T0255649

OPERATIONS RESEARCH

Introduction to Models and Methods

OPERATIONS RESEARCH
Introduction to Models and Methods

Richard J Boucherie
University of Twente, The Netherlands

Aleida Braaksma
University of Twente, The Netherlands

Henk Tijms
Vrije University Amsterdam, The Netherlands

 **World Scientific**

NEW JERSEY · LONDON · SINGAPORE · BEIJING · SHANGHAI · HONG KONG · TAIPEI · CHENNAI · TOKYO

Published by

World Scientific Publishing Co. Pte. Ltd.

5 Toh Tuck Link, Singapore 596224

USA office: 27 Warren Street, Suite 401-402, Hackensack, NJ 07601

UK office: 57 Shelton Street, Covent Garden, London WC2H 9HE

Library of Congress Cataloging-in-Publication Data
Names: Boucherie, R. J. (Richard J.), 1964– author. | Braaksma, Aleida, author. | Tijms, H. C., author.
Title: Operations research : introduction to models and methods / Richard J. Boucherie,
  Aleida Braaksma, Henk Tijms.
Description: Hackensack, New Jersey : World Scientific, [2022] | Includes bibliographical references and index.
Identifiers: LCCN 2021045245 | ISBN 9789811239342 (hardcover) | ISBN 9789811239816 (paperback) |
  ISBN 9789811239359 (ebook for institutions) | ISBN 9789811239366 (ebook for individuals)
Subjects: LCSH: Operations research--Textbooks. | Management science--Textbooks. |
  System analysis--Textbooks.
Classification: LCC T57.6 .B68 2022 | DDC 658.4/034--dc23/eng/20211101
LC record available at https://lccn.loc.gov/2021045245

British Library Cataloguing-in-Publication Data
A catalogue record for this book is available from the British Library.

Cover image: Photo by Hans Braxmeier, Brienzer Rothorn Ski Area Lift, pixabay.com

For any available supplementary material, please visit
https://www.worldscientific.com/worldscibooks/10.1142/12343#t=suppl

Printed in Singapore

Preface

Why this textbook when there is an abundance of textbooks on operations research on the market? We felt that there is a need for a manageable textbook that is not a thousand pages long, a length that is daunting for beginning students, especially when it takes them an hour to digest what's on one page. This motivational textbook provides a comprehensive yet concise introduction to operations research for students in a variety of fields such as engineering, business analytics, mathematics and statistics, computer science, and econometrics. It is the result of many years of teaching and feedback from many students. The didactic premise of the book is "motivation and insight first, before getting into details". The objective is to present the material in a way that would immediately make sense to a beginning student. The book covers a wide range of topics from both deterministic and stochastic operations research. The emphasis is on useful models and interpreting the solutions in the context of concrete applications.

The text is divided into several parts. The first three chapters deal exclusively with deterministic models, such as linear programming and sensitivity analysis, integer programming and heuristics, and network analysis including shortest paths, maximum flows and minimum cost flows. The next three chapters deal primarily with basic stochastic models and techniques, such as decision trees, dynamic programming and optimal stopping, production planning and inventory control. The first six chapters can be seen as a springboard to specialized texts, either practical or theoretical. The last five chapters contain more advanced material, including discrete-time and continuous-time Markov chains, queueing systems and networks of queues, Markov decision processes, and discrete-event simulation. Each chapter has many exercises and answers to a large selection of exercises are included. A solutions manual is available to qualified instructors who have adopted the book as required reading for their courses.

The book is an adaptation of an earlier Dutch version that was used at several Dutch universities. We would like to thank Reinie Erné for doing an excellent job in translating the Dutch text.

Contents

Chapter 1

Linear Programming

1.1 Introduction

Any future book on the history of mathematics will undoubtedly state: 1947, the invention of the simplex method for linear programming by George B. Dantzig. The year 1947 is a memorable year for applied mathematics. Nowadays, linear programming (LP) is one of the mathematical methods most frequently applied in practice. In LP, the optimal value of an objective function in several decision variables is determined. These variables must satisfy a number of linear constraints. LP models are used in many different decision situations, such as

- production and inventory level planning,
- blending problems in the oil or food industry,
- work schedules in public transportation and hospitals,
- route planning in distribution networks,
- stock portfolios selection.

Broadly speaking, it can be said that linear programming concerns the optimal distribution of a limited quantity of raw materials among competing production activities, subject to a number of constraints. The adjective "linear" refers to the use of linear equations and linear inequalities when formulating an LP model. The word "programming" should not immediately bring computers to mind; it rather refers to planning. Nevertheless, in real-world situations, the processing power of a computer is always necessary to find a numerical solution to an LP model.

   It is no coincidence that the simplex method for the numerical resolution of LP problems was developed right after the Second World War. During that war was the first time when scientific methods were used on a large scale to analyze all kinds of military logistic problems. The name "Operations Research" originates from the scientific activities developed then. After the war, people realized that the planning methods developed for military purposes could also be used for all sorts of planning problems in industry and the nonprofit sector. Research in the field of operations research flourished after the war, and this quickly led to the invention of the simplex method, which to this day has an enormous practical significance. The application possibilities of linear programming have only further increased through the rise of the personal computer.

In Section 1.2, we give the mathematical formulation of an LP model. We discuss a simple but illustrative example before giving the formulation in general terms. LP models cannot easily be solved using pen and paper, except in the case of two decision variables. In the particular case of two decision variables, the LP model can be solved using a graphical method. We discuss the graphical method only briefly as it cannot be applied to the general LP problem with more than two variables. No real-world problem has only two variables. Several LP software packages exist; these packages differ in their ease of use and the range of additional features they offer. All packages use a form of the simplex method to find the optimal solution to the LP model. We do not go into too much detail about the simplex method, but instead focus on understanding, formulating, and interpreting LP models. One does not need to know the technical details of the simplex method to apply LP (in the same way that one does not need to know how $2^{1/3}$ or $\ln(5)$ is computed to use a calculator). Nevertheless, the chosen approach with an emphasis on modeling and interpreting requires access to software to carry out the calculations on a computer. Nowadays, such software is readily available.

In Section 1.3, we discuss a number of applications of LP, including the composition of a stock portfolio and the fitting of a curve to given data points. We explain the basic ideas of the simplex method in Section 1.4. We also briefly touch upon an alternative solution method, the interior-point method developed by Karmarkar in 1984.

In real-world applications, it is often important not only to know the optimal solution but also to have an idea of the effects of changes to the data of the LP model on the optimal solution. These effects can be determined using sensitivity analysis. In real-world applications, being able to carry out a sensitivity analysis is often more important than solving the LP model itself. To understand sensitivity analysis well, some background knowledge of the duality theory of linear programming is useful. The dual of an LP problem and an economic interpretation of it are discussed in Section 1.5. This discussion also brings up the important concepts of shadow prices and reduced costs. In Section 1.6, the use of sensitivity analysis in practice is illustrated through two examples. Finally, for the more mathematically literate reader, we give a mathematical treatment of duality theory and sensitivity analysis in Section 1.7.

1.2   The Formulation of LP Models

To a large extent, how to formulate an LP model from a verbal description of a decision problem is an art that one can only develop through practice. Every beginning is difficult: how should one first choose the decision variables and then formulate the objective function and constraints? Here is some advice that is as simple as it is effective: put yourself in the shoes of the decision maker, and from that position, ask yourself what decision variables that person needs. In this section and Section 1.3, we provide detailed solutions to some typical examples of linear programming. After studying these, try a few exercises. Actively working through an exercise is completely different from passively understanding a ready-made solution.

1.2.1 A Production Planning Problem

Problem statement. A company manufactures four different products: A, B, C, and D. The production manager must draw up a production plan for the coming week. Each product undergoes three stages: assembly, finishing, and packaging. Each product requires a certain amount of time to undergo each of the processes. The amounts of time in hours per unit of product per operation are given in Table 1.1. For example, each unit of product A requires 0.70 hours for assembly, 0.55 hours for finishing, and 0.24 hours for packaging. Every week, 400 hours are available in the assembly department, 480 hours in the finishing department, and 220 hours in the packaging department. The same raw material is used for each of the products. The raw material requirements of the products A, B, C, and D are, respectively, 1.9, 2.5, 1.8, and 2 units per unit of product. In all, 1500 units of raw material are available for production in one week. After deducting the costs of raw materials, the products A, B, C, and D contribute 4.80 euros, 12 euros, 6 euros, and 7.20 euros per unit to the company's profit. The demand for each of the products is very high, and the company can sell as many products as it manufactures. The company has already accepted an order of 100 units of product A for the coming week. How should the production manager choose his production plan to maximize the total profit subject to the constraints mentioned above?

Table 1.1 Data for the production planning problem.

	Prod. A	Prod. B	Prod. C	Prod. D	Available
	\multicolumn{5}{c}{Processing Time (Hours)}				
Assembly	0.70	0.75	0.55	0.34	400
Finishing	0.55	0.82	0.80	0.55	480
Packaging	0.24	0.32	0.45	0.27	220

Formulation. The first step in the formulation process consists of the question "What decision variables are needed?". To answer this question, it might be useful to put oneself in the decision maker's shoes. In this case, one can quickly conclude that the decision variables are the numbers of units of each product to make. We use the following symbols for the decision variables:

$$x_A = \text{number of units of product } A \text{ to make,}$$
$$x_B = \text{number of units of product } B \text{ to make,}$$
$$x_C = \text{number of units of product } C \text{ to make,}$$
$$x_D = \text{number of units of product } D \text{ to make.}$$

The next step in the formulation process is to specify the objective function and the constraints in terms of the decision variables. The aim is to maximize the profit, which can be written as

$$\text{profit} = 4.8x_A + 12x_B + 6x_C + 7.2x_D.$$

Using the data in the table, we can express the constraints on the capacity of the production stages as

$$0.70x_A + 0.75x_B + 0.55x_C + 0.34x_D \leq 400 \text{ (assembly)},$$
$$0.55x_A + 0.82x_B + 0.80x_C + 0.55x_D \leq 480 \text{ (finishing)},$$
$$0.24x_A + 0.32x_B + 0.45x_C + 0.27x_D \leq 220 \text{ (packaging)}.$$

The raw material constraint is given by

$$1.9x_A + 2.5x_B + 1.8x_C + 2x_D \leq 1500.$$

The already accepted order for product A gives the constraint

$$x_A \geq 100.$$

The *linear programming* model then reads as follows:

$$
\begin{aligned}
\text{Maximize}\quad & 4.8x_A + 12x_B + 6x_C + 7.2x_D \\
\text{subject to}\quad & 0.70x_A + 0.75x_B + 0.55x_C + 0.34x_D \leq 400 \\
& 0.55x_A + 0.82x_B + 0.80x_C + 0.55x_D \leq 480 \\
& 0.24x_A + 0.32x_B + 0.45x_C + 0.27x_D \leq 220 \\
& 1.9x_A \;\;+ 2.5x_B \;\;+ 1.8x_C \;\;+ 2x_D \;\;\leq 1500 \\
& x_A \geq 100 \\
\text{and}\quad & x_A, x_B, x_C, x_D \geq 0.
\end{aligned}
$$

We must include nonnegativity constraints because we cannot make a negative number of products. The objective function must be maximized over the region consisting of the points (x_A, x_B, x_C, x_D) that satisfy the constraints and the non-negativity requirements. This region is called the *feasible region*. The model above is a typical example of a linear programming model. If we use a software package to solve this LP model, we find the solution $x_A = 100.00$, $x_B = 330.15$, $x_C = 0$, and $x_D = 242.31$, with a maximum objective value of 6186.46 euros.

In the LP model above, we have tacitly made the following implicit assumptions:

(1) **Proportionality.** It is implicitly assumed that the use of the available raw material is proportional to the level of production. For example, every unit of product A requires 1.9 units of the raw material, regardless of the number of units of product A that are made. Likewise, it is assumed that every unit of product yields a constant marginal profit, regardless of the quantity that is sold.

(2) **Additivity.** This means that the combined use of raw material for two or more production activities is equal to the sum of the use for each activity individually. Moreover, the combined contribution of two or more production activities to the objective function is equal to the sum of the profits realized by each of the individual activities.

(3) **Divisibility.** It is assumed that the decision variables can take on fractional values. This assumption is acceptable if, for example, the products represent liters of certain fluids. If, on the other hand, the products represent numbers of machines, then the assumption is only acceptable if the fractional values of the variables in the solution are sufficiently large that we can safely round to the nearest integers. In many real-world applications, there is no reason not to assume that the variables may take on fractional values. If, however, the variables must be integral, then we need to use integer programming. Unfortunately, integer programming problems have higher costs in computation time than LP problems of comparable size. We come back to integer programming in Chapter 2.

The proportionality and additivity assumptions imply that the objective function is a linear function and that the constraints are linear equalities or inequalities. In many real-world problems, it is justified to use these assumptions to formulate an (approximate) LP model. We stress that a linear programming model may have only linear expressions. There may not be any nonlinear terms such as $x_1^{1/2}$ or $x_1 x_2$. In those cases, one needs to switch to nonlinear programming models. We give an example of such a model in Section 1.2.2.

1.2.2 *A Nonlinear Programming Problem*

Problem statement. In a remote area, there are three oil fields. Two supply points are to be built in the area, from which the oil fields are to be supplied by air. Oil fields 1, 2, and 3 require 30 tons, 17 tons, and 41 tons of supplies per month, respectively. Oil field 2 is located 90 miles west and 400 miles north of oil field 1, and oil field 3 is located 270 miles east and 48 miles south of oil field 1. Where should the supply points be located, and how should the supplies be delivered so that the total number of flown ton-miles is as low as possible?

Formulation. First, we choose coordinates for the different places in the region in question. If we choose the coordinates $(0,0)$ for the position of oil field 1, then the positions of oil fields 2 and 3 have coordinates $(-90, 400)$ and $(270, -48)$, respectively. The decision variables include the coordinates (x_1, y_1) and (x_2, y_2) of the positions of supply posts 1 and 2. Additionally, six decision variables w_{11}, w_{12}, w_{13}, w_{21}, w_{22}, and w_{23} are needed, where w_{11} is the number of tons of supplies flown per month from supply point 1 to oil field 1 and, more generally,

$$w_{ij} = \text{number of tons of supplies flown per}$$
$$\text{month from supply point } i \text{ to oil field } j$$

for $i = 1, 2$ and $j = 1, 2, 3$. To formulate the objective function, one needs the *Euclidean distance* between two points (a_1, b_1) and (a_2, b_2) in the Euclidean plane. This distance is given by (Pythagorean theorem):

$$\sqrt{(a_1 - a_2)^2 + (b_1 - b_2)^2}.$$

For example, $\sqrt{(x_2 - 270)^2 + (y_2 + 48)^2}$ is the flight distance between supply point 2 and oil field 3. Multiplying this by w_{23} gives the total number of ton-miles flown per month from supply point 2 to oil field 3. This leads to the following optimization problem:

$$
\begin{aligned}
\text{Minimize} \quad & w_{11}\sqrt{(x_1 - 0) + (y_1 - 0)^2} + w_{12}\sqrt{(x_1 + 90)^2 + (y_1 - 400)^2} \\
& + w_{13}\sqrt{(x_1 - 270)^2 + (y_1 + 48)^2} + w_{21}\sqrt{(x_2 - 0)^2 + (y_2 - 0)^2} \\
& + w_{22}\sqrt{(x_2 + 90)^2 + (y_2 - 400)^2} + w_{23}\sqrt{(x_2 - 270)^2 + (y_2 + 48)^2}
\end{aligned}
$$

subject to
$$
\begin{aligned}
w_{11} + w_{21} &= 30 \\
w_{12} + w_{22} &= 17 \\
w_{13} + w_{23} &= 41
\end{aligned}
$$

and $w_{11}, w_{12}, w_{13}, w_2, w_{22}, w_{23} \geq 0.$

This is a nonlinear programming problem with a nonlinear objective function and linear constraints.

In the remainder of this chapter, we restrict ourselves to LP problems. An important class of LP problems consists of the so-called distribution problems, which are common in practice.

1.2.3   A Distribution Problem

Problem statement.   A company has one factory and three depots. The company ships part of what it produces from the factory to the depots, from where the goods are further distributed to the customers. The customers can also purchase goods directly from the factory. The company sells its product to four customers. The monthly demand from the customers C_1, C_2, C_3, and C_4 is 50 thousand tons, 70 thousand tons, 45 thousand tons, and 30 thousand tons, respectively. The production capacity of the factory is 250 thousand tons per month, and the storage capacities of the depots D_1, D_2, and D_3 are 80 thousand tons, 40 thousand tons, and 75 thousand tons, respectively. Table 1.2 gives the distribution costs in euros per ton. How should the goods be distributed among the clients to minimize the total distribution costs?

Table 1.2   Distribution costs (in euros per ton).

From\To	D_1	D_2	D_3	C_1	C_2	C_3	C_4
Factory	4.5	4	7	12	10	15	18
Depot 1	–	–	–	10	5	10	12
Depot 2	–	–	–	7	5	8	12
Depot 3	–	–	–	7	8	5	10

Formulation.   By putting oneself in the shoes of the decision maker, it becomes clear that the following decision variables are needed:

$$
\begin{aligned}
FD_i &= \text{number of tons to send from the factory to depot } i, \\
FC_j &= \text{number of tons to send from the factory to customer } j, \\
DC_{ij} &= \text{number of tons to send from depot } i \text{ to customer } j,
\end{aligned}
$$

where $i = 1, 2, 3$ and $j = 1, 2, 3, 4$. The index i indicates the depot number ($i = 1, 2, 3$), and the index j the customer number ($j = 1, 2, 3, 4$). For example, FD_2 is the number of tons to send from the factory to depot 2, and DC_{24} is the number of tons to send from depot 2 to customer 4. This notation of variables with indices takes some getting used to in the beginning but is indispensable for the formulation of realistic LP problems. The aim is to minimize the total distribution costs, given the capacity constraints and the condition that the customers' demand must be met. This leads to the following LP model:

$$
\begin{aligned}
\text{Minimize} \quad & 4.5FD_1 + 4FD_2 + 7FD_3 + 12FC_1 + 10FC_2 \\
& + 15FC_3 + 18FC_4 + 10DC_{11} + 5DC_{12} + 10DC_{13} \\
& + 12DC_{14} + 7DC_{21} + 5DC_{22} + 8DC_{23} + 12DC_{24} \\
& + 7DC_{31} + 8DC_{32} + 5DC_{33} + 10DC_{34}
\end{aligned}
$$

$$
\begin{aligned}
\text{subject to} \quad & FD_1 + FD_2 + FD_3 + FC_1 + FC_2 + FC_3 + FC_4 \leq 250\,000 \\
& FD_1 \leq 80\,000 \\
& FD_2 \leq 40\,000 \\
& FD_3 \leq 75\,000 \\
& FC_1 + DC_{11} + DC_{21} + DC_{31} = 50\,000 \\
& FC_2 + DC_{12} + DC_{22} + DC_{32} = 70\,000 \\
& FC_3 + DC_{13} + DC_{23} + DC_{33} = 45\,000 \\
& FC_4 + DC_{14} + DC_{24} + DC_{34} = 30\,000 \\
& DC_{11} + DC_{12} + DC_{13} + DC_{14} = FD_1 \\
& DC_{21} + DC_{22} + DC_{23} + DC_{24} = FD_2 \\
& DC_{31} + DC_{32} + DC_{33} + DC_{34} = FD_3
\end{aligned}
$$

and $\quad FD_i, FC_j, DC_{ij} \geq 0 \quad$ for all i, j.

The last three constraints ensure that the depots do not forward more products than they have received from the factory. The solution to the LP model is

$$
\begin{aligned}
FD_1 &= 80\,000, \; FD_2 = 40\,000, \; FD_3 = 45\,000, \\
FC_1 &= 30\,000, \; DC_{21} = 20\,000, \; DC_{12} = 70\,000, \\
DC_{33} &= 45\,000, \; DC_{14} = 10\,000, \; DC_{24} = 20\,000,
\end{aligned}
$$

and the other variables have value zero. The minimum costs are 2 270 million euros.

1.2.4 *Graphical Solution Method for LP*

An LP problem with two variables can be solved graphically. The graphical approach can only be applied in the case of two variables and therefore has no practical usefulness. Nevertheless, we give the method because it provides insight into the nature of the solution, insight that can be used for problems with more than two variables. Consider the LP problem

$$\text{Maximize} \quad 12x_1 + 10x_2$$
$$\text{subject to} \quad x_1 \quad + \quad x_2 \le 20$$
$$0.5x_1 + \quad x_2 \le 12$$
$$\tfrac{1}{16}x_1 + \tfrac{1}{24}x_2 \le 1$$
$$12x_1 - \quad 8x_2 \ge 0$$
$$\text{and} \quad x_1, x_2 \ge 0.$$

Fig. 1.1 The feasible region of the LP example.

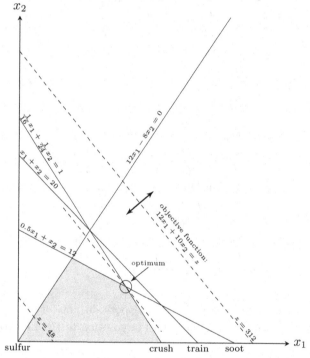

This LP model is the basis for the coal-fueled power plant problem in Exercise 1.2 at the end of the chapter. In Figure 1.1, the shaded area is the feasible region. This region consists of all points (x_1, x_2) that satisfy the constraints, including the nonnegativity constraints. It can be found by drawing the lines $x_1 + x_2 = 20$, $0.5x_1 + x_2 = 12$, $\tfrac{1}{16}x_1 + \tfrac{1}{24}x_2 = 1$, and $12x_1 - 8x_2 = 0$ in the (x_1, x_2)-plane. For example, the points that satisfy the constraint $x_1 + x_2 \le 20$ are either on or below the line $x_1 + x_2 = 20$. This is how to construct the shaded area of feasible solutions. How does one find the maximum of the objective function $12x_1 + 10x_2$ in the feasible region? For this, consider the line $12x_1 + 10x_2 = z$ for different values of z. These lines are all parallel to one another. Figure 1.1 shows the line $12x_1 + 10x_2 = z$ for the values $z = 48$ and $z = 312$. Now, slide the line $12x_1 + 10x_2 = z$ parallel to the lines drawn for $z = 48$ and $z = 312$, in such a way that the value of z increases and the line still has points in common with the feasible region. The line with the highest value of z that still has a point in common with the feasible region is the line that meets the region at the corner point indicated by a small circle in the figure. In the feasible region, the objective function is therefore maximum at this corner point. The corner point is the intersection point of the two lines $0.5x_1 + x_2 = 12$ and $\frac{1}{16}x_1 + \frac{1}{24}x_2 = 1$ and is therefore given by

$$(x_1, x_2) = (12, 6).$$

This corner point gives the optimal solution to the LP problem, and the maximum objective value is $z = 204$. One can immediately see from the geometric construction that for an LP problem with two variables, the optimum is always reached at a corner point. Fortunately, this property remains valid in dimensions higher than two. In LP problems with more than two variables, the optimal corner point is found using an algebraic method, namely the *simplex method*, which is discussed in Section 1.4.

1.2.5 *General Formulation of an LP Model*

In the previous subsections, examples have shown that an LP model maximizes (or minimizes) a linear objective function subject to a number of linear constraints. The following economic context leads to a general LP model:

- A company can make n products numbered $j = 1, 2, \ldots, n$ from m raw materials numbered $i = 1, 2, \ldots, m$, where n and m are given numbers.
- Of each raw material i, a given quantity b_i is available.
- On each unit of product j that is made, the profit is c_j.
- Of each raw material i, a given quantity a_{ij} is needed to make one unit of product j (for example, making one unit of product 7 requires the quantities $a_{17}, a_{27}, \ldots, a_{m7}$ of raw materials $1, 2, \ldots, m$, respectively).

Question: How much of each product should the company make to maximize the total profit?

To answer this question, define the following decision variables:

$$x_1 = \text{quantity to make of product 1,}$$
$$x_2 = \text{quantity to make of product 2,}$$

and so on. In general, define

$$x_j = \text{quantity to make of product } j$$

for $j = 1, 2, \ldots, n$. The aim is to maximize the total profit. This leads to the objective function

$$\text{total profit} = c_1 x_1 + c_2 x_2 + \cdots + c_n x_n.$$

Restrictions on the decision variables arise because only a limited quantity of each raw material is present. To make the quantities $x_1, x_2, \ldots, x_n$ of the products $1, 2, \ldots, n$, the respective quantities $a_{11}x_1, a_{12}x_2, \ldots, a_{1n}x_n$ of raw material 1 are required. This leads to the constraint

$$a_{11}x_1 + a_{12}x_2 + \cdots + a_{1n}x_n \leq b_1.$$

More generally, the limited quantity of raw material i gives the constraint

$$a_{i1}x_1 + a_{i2}x_2 + \cdots + a_{in}x_n \leq b_i$$

for $i = 1, 2, \ldots, m$. Additionally, we have the nonnegativity constraints

$$x_1 \geq 0, \ x_2 \geq 0, \ldots, x_n \geq 0.$$

Together, these lead to the general LP problem

$$\text{Maximize } c_1x_1 + c_2x_2 + \cdots + c_nx_n$$
$$\text{subject to} \quad a_{11}x_1 + a_{12}x_2 + \cdots + a_{1n}x_n \leq b_1$$
$$a_{21}x_1 + a_{22}x_2 + \cdots + a_{2n}x_n \leq b_2$$
$$\vdots$$
$$a_{m1}x_1 + a_{m2}x_2 + \cdots + a_{mn}x_n \leq b_m$$
$$\text{and} \quad x_1, x_2, \ldots, x_n \geq 0.$$

Indices are a useful tool to state LP problems clearly. In large-scale LP problems, double indices are often used to represent the decision variables; see the distribution problem in Section 1.2.3. Insightful notation is the basis for the formulation of real-world problems as LP models. The LP model above can be represented more clearly using the sum notation $\sum$:

$$\text{Maximize } \sum_{j=1}^{n} c_j x_j$$
$$\text{subject to } \sum_{j=1}^{n} a_{ij}x_j \leq b_i \ \text{ for } i = 1, 2, \ldots, m$$
$$\text{and} \quad x_j \geq 0 \qquad \text{for } j = 1, 2, \ldots, n$$

(a common abbreviation for maximize is "max"). In the LP formulation above, the constraints consist of linear inequalities $\leq$. Of course, linear inequalities $\geq$ and linear equalities $=$ can also occur as constraints in an LP formulation. Moreover, the objective function can be minimized in an LP model instead of being maximized. Needless to say,

$$\text{Minimize } \quad c_1x_1 + c_2x_2 + \cdots + c_nx_n$$

is equivalent to

$$\text{Maximize } \quad -c_1x_1 - c_2x_2 - \cdots - c_nx_n.$$

The constraints are linear equalities or inequalities. The inequalities must be of the form $\leq$ or $\geq$. Strict inequalities with $<$ or $>$ are not allowed. We do not elaborate on the mathematical reason for this. In practice, it is not an impediment that inequalities must be of the form $\leq$ or $\geq$. A $\leq$ constraint generally corresponds to a capacity or raw material restriction. A $\geq$ constraint generally indicates that the solution must meet a certain minimum requirement.

Nonnegativity Constraints

In the formulation of an LP model, all variables must be nonnegative. This is necessary due to the nature of the simplex method for solving LP models (in any LP software package, the nonnegativity constraints are automatically taken into account so that they do not need to be entered separately). If a variable x can take on both positive and negative values, then there is an easy trick to replace this variable with two nonnegative variables x^+ and x^-. Just as any number can be written as the difference between two nonnegative numbers (for example, $-3 = 0 - 3$), the variable x can be written as

$$x = x^+ - x^-, \quad \text{where } x^+, x^- \geq 0.$$

This transformation increases the number of variables but does not disturb the linearity of the model. In the objective function and the constraints, replace x everywhere with $x^+ - x^-$, and to the nonnegativity constraints, add $x^+, x^- \geq 0$. As an illustration, consider the system of linear equations

$$2x_1 - x_2 + x_3 = 2$$
$$3x_1 + 5x_2 + 5x_3 = 3.5$$
$$x_1 + x_2 + x_3 = 0.5.$$

The solution to this system can be determined using linear programming when each variable x_i is replaced with two nonnegative auxiliary variables x_i^+ and x_i^- and a suitable linear auxiliary function is chosen as objective function. An appropriate choice is to *minimize* the sum of the nonnegative auxiliary variables. Check that one finds a solution to the system of linear equations above by solving the following LP problem:

$$
\begin{aligned}
\text{Minimize} \quad & x_1^+ + x_1^- + x_2^+ + x_2^- + x_3^+ + x_3^- \\
\text{subject to} \quad & 2x_1^+ - 2x_1^- - x_2^+ + x_2^- + x_3^+ - x_3^- = 2 \\
& 3x_1^+ - 3x_1^- + 5x_2^+ - 5x_2^- + 5x_3^+ - 5x_3^- = 3.5 \\
& x_1^+ - x_1^- + x_2^+ - x_2^- + x_3^+ - x_3^- = 0.5 \\
\text{and} \quad & x_1^+, x_1^-, x_2^+, x_2^-, x_3^+, x_3^- \geq 0.
\end{aligned}
$$

The LP solution is $x_1^+ = 0$, $x_1^- = 0.5$, $x_2^+ = 0$, $x_2^- = 1$, $x_3^+ = 2$, $x_3^- = 0$. This gives the solution $x_1^* = -0.5$, $x_2^* = -1$, and $x_3^* = 2$ for the system of linear equations. Note that the maximization of the sum of the nonnegative auxiliary variables would have led to an unbounded objective value (why?).

A *feasible solution* to an LP model is a set of values for $x_1, \ldots, x_n$ that satisfy all constraints, including the nonnegativity constraints. An *optimal solution* to the LP model is a feasible solution that maximizes (or minimizes) the objective function over the feasible region. The following situations can occur in an LP model:

(i) **No feasible solution.** The reason is often that the decision problem was not formulated correctly as an LP model or that a typing error was made when entering the data into the computer program. As an illustration, the LP formulation

$$\text{Maximize} \quad x_1 + x_2$$
$$\text{subject to} \quad 2x_1 + x_2 \leq 1$$
$$x_1 + x_2 \geq 2$$
$$\text{and} \quad x_1, x_2 \geq 0$$

has no feasible solution. The nonnegativity constraints and the first constraint imply $x_1 + x_2 \leq 1$, which contradicts the second constraint.

(ii) **An unbounded solution.** For a maximization (respectively, minimization) problem, this means that the objective function can take on any arbitrarily large positive (respectively, negative) value within the set of feasible solutions. As an illustration, consider the LP formulation

$$\text{Maximize} \quad 3x_1 + x_2$$
$$\text{subject to} \quad 2x_1 + x_2 \geq 1$$
$$\text{and} \quad x_1, x_2 \geq 0.$$

The variables x_1 and x_2 can be made arbitrarily large, and thus the objective value can also be made arbitrarily large.

In real-world applications, an unbounded solution often indicates that one or more constraints were forgotten in the LP formulation. Both the absence of a feasible solution and the existence of an unbounded solution are detected by the simplex method.

(iii) **An optimal solution.** For the particular case of two decision variables, we saw in Section 1.2.4 that an optimal solution is reached at a corner point of the feasible region. This property holds in general. If an LP problem is solvable, then there exists an optimal solution that corresponds to a corner point of the feasible region. Roughly speaking, a corner point is a feasible solution that does not lie between two other feasible solutions. Moreover, it can be proved that at every corner point, at most m of the decision variables $x_1, \ldots, x_n$ can take on a *positive* value, where m is the number of "true" constraints (excluding the nonnegativity constraints). This means that the following important result holds for an LP problem:

> *There is an optimal solution in which at most m of the decision variables have a positive value, where m is the number of constraints (excluding the nonnegativity constraints).*

In case the LP problem has more than one optimal solution, the property does not necessarily hold for every optimal solution. This theoretical property has significant practical implications. For example, suppose that a diet-composition problem with 50 possible ingredients is formulated as an LP model with 10 constraints for the composition of the diet. Then it is clear, a priori, that a diet with minimum cost does not need to contain more than 10 different ingredients. The simplex method always leads to an optimal solution with the property stated above.

1.3 Linear Programming Applications

In this section, we discuss some applications of linear programming and show how LP models can be solved using a software module.

1.3.1 *An Investment Problem*

Problem statement. A bank has 100 thousand euros at its disposal to invest during the current year. The bank's financial analysts have selected the following investment options: corporate loans, personal loans, preferred shares, common shares, and government bonds. The annual returns are estimated at 12% for corporate loans, 17% for personal loans, 10.5% for preferred shares, 11.5% for common shares, and 9% for government bonds. In order to reduce the risk, the financial analysts have imposed the following three restrictions on the bank's portfolio:

- No more than 50% of the available amount may be invested in loans or shares.
- The investment in government bonds must be equal to at least 30% of the investment in loans.
- The personal loans may not account for more than 40% of the total investment in loans.

How should the bank invest its money in order to maximize the annual return on its portfolio?

Formulation. The decision variables are defined by:

$$loanc = \text{thousands of euros invested in corporate loans,}$$
$$loanp = \text{thousands of euros invested in personal loans,}$$
$$sharep = \text{thousands of euros invested in preferred shares,}$$
$$sharec = \text{thousands of euros invested in common shares,}$$
$$bond = \text{thousands of euros invested in government bonds.}$$

The unit of 1000 euros was chosen to prevent that the LP model contain both very large and very small numbers. For the investment problem, the LP model is

Maximize $\quad 0.12\, loanc + 0.17\, loanp + 0.105\, sharep + 0.115\, sharec + 0.09\, bond$
subject to $\quad loanc + loanp + sharep + sharec + bond \leq 100$
$\qquad\qquad loanc + loanp \leq 50$
$\qquad\qquad sharep + sharec \leq 50$
$\qquad\qquad bond \geq 0.30\, loanc + 0.30\, loanp$
$\qquad\qquad loanp \leq 0.40\, loanc + 0.40\, loanp$
and $\qquad loanc, loanp, sharep, sharec, bond \geq 0.$

The computer solution to this problem is given in Figure 1.2. This solution was obtained using the LP module of the software package ORSTAT2000. The computer output shows that the maximum annual revenue is 12.375 thousand euros. This return is obtained by investing 30 thousand euros in corporate loans, 20 thousand euros in personal loans, nothing in preferred shares, 35 thousand euros in common shares, and 15 thousand euros in government bonds. The column "Slack or Surplus" shows that all constraints except the third are *binding*. A linear inequality is called binding if equality holds for the optimal solution. We discuss the additional information in the computer output in Sections 1.5 and 1.6.

Fig. 1.2   Computer output for the investment problem.

```
SIMOPT Version 2.1                        IEOR  VU Amsterdam

The following model was read:

Objective Function :
   MAX  0.1200 LOANC +0.1700 LOANP +0.1050 SHAREP
        +0.1150 SHAREC +0.0900 BOND
Subject to :
   1.   1.0000 LOANC +1.0000 LOANP +1.0000 SHAREP
        +1.0000 SHAREC +1.0000 BOND <= 100.0000
   2.   1.0000 LOANC +1.0000 LOANP <= 50.0000
   3.   1.0000 SHAREP +1.0000 SHAREC <= 50.0000
   4.   1.0000 BOND >=  0.3000 LOANC +0.3000 LOANP
   5.   1.0000 LOANP <=  0.4000 LOANC +0.4000 LOANP

Summary of Results

Value Objective Function :        12.3750


Variable            Activity Level   Reduced Cost
--------            --------------   ------------

LOANC    :            30.0000         0.0000
LOANP    :            20.0000         0.0000
SHAREP   :             0.0000         0.0100
SHAREC   :            35.0000         0.0000
BOND     :            15.0000         0.0000


                   Slack or Surplus   Shadow Prices
                   ----------------   -------------

Constraint 1            0.0000          0.1150
Constraint 2            0.0000          0.0175
Constraint 3           15.0000          0.0000
Constraint 4            0.0000         -0.0250
Constraint 5            0.0000          0.0500


Accuracy Check Passed.
```

1.3.2 A Currency Problem

Problem statement. The exchange rates on the international money market are continually fluctuating. Small differences in exchange rates can allow one to make much profit by exchanging one currency to another. Suppose that one has a current working capital of 1 million US dollars. The current exchange rates are shown in Table 1.3 (next page), where we have taken exchange rates from the old days, with Dutch guilders (Gld) and German marks (DM). For example, every dollar gives 2.1421 guilders. There is a rule that one may not exchange more than 1.2 million US dollars, 0.2 million British pounds, 0.8 million German marks, 60 million Japanese yen, or 2 million Dutch guilders. How should one convert the dollars into other currencies and then back into dollars to maximize the overall profit?

Formulation. We introduce five nodes 1, 2, 3, 4, and 5, where node 1 corresponds to US dollars, node 2 to British pounds, node 3 to German marks, node 4 to Japanese yen, and node 5 to Dutch guilders. One can imagine a network consisting of nodes and arcs between the nodes, in which money flows from one node to another. The goal is to find a flow that begins and ends at node 1 and maximizes the profit. For this, we define the decision variables

$$x_{ij} = \text{amount in currency } i \text{ converted into currency } j$$

for $i, j = 1, \ldots, 5$, with $i \neq j$. The unit for the amounts is 10 thousand for each currency. After some thought, one can easily see that after all transactions, the size of the capital (in units of 10 thousand dollars) is equal to

$$out = 100 + 1.4367x_{21} + 0.3721x_{31} + 0.0065x_{41} + 0.4667x_{51} - x_{12} - x_{13} - x_{14} - x_{15}.$$

The objective is to maximize the outflow. Since the equation *"inflow = outflow"* has to hold at every node, we find the following LP model for the currency problem:

$$\begin{aligned}
\text{Maximize} \quad & out \\
\text{subject to} \quad & 100 + 1.4367x_{21} + 0.3721x_{31} + 0.0065x_{41} \\
& \quad + 0.4667x_{51} = x_{12} + x_{13} + x_{14} + x_{15} + out \\
& 0.6987x_{12} + 0.2702x_{32} + 0.0040x_{42} + 0.3266x_{52} \\
& \quad = x_{21} + x_{23} + x_{24} + x_{25} \\
& 2.6396x_{13} + 3.7240x_{23} + 0.0163x_{43} + 1.2326x_{53} \\
& \quad = x_{31} + x_{32} + x_{34} + x_{35} \\
& 155.9200x_{14} + 215.7520x_{24} + 62.5220x_{34} \\
& \quad + 76.0460x_{54} = x_{41} + x_{42} + x_{43} + x_{45} \\
& 2.1421x_{15} + 3.0045x_{25} + 0.8206x_{35} + 0.0136x_{45} \\
& \quad = x_{51} + x_{52} + x_{53} + x_{54} \\
& x_{12} + x_{13} + x_{14} + x_{15} \leq 120 \\
& x_{21} + x_{23} + x_{24} + x_{25} \leq 20 \\
& x_{31} + x_{32} + x_{34} + x_{35} \leq 80 \\
& x_{41} + x_{42} + x_{43} + x_{45} \leq 6000 \\
& x_{51} + x_{52} + x_{53} + x_{54} \leq 200 \\
\text{and} \quad & x_{ij} \geq 0 \text{ for all } i, j.
\end{aligned}$$

Table 1.3	Data for the current exchange rates.				
Sell\Buy	$	GBP	DM	Yen	Gld
$	—	0.6987	2.6396	155.9200	2.1421
GBP	1.4367	—	3.7240	215.7520	3.0045
DM	0.3721	0.2702	—	62.5220	0.8206
Yen	0.0065	0.0040	0.0163	—	0.0136
Gld	0.4667	0.3266	1.2326	76.0460	—

The computer solution shows that the maximum outflow is equal to 1 million 28 977 thousand dollars. So for the choices $x_{15}^* = 64.8363$, $x_{21}^* = 20$, $x_{32}^* = 74.0192$, $x_{34}^* = 5.9808$, $x_{41}^* = 6000$, $x_{53}^* = 64.9035$, $x_{54}^* = 73.9825$, and $x_{ij} = 0$ for the remaining variables, the currency transactions yield a profit of 28 977 thousand dollars.

1.3.3   *A Cutting Stock Problem*

Problem statement.   A paper mill has been commissioned to supply rolls of paper measuring a total of 550 m of width 1.70 m, 800 m of width 1.25 m, and 500 m of width 0.50 m. For each width, the required number of meters can be supplied in rolls of different lengths. The widths are cut out of standard 3 m wide reels of paper. The factory does not keep any residual rolls in stock. How should the standard reels be cut so that the number of m^2 of trim loss plus nonusable residual rolls is as small as possible? The cutting stock problem is a classic application of linear programming.

Formulation.   To set up an LP model for this problem, we first observe that a standard reel can be cut up in five different ways to give the desired widths. The five possible combinations are:

C1.	1 × 1.70 m; 1 × 1.25 m;		trim loss 0.05 m
C2.	1 × 1.70 m;	2 × 0.50 m;	trim loss 0.30 m
C3.		2 × 1.25 m; 1 × 0.50 m;	trim loss 0 m
C4.		1 × 1.25 m; 3 × 0.50 m;	trim loss 0.25 m
C5.		6 × 0.50 m;	trim loss 0 m.

Define the decision variables

$$x_i = \text{number of meters of standard reel being cut up}$$
$$\text{following combination } Ci \ (i = 1, \ldots, 5).$$

The requirement that 550 m of width 1.70 m be supplied leads to the constraint $x_1 + x_2 \geq 550$. Likewise, we find the constraints $x_1 + 2x_3 + x_4 \geq 800$ and $2x_2 + x_3 + 3x_4 + 6x_5 \geq 500$. In addition to trim loss, nonusable residual rolls are produced of length $x_1 + x_2 - 550$ m and width 1.70 m, length $x_1 + 2x_3 + x_4 - 800$ m and width 1.25 m, and length $2x_2 + x_3 + 3x_4 + 6x_5 - 500$ m and width 0.50 m. The cutting stock problem can therefore be formulated as the following LP model:

$$\text{Minimize} \quad 0.05x_1 + 0.30x_2 + 0.25x_4 + 1.7(x_1 + x_2 - 550)$$
$$+ 1.25(x_1 + 2x_3 + x_4 - 800)$$
$$+ 0.5(2x_2 + x_3 + 3x_4 + 6x_5 - 500)$$
$$\text{subject to} \quad x_1 + x_2 \qquad\qquad\qquad\qquad \geq 550$$
$$x_1 + \qquad + 2x_3 + x_4 \qquad\qquad \geq 800$$
$$2x_2 + x_3 + 3x_4 + 6x_5 \geq 500$$
$$\text{and} \qquad x_1, x_2, x_3, x_4, x_5 \qquad\qquad\qquad \geq 0.$$

Fig. 1.3   Computer output for the cutting stock problem.

```
SIMOPT Version 2.1                        IEOR   VU Amsterdam

The following model was read:

Objective Function :
   MIN  0.0500 X1 +0.3000 X2 +0.2500 X4 +1.700 X1 +1.700 X2 +1.2500 X1
        +2.5000 X3 +1.2500 X4 +1.000 X2 +0.5000 X3 +1.5000 X4 +3.000 X5
Subject to :
   1.  1.0000 X1 +1.0000 X2 >= 550.0000
   2.  1.0000 X1 +2.0000 X3 +1.0000 X4 >= 800.0000
   3.  2.0000 X2 +1.0000 X3 +3.0000 X4 +6.0000 X5 >= 500.0000


Summary of Results

Value Objective Function :              2212.5000


Variable        Activity Level   Reduced Cost
--------        --------------   ------------

X1        :       550.0000         0.0000
X2        :         0.0000         0.2500
X4        :         0.0000         0.2500
X3        :       125.0000         0.0000
X5        :        62.5000         0.0000


                Slack or Surplus  Shadow Prices
                ----------------  -------------

Constraint 1          0.0000         1.7500
Constraint 2          0.0000         1.2500
Constraint 3          0.0000         0.5000


Accuracy Check Passed.
```

Figure 1.3 shows the computer solution to the LP model. In the computer input, the constant -2185 $(= -1.7 \times 550 - 1.25 \times 800 - 0.5 \times 500)$ is not included in the objective function. The computer output shows that the optimal cutting pattern is to cut up 550 m of standard reel according to combination C1, 125 m according to combination C3, and 62.5 m according to combination C5. The total trim loss is $2212.5 - 2185 = 27.5$ m^2 of paper.

1.3.4   *Two-Finger Morra*

Problem statement.   A well-known street scene in Italy is excitedly gesticulating Italians playing a game in which marbles are moved back and forth incredibly fast. This game is called two-finger Morra. The game is played by two players A and B and boils down to the following. The players A and B simultaneously show 1 or 2 fingers as desired and say the number of fingers the opponent is going to show. Depending on what each of the two players does, a payment is made from one player to the other. Table 1.4 shows the reward matrix, where the positive numbers represent a payment from player B to player A and the negative numbers a payment from A to B. For example, if (i, j) indicates the decision to show i fingers and say j fingers, then player A receives 3 euros from player B if player A makes the decision $(2, 1)$ and player B makes the decision $(1, 1)$, while A pays B 2 euros if A makes the decision $(1, 2)$ and B makes the decision $(1, 1)$.

Table 1.4   Reward matrix for two-finger Morra.

A	B shows 1 says 1	shows 1 says 2	shows 2 says 1	shows 2 says 2
shows 1, says 1	0	2	−3	0
shows 1, says 2	−2	0	0	3
shows 2, says 1	3	0	0	−4
shows 2, says 2	0	−3	4	0

What strategy should the players A and B follow? In theory, neither of the players has the advantage in this game since the situation is completely symmetrical. Suppose, however, that in a specific case, player A loses the game more frequently than player B because in making decisions, player A is rather predictable for the opponent, who uses this to his advantage. Player A can only cancel out the psychological advantage of player B by keeping his opponent completely in the dark about what he plans to do. The only way a player can prevent his opponent from finding out what he is going to do is by not knowing himself what he is going to do until the last moment. This can be achieved by choosing a so-called *mixed strategy,* which means that at each round of the game, the decision is chosen randomly. This way, the player ensures that there is no pattern to his decisions that the opponent could take advantage of. Mixed strategy is a basic concept in game theory.

Formulation. Let us develop the idea of a mixed strategy for player A. The starting point is that player A wants to arm himself as well as possible against the most unfavorable situation that can occur for him. Suppose that player A makes the decision (i, j) with probability x_{ij} for $i, j \in \{1, 2\}$ (by, for example, first throwing a special die under the table). How should the probabilities x_{ij} be chosen? Of course, the numbers x_{ij} must satisfy $x_{ij} \geq 0$ and $x_{11} + x_{12} + x_{21} + x_{22} = 1$. Player A wants to choose the x_{ij} in such a way that the lowest value of his expected profit is as high as possible. If player A determines his decision according to the mixed strategy and player B makes decision (r, s), then the expected profit of player A equals

$$
\begin{array}{ll}
0x_{11} - 2x_{12} + 3x_{21} + 0x_{22} & \text{if } (r, s) = (1, 1), \\
2x_{11} + 0x_{12} + 0x_{21} - 3x_{22} & \text{if } (r, s) = (1, 2), \\
-3x_{11} + 0x_{12} + 0x_{21} + 4x_{22} & \text{if } (r, s) = (2, 1), \\
0x_{11} + 3x_{12} - 4x_{21} + 0x_{22} & \text{if } (r, s) = (2, 2).
\end{array}
$$

Now, suppose that the objective of player A is to maximize the least of these four payouts. Define the variable w by

$$
w = \min\{-2x_{12} + 3x_{21}, 2x_{11} - 3x_{22}, -3x_{11} + 4x_{22}, 3x_{12} - 4x_{21}\};
$$

player A aims to maximize the quantity w. This seems very complicated. Nevertheless, we can give an LP formulation for the optimization problem of player A. The key to this is the observation that $a = \min(a_1, a_2, a_3, a_4)$ can be written as the linear inequalities $a \leq a_1, a \leq a_2, a \leq a_3$, and $a \leq a_4$, with the additional requirement that a equals at least one of the a_i. This leads to the following LP problem for the maximization of w as a function of the x_{ij}:

$$
\begin{array}{ll}
\text{Maximize} & w \\
\text{subject to} & w \leq -2x_{12} + 3x_{21} \\
& w \leq 2x_{11} - 3x_{22} \\
& w \leq -3x_{11} + 4x_{22} \\
& w \leq 3x_{12} - 4x_{21} \\
& x_{11} + x_{12} + x_{21} + x_{22} = 1 \\
\text{and} & x_{11}, x_{12}, x_{21}, x_{22}, w \geq 0.
\end{array}
$$

From physical considerations, it should be clear that in the optimal solution, we have $w \geq 0$. That is why we have set $w \geq 0$ in the formulation of the LP model. In the literature on game theory, w is usually unrestricted in sign in the formulation, although in the optimum, $w \geq 0$ (and even $w = 0$).

Maximizing w automatically ensures that equality holds in at least one of the four inequalities (after all, suppose that the optimal solution has the sign $<$ for the first four inequalities; then the value of w would not be maximum because it can be increased by the least value of the differences between the right-hand and left-hand sides of these inequalities). The values of the variables in the optimal solution to this LP problem are

$$
w = 0, \ x_{11} = 0, \ x_{12} = \tfrac{4}{7}, \ x_{21} = \tfrac{3}{7}, \ x_{22} = 0.
$$

In words: If, each time, player A makes the decision $(1, 2)$ with probability $\frac{4}{7}$ and the decision $(2, 1)$ with probability $\frac{3}{7}$, then player A knows that the expected value of the payout to him in a given round of the game can never be negative, regardless of how player B plays. If player B's game is not optimal, then player A can be assured of a positive average profit per game for a large number of rounds of the game.

1.3.5   *Fitting Curves*

Problem statement.   Suppose that we want to draw a straight line $y = ax + b$ through n given data points $(x_i^{(0)}, y_i^{(0)})$, for $i = 1, \ldots, n$, in such a way that the sum of the vertical distances of the given y-values to the line $y = ax + b$ is as small as possible. This problem, which occurs frequently in applications, can directly be translated into the following optimization problem

$$\text{Minimize } \left| y_1^{(0)} - (ax_1^{(0)} + b) \right| + \cdots + \left| y_n^{(0)} - (ax_n^{(0)} + b) \right|$$

for the decision variables a and b. This nonlinear optimization problem with the two variables a and b can be transformed into an LP model. How does this work?

Before doing it, we first note that every number d can be represented as the difference between two *nonnegative* numbers. This representation is not unique. For example, $d = -3$ can be written as the difference between 0 and 3 but also as the difference between 5 and 8. We obtain a special representation of a given number d as the difference between two nonnegative numbers by defining the nonnegative numbers d^+ and d^- to be

$$d^+ = \begin{cases} d \text{ if } d \geq 0 \\ 0 \text{ if } d < 0 \end{cases}$$

and

$$d^- = \begin{cases} -d \text{ if } d < 0 \\ 0 \quad \text{if } d \geq 0. \end{cases}$$

In words, d^+ is the distance that d is above zero and d^- is the distance that d is below 0. For this representation of d, not only do we have

$$d = d^+ - d^- \text{ with } d^+, d^- \geq 0,$$

this is also the only representation of d as the difference between two *nonnegative* numbers that has the property that

$$|d| = d^+ + d^-.$$

This property holds purely because d^+ and d^- cannot both be positive at the same time; one of the two is always zero.

Formulation. Returning to the optimization problem, we define the (auxiliary) variable d_i by

$$d_i = y_i^{(0)} - ax_i^{(0)} - b \quad \text{for } i = 1, \dots, n.$$

We can then rewrite the optimization problem above in the equivalent form

Minimize $|d_1| + |d_2| + \cdots + |d_n|$

subject to $y_i^{(0)} - ax_i^{(0)} - b = d_i \quad \text{for } i = 1, \dots, n.$

To transform this problem into an LP problem, we define, for each variable d_i, the two nonnegative variables d_i^+ and d_i^- in the same way as we defined the two nonnegative numbers d^+ and d^- for the number d. This gives

$$y_i^{(0)} - ax_i^{(0)} - b = d_i^+ - d_i^- \quad \text{for } i = 1, \dots, n$$

and

$$\left| y_i^{(0)} - ax_i^{(0)} - b \right| = d_i^+ + d_i^- \quad \text{for } i = 1, \dots, n,$$

where for each index i, the two variables d_i^+ and d_i^- are nonnegative but cannot simultaneously be greater than zero. In words, d_i^+ is the distance that $y_i^{(0)}$ lies above the line $y = ax + b$ and d_i^- is the distance that $y_i^{(0)}$ lies below the line $y = ax + b$. Except for the requirement that the variables d_i^+ and d_i^- not both have positive values at the same time, we can convert the optimization problem above into the LP model

Minimize $d_1^+ + d_1^- + d_2^+ + d_2^- + \cdots + d_n^+ + d_n^-$

subject to $y_i^{(0)} - ax_i^{(0)} - b = d_i^+ - d_i^-, \quad i = 1, \dots, n$

and $d_i^+, d_i^- \geq 0, \quad i = 1, \dots, n.$

Luck is now on our side because the form of this LP model automatically ensures that in the optimal solution, d_i^+ and d_i^- are not both positive at the same time. The proof of this statement is by contradiction; that is, a contradiction is deduced by assuming that the statement is not true. Suppose that for a certain i, both d_i^+ and d_i^- have a positive value in the optimal solution, for example $d_1^+ = 5$ and $d_1^- = 3$. If we then set $d_1^+ = 2$ and $d_1^- = 0$, we still have $d_1^+ - d_1^- = 2$ while the objective value decreases by $3 + 3 = 6$. This is in contradiction with the fact that the solution has minimum objective value, thus proving the statement.

Since the parameters a and b do not necessarily have to be nonnegative, the linear minimization problem still needs a small adjustment before it can be entered into an LP software package. The variables a and b must be replaced by the variables a^+, a^-, b^+, and b^- with $a = a^+ - a^-$, and $b = b^+ - b^-$, where $a^+, a^-, b^+, b^- \geq 0$. The linear constraint $y_i^{(0)} - ax_i^{(0)} - b = d_i^+ - d_i^-$ is therefore rewritten as the linear constraint

$$y_i^{(0)} - a^+ x_i^{(0)} + a^- x_i^{(0)} - b^+ + b^- = d_i^+ - d_i^-.$$

Table 1.5   Data for the fitting curve.

	1	2	3	4	5	6	7	8	9	10
$x_i^{(0)}$	0.1	0.6	1.5	2.0	3.5	4.8	6.0	7.5	8.6	10.0
$y_i^{(0)}$	1.3	1.0	-0.3	-0.2	1.2	1.9	3.9	4.6	6.4	7.5

As a numerical illustration, consider the data given in Table 1.5. In the computer input for the LP model, we denote d_1^+ by d_{1p}, d_1^- by d_{1m}, and so on. The LP model corresponding to the data in Table 1.5 is

$$\begin{aligned}
\text{Minimize} \quad & d_{1p} + d_{1m} + d_{2p} + d_{2m} + d_{3p} + d_{3m} + d_{4p} + d_{4m} \\
& + d_{5p} + d_{5m} + d_{6p} + d_{6m} + d_{7p} + d_{7m} + d_{8p} \\
& + d_{8m} + d_{9p} + d_{9m} + d_{10p} + d_{10m} \\
\text{subject to} \quad & 1.3 - 0.1a_p + 0.1a_m - b_p + b_m = d_{1p} - d_{1m} \\
& 1.0 - 0.6a_p + 0.6a_m - b_p + b_m = d_{2p} - d_{2m} \\
& -0.3 - 1.5a_p + 1.5a_m - b_p + b_m = d_{3p} - d_{3m} \\
& -0.2 - 2.0a_p + 2.0a_m - b_p + b_m = d_{4p} - d_{4m} \\
& 1.2 - 3.5a_p + 3.5a_m - b_p + b_m = d_{5p} - d_{5m} \\
& 1.9 - 4.8a_p + 4.8a_m - b_p + b_m = d_{6p} - d_{6m} \\
& 3.9 - 6.0a_p + 6.0a_m - b_p + b_m = d_{7p} - d_{7m} \\
& 4.6 - 7.5a_p + 7.5a_m - b_p + b_m = d_{8p} - d_{8m} \\
& 6.4 - 8.6a_p + 8.6a_m - b_p + b_m = d_{9p} - d_{9m} \\
& 7.5 - 10.0a_p + 10.0a_m - b_p + b_m = d_{10p} - d_{10m} \\
\text{and} \quad & a_p, a_m, b_p, b_m, d_{ip}, d_{im} \geq 0 \quad \text{for all } i.
\end{aligned}$$

The computer solution to the LP model gives the minimum objective value 7.394 with $a_p = 0.918$, $a_m = 0$, $b_p = 0$, and $b_m = 1.676$. The coefficients of the fitting line $y = ax + b$ are therefore $a = 0.918$ and $b = -1.676$. The minimum sum of the distances between the data points and the fitted line is 7.394. The method from this subsection can also be applied to fit a quadratic curve $y = ax^2 + bx + c$ to given points $\left(x_i^{(0)}, y_i^{(0)}\right)$ in the plane or a plane $z = ax + by + c$ to given points $\left(x_i^{(0)}, y_i^{(0)}, z_i^{(0)}\right)$ in three-dimensional space.

1.4   The Simplex Method

To solve an LP model on the computer, one needs a *systematic* procedure. The simplex method is such a procedure. The basic idea of the simplex method for solving LP models shows a certain similarity to that of the famous elimination method of Gauss for solving systems of linear equations. The latter is a systematic method that works in the same way whether one has to solve two linear equations in two unknowns or a thousand linear equations in a thousand unknowns. Before describing the principle of the method, we begin with a seemingly trivial remark. The system $x_1 = 1.5$ and $x_2 = 4$ is a system of two equations in two unknowns whose solution has already been given. If we add $x_1 = 1.5$ twice to $x_2 = 4$ and subtract $x_2 = 4$ from $x_1 = 1.5$, we obtain the two linear equations $2x_1 + x_2 = 7$ and $x_1 - x_2 = -2.5$ with, of course, solution $x_1 = 1.5$ and $x_2 = 4$. The Gaussian elimination method works in the opposite direction to calculate the solution to a system of linear equations.

We explain the elimination method of Gauss (one of the greatest mathematicians of all time) using the following system of three linear equations in three unknowns:

$$2x_1 + x_2 + x_3 = 1$$
$$4x_1 + x_2 \qquad = -2$$
$$-2x_1 + 2x_2 + x_3 = 7.$$

The Gaussian elimination algorithm begins by subtracting a multiple of the first equation from the other equations to eliminate the variable x_1 from those other equations. This means that we

a. subtract 2 times the first equation from the second and
b. add 1 time the first equation to the third.

This gives the system of linear equations

$$2x_1 + x_2 + x_3 = 1$$
$$- x_2 - 2x_3 = -4$$
$$3x_2 + 2x_3 = 8.$$

This system is equivalent to the first system: every solution to the first system is a solution to the second system and vice versa.

In the second step of the Gaussian elimination algorithm, we leave the first equation of the new system as is. The last two equations contain only x_2 and x_3, and the same elimination idea is applied to these two equations: subtract a multiple of the second equation from the third equation to eliminate x_2. If we add 3 times the second equation to the third, we obtain the equivalent system of linear equations

$$2x_1 + x_2 + x_3 = 1$$
$$-x_2 - 2x_3 = -4$$
$$-4x_3 = -4.$$

It is clear how to solve this last system. The last equation gives $x_3 = 1$, substituting $x_3 = 1$ in the second equation gives $x_2 = 2$, and, finally, the first equation gives $x_1 = -1$. The solution to the original system of equations is therefore ($x_1 = -1$, $x_2 = 2$, $x_3 = 1$).

The elimination procedure only works if the system consists of equalities. In a system of inequalities, the procedure can go wrong. To see this, consider the two inequalities

$$x_1 + x_2 \leq 8$$
$$-2x_1 + x_2 \leq 5.$$

If we rewrite the second inequality by adding twice the first, we obtain the inequalities

$$x_1 + x_2 \leq 8$$
$$3x_2 \leq 21.$$

The point $(x_1 = 0, x_2 = 7)$ satisfies the second system of inequalities but not the first.

In an LP model, the constraints can be either equalities or inequalities. The simplex method for solving LP models requires that the LP model first be transformed into an equivalent LP model with only equalities (subject to the nonnegativity requirements). We will show how to do this shortly. In the equivalent LP model, called the *standard form*, all constraints are equalities. In the standard form, the variables in each equality constraint are on the left-hand side of the equation and the constant on the right-hand side is nonnegative. This transformation into the standard form is carried out automatically in LP software modules. First, nonnegative coefficients are created on the right-hand side by multiplying every constraint with a negative coefficient on the right-hand side by -1. Note that a $\leq$ constraint then becomes a $\geq$ constraint and vice versa.

Slack and Surplus Variables

Slack and surplus variables are used to transform linear inequalities into equivalent linear equalities. We illustrate this using two inequalities from the LP model in Section 1.2.1. Consider the inequality

$$0.70x_A + 0.75x_B + 0.55x_C + 0.34x_D \leq 400$$

that gives the restriction on the number of available assembly hours. If we introduce a new variable v, this inequality is equivalent to

$$0.70x_A + 0.75x_B + 0.55x_C + 0.34x_D + v = 400$$
$$\text{and} \qquad\qquad\qquad\qquad v \geq 0.$$

After all, if $(x_A = x_A^{(0)}, x_B = x_B^{(0)}, x_C = x_C^{(0)}, x_D = x_D^{(0)})$ satisfies the inequality, then the equality with nonnegativity constraint $v \geq 0$ is satisfied by $(x_A = x_A^{(0)}, x_B = x_B^{(0)}, x_C = x_C^{(0)}, x_D = x_D^{(0)}, v = v^{(0)})$ with $v^{(0)} = 400 - 0.70x_A^{(0)} - 0.75x_B^{(0)} - 0.55x_C^{(0)} - 0.34x_D^{(0)}$. Conversely, if $(x_A = x_A^{(1)}, x_B = x_B^{(1)}, x_C = x_C^{(1)}, x_D = x_D^{(1)}, v = v^{(1)})$ satisfies the equality with $v^{(1)} \geq 0$, then $(x_A = x_A^{(1)}, x_B = x_B^{(1)}, x_C = x_C^{(1)}, x_D = x_D^{(1)})$ satisfies the inequality.

The variable v is called a *slack variable*. In the example above, it gives the number of assembly hours remaining in the solution out of the available 400 hours. The method is similar for a $\geq$ constraint. Consider the constraint

$$x_A \geq 100$$

for the production quantity of product A. If we introduce the new variable w, then this constraint is equivalent to

$$x_A - w = 100$$
$$\text{and} \qquad w \geq 0.$$

The variable w is called a *surplus variable* and gives the quantity by which the production of product A exceeds the minimum requirement of 100 units.

Basic Solution and Basic Variables

For LP models in two variables, using the graphical method, we have already shown that an optimal solution is reached at a corner point of the feasible region. This property remains valid in general for LP problems with more than two decision variables (in higher dimension, the concept of corner point is difficult to visualize: a corner point is then a feasible solution that cannot be written as a linear combination of two other feasible solutions). In a feasible solution that corresponds to a corner point in the feasible region, *at most* m of the decision variables can take on a *positive* value, where m is the number of constraints in the LP problem (excluding the nonnegativity constraints). This result is proved in Section 1.7. A feasible solution that corresponds to a corner point is called a *basic solution*. If, for the sake of convenience, we take a feasible basic solution for which exactly m variables have a *positive* value, then those m variables are called the *basic variables* of the specific basic solution.

The Simplex Method

The simplex method runs through the set of all possible feasible solutions of the LP model in standard form in a "smart" way (this set can be enormous: in principle, there are $\binom{n}{m}$ possible basic solutions if n is the number of variables and m is the number of constraints, with $m \leq n$). The core idea behind the simplex method is as follows:

1. First, construct a basic feasible solution as initial solution.
2. From the current basic solution, construct a new basic solution that differs from the current basic solution by only one basic variable. For this, one of the current nonbasic variables is made positive while one of the current basic variables is set to zero. The choice of the "entering" and "exiting" variables is determined by the following criteria:
   (i) The objective value must be improved.
   (ii) The new solution must be a basic feasible solution.
3. Continue the construction of improved basic solutions until the objective value cannot be improved. The simplex method has then found an optimal solution.

Every algorithm requires an initial solution. Constructing an initial solution is not always easy. For LP models, it is only easy if every constraint in the original formulation is a $\leq$ constraint. For the explanation of the simplex method, we restrict ourselves to this case. The easiest way to explain the simplex method is by using a numerical example. Consider the following LP model:

$$
\begin{array}{lll}
\text{Maximize} & 3x_1 + 2x_2 - 0.5x_3 & \\
\text{subject to} & 4x_1 + 3x_2 + x_3 & \leq 10 \\
& 3x_1 + x_2 - 2x_3 & \leq 8 \\
\text{and} & x_1, x_2, x_3 & \geq 0.
\end{array}
$$

If we introduce the slack variables x_4 and x_5, this LP model is equivalent to the LP model

$$
\begin{aligned}
\text{Maximize} \quad & 3x_1 + 2x_2 - 0.5x_3 \\
\text{subject to} \quad & 4x_1 + 3x_2 + x_3 + x_4 && = 10 \\
& 3x_1 + x_2 - 2x_3 && + x_5 = 8 \\
\text{and} \quad & x_1, x_2, x_3, x_4, x_5 && \geq 0.
\end{aligned}
$$

This LP model in standard form already has the so-called *canonical form*; that is, in every equality constraint, there is a variable with coefficient 1 in that constraint and coefficient 0 in the other equality constraints. In the representation above, these are the variables x_4 and x_5. If the LP model is written in canonical form, one can immediately see a basic feasible solution. In the representation above, one can find this by setting x_4 equal to 10, x_5 equal to 8, and the other variables equal to zero. This gives the basic feasible solution

$$ x_4 = 10, \quad x_5 = 8, \quad x_1 = x_2 = x_3 = 0 $$

with basic variables x_4 and x_5. To find a better basic solution, we are going to rewrite the system of linear equations as an equivalent system of linear equations that is also in *canonical form*, that is, contains a subsystem whose matrix is the identity. This is the core idea of the simplex method. To rewrite the system, we use a method similar to Gaussian elimination:

a. Divide both sides of an equation by a nonzero number.
b. Add multiples of one equation to the other equations.

To explain the simplex step to improve the existing basic feasible solution, it is didactically useful to introduce a variable z for the objective function. The previous representation of the LP model in canonical form can be written as

$$
\begin{aligned}
\text{Maximize} \quad & z = 3x_1 + 2x_2 - 0.5x_3 \\
\text{subject to} \quad & 4x_1 + 3x_2 + x_3 + x_4 && = 10 \\
& 3x_1 + x_2 - 2x_3 && + x_5 = 8 \\
\text{and} \quad & x_1, x_2, x_3, x_4, x_5 && \geq 0.
\end{aligned}
$$

Once again, the basic feasible solution associated with this canonical form is $x_4 = 10$, $x_5 = 8$, and $x_i = 0$ for the remaining variables, with corresponding objective value $z = 0$. Note that the auxiliary equation for the variable z contains only the nonbasic variables x_1, x_2, x_3 of the current basic solution. This fact is essential when constructing an improved basic solution. Now, ask yourself the following question.

Question: If you could make exactly one of the nonbasic variables positive (at the expense of the values of the current basic variables), which variable would you choose, and how big would you make that variable?

The first part of the answer lies in the z-equation. In the current canonical form, the z-equation contains several nonbasic variables with *positive* coefficients. It seems obvious that one should change the nonbasic variable with the greatest positive coefficient. Doing this gives the greatest "short-term" improvement in the z-value. The nonbasic variable x_1 has the greatest positive coefficient in the current z-equation. If one chooses to make x_1 positive, the values of the current basic variables x_4 and x_5 change. How big can one make x_1? The equality constraint $4x_1 + 3x_2 + x_3 + x_4 = 10$ implies that x_1 should not exceed $\frac{10}{4}$, because otherwise x_4 would be negative. Likewise, the equality constraint $3x_1 + x_2 - 2x_3 + x_5 = 8$ implies that x_1 should not exceed $\frac{8}{3}$. Hence, one cannot increase x_1 to more than $\frac{10}{4}$, namely the least of the two bounds $\frac{10}{4}$ and $\frac{8}{3}$. The first equality constraint with basic variable x_4 imposes the upper bound $\frac{10}{4}$. If one increases x_1 up to this bound, the basic variable x_4 is pushed down to zero. In other words, x_1 takes the place of x_4 in the basis, and the new basic variables are x_1 and x_5. In the next step of the simplex method, the LP model is rewritten in the canonical form corresponding to the new basic variables x_1 and x_5. Above, we have already given a global explanation of how to do this. To rewrite the equality constraints in the new canonical form, first divide both sides of the equation $4x_1 + 3x_2 + x_3 + x_4 = 10$ by 4 to make the coefficient of x_1 equal to 1. The resulting equation $x_1 + 0.75x_2 + 0.25x_3 + 0.25x_4 = 2.5$ is subtracted 3 times from the equation $3x_1 + x_2 - 2x_3 + x_5 = 8$ to make the coefficient of x_1 in that equation equal to zero. This gives the equality $-1.25x_2 - 2.75x_3 - 0.75x_4 + x_5 = 0.5$. The z-equation is also adjusted to include only the new nonbasic variables x_2, x_3, x_4. Substituting $x_1 = 2.5 - 0.75x_2 - 0.25x_3 - 0.25x_4$ in $z = 3x_1 + 2x_2 + 0.5x_3$ gives

$$z = 3(2.5 - 0.75x_2 - 0.25x_3 - 0.25x_4) + 2x_2 - 0.5x_3$$
$$= 7.5 - 0.25x_2 - 1.25x_3 - 0.75x_4.$$

The canonical form of the LP model corresponding to the new basic variables x_1 and x_5 is therefore

$$\begin{array}{lll} \text{Maximize} & z = 7.5 - 0.25x_2 - 1.25x_3 - 0.75x_4 \\ \text{subject to} & x_1 + 0.75x_2 + 0.25x_3 + 0.25x_4 & = 2.5 \\ & \quad\quad -1.25x_2 - 2.75x_3 - 0.75x_4 + x_5 = 0.5 \\ \text{and} & x_1, x_2, x_3, x_4, x_5 & \geq 0. \end{array}$$

The corresponding basic solution is

$$x_1 = 2.5, \ x_5 = 0.5, \ x_2 = x_3 = x_4 = 0$$

with objective value $z = 7.5$. This is indeed an improved basic solution. However, in the z-equation $z = 7.5 - 0.25x_2 - 1.25x_3 - 0.75x_4$, all nonbasic variables have a *negative* coefficient, so that the value of z would decrease if one of these variables were made positive. The most recent canonical form of the LP model shows that the corresponding basic solution must be optimal. After all, based on the nonnegativity constraints for the variables, it immediately follows from the z-equation that $z \leq 7.5$ for every feasible solution. For the basic solution $x_1 = 2.5$, $x_2 = 0$, $x_3 = 0$, $x_4 = 0$,

$x_5 = 0.5$, we have $z = 7.5$, so that this solution is indeed optimal for the LP model in standard form. The optimal solution to the original LP model with the $\leq$ constraints is of course $x_1 = 2.5$, $x_2 = x_3 = 0$ with maximum objective value 7.5.

As further clarification, we repeat the steps for the following LP example:

$$\begin{aligned}
\text{Maximize} \quad & 3x_1 + 2x_2 + x_3 \\
\text{subject to} \quad & 2x_1 + x_2 + x_3 \leq 40 \\
& x_1 + 3x_2 + 2x_3 \leq 70 \\
\text{and} \quad & x_1, x_2, x_3 \geq 0.
\end{aligned}$$

Initialization. Introduction of the slack variables x_4 and x_5 gives the equivalent LP model

$$\begin{aligned}
\text{Maximize} \quad & 3x_1 + 2x_2 + x_3 \\
\text{subject to} \quad & 2x_1 + x_2 + x_3 + x_4 = 40 \\
& x_1 + 3x_2 + 2x_3 + x_5 = 70 \\
\text{and} \quad & x_1, x_2, x_3 \geq 0.
\end{aligned}$$

For this canonical form, x_4 and x_5 are the basic variables. The basic solution is $x_4 = 40$, $x_5 = 70$, and $x_j = 0$ for the remaining variables, with $z = 0$.

Iteration 1. (a) In the z-equation, x_1 has the greatest positive coefficient. So x_1 becomes a new basic variable. Since $\min(\frac{40}{2}, \frac{70}{1}) = \frac{40}{2}$, the variable x_4 leaves the basis.

(b) Rewrite the two constraints for the new basic variables x_1 and x_5 in the corresponding canonical form:

$$\begin{aligned}
x_1 + 0.5x_2 + 0.5x_3 + 0.5x_4 &= 20 \\
2.5x_2 + 1.5x_3 - 0.5x_4 + x_5 &= 50.
\end{aligned}$$

Rewrite the z-equation as

$$\begin{aligned}
z &= 3(20 - 0.5x_2 - 0.5x_3 - 0.5x_4) + 2x_2 + x_3 \\
&= 60 + 0.5x_2 - 0.5x_3 - 1.5x_4,
\end{aligned}$$

so that this includes only the nonbasic variables.

The new canonical form is

$$\begin{aligned}
\text{Maximize} \quad z = \; & 60 + 0.5x_2 - 0.5x_3 - 1.5x_4 \\
\text{subject to} \quad & x_1 + 0.5x_2 + 0.5x_3 + 0.5x_4 = 20 \\
& 2.5x_2 + 1.5x_3 - 0.5x_4 + x_5 = 50.
\end{aligned}$$

The basic variables are x_1 and x_5. The basic solution is $x_1 = 20$, $x_5 = 50$, and $x_j = 0$ for the remaining variables, with z-value $z = 60$.

Iteration 2. (a) In the z-equation, x_2 has the greatest positive coefficient. So x_2 becomes the new basic variable. Since $\min(\frac{20}{0.5}, \frac{50}{2.5}) = \frac{50}{2.5}$, the variable x_5 leaves the basis.

(b) Rewrite the two constraints in the canonical form corresponding to the basic variables x_1 and x_2 (the new basic variable x_2 has coefficient $+1$ in the equation that contains the variable x_5 leaving the basis; therefore, divide this equation by 2.5):

$$x_1 \quad + 0.2x_3 + 0.6x_4 - 0.2x_5 = 10$$
$$x_2 + 0.6x_3 - 0.2x_4 + 0.4x_5 = 20.$$

Rewrite the z-equation as

$$z = 60 + 0.5(20 - 0.6x_3 + 0.2x_4 - 0.4x_5) - 0.5x_3 - 1.5x_4$$
$$= 70 - 0.8x_3 - 1.4x_4 - 0.2x_5.$$

In the z-equation, all nonbasic variables have a negative coefficient. This means that the objective value of 70 cannot be improved. The optimal basic solution for the basic variables x_1 and x_2 is therefore $x_1 = 10$, $x_2 = 20$, and $x_j = 0$ for the remaining variables, with maximum objective value 70.

The example above explains the essence of the calculations of the simplex method. Most educational LP software packages perform the calculations following the procedure outlined above. This procedure is simple and has didactic advantages, but it is not the most efficient implementation of the simplex method. The calculations of the simplex method can be significantly reduced using matrix algebra. This subject is briefly discussed in Section 1.7.

The Construction of an Initial Solution

When the LP model contains $\geq$ constraints and/or $=$ constraints, constructing an initial solution is not as easy as in the case of only $\leq$ constraints. We illustrate how to proceed in such a case using the following LP example:

$$
\begin{array}{llll}
\text{Maximize} & 3x_1 + 5x_2 & + 2x_3 & \\
\text{subject to} & 2x_1 & - x_2 + 0.5x_3 & \geq 7 \\
& 5x_1 + 2x_2 & + x_3 & = 20 \\
\text{and} & & x_1, x_2, x_3 & \geq 0.
\end{array}
$$

First, a surplus variable x_4 is introduced to rewrite the LP model in the following equivalent standard form:

$$
\begin{array}{llll}
\text{Maximize} & 3x_1 + 5x_2 & + 2x_3 & \\
\text{subject to} & 2x_1 & - x_2 + 0.5x_3 - x_4 = & 7 \\
& 5x_1 + 2x_2 & + x_3 & = 20 \\
\text{and} & & x_1, x_2, x_3, x_4 & \geq 0.
\end{array}
$$

This LP formulation is not in canonical form. To construct an initial solution for an LP formulation that is not in canonical form, we need so-called *artificial variables*. In the standard form, an artificial variable is added to every equality constraint that was not originally a $\leq$ constraint. This is given coefficient 1 in the equality constraint in question and coefficient 0 in all other equality constraints. Then, an auxiliary model is considered with the new linear equality constraints and with objective function the sum of the artificial variables. For the LP example above, two artificial variables u_1 and u_2 are introduced, and the following auxiliary model is considered:

$$
\begin{aligned}
\text{Minimize}\quad & u_1 \;+ u_2 \\
\text{subject to}\quad & 2x_1 \;- x_2 + 0.5x_3 - x_4 + u_1 \;= 7 \\
& 5x_1 + 2x_2 + x_3 + u_2 = 20 \\
\text{and}\quad & x_1, x_2, x_3, x_4, u_1, u_2 \geq 0.
\end{aligned}
$$

This is an LP model in canonical form. If the original LP model has a feasible solution, then the auxiliary model must have an optimal basic solution with $u_1 = u_2 = 0$ (why?). The auxiliary model can be solved using the simplex method as described previously. The values of the variables x_1, x_2, x_3, x_4 in the optimal basic solution to the auxiliary model then form a feasible initial solution for the standard form of the original LP model. For more details, the interested reader is referred to the book of Bradley, Hax, and Magnanti (1977).

Complexity of the Simplex Method

It is interesting to know how much work the simplex method requires. Does the simplex method always stop after a finite number of iterations? How quickly does the number of iterations increase when the number of constraints is increased?

The answer to the first question is "no." Examples have been constructed for which the simplex method does not stop after a finite number of steps. These examples are artificial ones that do seldom occur in real-world problems. Consider the following example:

$$
\begin{aligned}
\text{Maximize}\quad & 0.75x_1 - 20x_2 + 0.5x_3 - 6x_4 \\
\text{subject to}\quad & 0.25x_1 - 8x_2 - x_3 + 9x_4 \leq 0 \\
& 0.5x_1 - 12x_2 - 0.5x_3 + 3x_4 \leq 0 \\
& x_3 \leq 1 \\
\text{and}\quad & x_1, x_2, x_3, x_4 \geq 0.
\end{aligned}
$$

In this example, the phenomenon called "cycling" occurs: the series of feasible solutions generated by the simplex method ends up in a loop that keeps repeating itself and where the objective value remains the same. One can check this experimentally using an LP software module. Fortunately, in economic planning models, cycling is rare. Therefore, LP computer programs do not include any provisions against cycling. Such provisions would slow down the simplex method too much. We may therefore assume that in real-world cases, the simplex method stops after a finite number of iterations. Let us proceed to the second question. Again, an artificial example can be constructed that shows that the number of required iterations can explode. Consider the following example:

$$\text{Maximize} \quad \sum_{j=1}^{m} 10^{m-j} x_j$$

$$\text{subject to} \quad 2 \sum_{j=1}^{i-1} 10^{i-j} x_j + x_i \leq 100^{i-1} \quad \text{for } i = 1, \ldots, m$$

$$\text{and} \quad x_j \geq 0 \quad \text{for } j = 1, \ldots, m.$$

For this example, it has been shown that the simplex method needs 2^m iterations; see the book of Chvátal (1983). This number of iterations grows exponentially above any practical bound as m increases. However, many years of experience with the simplex method have shown that the example above is an abnormal one. Empirical findings indicate that for real-world problems, the number of iterations is generally less than $\frac{3}{2}m$ and rarely exceeds $3m$, where m is the number of constraints of the problem. The number of constraints is the dominant factor for the number of required iterations. With regard to the computation time, experience shows that the total computation time roughly increases as the third power of the number of constraints. The computation time of an LP problem with 200 constraints is probably 8 times as much as the computation time of a similar LP problem with 100 constraints.

Karmarkar's Interior-Point Method

As shown by the example above, the simplex method is not a *polynomial-time algorithm*; that is, the computation time is not guaranteed to be bounded from above by a polynomial function of the size of the problem. Nevertheless, the simplex method works excellently in practice. The average performance of an algorithm is more important than performing less well in an abnormal case. The fact that the simplex method is not a polynomial-time algorithm has led to research into alternative solution methods for LP problems. Nowadays, the simplex method has a serious competitor in the form of the interior-point method of Karmarkar, which is a polynomial-time algorithm. In contrast to the simplex method, which explores the corner points of the feasible region in a clever manner, the interior-point method cuts through the interior of the feasible region to approach the optimal corner point closer and closer. The publication of Karmarkar's method was front-page news in the *New York Times* because it was thought that this method represented a breakthrough that would make it possible to solve all sorts of hitherto unsolvable optimization problems in, for example, aviation, the oil industry, and the military. The high expectations of the new method have now been adjusted. Nevertheless, for very large LP problems, the interior-point method is often significantly faster than the simplex method. There are two explanations for the speed of the interior-point method. On the one hand, the system of linear equations that must be solved in each iteration step has a special property (a symmetric positive-definite coefficient matrix) that allows the use of an extremely efficient solution method for linear equations. On the other hand, empirically, it turns out that the number of iterations of the interior-point method hardly depends on the problem size and typically lies between 20 and 80. The simplex method was not driven out by Karmarkar's interior-point method. The development of the latter has led to renewed research into the simplex method, resulting in significantly faster versions of the simplex method. Competition is the best guarantee for improvement. An excellent introduction to Karmarkar's method can be found in the book of G. L. Strang (1986), *Introduction to Applied Mathematics*, Wellesley-Cambridge Press.

1.5   The Dual LP Problem

The computer output of an LP software package provides not only the solution to the original LP problem but also the solution to another LP problem. The other LP problem is called the dual LP problem, or dual, and is naturally related to the original LP problem. Duality is a fundamental concept in applied mathematics and also plays an essential role in linear programming. The original LP problem we want to solve is called the primal LP problem, or primal. While the solution to the primal gives the optimal values of the production activities, the solution to the dual gives us information about the economic value of each of the required raw materials. The solution to the dual is the basis for the so-called shadow prices for the constraints of the primal given in the computer output for the primal. If the constraint concerns a limit on the available quantity of a specific raw material, then the shadow price associated with that constraint gives us the marginal value of an extra unit of the raw material. Not only does duality in linear programming have an important economic interpretation, it is also of great practical importance for sensitivity analysis. The theory of duality for linear programming provides the basis for sensitivity analysis. Sensitivity analysis allows us to study the influence of changes in the model's data on the optimal solution. In practice, people are very often interested in this because in many cases, there is a margin of uncertainty in the data. In sensitivity analysis, one can, for example, check how much a cost or profit coefficient of a decision variable may change without changing the optimal values of the decision variables. In Section 1.6, we go into more detail about sensitivity analysis and show how the simplex method is ideally suited to perform sensitivity analysis.

In the next subsection, we clarify the significance of shadow prices and dual variables intuitively through an appealing example, with the help of economic "common sense" principles. Section 1.7 provides a mathematical treatment of duality.

1.5.1   *An Illustrative Example*

Problem statement.   Using a simple LP problem, we explain the significance of the dual and shadow prices intuitively with economic arguments. Suppose that a company produces three types of fertilizer, called Super-G, Quick-G, and Normal-G. Two basic raw materials are used in the production process. Table 1.6 gives the quantities of raw materials required for the various types of fertilizer, the availability of each raw material, and the selling prices of the various types of fertilizer. The purchase costs of the raw materials have been included in the selling prices. We assume that any production costs for making the fertilizer are negligible. This assumption is made in order to avoid ambiguities in the economic arguments justifying the dual problem.

Table 1.6   Data for the fertilizer example.

Raw Material	Super-G	Quick-G	Normal-G	Available (Tons)
	Raw Material Requirements (Tons per Ton)			
1	3	2.5	2	1250
2	4	2	1	1000
Selling Price (per Ton)	275 euros	210 euros	175 euros	

Formulation.   The decision variables are

$$x_1 = \text{number of tons of Super-G to produce}$$
$$x_2 = \text{number of tons of Quick-G to produce}$$
$$x_3 = \text{number of tons of Normal-G to produce.}$$

The LP model for the fertilizer company is given by

$$\text{Maximize } 275x_1 + 210x_2 + 175x_3$$
$$\text{subject to } \quad 3x_1 + 2.5x_2 + 2x_3 \le 1250$$
$$4x_1 + 2x_2 + x_3 \le 1000$$
$$\text{and} \quad x_1, x_2, x_3 \ge 0.$$

We provide an explicit evaluation of this LP using the simplex method in Section 1.6.4. Here we continue with the computer output for the LP model of the fertilizer company:

Fig. 1.4   Computer output for the LP model of the factory.

```
Summary of Results

Value Objective Function :      111250.0000

Variable            Activity Level    Reduced Cost
--------            --------------    ------------
X1          :           150.0000         0.0000
X2          :             0.0000        12.5000
X3          :           400.0000         0.0000
```

	Slack or Surplus	Shadow Prices
Constraint 1	0.0000	85.0000
Constraint 2	0.0000	5.0000

The optimal solution shows that the available quantities of both raw materials are fully used and that the selling price of product 2 is too low to make this product. This leads to the following questions:

1. By how much would the total revenue increase if one extra ton of raw material 1 or 2 were available?
2. How much higher should the selling price of product 2 be to make it worthwhile to manufacture this product?

The answers to these questions are contained in the computer output. These answers are obtained using the so-called *dual* of the LP problem. To motivate the dual of the fertilizer problem, we conduct a *thought experiment*, in which it is important to make the following assumptions: (a) the production costs for making the fertilizer are negligible and (b) the purchase costs of the raw materials are already included in the selling prices. The thought experiment is as follows. Suppose that a trader wants to buy all the company's raw materials. How does the trader choose the prices she is willing to pay for the raw materials? For this, define

$$y_1 = \text{price to pay for raw material 1}$$
$$y_2 = \text{price to pay for raw material 2,}$$

where the prices y_1 and y_2 are expressed in euros per ton. The trader's objective is, of course, to minimize the total payment, that is,

$$\text{Minimize} \quad 1250y_1 + 1000y_2.$$

On the other hand, the trader need to set the prices sufficiently high that the company is willing to sell her the raw materials. The table shows that the company can obtain a revenue of 275 euros by using 3 tons of raw material 1 and 4 tons of raw material 2 to make 1 ton of Super-G fertilizer. The trader's prices y_1 and y_2 must therefore satisfy

$$3y_1 + 4y_2 \geq 275.$$

Likewise, the prices y_1 and y_2 must satisfy

$$2.5y_1 + 2y_2 \geq 210;$$

otherwise, it is more profitable for the company to make the fertilizer Quick-G than to sell the trader the raw materials. The third activity, Normal-G, gives the constraint

$$2y_1 + y_2 \geq 175.$$

Finally, the prices y_1 and y_2 must be nonnegative since the company is, of course, not willing to give away money. One can then determine the optimal prices y_1 and y_2 by solving the following LP problem for the trader:

$$
\begin{aligned}
\text{Minimize} \quad & 1250y_1 + 1000y_2 \\
\text{subject to} \quad & 3y_1 + 4y_2 \geq 275 \\
& 2.5y_1 + 2y_2 \geq 210 \\
& 2y_1 + y_2 \geq 175 \\
\text{and} \quad & y_1, y_2 \geq 0.
\end{aligned}
$$

The computer output for the LP solution to this minimization problem is given by

Fig. 1.5   Computer output for the LP model for the trader.

```
Summary of Results


Value Objective Function :      111250.0000


Variable          Activity Level    Reduced Cost
--------          --------------    ------------

Y1        :           85.0000          0.0000
Y2        :            5.0000          0.0000


                  Slack of Surplus  Shadow prices
                  ----------------  -------------

Constraint 1           0.0000          150.0000
Constraint 2          12.5000            0.0000
Constraint 3           0.0000          400.0000
```

The optimal values of the prices y_1 and y_2 are therefore 85 euros and 5 euros. Because of the significance of the dual variables y_1 and y_2, the optimal values of y_1 and y_2 are called the *shadow prices* of the raw materials. Moreover, these optimal values do not require a separate resolution of the trader's LP problem, they are automatically calculated when the simplex method is applied to the (primal) LP model for the factory. For this LP problem, the final z-equation, representing the objective function, is given by $z = 111\,250 - 12.50x_2 - 85s_1 - 5s_2$, where s_1 and s_2 are the slack variables. So one sees that the negatives of the values of the coefficients of the two slack variables in the final z-equation give the optimal values of the dual variables y_1 and y_2. This result holds in general for the (primal) LP problem: maximize $\sum_{j=1}^{n} c_j x_j$ subject to $\sum_{j=1}^{n} a_{ij} x_j \leq b_i$ for $i = 1, \ldots, m$ and $x_j \geq 0$ for all j; see Section 1.6.4.

The LP problem for the trader is called the *dual* of the original LP problem for the factory. For the sake of convenience, the original problem is called the *primal*. The discussion above shows that the dual concerns the economic value of the raw materials. There is an interesting relation between the optimal objective value of the primal and that of the dual. Needless to say, one cannot pay the company less than the maximum revenue the company can achieve by using the raw materials for the production of fertilizer. On the other hand, common sense says that there is no reason to pay more than that maximum return. From this, we can conclude that the *minimum objective value of the dual* must equal the *maximum objective value of the primal*. This important result can indeed be proved mathematically, but the mathematical proof of this intuitively explained result is far from easy.

Interpretation of the Dual Solution

Solving the LP problem for the fertilizer factory (the primal) gives the optimal solution

$$x_1^* = 150, \ x_2^* = 0, \ \text{and} \ x_3^* = 400,$$

with maximum objective value 111 250 euros (in optimization, it is customary to indicate the optimal numerical values of the decision variables by adding an asterisk). Solving the LP problem for the trader (the dual) gives the optimal solution

$$y_1^* = 85 \ \text{and} \ y_2^* = 5,$$

with minimum objective value 111 250 euros. Indeed, the minimum of the dual is equal to the maximum of the primal. The dual solution (y_1^*, y_2^*) is not only important for the trader; it also contains a wealth of information for the fertilizer factory. For example, in the primal, the constraint for raw material 1 is binding in the optimal solution; that is, equality holds. This means that the fertilizer factory could make a higher profit if it had access to an extra quantity of raw material 1. The dual value y_1^* gives an indication of the increase in profit if the factory had access to one extra ton of raw material 1 in addition to the currently available quantity. If y_1^* is the amount the trader is willing to pay for each ton of the available quantity of raw material 1, then for economic reasons, it makes sense for the factory to also set the value of one extra ton of raw material 1 equal to y_1^*. In other words,

> y_1^* *gives an indication of the increase in the company's total revenue if the factory has 1 additional ton of raw material 1 at its disposal.*

Likewise, y_2^* gives an indication of the increase in the total profits of the company if it has access to 1 extra ton of raw material 2 in addition to the currently available quantity. These *intuitive* interpretations of the dual values are substantiated mathematically in Sections 1.6.4 and 1.7. The dual values y_1^* and y_2^* can be found in the computer output as the *shadow prices* of constraints 1 and 2.

In the optimal solution to the LP problem for the fertilizer company, we have $x_2^* = 0$. This means that the profit coefficient of product 2 must be greater for it to be profitable to manufacture this product. The amount by which the profit coefficient of product 2 should at least increase can also be related to the dual values y_1^* and y_2^*. For this, we first derive the economic cost of product 2. To manufacture

1 ton of product 2, we require 2.5 tons of raw material 1 and 2 tons of raw material 2. The economic value of 1 ton of raw material i is equal to y_i^*, so that

$$\text{economic cost of product } 2 = 2.5y_1^* + 2y_2^*.$$

Substitution gives the economic cost of $2.5y_1^* + 2y_2^* = 222.50$ euros per ton of product 2, while the selling price is 210 euros per ton. The selling price per ton of product 2 must then increase by at least $2.5y_1^* + 2y_2^* - 210 = 12.50$ euros for it to be worthwhile to manufacture product 2. The numerical value 12.50 is included in the computer output as the *reduced cost* for variable x_2. The significance we intuitively give to the reduced costs can, of course, also be substantiated mathematically; see Section 1.6.4.

1.5.2 *Economic Interpretation of the Dual: Shadow Prices and Reduced Costs*

Consider the (primal) LP problem

$$\text{Maximize} \quad \sum_{j=1}^{n} c_j x_j$$

$$\text{subject to} \quad \sum_{j=1}^{n} a_{ij} x_j \le b_i \quad \text{for } i = 1, \dots, m$$

$$\text{and} \quad x_j \ge 0 \quad \text{for } j = 1, \dots, n.$$

For this LP problem, the associated dual is defined by

$$\text{Minimize} \quad \sum_{i=1}^{m} b_i y_i$$

$$\text{subject to} \quad \sum_{i=1}^{m} a_{ij} y_i \ge c_j \quad \text{for } j = 1, \dots, n$$

$$\text{and} \quad y_i \ge 0 \quad \text{for } i = 1, \dots, m.$$

This definition did not come out of thin air; it complies with the findings in the previous subsection, in which it was made plausible that the LP problem

$$
\begin{aligned}
\text{Maximize} \quad & 275x_1 + 210x_2 + 175x_3 \\
\text{subject to} \quad & 3x_1 + 2.5x_2 + 2x_3 \le 1250 \\
& 4x_1 + 2x_2 + x_3 \le 1000 \\
\text{and} \quad & x_1, x_2, x_3 \ge 0
\end{aligned}
$$

is naturally connected to the dual LP problem

$$
\begin{aligned}
\text{Minimize} \quad & 1250y_1 + 1000y_2 \\
\text{subject to} \quad & 3y_1 + 4y_2 \ge 275 \\
& 2.5y_1 + 2y_2 \ge 210 \\
& 2y_1 + y_2 \ge 175 \\
\text{and} \quad & y_1, y_2 \ge 0.
\end{aligned}
$$

For the discussion below, it is enlightening to give an economic interpretation to the primal and dual. The primal arises when a company wants to choose the production levels $x_1, \ldots, x_n$ of the n different production activities in such a way that the total revenue is the greatest at a revenue of c_j per unit of product j. The production activities use m raw materials; b_i is the available quantity of raw material i, and a_{ij} is the quantity of raw material i required to manufacture one unit of product j. The dual arises when someone wants to buy all of the company's raw materials. The dual variable y_i represents the "fair price" that person must pay per unit of raw material i. Since the company can obtain a revenue of c_j by using a quantity a_{ij} of each raw material i, the buyer's prices must satisfy the constraints $\sum_{i=1}^{m} a_{ij} y_i \geq c_j$ for all j. The buyer wants to minimize the total payment $\sum_{i=1}^{m} b_i y_i$ subject to these constraints. Common sense says that in the event of a supply shortage, the total value of all raw materials must equal the total revenue that can be generated through the production activities. It is indeed possible to provide a mathematical proof of this fundamental result, which we state as

optimal objective value of the dual = optimal objective value of the primal.

If we denote the optimal numerical values of the dual variables $y_1, \ldots, y_m$ by $y_1^*, \ldots, y_m^*$, then those numerical values are called the *shadow prices* of the raw materials $1, \ldots, m$ from the primal. The shadow price y_i^* can be viewed as the economic value of one additional unit of raw material i given that quantities $b_1, \ldots, b_m$ of the raw materials $1, \ldots, m$ are already present. In other words,

the shadow price y_i^* gives an indication of the increase in the maximum revenue of the company if 1 additional unit of raw material i is available in addition to the already present quantities of the raw materials.

We prove this important interpretation of y_i^* mathematically in Sections 1.6.4 and 1.7.

The economic costs can be determined for every product in the primal. To make one unit of a given product j, we need the quantities $a_{1j}, \ldots, a_{mj}$ of the raw materials $1, \ldots, m$. Consequently, we can say that

$$\text{economic cost per unit of product } j = \sum_{i=1}^{m} a_{ij} y_i^*.$$

Of course, product j is not made if the economic cost is higher than the revenue c_j per unit. This leads to the following result:

the reduced cost $\sum_{i=1}^{m} a_{ij} y_i^* - c_j$ gives the amount by which the cost coefficient of product j must at least increase to manufacture product j in an optimal solution.

This interpretation explains why, in the computer output, one sees that a variable with a positive value for the reduced cost is always zero in the optimal solution.

The reduced costs in fact only have significance for a variable with value zero in the optimal solution. For such a variable x_r, the reduced cost indicates by how much the profit coefficient c_r must at least increase for it to become worthwhile to manufacture product r.

Multiple Optimal Solutions

The reduced costs can also help one discover an *alternative optimal solution* (if it exists). It may happen that an original decision variable x_j has reduced costs with value zero, while the variable's value in the optimal solution is also zero. Generally, this means that there is an alternative optimal solution in which x_j is positive. One might find the alternative optimum by solving the LP model once more, replacing the relevant profit coefficient with $c_j + \varepsilon$, with $\varepsilon > 0$ small (for example, $\varepsilon = |c_j| \times 10^{-4}$). Use a software module to verify that the LP problem

$$\text{Maximize} \quad 4x_1 + 10x_2$$
$$\text{subject to} \quad x_1 + x_2 \le 5$$
$$2x_1 + 5x_2 \le 15$$
$$\text{and} \quad x_1, x_2 \ge 0$$

has the alternative optima $(x_1^*, x_2^*) = (0, 3)$ and $(x_1^*, x_2^*) = (\frac{10}{3}, \frac{5}{3})$.

1.5.3 *The General Dual LP Problem*

To understand the information the computer output gives for sensitivity analysis for a general LP problem that does not have only $\le$ constraints, it is useful to have insight into how the dual of an arbitrary LP problem looks. Section 1.5.2 showed how the dual of an LP problem is defined in the case of a maximization problem with only $\le$ constraints. Since a $\le$ constraint can be seen as a resource constraint, in that case, a physical interpretation of the dual exists, and it is intuitively clear that in a maximization problem, the shadow price of a $\le$ constraint is always nonnegative. The dual defined in Section 1.5.2 also determines the dual of an arbitrary LP problem with $\le$, $=$, and $\ge$ constraints. For this, we first observe that any minimization problem can be transformed into a maximization problem by multiplying the objective function by -1. Moreover, every constraint can be transformed into a $\le$ constraint. Any $\ge$ constraint can be written as a $\le$ constraint by multiplying both sides of the constraint by -1. An equality constraint $\sum a_j x_j = b$ is equivalent to the two inequality constraints $\sum a_j x_j \le b$ and $\sum a_j x_j \ge b$, where the latter can be rewritten as $-\sum a_j x_j \le -b$. Once all constraints have been transformed into $\le$ constraints, the dual can be written as in Section 1.5.2. So a unique dual is associate with every LP problem. As an illustration, the (primal) LP problem

$$\text{Maximize} \quad 7x_1 + 5x_2 + 4x_3$$
$$\text{subject to} \quad 2x_1 + 3x_2 + 5x_3 \ge 8$$
$$x_1 + x_2 + x_3 = 2$$
$$\text{and} \quad x_1, x_2, x_3 \ge 0$$

can be rewritten as

$$\begin{aligned}
\text{Maximize} \quad & 7x_1 + 5x_2 + 4x_3 \\
\text{subject to} \quad & -2x_1 - 3x_2 - 5x_3 \le -8 \\
& x_1 + x_2 + x_3 \le 2 \\
& -x_1 - x_2 - x_3 \le -2 \\
\text{and} \quad & x_1, x_2, x_3 \ge 0.
\end{aligned}$$

This LP problem has dual

$$\begin{aligned}
\text{Minimize} \quad & -8w_1 + 2w_2 - 2w_3 \\
\text{subject to} \quad & -2w_1 + w_2 - w_3 \ge 7 \\
& -3w_1 + w_2 - w_3 \ge 5 \\
& -5w_1 + w_2 - w_3 \ge 4 \\
\text{and} \quad & w_1, w_2, w_3 \ge 0.
\end{aligned}$$

We can rewrite this dual in such a way that the number of dual variables is equal to the number of constraints in the primal and that the coefficients of the dual variables in the objective function are equal to the right-hand-side coefficients of the constraints of the primal. For this, we replace $-w_1$ everywhere with the variable y_1 and the difference $w_2 - w_3$ with the variable y_2. In other words, a nonpositive dual variable y_1 belongs to the $\ge$ constraint and the unrestricted-in-sign dual variable y_2 belongs to the $=$ constraint in the primal. The dual of the primal LP problem can therefore be written as

$$\begin{aligned}
\text{Minimize} \quad & 8y_1 + 2y_2 \\
\text{subject to} \quad & 2y_1 + y_2 \ge 7 \\
& 3y_1 + y_2 \ge 5 \\
& 5y_1 + y_2 \ge 4 \\
\text{and} \quad & y_1 \le 0, \ y_2 \text{ unrestricted in sign.}
\end{aligned}$$

In the same way as this dual formulation was found for the numerical example above, we can show more generally that the dual of the general primal

$$\begin{aligned}
\text{Maximize} \quad & \sum_{j=1}^{n} c_j x_j \\
\text{subject to} \quad & \sum_{j=1}^{n} a_{ij} x_j \ \{\le, \text{ or } =, \text{ or } \ge\} \ b_i \quad \text{for } i = 1, \ldots, m \\
\text{and} \quad & x_j \ge 0 \quad \text{for } j = 1, \ldots, n
\end{aligned}$$

can be written as

$$\begin{aligned}
\text{Minimize} \quad & \sum_{i=1}^{m} b_i y_i \\
\text{subject to} \quad & \sum_{i=1}^{m} a_{ij} y_i \ge c_j \quad \text{for } j = 1, \ldots, n,
\end{aligned}$$

where

- the dual variable y_i is *nonnegative* if constraint i of the primal is a $\leq$ constraint,
- the dual variable y_i is *nonpositive* if constraint i of the primal is a $\geq$ constraint,
- the dual variable y_i can take on *both positive and negative* values if constraint i of the primal is an equality constraint.

This explains the following property of the shadow prices in the computer output. Suppose that we use an LP software package to solve an LP-*maximization problem*. If we assume that in the input, the right-hand-side coefficients of the constraints are nonnegative, then we see that in the output of the computer solution,

- the shadow price of a $\leq$ constraint is always nonnegative,
- the shadow price of a $\geq$ constraint is always nonpositive,
- the shadow price of an $=$ constraint can be positive, zero, or negative.

Considering that min z is the same as max $-z$, it is not difficult to verify that the primal

$$\text{Minimize} \quad \sum_{j=1}^{n} c_j x_j$$

$$\text{subject to} \quad \sum_{j=1}^{n} a_{ij} x_j \ \{\leq, \text{ or } =, \text{ or } \geq\} \ b_i \quad \text{for } i = 1, \ldots, m$$

$$x_j \geq 0 \quad \text{for } j = 1, \ldots, n$$

has dual

$$\text{Maximize} \quad \sum_{i=1}^{m} b_i y_i$$

$$\text{subject to} \quad \sum_{i=1}^{m} a_{ij} y_i \leq c_j \quad \text{for } j = 1, \ldots, n,$$

where

- $y_i \geq 0$ if the ith primal constraint is a $\geq$ constraint
- $y_i \leq 0$ if the ith primal constraint is a $\leq$ constraint
- y_i is unrestricted in sign if the ith primal constraint is an $=$ constraint.

For linear minimization problems, some codes and textbooks use the rather unfortunate convention to define shadow prices as the negative of the dual variables. This convention does not conform with duality theory and creates more confusion than clarity. For example, if the LP module of the software package ORSTAT2000 is used to solve a linear minimization problem with nonnegative right-hand-side coefficients, then the computer output satisfies

- the shadow price of a $\leq$ constraint is always nonpositive,
- the shadow price of a $\geq$ constraint is always nonnegative,
- the shadow price of an $=$ constraint can be positive, zero, or negative.

Finally, we note that a shadow price does not always have a physical significance. For example, if the decision variable x_j represents the fraction of a specific ingredient j in a mixture, then the shadow price of the constraint $\sum x_j = 1$ has no meaningful interpretation.

1.6   Sensitivity Analysis

In linear programming models for real-life problems, it is rare for the data of a model to be known with absolute certainty. In many cases, the data have been obtained using statistical estimation procedures. On the other hand, there may be certain coefficients that are known precisely, for example the availability of raw materials or labor. The analysis of the influence of changes in the data on the solution to the model is known as *sensitivity analysis* or *post-optimality analysis*. In many cases, the effect of changes can be studied directly using the computer output of the LP model. Hence, sensitivity analysis provides us with additional information without having to solve a new LP model. In general, sensitivity analysis is limited to situations in which *only one* parameter in the original problem is changed. We restrict ourselves to the situations where the parameter is either a coefficient of the objective function or a right-hand-side coefficient. Typical post-optimality questions are "what if" questions of the type "What happens to the maximum profit if the available quantity of a certain raw material is increased?" or "What happens to the optimal solution if the profit coefficient of a certain variable is decreased?". In many cases, the problem does not need to be solved again to answer these questions. In this section, we clarify the practical significance of sensitivity analysis using two illustrative examples. Before we begin with the examples, we first discuss in general terms the significance of the information given in the computer output for sensitivity analysis. The insights gained in the previous subsection are useful in the discussion of sensitivity analysis.

1.6.1   *Computer Output for Sensitivity Analysis*

The output of an LP software module gives not only an optimal solution to the LP model but also additional information for sensitivity analysis. The computer output of LP software modules includes not only the shadow prices and the reduced costs, but generally also ranges for the profit coefficients and right-hand-side coefficients.

Right-Hand-Side Coefficient Ranges

In the computer output, a shadow price is given for each constraint s. Let us assume that the right-hand-side coefficient b_s of this constraint represents the available quantity of a specific raw material. In Section 1.5.2, we made it plausible that the shadow price of constraint s gives the *marginal* increase in the current maximum revenue if a *small additional* quantity of raw material s is available. The reason we speak of a "small" additional quantity is that the shadow price of a raw material naturally depends on the quantity of that raw material that is already present. The more of a raw material that is available, the lower the value of an extra unit of that raw material generally is. This is made precise in the following Rule.

Rule 1.1. *If the right-hand-side coefficient b_s of only one constraint s is changed to $b_s + \Delta$ while the other data of the LP model remain the same, then*

$$new\ optimal\ objective\ value = old\ optimal\ objective\ value$$
$$+ \Delta \times (shadow\ price\ of\ constraint\ s)$$

provided that $b_s + \Delta$ lie inside the range for the right-hand-side coefficient of constraint s given in the computer output.

In Section 1.6.2, we show how Rule 1.1 can be applied in the context of concrete examples. The examples focus on the use of the ranges for the right-hand-side coefficients and not on their calculation. The reader interested in how to calculate the ranges can find background material on this in Section 1.6.4. It should be noted that the ranges for the right-hand-side coefficients do not provide any information on how the optimal values of the decision variables change. In most applications, this is not a problem because people are generally interested only in the effect the change in a right-hand-side coefficient has on the maximum value of the objective function.

How does the optimal objective value change if the new right-hand-side coefficient of constraint s is outside the allowable range? The answer to this question is more difficult and depends on the situation of the problem. The general principle is that *"loosening"* a constraint in an optimization model leaves the optimal objective value unchanged or *improves* it, while *"sharpening"* the constraint leaves the optimal objective value unchanged or *worsens* it. Another principle that holds for an LP model is that "loosening" a constraint has less and less effect on the improvement of the optimal objective value the further the constraint is loosened, while "sharpening" the constraint has an increasingly strong effect on the worsening of the optimal objective value the more stringent the constraint becomes. Intuitively, this should be clear: for example, the greater the availability of a raw material, the smaller the marginal value of one extra unit of the raw material.

Returning to the question above, consider, for example, a maximization problem in which the sth constraint is a $\leq$ constraint and its right-hand-side coefficient is changed from b_s to $b_s + \Delta$ with $b_s + \Delta$ outside the allowable range. If $\Delta > 0$ (the sth constraint has been "loosened"), it can be shown that the maximum objective value increases by *at most* $|\Delta \times (shadow\ price\ of\ constraint\ s)|$. If $\Delta < 0$ (the constraint has been "sharpened"), the maximum objective value decreases by *at least* $|\Delta \times (shadow\ price\ of\ constraint\ s)|$. If we restrict ourselves to the situation where the sth constraint is an inequality constraint, then the following rule holds when the new right-hand-side coefficient $b_s + \Delta$ lies outside the allowable range.

Rule 1.2.

- *If the sth constraint is a $\leq$ constraint and $\Delta > 0$, then for a maximization (resp., minimization) problem, the optimal objective value increases (resp., decreases) by at most*

$$|\Delta \times (shadow\ price\ of\ constraint\ s)|.$$

  This statement also holds in the case of a $\geq$ constraint with $\Delta < 0$.
- *If the sth constraint is a $\leq$ constraint and $\Delta < 0$, then for a maximization (resp., minimization) problem, the optimal objective value decreases (resp., increases) by at least*

$$|\Delta \times (shadow\ price\ of\ constraint\ s)|.$$

  This statement also holds in the case of a $\geq$ constraint with $\Delta > 0$.

In many cases, the information obtained from Rule 2 suffices for management purposes. To know the new objective value exactly, we need to solve the LP problem again with the changed data. We conclude with the following remark. The formulation of this rule is limited to inequality constraints because in the case of an equality constraint, the sign of the corresponding shadow price is not known beforehand. Once we know the sign of the shadow price from the computer output, then in the case of a maximization problem, for example, we may conclude that the equality constraint behaves "physically" as a $\leq$ constraint or $\geq$ constraint depending on whether the shadow price is positive or negative, respectively. The correctness of this conclusion follows from duality theory as stated in Section 1.5.3.

Objective Function Coefficient Ranges

As mentioned earlier, the reduced costs are only significant for variables with value zero in the optimal solution. The reduced cost of such a variable indicates the amount by which the profit or cost coefficient of the variable must at least be increased or decreased, respectively, for the variable to have a positive value in the optimal solution. Related to the reduced costs are the ranges for the objective function coefficients.

Rule 1.3. *Suppose that the objective function coefficient c_r of only one variable x_r is changed to $c_r + \Delta$ while the other coefficients in the LP model remain unchanged. Then*

> *the optimal values of the decision variables remain the same provided that $c_r + \Delta$ lie within the allowable range given for the objective function coefficient of variable x_r in the computer output.*

This result may be surprising. It is not entirely clear intuitively. A mathematical explanation is given in Section 1.6.4. One can check the result experimentally using an LP software package. Although the optimal values of the decision variables remain unchanged, the optimal objective value changes by $x_r^* \Delta$, where x_r^* is the value of x_r in the optimal solution. An intuitive explanation for why the optimal values of the decision variables do not change lies in the fact that for an LP problem, the optimum is reached at a corner point of the feasible region. Consider Figure 1.1, which gives the feasible region of an LP problem with two variables. If we do not vary the slope of the line that represents the objective function too much, then the same corner point continues to give the optimal solution.

Simultaneous Changes

The ranges for the right-hand-side and objective function coefficients hold for changes in a single right-hand-side or objective function coefficient, with the other data remaining unchanged. Nevertheless, it is possible to say something about the effect of simultaneous changes in objective or right-hand-side coefficients. For this, the 100% rule, taken from the book of Bradley, Hax, and Magnanti (1977), can be used.

Objective Function Coefficients

Suppose that the current objective function coefficient c_j is changed to $c_j + \Delta c_j$ for $j = 1, \ldots, n$. In the computer output, an allowable range $[l_j, h_j]$ is given for each objective function coefficient c_j. Use this to compute the numbers S_j and D_j given by

$$S_j = h_j - c_j \quad (= \text{the allowable increase of } c_j),$$
$$D_j = c_j - l_j \quad (= \text{the allowable decrease of } c_j).$$

Then, for each of the decision variables x_j, calculate the number given by

$$r_j = \begin{cases} \Delta c_j / S_j & \text{if } \Delta c_j \geq 0, \\ -\Delta c_j / D_j & \text{if } \Delta c_j < 0. \end{cases}$$

The number r_j measures the ratio of the actual change of c_j to the maximum allowable change of c_j.

The 100% rule for the c_j. *The current optimal values of the decision variables remain the same if* $\sum_{j=1}^{n} r_j \leq 1$.

Right-Hand-Side Coefficients

Suppose that the right-hand-side coefficients b_i are changed to $b_i + \Delta b_i$ for $i = 1, \ldots, m$. In the computer output, an allowable range $[l_i, h_i]$ is given for each right-hand-side coefficient b_i. Use this to compute the numbers S_i and D_i given by

$$S_i = h_i - b_i \quad (= \text{the allowable increase of } b_i),$$
$$D_i = b_i - l_i \quad (= \text{the allowable decrease of } b_i).$$

Then, for each constraint i, calculate the number r_i given by

$$r_i = \begin{cases} \Delta b_i / S_i & \text{if } \Delta b_i \geq 0, \\ -\Delta b_i / D_i & \text{if } \Delta b_i < 0. \end{cases}$$

The 100% rule for the b_i. *If $\sum_{i=1}^{m} r_i \leq 1$, then the optimal objective value changes by*

$$\sum_{i=1}^{m} \Delta b_i \times (shadow\ price\ of\ constraint\ i).$$

The 100% rule is a rule that is on the safe side. Nevertheless, it is advisable to solve the LP model again when $\sum_{i=1}^{m} r_i > 1$.

1.6.2   *A Production Planning Problem*

We return to the production planning example introduced in Section 1.2.1. The formulation and the computer solution to the LP model are given in Figure 1.6. The computer output contains the optimal values of the decision variables and data for sensitivity analysis. The optimal production plan is to make 100 units of product A, 330.154 units of product B, no units of product C, and 242.308 units of product D. The optimal objective value is 6186.46 euros.

Before a production plan is implemented, the management team meets to discuss it. Suppose that one is required to answer the following questions that were asked during that meeting:

Questions:

(1) The sales manager thinks that it may be possible to increase the profit on product C from 6 euros to 8 euros per unit. What happens to the optimal solution?

(2) The purchasing manager can buy 100 additional units of raw material from another supplier. However, this supplier's price is 1.50 euros more per unit of raw material than the current supplier's unit price. It is not possible to buy more raw material from the current supplier. What is the advice?

(3) The customer to whom 100 units of product A had already been promised is willing to cancel the order in exchange for a compensation of 575 euros. What is the advice to the sales manager?

(4) The production manager thinks that 14 production hours can be transferred from finishing to assembly. The cost of this transfer is 140 euros. What are the comments on this?

(5) The marketing manager proposes to introduce a new product E, which contributes 7 euros per unit to the total profit. Every unit of the new product requires 0.40 hours of assembly, 0.50 hours of finishing, 0.20 hours of packaging, and 2 units of raw material. What are the comments on this?

Fig. 1.6   Computer solution for the production example.

```
SIMOPT Version 2.1                    IEOR   VU Amsterdam

The following model was read:

Objective Function :
  MAX  4.8000 XA +12.0000 XB +6.0000 XC +7.2000 XD
Subject to :
  1.  0.7000 XA +0.7500 XB +0.5500 XC +0.3400 XD <= 400.0000
  2.  0.5500 XA +0.8200 XB +0.8000 XC +0.5500 XD <= 480.0000
  3.  0.2400 XA +0.3200 XB +0.4500 XC +0.2700 XD <= 220.0000
  4.  1.9000 XA +2.5000 XB +1.8000 XC +2.0000 XD <= 1500.0000
  5.  1.0000 XA >= 100.0000


Summary of Results

Value Objective Function :        6186.4615


Variable        Activity Level  Reduced Cost
--------        --------------  ------------

XA         :        100.0000       0.0000
XB         :        330.1538       0.0000
XC         :          0.0000       2.7323
XD         :        242.3077       0.0000


                Slack or Surplus  Shadow Prices
                ----------------  -------------

Constraint 1          0.0000         9.2308
Constraint 2         21.0046         0.0000
Constraint 3         24.9277         0.0000
Constraint 4          0.0000         2.0308
Constraint 5          0.0000        -5.5200

Objective Coefficient Ranges

                Current
Variable        Coefficient  Allowed Interval
--------        -----------  ----------------

XA               4.8000      [ -Infinity ,   10.3200 ]
XB              12.0000      [    9.0000 ,   15.8824 ]
XC               6.0000      [ -Infinity ,    8.7323 ]
XD               7.2000      [    5.4400 ,    9.6000 ]
```

```
Right-Hand-Side Ranges


                    Current
     Constraint      RHS              Allowed Interval
     ----------      ---              ----------------
     1.            400.0000        [  292.7000 ,    451.5208 ]
     2.            480.0000        [  458.9954 ,    Infinity ]
     3.            220.0000        [  195.0723 ,    Infinity ]
     4.           1500.0000        [ 1290.0000 ,   1602.1167 ]
     5.            100.0000        [    0.0000 ,    384.6154 ]


Accuracy Check Passed.
```

Answers:

(1) No units of product C are manufactured in the current production plan. The reduced cost of product C indicates that the unit price of this product must increase by at least 2.73 euros for product C to become profitable. The optimal production plan therefore remains unchanged if the unit price of product C increases by 2 euros.

(2) Constraint 4 concerns the raw material. This constraint is binding in the optimal solution, so it could be worthwhile to buy additional raw material. The shadow price of the raw material constraint is 2.031, and the range for the right-hand-side coefficient shows that this shadow price remains valid for increases in the quantity of raw material of up to $1602.117 - 1500 = 102.117$ units. The proposed increase is 100 units. The optimal objective value therefore increases by $100 \times 2.031 = 203.10$ euros. This increase is greater than the additional cost of 150 euros, so the advice is to buy those 100 additional units of raw material. To determine the new optimal production plan, one needs to solve the new LP model. The new optimal production plan is to make 100 units of product A, 277.846 units of product B, no units of product C, and 357.692 units of product D.

(3) The lower bound constraint $x_A \geq 100$ is binding in the optimal solution and has a shadow price -5.52. The objective value therefore increases if this constraint is weakened. The range for the right-hand-side coefficient of constraint 5 indicates that the shadow price of -5.52 remains valid for decreases of the right-hand-side coefficient of up to $100 - 0 = 100$ units. The optimal objective value therefore increases by $100 \times 5.52 = 552$ euros if the order of product A is canceled. However, the cost of buying off the order of 100 units of product A is higher than the increase in profit. The advice would therefore be to deliver the order of 100 units.

(4) Constraint 1, which concerns the available hours for assembly, is binding in the optimal solution, while constraint 2, which concerns the available hours for

finishing is not binding. It could therefore be profitable to shift production hours from finishing to assembly. Since this concerns simultaneous changes, we use the 100% rule. For the assembly constraint 1 and the finishing constraint 2, the computer output shows that the allowable increase of $b_1 = 400$ gives $S_1 = 51.52$ and the allowable decrease of $b_2 = 480$ gives $D_2 = 21.00$. This means that for constraints 1 and 2, the ratios of the actual changes in the right-hand-side coefficients to the greatest allowable changes are given by $r_1 = 14/51.52 = 0.272$ and $r_2 = 14/21 = 0.667$. The sum $r_1 + r_2 = 0.939$ is not greater than 1. Since the shadow prices of constraints 1 and 2 are 9.231 and 0, the optimal objective value would change by $14 \times 9.231 + 14 \times 0 = 129.23$ euros if 14 production hours were shifted from finishing to assembly. This increase in profit is less than the transfer cost of 140 euros. The advice is to not transfer the 14 production hours from finishing to assembly.

(5) There are two ways to answer this question. The first uses the concept of economic cost $\sum_{i=1}^{m} a_{ij} y_i$ introduced in Section 1.5.2. The underlying economic cost of the new product E is computed to be $0.40 \times 9.231 + 0.50 \times 0 + 0.20 \times 0 + 2 \times 2.031 - 0 \times 5.52 = 7.75$ euros per unit. The economic cost of the product is higher than the revenue. The observation is therefore that there is no point in introducing the new product. This can also be seen using the 100% rule for the right-hand-side coefficients. Namely, suppose that in the LP formulation of the problem, one introduces the variable x_E with coefficient a_{iE} in constraint i for $i = 1, \ldots, 4$ and that this variable is subsequently set equal to δ with δ a small positive number. This is the same as replacing the right-hand-side coefficient b_i in the LP formulation with $b_i - a_{iE}\delta$ for $i = 1, \ldots, 4$. Applying the 100% rule with $\Delta b_i = -a_{iE}\delta$ gives that the optimal objective value changes by $-0.40\delta \times 9.231 - 0.50\delta \times 0 - 0.20\delta \times 0 - 2\delta \times 2.031 + 0\delta \times 5.52 = -7.75\delta$ euros. The decrease of 7.75δ euros does not outweigh the increase in profit of 7δ euros.

1.6.3 *A Feed Mixing Problem*

Problem statement. Suppose that we wish to produce a feed mix for cattle at the lowest possible cost, with the mix having to meet certain composition requirements. The feed mix is made out of four different raw materials 1, 2, 3, and 4. The mix must contain at least 25% protein, at least 10% fat, and exactly 25% carbohydrates. In addition, the fat percentage must not exceed 65% of the protein percentage. Table 1.7 gives the compositions of the different raw materials in percentages and the costs of those raw materials.

Formulation. A reasonable objective is to minimize the cost per ton of feed mix. In view of this objective, the decision variables have been chosen as

$$x_j = \text{proportion of raw material } j \text{ in the feed mix } (j = 1, 2, 3, 4).$$

Table 1.7   Data for the feed mix problem.

Raw Materials	1	2	3	4
Protein	15	50	25	5
Fat	5	20	10	15
Carbohydrates	50	10	5	50
Costs per Ton	€80	€140	€100	€60

We find the following LP model for the feed mix problem:

$$
\begin{aligned}
\text{Minimize} \quad & 80x_1 + 140x_2 + 100x_3 + 60x_4 \\
\text{subject to} \quad & 15x_1 + 50x_2 + 25x_3 + 5x_4 \geq 25 \\
& 5x_1 + 20x_2 + 10x_3 + 15x_4 \geq 10 \\
& 50x_1 + 10x_2 + 5x_3 + 50x_4 = 25 \\
& -4.75x_1 - 12.5x_2 - 6.25x_3 + 11.75x_4 \leq 0 \\
& x_1 + x_2 + x_3 + x_4 = 1 \\
\text{and} \quad & x_1, x_2, x_3, x_4 \geq 0.
\end{aligned}
$$

The fourth constraint comes from rewriting the constraint $(5x_1 + 20x_2 + 10x_3 + 15x_4)/(15x_1 + 50x_2 + 25x_3 + 5x_4) \leq 0.65$.

The computer solution to the LP model is given in Figure 1.7. The computer output contains the optimal solution and the data for sensitivity analysis. The minimum costs are 96.73 euros per ton, and the composition of the optimal feed mix is 0% of ingredient 1, 32.7% of ingredient 2, 26.5% of ingredient 3, and 40.8% of ingredient 4.

Fig. 1.7   Computer output for the feed mix example.

```
SIMOPT Version 2.1                      IEOR  VU Amsterdam

The following model was read:

Objective Function :
  MIN 80.0000 X1 +140.0000 X2 +100.0000 X3 +60.0000 X4
Subject to :
  1. 15.0000 X1 +50.0000 X2 +25.0000 X3 +5.0000 X4 >= 25.0000
  2.  5.0000 X1 +20.0000 X2 +10.0000 X3 +15.0000 X4 >= 10.0000
  3. 50.0000 X1 +10.0000 X2 +5.0000 X3 +50.0000 X4 = 25.0000
  4. -4.7500 X1 -12.5000 X2 -6.2500 X3 +11.7500 X4 <= 0.0000
  5.  1.0000 X1 +1.0000 X2 +1.0000 X3 +1.0000 X4 = 1.0000


Summary of Results

Value Objective Function :            96.7347
```

Variable		Activity Level	Reduced Cost
X1	:	0.0000	3.6735
X2	:	0.3265	0.0000
X3	:	0.2653	0.0000
X4	:	0.4082	0.0000

	Slack or Surplus	Shadow Prices
Constraint 1	0.0000	1.6327
Constraint 2	5.3061	0.0000
Constraint 3	0.0000	-0.1633
Constraint 4	0.9439	0.0000
Constraint 5	0.0000	60.0000

Objective Coefficient Ranges

Variable	Current Coefficient	Allowed Interval	
X1	80.0000	[76.3265 ,	Infinity]
X2	140.0000	[95.5556 ,	150.0000]
X3	100.0000	[88.7500 ,	150.0000]
X4	60.0000	[-Infinity ,	63.8298]

Right-Hand-Side Ranges

Constraint	Current RHS	Allowed Interval	
1.	25.0000	[21.8855 ,	33.1250]
2.	10.0000	[-Infinity ,	15.3061]
3.	25.0000	[5.0000 ,	28.5577]
4.	0.0000	[-0.9439 ,	Infinity]
5.	1.0000	[0.8673 ,	1.3265]

Accuracy Check Passed.

Questions:

(1) By how much do the minimum costs of the feed mix decrease or increase if the percentage of protein in the mix must be at least 30% instead of 25%?

(2) What happens to the minimum costs if the percentage of carbohydrates in the feed mix must be 27% instead of 25%?

(3) What happens to the optimal solution if the prices of ingredients 3 and 4 increase by 10 euros and 2 euros per ton, respectively?
(4) Give an interpretation of the shadow price of constraint 5.
(5) How does the optimal solution change if it is required that the fat percentage not be higher than half of the protein percentage?

Answers:

(1) The shadow price of constraint 1 is 1.633. According to the range for the right-hand-side coefficient in the computer output, this shadow price remains valid for an increase in the right-hand-side coefficient of up to 33.125. The new protein percentage is within the allowable range. The optimal objective value therefore changes by $5 \times 1.633 = 8.165$ euros if the protein percentage must be at least 30% instead of 25%. The change in the objective value means that the costs per ton of feed mix increase by 8.165 euros.

(2) The shadow price of the equality constraint 3 is -0.163, and this shadow price remains valid for an increase in the right-hand-side coefficient of up to 28.558 (the minus sign of the shadow price of the equality constraint 3 in the minimization problem indicates that, physically, this constraint behaves like a $\leq$ constraint). The optimal objective value therefore changes by $2 \times (-0.163) = -0.326$ if the carbohydrates percentage changes from 25% to 27%. In other words, the costs per ton of feed mix is then reduced by 0.326 euro.

(3) The computer output shows that the allowable increases in the objective function coefficients $c_3 = 100$ and $c_4 = 60$ correspond to $S_3 = 50$ and $S_4 = 3.83$. This leads to the ratios $r_3 = 10/50 = 0.2$ and $r_4 = 2/3.83 = 0.52$. Since $r_3 + r_4$ is not greater than 1, the 100% rule says that the optimal values of the decision variables remain unchanged when the prices of ingredients 3 and 4 simultaneously increase by 10 euros and 2 euros per ton, respectively. Even though the optimal composition of the feed mix remains the same, the minimum costs increase by $0.265 \times 10 + 0.408 \times 2 = 3.466$ euros per ton.

(4) The shadow price of constraint 5 has no significant meaning because changing the right-hand side of this constraint is not possible physically (the fractions must always add up to 1). The shadow price of constraint 4 also has no practical meaning.

(5) The column "Slack or Surplus" in the output tells us that in the current optimal feed mix, the fat percentage is 15.306% and the protein percentage is 25%. The fat percentage is more than half of the protein percentage. In general, a sensitivity analysis cannot help us answer questions concerning the change of a technological coefficient in a constraint. To answer question 5, we need to solve a new LP model, where constraint 4 is replaced by the new linear constraint

$$-2.5x_1 - 5x_2 - 2.5x_3 + 12.5x_4 \leq 0.$$

The new solution can easily be found using an LP software package. The new LP problem has optimal solution $x_1 = 5/24$, $x_2 = 1/4$, $x_3 = 1/3$, and $x_4 = 5/24$. The costs of the new feed mix are 97.50 euros per ton.

1.6.4 *Computational Method for Sensitivity Analysis*[1]

This section provides some insight into calculating the shadow prices, reduced costs, and ranges for the coefficients. The calculations are best understood in the context of an example. We therefore begin by applying the simplex method to the fertilizer example from Section 1.5.1. By carefully studying the functioning of the simplex method, we see how to carry out the sensitivity calculations. The explanation is limited to the case of a maximization problem with only $\leq$ constraints. The sensitivity calculations for other situations are similar. In this section, it is assumed that the reader is familiar with the simplex method as discussed in Section 1.4.

Simplex Iterations for the Fertilizer Example

We first rewrite the LP model for the fertilizer problem. As argued in Section 1.4, in the simplex method, it is useful to represent the objective function by a variable z. The LP model for the fertilizer problem is therefore equivalent to

$$
\begin{aligned}
\text{Maximize} \quad & z = 275x_1 + 210x_2 + 175x_3 \\
\text{subject to} \quad & 3x_1 + 2.5x_2 + 2x_3 + s_1 \quad\;\; = 1250 \\
& 4x_1 + \;\; 2x_2 + \;\; x_3 \quad\;\; + s_2 = 1000 \\
\text{and} \quad & \qquad\qquad x_1, x_2, x_3, s_1, s_2 \geq 0.
\end{aligned}
$$

Here, s_1 and s_2 are slack variables introduced to transform the inequalities into equalities. The objective function is used as an additional linear constraint in order to better understand how the simplex method works. The linear equality constraints are already in canonical form. An initial solution for the simplex method can therefore be given immediately. The feasible basic solution we begin with is $s_1 = 1250$, $s_2 = 1000$, $x_1 = x_2 = x_3 = 0$, with an objective value of $z = 0$. In the next step of the simplex method, the variable x_1 is made basic, because x_1 has the greatest positive coefficient in the z-equation. By making this variable positive, we obtain the greatest immediate increase in z. We can increase x_1 up to $\min(1250/3, 1000/4) = 1000/4$ without making another variable negative. This results in the variable s_2 becoming zero and leaving the basis. Next, the linear equations are rewritten into an equivalent canonical form. In the new canonical form, s_1 and x_1 are the basic variables and the linear equation for z is expressed in terms of the nonbasic variables x_2, x_3, s_2. To accomplish this, we first divide both sides of the equation $4x_1 + 2x_2 + x_3 + s_2 = 1000$ by 4 to make the coefficient of x_1 equal to 1. We subtract the resulting equation $x_1 + 0.5x_2 + 0.25x_3 + 0.25s_2 = 250$ three times from the equation $3x_1 + 2.5x_2 + 2x_3 + s_1 = 1250$ to make the coefficient

---

[1] Section 1.4 is a prerequisite for this optional section.

of x_1 equal to 0 in the latter. Then, we substitute $x_1 = 250 - 0.5x_2 - 0.25x_3 - 0.25s_2$ in the equation $z = 275x_1 + 210x_2 + 175x_3$ to obtain the equation for z in the new nonbasic variables. This leads to the following equivalent system of linear equations:

$$z = 68\,750 + 72.5x_2 + 106.25x_3 - 68.75s_2$$
$$x_2 + 1.25x_3 + s_1 - 0.75s_2 = 500$$
$$x_1 + 0.5x_2 + 0.25x_3 \qquad + 0.25s_2 = 250.$$

The new basic solution is $s_1 = 500$, $x_1 = 250$, $x_2 = x_3 = s_2 = 0$, with an objective value of $z = 68\,750$. Since the z-equation contains a nonbasic variable with positive coefficient, one can immediately conclude that the objective value can be improved. The variable x_3 has the greatest positive coefficient in the current z-equation and is therefore made positive in the next iteration. The variable x_3 can be increased to $\min(500/1.25, 250/0.25) = 500/1.25$. The variable s_1 is therefore made zero and leaves the basis. Both sides of the equation $x_2 + 1.25x_3 + s_1 - 0.75x_2 = 500$ are now divided by 1.25, and the resulting equation $0.8x_2 + x_3 + 0.8s_1 - 0.6s_2 = 400$ is subtracted 0.25 times from the equation $x_1 + 0.5x_2 + 0.25x_3 + 0.25s_2 = 250$. Finally, the equation $x_3 = 400 - 0.8x_2 - 0.8s_1 + 0.6s_2$ is substituted into the equation for z. This leads to the following equivalent system of linear equations:

$$z = 111\,250 - 12.50x_2 - 85s_1 - 5s_2$$
$$0.8x_2 + x_3 + 0.8s_1 - 0.6s_2 = 400$$
$$x_1 + 0.3x_2 \qquad - 0.2s_1 + 0.4s_2 = 150.$$

The corresponding basic solution is $x_3 = 400$, $x_1 = 150$, $x_2 = s_1 = s_2 = 0$, with an objective value of $z = 111\,250$. The objective value cannot be further increased because every variable in the z-equation has a negative coefficient. The solution $x_1 = 150$, $x_2 = 0$, $x_3 = 400$ must therefore be optimal.

We have executed the calculations above in preparation for the explanation of the calculations for sensitivity analysis. As we have seen above, in every iteration, the simplex method rewrites the linear equations to an equivalent form by dividing a specific equality constraint by a nonzero number, adding a multiple of this equivalent constraint to another equality constraint, and, finally, substituting the chosen equality constraint into the equation for the objective function z. This observation allows us to analyze how a change in a particular coefficient of the LP model influences the steps of the simplex method.

Right-Hand-Side Coefficient Ranges

Suppose that the coefficient $b_1 = 1250$ is changed to $b_1 = 1250 + \Delta_1$. What is the range of Δ_1 for which the same basic variables continue to be part of the optimal solution? To answer this question, let us look at what happens if we apply the same series of simplex operations as above to the new equations

$$3x_1 + 2.5x_2 + 2x_3 + s_1 \qquad = 1250 + \Delta_1$$
$$4x_1 + 2x_2 + x_3 \qquad + s_2 = 1000.$$

The equations would then end up being transformed into the equivalent system

$$0.8x_2 + x_3 + 0.8s_1 - 0.6s_2 = 400 + 0.8\Delta_1$$
$$x_1 + 0.3x_2 \quad - 0.2s_1 + 0.4s_2 = 150 - 0.2\Delta_1.$$

Why is this the case? The reason is just that Δ_1 and the slack variable s_1 have the same coefficients in the original system of equations. The coefficients are therefore transformed in the same way. By looking carefully at how the equation for z is transformed and observing that in the original z-equation, the "variable" Δ_1 and slack variable s_1 both have coefficient zero, we also see that the linear equation for z is transformed into

$$z = 111\,250 + 85\Delta_1 - 12.50x_2 - 85s_1 - 5s_2.$$

The coefficient of Δ_1 is the negative of the coefficient of s_1 because the coefficients of Δ_1 and s_1 have opposite signs in the equations that have been substituted into the z-equation. It is important to note that the change in the transformed z-equation is in the term $111\,250 + 85\Delta_1$ but not in the coefficients of the nonbasic variables x_2, s_1, s_2. Consequently, the basic solution

$$x_3 = 400 + 0.8\Delta_1, \qquad x_1 = 150 - 0.2\Delta_1, \qquad x_2 = s_1 = s_2 = 0$$

is optimal for the values of Δ_1 that satisfy the linear inequalities

$$400 + 0.8\Delta_1 \geq 0$$
$$150 - 0.2\Delta_1 \geq 0.$$

These inequalities give the interval

$$-500 \leq \Delta_1 \leq 750.$$

In other words, the same basic variables x_1 and x_3 continue to be part of the optimal solution as long as the right-hand-side coefficient b_1 lies within the range $750 \leq b_1 \leq 2000$.

At the same time, the analysis above shows how to calculate the shadow prices. The objective value changes by an amount of $85\Delta_1$ if $b_1 = 1250$ is changed to $b_1 = 1250 + \Delta_1$, provided that $-500 \leq \Delta_1 \leq 750$. In Section 1.5.2, we deduced that the objective value changes by an amount of $y_1\Delta_1$, where y_1 is the shadow price of constraint 1. The shadow price of constraint 1 must therefore equal 85. Likewise, we can use the coefficient of the slack variable s_2 in the final z-equation to deduce that the shadow price of constraint 2 is equal to 5. We have now reached the following important conclusion:

The shadow prices of the constraints are given by the negatives of the coefficients of the corresponding slack variables in the final equation for z.

This conclusion was reached for an LP model in which all original equations are $\leq$ equations. The situation becomes somewhat more complicated when the original LP model also contains $\geq$ and $=$ constraints. In that case, one also needs to look at the coefficients of the surplus variables and auxiliary variables in the final z-equation. We do not discuss these technical details further.

Objective Function Coefficient Ranges

A closer look at the simplex operations shows us that the transformations of the constraints are not influenced by a change in one of the profit coefficients c_j. The linear equality constraints are still being transformed into

$$0.8x_2 + x_3 + 0.8s_1 - 0.6s_2 = 400$$
$$x_1 + 0.3x_2 \quad\quad - 0.2s_1 + 0.4s_2 = 150.$$

What happens to the equation for the objective function z? The easiest way to find the answer is by substituting

$$x_3 = 400 - 0.8x_2 - 0.8s_1 + 0.6s_2$$
$$x_1 = 150 - 0.3x_2 + 0.2s_1 - 0.4s_2$$

into the original equation for z. In doing so, we distinguish between a change in the coefficient c_j of a nonbasic variable and a change in the coefficient c_j of a basic variable.

We first consider the case where the coefficient of a nonbasic variable is changed. Suppose that $c_2 = 210$ is changed to $c_2 = 210 + \Delta_2$. Then the z-equation is transformed into the equivalent equation

$$z = 275(150 - 0.3x_2 + 0.2s_1 - 0.4s_2) + (210 + \Delta_2)x_2$$
$$+175(400 - 0.8x_2 - 0.8s_1 + 0.6s_2).$$

This equation can be rewritten as

$$z = 111\,250 - (12.50 - \Delta_2)x_2 - 85s_1 - 5s_2.$$

This means that a change in the profit coefficient of a nonbasic variable influences only the coefficient of this variable in the final z-equation, and not the other coefficients. The current basic solution $x_1 = 150$, $x_3 = 400$, $x_2 = s_1 = s_2 = 0$ remains the same as long as $12.50 - \Delta_2 \geq 0$ or, equivalently,

$$\Delta_2 \leq 12.50.$$

This means that the profit coefficient of the nonbasic variable x_2 must increase by at least 12.50 for it to become profitable to make x_2 positive. The reduced cost of the nonbasic variable x_2 is therefore 12.50. So, we find that

the reduced cost of a nonbasic variable is the negative of the coefficient of that variable in the final z-equation.

We now consider the case where the coefficient of a basic variable is changed. Suppose that $c_1 = 275$ is changed to $c_1 = 275 + \Delta_1$. The final z-equation becomes

$$z = (275 + \Delta_1)(150 - 0.3x_2 + 0.2s_1 - 0.4s_2) + 210x_2$$
$$+175(400 - 0.8x_2 - 0.8s_1 + 0.6s_2).$$

If we rewrite this equation as

$$z = 111\,250 + 150\Delta_1 - (12.50 + 0.3\Delta_1)x_2 - (85 - 0.2\Delta_1)s_1$$
$$-(5 + 0.4\Delta_1)s_2,$$

then we see that a change in the profit coefficient of a basic variable influences the coefficient of every variable in the final z-equation. The current optimal basic solution $x_1 = 150$, $x_3 = 400$, $x_2 = s_1 = s_2 = 0$ remains optimal for the values of Δ_1 that satisfy the linear inequalities

$$12.50 + 0.3\Delta_1 \geq 0$$
$$85 - 0.2\Delta_1 \geq 0$$
$$5 + 0.4\Delta_1 \geq 0.$$

The optimal objective value then changes by $150\Delta_1$. The inequalities provide the range $-12.5 \leq \Delta_1 \leq 425$. In other words, the current optimal basic solution remains optimal as long as the profit coefficient c_1 lies within the range $262.5 \leq c_1 \leq 700$.

We conclude this section with the following remark. The procedure described above for calculating shadow prices and reduced costs may be elementary, but it is not very efficient from a calculatory point of view. By using a modest amount of matrix algebra, the simplex calculations can be streamlined significantly. In this book, the emphasis is on formulating models and interpreting the computer output, not on mathematical proofs. This does not mean that the mathematics underlying the LP model is not important. Hopefully, the discussion above motivates the reader with mathematical interest to study the beautiful piece of mathematics hidden behind linear programming, which is discussed in Section 1.7.

1.7 Linear Programming with Matrix Algebra[2]

This section assumes that the reader has some basic knowledge of matrix algebra. The aim of this section is to show that the calculations of the simplex method can be made more transparent using a modest amount of matrix algebra. We also deduce the various results that were presented without proof in Sections 1.5 and 1.6.

We use the following notation. Capital letters in bold represent matrices and lowercase letters in bold represent vectors. For the sake of clarity, we do not distinguish between row and column vectors when matrices and vectors are multiplied with each other. The scalar product $\mathbf{xy}$ of two n-dimensional vectors $\mathbf{x} = (x_1, \ldots, x_n)$ and $\mathbf{y} = (y_1, \ldots, y_n)$ is the number $\sum_{i=1}^{n} x_i y_i$. The product $\mathbf{Ax}$ of an $m \times n$ matrix $\mathbf{A} = (a_{ij})$ and an n-dimensional vector $\mathbf{x} = (x_i)$ is an m-dimensional vector with ith component $(\mathbf{Ax})_i = \sum_{j=1}^{n} a_{ij} x_j$. Likewise, the product $\mathbf{yA}$ of an $m \times n$ matrix $\mathbf{A} = (a_{ij})$ and an m-dimensional vector $\mathbf{y}$ is an n-dimensional vector with jth component $(\mathbf{yA})_j = \sum_{i=1}^{m} y_i a_{ij}$. We denote the jth column vector of the matrix $\mathbf{A}$ by $\mathbf{a}_j$.

The LP model in *standard form* can be expressed in matrix form:

$$
\begin{aligned}
&\text{Maximize} && \mathbf{cx} \\
&\text{subject to} && \mathbf{Ax} = \mathbf{b} \\
&\text{and} && \mathbf{x} \geq \mathbf{0},
\end{aligned}
$$

---

[2]This specialized section contains mathematically more advanced material.

where $\mathbf{A}$ is a given $m \times n$ matrix, $\mathbf{b}$ is a given n-dimensional vector, $\mathbf{c}$ is a given n-dimensional vector, and $\mathbf{x}$ is the n-dimensional vector of decision variables.

From now on, we not only assume that n (= the number of decision variables) is greater than or equal to m (= the number of constraints), but also make the stronger assumption that the rank of the matrix $\mathbf{A}$ is equal to m; that is, the matrix $\mathbf{A}$ has at least one set of m linearly independent column vectors. In practice, this assumption is not limiting. The use of slack, surplus, and auxiliary variables to create the standard form of the LP model automatically ensures that the number of variables in the standard form is always greater than the number of constraints and that the matrix $\mathbf{A}$ contains the $m \times m$ identity matrix $\mathbf{I}$, so that the matrix $\mathbf{A}$ has rank m.

Definition 1.1. *An $m \times m$ matrix $\mathbf{B}$ is called a* basis *for $\mathbf{Ax} = \mathbf{b}$ if $\mathbf{B}$ consists of m linearly independent column vectors of $\mathbf{A}$. The* basic variables *for $\mathbf{B}$ are those variables x_j for which the corresponding column vector $\mathbf{a}_j$ is contained in $\mathbf{B}$.*

A basis $\mathbf{B}$ is by definition nonsingular; that is, the inverse matrix $\mathbf{B}^{-1}$ exists. Suppose that we have a certain basis $\mathbf{B}$. Split up the matrix $\mathbf{A}$ as $\mathbf{A} = (\mathbf{B}, \mathbf{R})$, where $\mathbf{R}$ is an $m \times (n - m)$ matrix consisting of the column vectors $\mathbf{a}_j$ that do not belong to $\mathbf{B}$. The vector $\mathbf{x}$ is split up as $\mathbf{x} = (\mathbf{x_B}, \mathbf{x_R})$, where $\mathbf{x_B}$ is the vector with basic variables for $\mathbf{B}$ and $\mathbf{x_R}$ contains the variables x_j for which $\mathbf{a}_j$ belongs to $\mathbf{R}$. Likewise, the vector $\mathbf{c}$ is split up as $\mathbf{c} = (\mathbf{c_B}, \mathbf{c_R})$. Then the system $\mathbf{Ax} = \mathbf{b}$ can be rewritten as $\mathbf{Bx_B} + \mathbf{Rx_R} = \mathbf{b}$. If we multiply both sides of this equation by $\mathbf{B}^{-1}$, we find that the system $\mathbf{Ax} = \mathbf{b}$ is equivalent to the system

$$\mathbf{x_B} + \mathbf{B}^{-1}\mathbf{R}\,\mathbf{x_R} = \mathbf{B}^{-1}\mathbf{b}. \tag{1.1}$$

This system of linear equations is called the *canonical form* of $\mathbf{Ax} = \mathbf{b}$ for the basis $\mathbf{B}$. If we substitute $\mathbf{x_B} = \mathbf{B}^{-1}\mathbf{b} - \mathbf{B}^{-1}\mathbf{R}\mathbf{x_R}$ into the objective function $\mathbf{cx} = \mathbf{c_B}\mathbf{x_B} + \mathbf{c_R}\mathbf{x_R}$, then we find the following equivalent representation for the objective function:

$$\mathbf{cx} = \mathbf{c_B}\mathbf{B}^{-1}\mathbf{b} - (\mathbf{c_B}\mathbf{B}^{-1}\mathbf{R} - \mathbf{c_R})\mathbf{x_R} \tag{1.2}$$

for every $\mathbf{x}$ that satisfies $\mathbf{Ax} = \mathbf{b}$.

Illustration

Consider the LP problem

$$
\begin{array}{llllllllll}
\text{Maximize} & 2x_1 & + & 3x_2 & + & x_3 & + & x_4 & & \\
\text{subject to} & x_1 & + & 2x_2 & + & 2x_3 & + & x_4 & = & 8 \\
& x_1 & + & x_2 & + & 4x_3 & - & 2x_4 & = & 5 \\
\text{and} & & & & & x_1, x_2, x_3, x_4 & \geq & 0. &
\end{array}
$$

The matrix $\mathbf{A}$, the row vector $\mathbf{c}$, and the column vector $\mathbf{b}$ are given by

$$\mathbf{A} = \begin{pmatrix} 1 & 2 & 2 & 1 \\ 1 & 1 & 4 & -2 \end{pmatrix}, \quad \mathbf{c} = (2\ 3\ 1\ 1), \quad \text{and} \quad \mathbf{b} = \begin{pmatrix} 8 \\ 5 \end{pmatrix}.$$

The matrix $\mathbf{A}$ consists of four column vectors $\mathbf{a}_1$, $\mathbf{a}_2$, $\mathbf{a}_3$, and $\mathbf{a}_4$. If we take the basis $\mathbf{B} = (\mathbf{a}_1, \mathbf{a}_4)$, then $\mathbf{R} = (\mathbf{a}_2, \mathbf{a}_3)$. We then find

$$\mathbf{B}^{-1} = \begin{pmatrix} \frac{2}{3} & \frac{1}{3} \\ \frac{1}{3} & -\frac{1}{3} \end{pmatrix} \quad \text{and} \quad \mathbf{B}^{-1}\mathbf{b} = \begin{pmatrix} 7 \\ 1 \end{pmatrix}.$$

So the system of linear constraints can be rewritten in the equivalent canonical form

$$x_1 + \tfrac{5}{3}x_2 + \tfrac{8}{3}x_3 \qquad = 7$$
$$\tfrac{1}{3}x_2 - \tfrac{2}{3}x_3 + x_4 = 1.$$

Likewise, we can rewrite the objective function $z = 2x_1 + 3x_2 + x_3 + x_4$ in an equivalent form. It follows from $\mathbf{c_B} = (2, 1)$ and $\mathbf{c_R} = (3, 1)$ that $\mathbf{c_B}\,\mathbf{B}^{-1}\mathbf{b} = 15$ and $\mathbf{c_B}\mathbf{B}^{-1}\mathbf{R} - \mathbf{c_R} = (2/3, 11/3)$. Hence, for every solution (x_1, x_2, x_3, x_4) to the system of linear constraints, the objective function $z = 2x_1 + 3x_2 + x_3 + x_4$ can be written as

$$z = 15 - \frac{2}{3}x_2 - \frac{11}{3}x_3.$$

As an aside, note that the solution $x_1 = 7$, $x_4 = 1$, $x_2 = x_3 = 0$ generated by the basis $\mathbf{B} = (\mathbf{a}_1, \mathbf{a}_4)$ is optimal because in the corresponding representation of the objective function, the coefficients of the nonbasic variables x_2 and x_3 are negative.

The representation (1.1) leads to the following definition.

Definition 1.2. *For a given basis $\mathbf{B}$, the solution $(\mathbf{x_B} = \mathbf{B}^{-1}\mathbf{b}, \mathbf{x_R} = 0)$ to $\mathbf{Ax} = \mathbf{b}$ is called the basic solution corresponding to $\mathbf{B}$. This solution is feasible if $\mathbf{B}^{-1}\mathbf{b} \geq 0$.*

The following theorem gives sufficient conditions for the optimality of a basic solution to a maximization problem.

Theorem 1.1. *Let $\mathbf{B}$ be a given basis for $\mathbf{Ax} = \mathbf{b}$. The basic solution corresponding to $\mathbf{B}$ is an optimal solution to the LP model if the following two optimality criteria hold:*

C1. $\mathbf{B}^{-1}\mathbf{b} \geq 0$,
C2. $\mathbf{c_B}\mathbf{B}^{-1}\mathbf{a}_j - c_j \geq 0 \qquad$ *for all $j = 1, \ldots, n$.*

Proof. It follows from criterion C1 that the basic solution $(\mathbf{x_B} = \mathbf{B}^{-1}\mathbf{b}, \mathbf{x_R} = 0)$ is feasible for $\mathbf{Ax} = \mathbf{b}$. By (1.2), the corresponding value of the objective function is equal to $\mathbf{c_B}\mathbf{B}^{-1}\mathbf{b}$. To prove that the basic solution is optimal, we must verify $\mathbf{cx} \leq \mathbf{c_B}\mathbf{B}^{-1}\mathbf{b}$ for every feasible solution $\mathbf{x}$. From (1.2), we deduce

$$\mathbf{cx} = \mathbf{c_B}\mathbf{B}^{-1}\mathbf{b} - \sum_{j:\mathbf{a}_j \in R} (\mathbf{c_B}\mathbf{B}^{-1}\mathbf{a}_j - c_j)x_j$$
$$\leq \mathbf{c_B}\mathbf{B}^{-1}\mathbf{b},$$

using criterion C2 and the inequalities $x_j \geq 0$ for all j. $\qquad\square$

The representations (1.1) and (1.2) for the canonical form and the optimality criteria C1 and C2 are the key to the revised simplex method.

The Revised Simplex Method

The revised simplex method is a streamlined version of the simplex method discussed in Section 1.4 and is particularly well suited for implementation on a computer. The revised simplex method generates a series of bases such that

- every new basis differs by exactly one column vector from the previous basis;
- the objective value for a new basis is at least as good as that for the previous basis.

The algorithm ensures that the optimality criterion C1 is satisfied at every iteration and continues until optimality criterion C2 is also satisfied. How can we obtain the new basis $\mathbf{B}_{\text{new}}$ from the basis $\mathbf{B}$? A single column vector in $\mathbf{B}$ must be replaced by a column vector from $\mathbf{R}$. The representation (1.2) suggests that the best way to choose the new vector is as follows:

1. Add a vector $\mathbf{a}_s$ to $\mathbf{B}_{\text{new}}$, where $\mathbf{a}_s$ is a vector $\mathbf{a}_j \in R$ for which $\mathbf{c}_\mathbf{B}\mathbf{B}^{-1}\mathbf{a}_j - c_j$ has a negative value (if we then make the nonbasic variable x_s positive, the objective value increases). Commercial computer programs use a clever strategy to choose the entering vector $\mathbf{a}_s$ to reduce the computation time as much as possible.

The column vector that must be removed from $\mathbf{B}$ is the first vector $\mathbf{a}_j$ for which the variable x_j becomes zero as x_s increases. Without going into detail, we state that it follows from representation (1.1) that the exiting vector can be chosen as follows:

2. Remove the vector $\mathbf{a}_\ell$ from $\mathbf{B}$, where the index ℓ is equal to the index i for which the minimum is reached in

$$\min_{1 \le i \le m} \left\{ \frac{(\mathbf{B}^{-1}\mathbf{b})_i}{(\mathbf{B}^{-1}\mathbf{a}_s)_i} \mid (\mathbf{B}^{-1}\mathbf{a}_s)_i > 0 \right\}.$$

This technical formula is only given to demonstrate that the revised simplex method uses only $\mathbf{B}^{-1}$ and the original data of the LP model. The revised simplex method does not transform all data in every step as in the original simplex method. Moreover, the revised simplex method has better control over rounding errors during the calculations. It suffices to check the rounding errors when calculating the inverses of the successive bases. The inverse of the new basis $\mathbf{B}_{\text{new}}$ can easily be calculated using the inverse of the previous basis $\mathbf{B}$, since the bases differ by only one column vector. We do not give the formula for calculating $\mathbf{B}_{\text{new}}^{-1}$.

The Dual LP Problem

The primal LP problem

$$
\begin{array}{ll}
\text{Maximize} & \mathbf{cx} \\
\text{subject to} & \mathbf{Ax} = \mathbf{b} \\
\text{and} & \mathbf{x} \ge \mathbf{0}
\end{array}
$$

has dual

$$\begin{aligned}
&\text{Minimize} &&\mathbf{yb}\\
&\text{subject to} &&\mathbf{yA} \geq \mathbf{c}\\
&\text{and} &&\mathbf{y} \text{ unrestricted in sign.}
\end{aligned}$$

The (revised) simplex method calculates not only an optimal solution to the primal but also an optimal solution to the dual. To show this, we first prove the following essential theorem.

Theorem 1.2.

(1) For every pair of feasible solutions $\mathbf{x}$ *and* $\mathbf{y}$ *to the primal and dual, respectively, we have*

$$\mathbf{cx} \leq \mathbf{yb}.$$

(2) If $\mathbf{cx}^* = \mathbf{y}^*\mathbf{b}$ *holds for a pair of feasible solutions* $\mathbf{x}^*$ *and* $\mathbf{y}^*$ *to the primal and dual, respectively, then* $\mathbf{x}^*$ *is optimal for the primal and* $\mathbf{y}^*$ *is optimal for the dual.*

Proof. (1) If the vectors $\mathbf{x}$ and $\mathbf{y}$ are feasible solutions to the primal and dual, respectively, then we have $\mathbf{Ax} = \mathbf{b}$, $\mathbf{x} \geq \mathbf{0}$, and $\mathbf{yA} \geq \mathbf{c}$. Multiplying $\mathbf{Ax} = \mathbf{b}$ by $\mathbf{y}$ and $\mathbf{yA} \geq \mathbf{c}$ by $\mathbf{x} \geq \mathbf{0}$ gives

$$\mathbf{yAx} = \mathbf{yb} \qquad \text{and} \qquad \mathbf{yAx} \geq \mathbf{cx},$$

where in the second statement, the inequality $\mathbf{x} \geq \mathbf{0}$ has been used. This gives $\mathbf{yb} \geq \mathbf{cx}$.

(2) In part (a), we have shown that $\max \mathbf{cx} \leq \min \mathbf{yb}$; therefore, we have

$$\mathbf{cx}^* \leq \max \mathbf{cx} \leq \min \mathbf{yb} \leq \mathbf{y}^*\mathbf{b}.$$

If $\mathbf{cx}^* = \mathbf{y}^*\mathbf{b}$, then equality must hold in this inequality, so that $\mathbf{cx}^* = \max \mathbf{cx}$ and $\mathbf{y}^*\mathbf{b} = \min \mathbf{yb}$. □

Using Theorem 1.2, we can deduce the following result.

Theorem 1.3. *Let* $\mathbf{x}^*$ *and* $\mathbf{y}^*$ *be feasible solutions to the primal and dual, respectively, that satisfy the condition*

$$x_j^* \left[\sum_{i=1}^{m} a_{ij} y_i^* - c_j \right] = 0, \quad j = 1, \ldots, n. \tag{1.3}$$

Then $\mathbf{x}^*$ *is optimal for the primal and* $\mathbf{y}^*$ *is optimal for the dual.*

Proof. If, in the equality above, we sum over all j, we find

$$\sum_{j=1}^{n} c_j x_j^* = \sum_{j=1}^{n} x_j^* \sum_{i=1}^{m} a_{ij} y_i^*$$

$$= \sum_{i=1}^{m} y_i^* \sum_{j=1}^{n} a_{ij} x_j^* = \sum_{i=1}^{m} b_i y_i^*,$$

that is, $\mathbf{cx}^* = \mathbf{by}^*$. The result now follows from part (b) of Theorem 1.2. □

The conditions (1.3) in Theorem 1.3 are called the *complementary slackness conditions*. Not only are these conditions important theoretically, they can sometimes also be used to construct special algorithms for LP models with a particular structure, for example in network optimization.

Next, we prove the most important result.

Theorem 1.4 (Duality theorem). *Suppose that the (revised) simplex method is applied to the primal and that the resulting basis* $\mathbf{B}$ *satisfies the optimality criteria C1 and C2 from Theorem 1.1. Then*

$$\mathbf{y}^* = \mathbf{c_B}\mathbf{B}^{-1}$$

is an optimal solution to the dual. Moreover, we have the duality result $\max \mathbf{cx} = \min \mathbf{yb} \ (= \mathbf{c_B}\mathbf{B}^{-1}\mathbf{b})$.

Proof. Since criterion C2 is satisfied, we have $\mathbf{y}^*\mathbf{a}_j \geq c_j$ for all $j = 1, \ldots, n$ or, equivalently, $\mathbf{y}^*\mathbf{A} \geq \mathbf{c}$. Therefore, $\mathbf{y}^*$ is a feasible solution to the dual. The objective value is $\mathbf{y}^*\mathbf{b} = \mathbf{c_B}\mathbf{B}^{-1}\mathbf{b}$. The optimal solution $\mathbf{x}^* = (\mathbf{x_B^*}, \mathbf{x_R^*})$ to the primal with $\mathbf{x_B^*} = \mathbf{B}^{-1}\mathbf{b}$ and $\mathbf{x_R^*} = \mathbf{0}$ now has objective value $\mathbf{cx}^* = \mathbf{c_B}\mathbf{B}^{-1}\mathbf{b} = \mathbf{y}^*\mathbf{b}$. It then follows from part (b) of Theorem 1.2 that $\mathbf{y}^*$ is an optimal solution to the dual. We also find that the maximum objective value of the primal is equal to the minimum objective value of the dual and is given by $\mathbf{c_B}\mathbf{B}^{-1}\mathbf{b}$.   □

Not only is the duality theorem a beautiful mathematical result, it also has important practical consequences. Let $\mathbf{B}$ be a basis that satisfies the optimality criteria C1 and C2. Suppose that in the primal right-hand-side vector, we change $\mathbf{b}$ to $\mathbf{b} + \mathbf{\Delta b}$. The new basic solution corresponding to $\mathbf{B}$ then becomes $\mathbf{B}^{-1}(\mathbf{b} + \mathbf{\Delta b})$. It follows from $\mathbf{B}^{-1}\mathbf{b} \geq \mathbf{0}$ that $\mathbf{B}^{-1}(\mathbf{b} + \mathbf{\Delta b}) \geq \mathbf{0}$ for $\mathbf{\Delta b}$ sufficiently small. In other words, the new basic solution corresponding to $\mathbf{B}$ is only feasible if $\mathbf{\Delta b}$ is sufficiently small. Since $\mathbf{b}$ does not affect the constraints of the dual, the vector $\mathbf{y}^* = \mathbf{c_B}\mathbf{B}^{-1}$ is always feasible for the new dual, regardless of the value of $\mathbf{\Delta b}$. For this feasible solution, the dual objective function has value $\mathbf{c_B}\mathbf{B}^{-1}(\mathbf{b} + \mathbf{\Delta b})$. It now follows from part (a) of Theorem 1.2 that

new maximum objective value of the primal $\leq \mathbf{c_B}\mathbf{B}^{-1}(\mathbf{b} + \mathbf{\Delta b})$,

where equality holds if $\mathbf{\Delta b}$ is sufficiently small, in which case we have $\mathbf{B}^{-1}(\mathbf{b} + \mathbf{\Delta b}) \geq \mathbf{0}$. After all, in that case, $\bar{\mathbf{x}} = \mathbf{B}^{-1}(\mathbf{b} + \mathbf{\Delta b})$ and $\bar{\mathbf{y}} = \mathbf{c_B}\mathbf{B}^{-1}$ are feasible solutions to the new primal and dual, and these solutions have the same objective value. It follows from part (b) of Theorem 1.2 that equality indeed holds in the inequality above if $\mathbf{\Delta b}$ is sufficiently small. Note that in that case, $\mathbf{c_B}\mathbf{B}^{-1}\mathbf{b}$ is the old maximum objective value of the primal and $\mathbf{c_B}\mathbf{B}^{-1}\mathbf{\Delta b} = \sum_{i=1}^{m} y_i^* \Delta b_i$. This means that the inequality above can be reformulated as

new maximum objective value of the primal $\leq$

old maximum objective value of the primal $+ \sum_{i=1}^{m} y_i^* \Delta b_i,$

where equality holds if $\mathbf{\Delta b}$ is sufficiently small. So the optimal dual value y_i^* gives the marginal increase of the maximum objective value of the primal if the available quantity b_i of the ith raw material is increased by a limited amount. This explains why the optimal dual values can be interpreted as "shadow prices" for the raw materials. This interpretation is of great importance in LP applications.

Parametric Analysis

The optimality criteria C1 and C2 of Theorem 1.1 also give fundamental insight for parametric analysis. We briefly illustrate this for the right-hand-side coefficients. Suppose that the LP problem $\max \mathbf{cx}$ subject to $\mathbf{Ax} = \mathbf{b}$ and $\mathbf{x} \geq \mathbf{0}$ has been solved. The revised simplex method has provided an optimal basis $\mathbf{B}$. The right-hand-side coefficients b_i are now changed to $b_i + \lambda d_i$ for $i = 1, \ldots, m$, where λ is a parameter. The new LP problem is $\max \mathbf{cx}$ subject to $\mathbf{Ax} = \mathbf{b} + \lambda \mathbf{d}$ and $\mathbf{x} \geq 0$. The optimality criterion C2 is not affected by the change in the data. The basis $\mathbf{B}$ therefore stays optimal for all values of λ that satisfy $\mathbf{B}^{-1}(\mathbf{b} + \lambda \mathbf{d}) \geq \mathbf{0}$. Component-wise, this condition is $(\mathbf{B}^{-1}\mathbf{b})_i + \lambda(\mathbf{B}^{-1}\mathbf{d})_i \geq 0$ for $i = 1, \ldots, m$. So for every i, we obtain an upper and a lower bound for λ. Together, these bounds give a range of λ-values for which the basis $\mathbf{B}$ stays optimal. Of course, the values of the optimal basic variables do change. The parametric analysis for the profit coefficients c_j is similar. In this analysis, we also find that the values of the optimal basic variables stay the same if the optimality criterion C2 continues to be satisfied when the profit coefficients are changed.

Illustration

Consider the parametric LP problem

$$\text{Maximize } 2x_1 + 3x_2 + x_3 + x_4$$
$$\text{subject to } x_1 + 2x_2 + 2x_3 + x_4 = 8 + 2\lambda$$
$$x_1 + x_2 + 4x_3 - 2x_4 = 5 + 4\lambda$$
$$x_1, x_2, x_3, x_4 \geq 0,$$

where λ is a parameter. For the LP problem with $\lambda = 0$, we saw at the beginning of this section that the basis $\mathbf{B} = (\mathbf{a}_1, \mathbf{a}_4)$ leads to an optimal solution. The inverse of the basis $\mathbf{B}$ is

$$\mathbf{B}^{-1} = \begin{pmatrix} \frac{2}{3} & \frac{1}{3} \\ \frac{1}{3} & -\frac{1}{3} \end{pmatrix}.$$

The basis $\mathbf{B}$ continues to lead to an optimal solution to the parametric LP problem for all λ-values satisfying

$$\mathbf{B}^{-1} \begin{pmatrix} 8 + 2\lambda \\ 5 + 4\lambda \end{pmatrix} \geq \mathbf{0}.$$

This gives the two linear inequalities

$$7 + \frac{8}{3}\lambda \geq 0 \quad \text{and} \quad 1 - \frac{2}{3}\lambda \geq 0.$$

So the solution $x_1 = 7 + \frac{8}{3}\lambda$, $x_4 = 1 - \frac{2}{3}\lambda$, $x_2 = x_3 = 0$ is optimal for the parametric LP problem for all values of λ with

$$-\frac{21}{8} \leq \lambda \leq \frac{3}{2}.$$

As a function of λ, the maximum objective value is given by

$$z^*(\lambda) = 15 + \frac{14}{3}\lambda \quad \text{for} \quad -\frac{21}{8} \leq \lambda \leq \frac{3}{2}.$$

Verify the following results. For the LP problem with $\lambda = \frac{3}{2}$, the basis $\mathbf{B} = (\mathbf{a}_1, \mathbf{a}_3)$ is an alternative basis that leads to an optimal solution. The basis $\mathbf{B} = (\mathbf{a}_1, \mathbf{a}_3)$ leads to an optimal basic solution $x_1 = 11$, $x_3 = \lambda - \frac{3}{2}$, $x_2 = x_4 = 0$ for all $\lambda \geq \frac{3}{2}$. The maximum objective value is given by

$$z^*(\lambda) = 20.5 + \lambda \quad \text{for} \quad \lambda \geq \frac{3}{2}.$$

The maximum objective value $z^*(\lambda)$ is a piecewise linear function whose slope decreases as λ increases. The result that the marginal increase of $z^*(\lambda)$ decreases as λ increases is not surprising. The greater the available quantity of raw material, the smaller the marginal value of an additional unit of raw material. In mathematical terms, the function $z^*(\lambda)$ is called *concave*. If, in the LP problem, we had varied the profit coefficients instead of the right-hand-side coefficients by changing c_j to $c_j + \lambda d_j$, we would have found that the maximum objective value $z^*(\lambda)$ is a piecewise linear function whose slope increases as λ increases (a *convex* function).

The Dual Simplex Method

This simplex method generates a series of basic solutions such that optimality criterion C1 is always satisfied (that is, the basic solutions are feasible for the primal problem) and continues until optimality criterion C2 is also satisfied. The dual simplex method described below generates a series of basic solutions such that optimality criterion C2 is always satisfied (that is, the basic solutions are feasible for the dual problem) and continues until optimality criterion C1 is also satisfied. The dual simplex method is especially important for post-optimization when after solving the LP problem, one extra constraint is added, and it turns out that the previously found optimal basic solution does not satisfy this extra constraint. In this case, the new LP problem does not need to be solved from the beginning; instead, the already calculated optimal basic solution to the old problem can be used as initial solution to the new problem.[3] This is something we can easily explain. Suppose that the original LP problem $\max \mathbf{cx}$ subject to $\mathbf{Ax} = \mathbf{b}$ and $\mathbf{x} \geq 0$ has been solved with $\mathbf{B}_0 = (\mathbf{a}_{i_1}, \ldots, \mathbf{a}_{i_m})$ the basis of the optimal basic solution that was found. Afterward, we add the extra constraint

$$\sum_{j=1}^{n} a_{m+1,j} x_j \leq d_{m+1},$$

---

[3] This fact is of great importance in the branch-and-bound method for solving integer LP problems, in which a series of LP problems is solved, each time with an additional constraint.

and it turns out that the previously found optimal basic solution $\mathbf{x_B} = \mathbf{B_0^{-1}b}$ does not satisfy this constraint. After introducing the slack variable v_{n+1} for the extra constraint, we can represent the new LP problem as

$$\text{Maximize } \mathbf{cx} + 0v_{n+1}$$

$$\text{subject to } \begin{pmatrix} \mathbf{A} & \mathbf{0} \\ \mathbf{a}_{m+1} & 1 \end{pmatrix} \begin{pmatrix} \mathbf{x} \\ v_{n+1} \end{pmatrix} = \begin{pmatrix} \mathbf{b} \\ d_{m+1} \end{pmatrix}$$

with $\mathbf{a}_{m+1} = (a_{m+1,1}, \ldots, a_{m+1,n})$. The basis $\mathbf{B_0} = (\mathbf{a}_{i_1}, \ldots, \mathbf{a}_{i_m})$ corresponding to the optimal basic solution to the old LP problem immediately leads to the basis

$$\overline{\mathbf{B}} = \begin{pmatrix} \mathbf{B_0} & \mathbf{0} \\ \gamma & 1 \end{pmatrix} \quad \text{with } \gamma = (a_{m+1,i_1}, \ldots, a_{m+1,i_m})$$

for the new LP problem. The basis $\overline{\mathbf{B}}$ has inverse (verify this!)

$$\overline{\mathbf{B}}^{-1} = \begin{pmatrix} \mathbf{B_0^{-1}} & \mathbf{0} \\ -\gamma \mathbf{B_0^{-1}} & 1 \end{pmatrix}.$$

For the basic solution corresponding to $\overline{\mathbf{B}}^{-1}$, we have

$$\overline{z}_j - c_j = (\mathbf{c}_{\overline{\mathbf{B}}_0}, 0) \begin{pmatrix} \mathbf{B_0^{-1}} & \mathbf{0} \\ -\gamma \mathbf{B_0^{-1}} & 1 \end{pmatrix} \begin{pmatrix} \mathbf{a}_j \\ a_{m+1,j} \end{pmatrix} - c_j$$

$$= \mathbf{c_B} \mathbf{B_0^{-1}} \mathbf{a}_j - c_j = z_j - c_j \geq 0$$

for $j = 1, \ldots, n$, while $z_{n+1} - c_{n+1} = 0$ because x_{n+1} is a basic variable for $\overline{\mathbf{B}}$. So the situation is that we can begin with a basic solution $\overline{\mathbf{B}}^{-1} \binom{\mathbf{b}}{d_{m+1}}$ that does not satisfy optimality criterion C1 but does satisfy C2. The dual simplex method now proceeds as follows, starting with the LP formulation $\max \mathbf{cx}$ subject to $\mathbf{Ax} = \mathbf{b}$ and $\mathbf{x} \geq \mathbf{0}$.

1. For the current basis $\mathbf{B}$, let the index r be the index i for which $(\mathbf{B^{-1}b})_i$ has the most negative value.
2. The new basis $\mathbf{B}_{\text{new}}$ is obtained by replacing the rth column vector of $\mathbf{B}$ with $\mathbf{a}_k$, where k is determined by

$$\max_j \left\{ \frac{\mathbf{c_B} \mathbf{B^{-1}} \mathbf{a}_j - c_j}{(\mathbf{B^{-1}} \mathbf{a}_j)_r} \mid (\mathbf{B^{-1}} \mathbf{a}_j)_r < 0 \right\}.$$

1.8 Exercises

1.1 Every week, McRonald makes two types of steak sauce for a sister company: Spicy Gonzales and Cool Gringo. Each of the sauces is made by mixing two ingredients, A and B. Although a certain degree of freedom is permitted when making the sauces, the following constraints must be satisfied: (1) Cool Gringo must not consist for more than 75% out of ingredient A, (2) Spicy Gonzales must consist for at least 25% out of ingredient A and at least 50% out of ingredient B. Each week, at most 40 liters of ingredient A and at most 30 liters of ingredient B can be purchased. McRonald can sell as much sauce as it produces at the price of €6.70 per liter of Spicy Gonzales and a price of €5.70 per liter of Cool Gringo. The ingredients A and B cost €3.20 and €4.10 per liter, respectively. How can McRonald maximize the net profit on the sauces?

1.2 A coal-fired power plant processes two types of coal, A and B, to generate steam energy. The coal is first crushed and then burned to produce steam energy to power steam turbines. The question is in what ratio the coal types A and B must be processed to maximize the production of steam energy given a number of limiting conditions. For example, environmental requirements have been imposed on gas and soot emissions. Soot emission levels may not exceed 12 kg per hour, while the requirement for gas emission is that the number of particles of sulfur per million particles of gas not exceed 3000. Burning 1 ton of coal type A gives a soot emission level of 0.5 kg per hour, and 1 ton of coal type B gives a soot emission level of 1 kg per hour. Coal type A has the property that for every million particles that are released after combustion, 1800 are sulfur particles, while for coal type B, the number of sulfur particles per million of released particles is 3800. Per ton burned, the coal types A and B emit the same number of particles. The coal is transported by train, and the supply is limited to 20 tons an hour. The capacity of the coal-crushing installation is also limited: the crushing capacity is 16 tons of coal type A per hour if A is the only type of coal and 24 tons of coal type B per hour if B is the only one. Finally, after combustion, 1 ton of coal type A gives 12 000 kg of steam energy, and 1 ton of coal type B gives 10 000 kg. Formulate this as an LP problem.

1.3 **(a)** Billy Smart is the head of a one-man company that manages the stock portfolios of a number of wealthy clients. A new client has just commissioned the creation of a stock portfolio worth 1 million euros. The client wants to keep things simple and only wants shares in the three funds Vilyps, ThickSteel, and MultiDelivery, with the extra requirement that at least 30% be invested in Vilyps and that the amount invested in MultiDelivery is at least three times the amount invested in ThickSteel. For the funds Vilyps, ThickSteel, and MultiDelivery, the prices per share are €50, €20, and €75, while the expected annual dividends per share are €7, €6, and €9. How many shares of each fund should be bought to maximize the total expected dividend? Formulate this as an LP model and solve it using a software package (obviously, one cannot buy fractional quantities of shares, but for sufficiently large values, say ≥ 20, of the decision variables, one can just use an LP model and if necessary round off the values of the decision variables that are found.)

(b) The law firm Right and Away has a separate department that deals with lawsuits against unscrupulous medical professionals who carry out unnecessary operations, as well as against hard-working professors who fail students who have not studied the course material. There are a plethora of such cases. A lawsuit against a physician requires 2.5 man-days of preparation and the hiring of 2 expert witnesses, while a lawsuit against a professor requires 2.5 man-days of preparation and the hiring of 1 expert witness. The law firm makes a profit of 17 thousand euros in a court case against a physician, while the profit in a court case against a professor is 9 thousand euros. The firm cannot allocate more than 75 man-days or hire more than 50 expert witnesses for these trials. How many court cases of each type must the firm accept to achieve maximum turnover? Solve the problem graphically.

1.4 Every Friday, a moneylender issues two loans: a loan of €3500 for one week with an interest of €1000 and a loan of €15000 for two weeks with an interest of €3000. The loans must be repaid on time, but unfortunately, not everyone does so. Twenty percent of the €3500 loans and ten percent of the €15000 loans are not repaid on time. In this case, an enforcer is used to collect the loan. The fine is 10% of the amount borrowed, half of which is for the moneylender's "gorilla." Collecting the money takes a week, and each week, 30 defaulters can be processed. (Defaulters' processing cannot be postponed.) The moneylender cannot issue more than 180 loans a week. Every week, there are sufficiently many candidates for taking out loans. The capital he has at his disposal for lending is 4 million euros, and each week he wants to have no more than 4 million euros in outstanding loans. He wants to use his ill-gotten gains to go on holiday in four weeks, which is why he is willing to issue loans of €3500 only in the next three weeks and loans of €15000 only in the next two weeks. How many loans of each type must the moneylender issue to maximize his vacation allowance? Ignore the fact that the numbers of loans should be integers.

1.5 A company produces rolls of textile with a standard width of 120 cm, each roll having a fixed length. A customer orders 50 rolls of width 40 cm and 60 rolls of width 25 cm. These smaller widths must be cut out of the standard rolls. The company can, for example, decide to cut up a 120 cm roll into one 40 cm roll and three 25 cm rolls. This gives 5 cm of trim loss. The company wants to meet the order by cutting up the standard rolls in such a way that the total amount of waste is minimized. In addition to the trim loss, the residual rolls in the ordered widths are also considered waste. Use linear programming to solve this cutting stock problem. (*Hint*: Use a decision variable for each of the different patterns the company can use to cut up the standard roll.)

1.6 (a) A small investor wants to enjoy her money in five years' time. She currently has 100 thousand euros at her disposal for investment. There are three investment opportunities. Investment A reaches maturity in one year and returns 1.25 euros at the end of the year for every euro invested at the beginning of the year. Investment A can only be made in the second, third, and fifth year. Investments B and C require that the invested money be fixed for a number of years and give returns for every invested euro of 1.35 euros after three years and 1.50 euros after four years, respectively. Formulate an LP model to determine how the investor should invest to maximize her capital after five years. (*Hint*: Use the resource variables $v_1, \ldots, v_5$, where the variable v_i indicates the number of euros not invested in year i).

(b) A company wants to draw up a production plan for the next five weeks. The demand for the product in each of the next five weeks is known and is equal to 1300, 1400, 1000, 800, and 1700 units in weeks 1, 2, 3, 4, and 5, respectively. Currently, there is a stock of 200 units. The product can be manufactured at the beginning of the week and kept in stock. The production time of each round of production may be ignored, and the maximum stock capacity is 250. The production costs per unit of product are 10 in weeks 1 and 2, 11 in weeks 3 and 4, and 12 in week 4. The stock charge for each unit in stock at the end of the week is 0.4. Use LP to draw up a production plan that minimizes the total cost and meets the demand every week.

1.7 Cheese manufacturer Tozzi produces mozzarella and ricotta daily; milk is the main raw material for this. The production of 1 kilo of mozzarella requires 7 liters of milk, and 1 kilo of ricotta requires 3 liters of milk. On any given day, no more than 12 thousand liters of milk, which can be purchased at 75 cents a liter, are available. During the production process, 1% of the mozzarella and 2% of the ricotta is lost. Retailers guarantee a daily acquisition of up to 1500 kilos of these types of cheese provided that at least 30% consist of ricotta. Tozzi sells the mozzarella and ricotta for 9 and 5 euros per kilo, respectively. Any leftover mozzarella is sold to a pizzeria at 7 euros per kilo. The pizzeria is willing to purchase 200 kilos of mozzarella per day. Tozzi is required to deliver at least 10% of its total net cheese production to the pizzeria. Leftover ricotta is sold to a feed company at 3 euros per kilo. Tozzi wants to maximize the daily net revenue. Formulate this as an LP problem and solve it using a software package.

1.8 Jake Nuts has been able to purchase 250 kg of walnuts, 75 kg of hazelnuts, 50 kg of cashews, and 37.5 kg of pistachios. Jake sells four types of mixed nuts at the market, in standard packages of 250 g. The table below gives the types of nuts, the requirements for the mixes, and the selling prices.

Type	Requirements for the Mix	Price
Jake Mix	$\geq$ 5% of each type of nut	€3.75
Party Mix	$\leq$ 50% walnuts, $\geq$ 15% hazelnuts	€5.00
Lulu Mix	$\geq$ 20% cashews, $\geq$ 20% pistachios	€6.25
Elite Mix	$\geq$ 30% cashews, $\geq$ 5% pistachios, $\geq$ 15% hazelnuts	€7.50

How many 250 g packages of each mix must Jake make, and what should the composition of each mix be to maximize Jake's total profit? Formulate this as an LP model and solve it using a software package.

1.9 (a) Use linear programming to fit a quadratic curve to the data in Table 1.3, where the aim is to minimize the sum of the distances between the observed and predicted values.

(b) Suppose given the following 8 data points: $(0.5, -1.24)$, $(1, 0.51)$, $(1, 1.52)$, $(1.5, 2.23)$, $(2.5, 2.78)$, $(3, 3.23)$, $(4, 3.91)$, $(5, 4.58)$. Fit a curve of the form $y = a + b \log(x)$ to the data, where the aim is to minimize the sum of the distances between the observed and predicted values.

1.10 You have received inside information about a specific horse race, namely that a deal has been made to let one of the four horses Lucky Luke, Joss Jamon, Ma Dalton, and Billy the Kid win. However, you do not know which of the four horses is going to win. If the horse Lucky Luke wins, the payment is 3.20 euros for every euro that was bet. For the horses Joss Jamon, Ma Dalton, and Billy the Kid, the payments are respectively 4.50 euros, 5 euros, and 7 euros for every euro that was bet. You have scraped together 2717 euros to bet on the race. Because you have no idea which of the four horses will win, you want to cover yourself against the worst possibility. How should you divide up your money over the four horses so that your minimum guaranteed payout is as high as possible?

1.11 (a) You are playing the following game with someone. You and your opponent both have three tokens, namely a red one, a white one, and a blue one. At the same time, each of you takes a token and places it under a beer mat without the other seeing which color the token has. Then, both players lift up the beer mat. The payout depends on the colors. The color red wins from blue with a payment of 5 euros, the color blue wins from white with a payment of 4 euros, and the color white wins from red with a payment of 3 euros. If the colors are the same, no payment occurs. What strategy maximizes the minimum guaranteed expected profit?

(b) Fatima and Eric repeatedly play the following game. Each round, they simultaneously write one of the numbers 1, 2, 3 on a piece of paper, in such a way that neither can see what the other has written down. Then, they show the numbers. The payout between the players depends on these numbers. If they have written down the same number, Eric receives 1 euro from Fatima if the number is 1 or 2, and Fatima receives 1 euro from Eric if the number is 3. The number 1 wins from 2 with a payment of 7 or 4 euros by the loser to the winner depending on whether Eric or Fatima is the winner, the number 2 wins from 3 with a payment of 4 or 5 euros by the loser to the winner depending on whether Eric or Fatima is the winner, and the number 3 wins from 1 with a payment of 1 or 2 euros by the loser to the winner depending on whether Eric or Fatima is the winner. Determine Fatima's strategy to maximize her minimum guaranteed expected profit.

(c) Two players A and B each choose a number from the list $10, 11, \ldots, 99$ at the same time. If the first digit of the product of the two numbers is 1, 2, or 3, then A wins; otherwise, B wins. What strategy should player A follow?

1.12 The West-Frisian Lottery has reached an agreement with the winner of the jackpot of the end-of-year lotto to pay the main prize spread out over 15 years instead of paying one sum of money at once. The Lottery faces the problem of financing the following amounts of money owed by the Lottery at the beginning of each of the next 15 years, in units of one thousand euros.

Year:	1	2	3	4	5	6	7	8
Needed:	800	900	1100	1200	1200	1400	1500	1500

Year:	9	10	11	12	13	14	15
Needed:	1700	1800	1900	2200	2300	2500	2500

The Lottery wonders how much money should be set aside now to meet the payment obligations when the following financial instruments are available:

- keep money on a separate savings account at 4% interest per year;
- purchase bonds at the beginning of year 1.

Bonds	Current Costs	Interest %	Duration
1	985 euros	6.0%	5 years
2	975 euros	6.5%	11 years
3	1025 euros	7.5%	14 years

Each bond has a nominal value of 1000 euros; that is, the bond pays 1000 euros at maturity and pays interest annually on its nominal value (this interest is transferred to the separate savings account).

How can the Lottery meet its payment obligations over the next 15 years against a minimum reservation? Formulate an LP model and solve it.

1.13 A computer chip manufacturer is facing the problem of planning the monthly production for the next three months. The expected demand for the chip is 600, 800, and 900 units for months 1, 2, and 3. The demand must be met on time. The production costs of each chip amount to 12.50 euros. The company can produce up to 750 chips per month during regular working hours. However, overtime may be used in the last two months. Overtime increases the monthly production capacity by 150 additional units. The production costs per chip are 4 euros higher for overtime than for regular working hours. Production surplus can be kept in stock at the cost of 2.50 euros per chip per month.

(a) How should production be planned to minimize the total cost? Formulate an LP model.

(b) Adapt the LP model to the case where there are additional costs of 0.45 euros for each unit of change in the total production level from one month to the next.

1.14 (a) Solve the following LP problem using the simplex method:

$$\begin{array}{ll}
\text{Maximize} & 2x_1 + 3x_2 + 1.5x_3 \\
\text{subject to} & x_1 \ \ + 4x_2 + x_3 \ \ \le 7 \\
& 3x_1 + 2x_2 + 2x_3 \le 8 \\
\text{and} & x_1, x_2, x_3 \ge 0.
\end{array}$$

(b) Solve the following LP problem using the simplex method:

$$\begin{array}{ll}
\text{Maximize} & 3x_1 \ + 2x_2 \\
\text{subject to} & 2x_1 \ + \ \ x_2 \le 40 \\
& -x_1 + 2x_2 \ \le 30 \\
\text{and} & x_1, x_2 \ge 0.
\end{array}$$

Note: If a new basic variable has a negative coefficient in an equation, then that equation does not impose any restrictions on how large the new basic variable can become.

(c) Give the dual of each of the LP problems above. What are the optimal objective values of these two duals?

1.15 A paper mill makes three types of paper from four different types of wood pulp. The table below summarizes for each type of paper how many tons of the different types of wood pulp are required to produce 1 ton of that type of paper. The available quantities of the four wood pulps $j = 1, 2, 3, 4$ are 1500, 800, 500, and 2000 tons, respectively, while the selling prices per ton of the three paper types $i = 1, 2, 3$ are 500, 400, and 325 euros, respectively. It is assumed that there are no production costs for making the paper from the wood pulp (the purchase costs of the wood pulp are included in the selling prices of the types of paper).

	Paper Type 1	Paper Type 2	Paper Type 3
Pulp 1	0.1	1.9	2.4
Pulp 2	0.2	0.5	0
Pulp 3	0.5	1.1	0.8
Pulp 4	2.5	0	0

First, formulate an LP problem for the paper mill that allows the mill to maximize the total revenue from the sale of the types of paper. Then, carry out the following thought experiment: A trader proposes that the mill sell him the available quantities of wood pulp instead of producing paper. What LP problem must the trader solve for his offer to be interesting for the paper mill? What is the significance of the variables in the trader's LP problem for the mill? Calculate only the solution to the LP problem for the trader and use the resulting knowledge to answer the following questions:

(a) What is the first estimate for the value for the mill of having 1 additional ton of wood pulp 3?

(b) Would paper type 2 be produced in the optimal solution to the mill's LP problem? If not, by how much would the selling price per ton of paper type 2 need to at least increase for it to become worthwhile for the mill to make this product?

1.16 Four ingredients are available for the composition of a menu that needs to meet certain requirements for the vitamin content. The menu must contain 7, 3, and 4 mg of vitamins A, B, and C, respectively. The costs of the ingredients $i = 1, 2, 3$, and 4 are 3.80, 1.25, 2.50, and 2 euros per unit, respectively. The table gives the quantities in mg of vitamins A, B, and C per unit of ingredient:

Ingredient	1	2	3	4
Vitamin A	3	1	2	1
Vitamin B	1.5	1	0.5	1
Vitamin C	2	0.5	1	2

(a) Formulate an LP problem for the composition of a menu that is the least expensive possible. (You do not have to solve this LP problem.)

Instead of making the menu, the possibility of marketing a vitamin pill containing 7 mg of vitamin A, 3 mg of vitamin B, and 4 mg of vitamin C is considered. The producer of this pill asks how she should set the price of the pill for this to be competitive with the costs of the menu.

(b) Formulate an LP problem for the pill manufacturer (you do not have to solve it) and compare this LP problem with the dual of the LP problem for composing the menu.

The optimal solution for the pill producer is to charge 1.20 euros for each mg of vitamin A in the pill, charge nothing for vitamin B, and charge 0.10 euros for each mg of vitamin C in the pill. Use this solution to answer the following questions concerning the cook's LP problem:

(c) What are the minimum costs of the menu?
(d) What is the first estimate for the increase in costs of the menu if the requirement that the menu contain at least 7 mg of vitamin A is changed to a requirement of at least 7.5 mg of vitamin A?
(e) Is ingredient 4 used in the optimal solution to the cook's LP problem?

1.17 Consider the investment problem from Section 1.3.1.

(a) What is the highest interest rate the bank is willing to pay a client who has increased its investment capital of 100 000 euros by 10000 euros through a deposit?
(b) By how much should the annual return on preferred shares be increased for it to become worthwhile to invest in them?
(c) By how much does the maximum annual return increase if up to 60% of the money may be invested in loans?
(d) What happens to the optimal investment plan if the investment in bonds must be at least 10% of the investment in loans?

1.18 The company Bingle makes two products, 1 and 2, by processing raw material. The company can purchase up to 90 kilos of raw material at a cost of 15 euros per kilo. One kilo of raw material suffices for either 1 kilo of product 1 or 0.3 kilos of product 2. To convert 1 kilo of raw material into 1 kilo of product 1 requires 2 hours of labor, while 3 hours of labor are needed to convert 1 kilo of raw material into 0.3 kilos of product 2. In all, 200 hours of labor are available. Furthermore, it is not possible to sell more than 25 kilos of product 2. The selling prices of products 1 and 2 are 20 and 70 euros per kilo, respectively. To maximize the net profit, Bingle solves an LP problem with three decision variables: I = number of kilos of raw material that are purchased, x_1 = number of kilos of raw material used to make product 1, and x_2 = number of kilos of raw material used to make product 2. The LP formulation and computer output are as follows:

```
Objective function
   MAX 20.0000 X1 +21.0000 X2 -15.0000 I
Subject to :
   1.   1.0000 X1 +1.0000 X2 -1.0000 I <=  0.0000
   2.   2.0000 X1 +3.0000 X2 <= 200.0000
   3.   1.0000 I <= 90.0000
   4.   0.3000 X2 <= 25.0000
```

```
Summary of Results

Value Objective Function :          470.0000

Variable          Activity Level    Reduced Cost
--------          --------------    ------------

X1        :           70.0000          0.0000
X2        :           20.0000          0.0000
I         :           90.0000          0.0000

                  Slack or Surplus    Shadow price
                  ----------------    ------------
Constraint 1            0.0000          18.0000
Constraint 2            0.0000           1.0000
Constraint 3            0.0000           3.0000
Constraint 4           19.0000           0.0000
```

```
Objective Coefficient Ranges

                  Current
Variable          Coefficient    Allowed Interval
--------          -----------    ----------------

X1                 20.0000        [  19.0000 ,    21.0000 ]
X2                 21.0000        [  20.0000 ,    22.5000 ]
I                 -15.0000        [ -18.0000 ,    Infinity ]
```

```
Right-Hand-Side Ranges

                  Current
Constraint        RHS            Allowed Interval
----------        ---            ----------------

   1.              0.0000        [ -23.3333 ,    10.0000 ]
   2.            200.0000        [ 180.0000 ,   263.3333 ]
   3.             90.0000        [  66.6667 ,   100.0000 ]
   4.             25.0000        [   6.0000 ,    Infinity ]
```

Answer the following questions (include your reasoning):

(a) What happens to the maximum net profit if only 87 kilos of raw material can be purchased? Also answer this question if only 65 kilos of raw material can be purchased.

(b) If the selling price of product 2 is changed to 68 euros per kilo, what happens to the values of the variables in the optimal solution?

(c) How much is the company Bingle willing to pay, at most, for 5 additional kilos of raw material?

(d) How much is the company Bingle willing to pay, at most, for 10 additional hours of labor?

(e) Suppose that the company is offered both 5 additional kilos of raw material and 10 extra hours of labor at a combined price of 90 euros. Is this offer attractive?

(f) Suppose that product 3 can also be made out of the raw material, where 0.9 kilos of product 3 can be obtained from 1 kilo of raw material, and 5 hours of labor are required to turn 1 kilo of raw material into 0.9 kilos of product 3. The selling price of product 3 is 25 euros per kilo. Is it worthwhile to make this product?

1.19 A steel manufacturer has been awarded a contract to supply 100 tons of a particular steel grade. The steel must meet the following requirements: it must consist of at least 3.5% nickel, at most 3% carbon, and exactly 4.5% manganese. The company receives 1000 euros per ton of this steel. Four alloys are available, in any desired quantity, for making the steel. The table below gives the percentages of nickel, carbon, and manganese for each of these four alloys.

	Alloy 1	Alloy 2	Alloy 3	Alloy 4
Nickel	6%	3%	2%	1%
Carbon	3%	2%	5%	6%
Manganese	8%	3%	2%	1%

The costs of the four alloys 1, 2, 3, and 4 are 600, 500, 400, and 300 euros per ton, respectively. Let x_i be the number of tons of alloy i to be used; the LP model and computer output are then as follows:

```
Objective Function :
   MAX 400.0000 X1 +500.0000 X2 +600.0000 X3 +700.0000 X4
Subject to :
   1.  0.0600 X1 +0.0300 X2 +0.0200 X3 +0.0100 X4 >=   3.5000
   2.  0.0300 X1 +0.0200 X2 +0.0500 X3 +0.0600 X4 <=   3.0000
   3.  0.0800 X1 +0.0300 X2 +0.0200 X3 +0.0100 X4 =    4.5000
   4.  1.0000 X1 +1.0000 X2 +1.0000 X3 +1.0000 X4 =  100.0000


Summary of Results

Value Objective Function :      49545.4545

Variable        Activity Level   Reduced Cost
--------        -------------    ------------
X1          :        36.3636        0.0000
X2          :        47.7273        0.0000
X3          :         0.0000       36.3636
X4          :        15.9091        0.0000
```

```
                        Slack or Surplus   Shadow prices
                        ----------------   -------------
        Constraint 1          0.2727          -0.0000
        Constraint 2          0.0000         3636.3636
        Constraint 3          0.0000        -2727.2727
        Constraint 4          0.0000          509.0909
```

Objective Coefficient Ranges

```
                       Current
        Variable       Coefficient   Allowed Interval
        --------       -----------   ----------------
        X1             400.0000       [    0.0000 ,   Infinity ]
        X2             500.0000       [  300.0000 ,   614.2857 ]
        X3             600.0000       [ -Infinity ,   636.3636 ]
        X4             700.0000       [  650.0000 ,   Infinity ]
```

Right-Hand-Side Ranges

```
                       Current
        Constraint     RHS           Allowed Interval
        ----------     ---           ----------------
        1.             3.5000        [ -Infinity ,     3.7727 ]
        2.             3.0000        [    2.3000 ,     4.5000 ]
        3.             4.5000        [    4.0714 ,     8.0000 ]
        4.           100.0000        [   81.2500 ,   150.0000 ]
```

Answer the following questions:

(a) What is the physical meaning of the shadow price of constraint 4?

(b) What is the effect on the values of the variables in the optimal solution if the cost of alloy 1 increases from 600 euros per ton to 750 euros per ton?

(c) What is the effect on the net profit if the requirement of at most 3% carbon is sharpened to at most 2.95% carbon, and what is the effect of sharpening it to at most 2.2% carbon?

(d) What happens to the net profit if the requirement of exactly 4.5% manganese is replaced by the requirement of exactly 4.55% manganese?

(e) What is the effect on the optimal solution if the selling price per ton of steel drops from 1000 euros per ton to 980 euros per ton?

(f) Suppose that alloy 5 is also available, for which the percentages of nickel, carbon, and manganese are 3%, 2%, and 10%. At what cost for alloy 5 does it become worthwhile to use it?

1.20 The farm Grain Harvest grows wheat and barley on its fields, which cover 45 hectares. Each cultivated hectare produces either 5 "bushels" of wheat or 4 bushels of barley (a "bushel" is a measure of capacity used for agricultural products). Wheat sells for 60 euros per bushel, and barley sells for 100 euros per bushel. The farm can sell at most 140 bushels of wheat and 120 bushels of barley. Six hours of labor are needed to harvest one hectare of wheat, and 10 to harvest one hectare of barley. At most 350 hours of labor are available at a cost of 20 euros per hour. To maximize the net profit, the farm solves an LP problem with the decision variables x_1 = number of hectares to cultivate with wheat, x_2 = number of hectares to cultivate with barley, and x_3 = number of hired hours of labor. The computer output for this LP problem is:

```
Objective Function :
  MAX   300 X1 + 400 X2 - 20 X3
Subject to :
  1.   1.0000 X1 +  1.0000 X2  <= 45.0000
  2.   6.0000 X1 + 10.0000 X2 - 1.0000 X3 <= 0.0000
  3.   5.0000 X1 <= 140.0000
  4.   4.0000 X2 <= 120.0000
  5.   1.0000 X3 <= 350.0000


Summary of Results

Value Objective Function :            8500.0000

  Variable          Activity Level   Reduced Cost
  --------          --------------   ------------

  X1    :             25.0000          0.0000
  X2    :             20.0000          0.0000
  X3    :            350.0000          0.0000

                    Slack or Surplus  Shadow Prices
                    ----------------  -------------

  Constraint 1          0.0000         150.0000
  Constraint 2          0.0000          25.0000
  Constraint 3         15.0000           0.0000
  Constraint 4         40.0000           0.0000
  Constraint 5          0.0000           5.0000


Objective Coefficient Ranges

                      Current
  Variable          Coefficient   Allowed Interval
  --------          -----------   ----------------

  X1                 300.0000     [  240.0000 ,   320.0000 ]
  X2                 400.0000     [  380.0000 ,   500.0000 ]
  X3                 -20.0000     [  -25.0000 ,   Infinity ]


Right-Hand-Side Ranges

                      Current
  Constraint          RHS           Allowed Interval
  ----------          ---           ----------------

  1.                 45.0000       [   38.3333 ,    46.2000 ]
  2.                  0.0000       [  -12.0000 ,    40.0000 ]
  3.                140.0000       [  125.0000 ,    Infinity ]
  4.                120.0000       [   80.0000 ,    Infinity ]
  5.                350.0000       [  338.0000 ,   390.0000 ]
```

Answer the following questions (include your reasoning):

(a) How much is the farm willing to pay, at most, to rent 1 additional hectare of land?

(b) How much is the farm willing to pay, at most, for 25 additional hours of labor?

(c) By how much would the net profit decrease if only 40 hectares of land were cultivated?

(d) What is the effect on the optimal solution if the selling price of wheat decreases to 56 euros per bushel and that of barley to 97 euros per bushel?

(e) Suppose that the farm can also grow hemp and that there is no limit to the quantity they can sell. Every hectare cultivated with hemp produces 4 bushels of hemp and requires 3 hours of labor. Would it be worthwhile to cultivate hemp if the selling price is 65 euros per bushel?

1.21 The company RAF has just signed a contract to produce 1000 RAF trucks. The company has four locations at its disposal for production. Owing to local conditions, the costs for producing a truck vary from location to location, as do the hours of labor and number of units material required for making a truck. The following table contains these data:

Location	Costs (in 1000 Euros)	Hours of Labor	Units of Material
1	15	2	3
2	10	3	4
3	9	4	5
4	7	5	6

For the four production locations together, a total of 3300 hours of labor and 4000 units of material are available. In their contract, the trade union has demanded that at least 400 trucks be produced at location 3. The computer output for this problem is:

```
Objective Function :
  MIN 15.0000 X1 +10.0000 X2 +9.0000 X3 +7.0000 X4
Subject to :
  1.  2.0000 X1 +3.0000 X2 +4.0000 X3 +5.0000 X4 <= 3300.0000
  2.  3.0000 X1 +4.0000 X2 +5.0000 X3 +6.0000 X4 <= 4000.0000
  3.  1.0000 X3 >= 400.0000
  4.  1.0000 X1 +1.0000 X2 +1.0000 X3 +1.0000 X4 =  1000.0000

  Summary of Results

  Value Objective Function :      11600.0000

  Variable          Activity Level   Reduced Cost
  --------          --------------   ------------
  X1        :          400.0000         0.0000
  X2        :          200.0000         0.0000
  X3        :          400.0000         0.0000
  X4        :            0.0000         7.0000
```

	Slack or Surplus	Shadow prices
Constraint 1	300.0000	-0.0000
Constraint 2	0.0000	-5.0000
Constraint 3	0.0000	4.0000
Constraint 4	0.0000	30.0000

Objective Coefficient Ranges

Variable	Current Coefficient	Allowed Interval
X1	15.0000	[11.5000 , Infinity]
X2	10.0000	[-Infinity , 12.0000]
X3	9.0000	[5.0000 , Infinity]
X4	7.0000	[0.0000 , Infinity]

Right-Hand-Side Ranges

Constraint	Current RHS	Allowed Interval
1.	3300.0000	[3000.0000 , Infinity]
2.	4000.0000	[3800.0000 , 4300.0000]
3.	400.0000	[0.0000 , 500.0000]
4.	1000.0000	[900.0000 , 1066.6667]

(a) What are the extra costs of making 50 additional trucks?
(b) What happens to the optimal solution if the production costs per truck at location 2 are equal to 12 000 euros instead of 10 000 euros?
(c) By how much would the costs decrease if the trade union dropped the demand that at least 400 trucks be produced at location 3? What happens to the costs if the demand is sharpened to 525 trucks being produced at location 3?
(d) Suppose that the company can have 100 additional units of material at its disposal by surrendering 200 hours of labor. What happens to the minimum costs?
(e) Suppose that the production costs per truck increase by 25% at every location. Does the optimal solution change?
(f) Suppose that there is also a location 5. At that location, the production costs of a truck are 7000 euros, and 4 hours of labor and 5 units of material are needed to make one truck. Is it worthwhile to use this location in the optimal solution?

1.22 An electricity company has installations at four different locations. For the coming month, the company needs the following quantities of coal for its installations: 75 000 tons at location 1, 150 000 tons at location 2, 250 000 tons at location 3, and 300 000 tons at location 4. The electricity company has three coal suppliers. The maximum quantities the coal companies can supply during the coming month are 400 000 tons for company 1, 300 000 tons for company 2, and 200 000 tons for company 3. The prices per tons (including transportation costs) are as follows:

	Location 1	Location 2	Location 3	Location 4
Coal Company 1	23 euros	18 euros	18 euros	21 euros
Coal Company 2	14 euros	15 euros	12 euros	19 euros
Coal Company 3	18 euros	17 euros	16 euros	17 euros

(a) Use an LP model to determine the supply schedule that gives the lowest costs.

(b) By how much do the minimum costs increase if the demand at location 1 increases from 75 000 tons to 90 000 tons?

(c) What happens to the optimal solution if coal company 2 increases its price by 3 euros per ton for location 3?

1.23 A company sells three products A, B, and C that are produced at five of the company's factories. The production capacities and the production costs per unit of product for the factories are given by

	Factory 1	Factory 2	Factory 3	Factory 4	Factory 5
A	€36	€40	€42	€40	€45
B	€42	€45	€50	€45	€45
C	€45	€75	€60	€60	€75
Cap.	300	400	500	250	550

The selling prices per unit of product A, B, and C are €125, €150, and €200. Each of these products can be sold in unlimited quantities. The sales department has already agreed to supply 400 units of product A, 300 units of product B, and 500 units of product C.

(a) Determine a production plan that maximizes the total net profit.

(b) Suppose that the capacity of one of the factories can be increased by 10%. Which factory should be extended?

(c) By how much does the total net profit decrease if the sales department has already agreed to sell 500 units of product A instead of 400.

(d) What happens to the optimal production plan if the production costs per unit increase by €2 for each of the products of factory 5?

1.24 A company can make two different products. The selling price of product 1 is 15 euros per ton up to 15 tons and 10 euros per ton for each additional ton. No more than 35 tons of product 1 may be sold. Product 2 can be sold without restriction. The selling price of product 2 is 12 euros per ton up to 10 tons and 9 euros for each additional ton. Two types of raw material are used in the production process. Each ton of product 1 requires 3 tons of raw material 1 and 1 ton of raw material 2. The raw material requirements for each ton of product 2 are 2.5 tons of raw material 1 and 1 ton of raw material 2. A total of 150 tons of raw material 1 and 75 tons of raw material 2 are available.

(a) Formulate an LP model to find the production plan that maximizes the total profit. *Hint:* Define the decision variable $prod_i^H$ ($prod_i^L$) as the number of tons of product i produced and sold at the high (low) price. Check whether the objective function ensures that $prod_i^L$ cannot be positive unless $prod_i^H$ is at its upper bound.

(b) Suppose that the company can buy an additional 50 tons of raw material 1 and 25 tons of raw material 2 for a combined price of 195 euros. Should the company accept this offer?

(c) The suggestion is made to lower all selling prices by the same percentage. How high can that percentage be without changing the optimal production plan?

1.25 An oil refinery sells three types of gasoline that are produced by blending two types of crude oil. The selling price per barrel is 85 euros for gasoline 1, 75 euros for gasoline 2, and 65 euros for gasoline 3. The purchase price per barrel is 55 euros for crude oil 1 and 35 euros for crude oil 2. The daily amount of crude oil 1 available is 10 000 barrels, and the daily amount of crude oil 2 available is 15 000 barrels. At most 22 000 barrels of crude oil can be processed per day. The company must deliver at least 2 500 barrels of gasoline 1, 4 000 barrels of gasoline 2, and 3 000 barrels of gasoline 3 per day. Every type of gasoline can be sold in unlimited quantities. The three types of gasoline differ in octane and lead content. Gasoline 1, 2, and 3 must have, respectively, octane levels of at least 82, 85, and 87. The lead content must be at most 1%, 1.5%, and 2%, respectively. The octane levels for crude oil 1 and 2 are 80 and 90, respectively, and the lead percentages are 0.5% and 2.5%. We may assume that both the octane and lead contents blend linearly by volume.

(a) How should the production be planned to maximize the profit?

(b) How much would the refinery be willing to pay for 1000 additional barrels of crude oil 1?

1.26 An investment plan for the next three years needs to be drawn up for the Holland Investment Company (HIC). Currently, HIC has 2.5 million euros available for investment. Furthermore, in the next three years, amounts of money become available at six-month intervals: 625 thousand euros (in six months), 500 thousand euros, 475 thousand euros, 450 thousand euros, 425 thousand euros, and 375 thousand euros (at the end of the third year). The investment company has three projects in mind for investment: the Euroborg project, the Zuiderzee project, and the Maas project. At 100% participation in the Euroborg project, HIC has the following cash flows at six-month intervals over the next three years: −3.75 million euros (now), −1.25 million euros, −2.25 million euros, 500 thousand euros, 2.25 million euros, 2.25 million euros, and 6.875 million euros (at the end of the third year), where a negative value indicates an investment in the project and a positive value a return on the project. For the Zuiderzee project, the cash flows at 100% participation are −2.5 million euros (now), −625 thousand euros, 1.875 million euros, 1.875 million euros, 1.875 million euros, 250 thousand euros, and −1.25 million euros (at the end of the third year), while the cash flows for the Maas project at 100% participation are −2.5 million euros (now), −2.5 million euros, −2.25 million euros, 1.25 million euros, 1.25 million euros, 1.25 million euros, and 7.5 million euros (at the end of the third year). HIC can also invest less than 100% in each of the three projects, in which case the cash flows mentioned above change proportionally. The investment company can also borrow money from the bank for its investments, for a period of six months each time, where the amount of the loan must be repaid with 3.5% interest after six months. The loan contracted at the bank can never exceed 2.5 million euros. Finally, HIC can deposit the noninvested cash at the bank, each time for a period of six months at 3% interest.

(a) Formulate an LP problem to determine an investment plan that maximizes the net amount of money HIC has in three years. Solve this LP problem using a software package.

(b) Suppose that in addition to borrowing money at the bank, where there is a credit limit of 2.5 million euros, HIC can also borrow money elsewhere for six-month periods. Calculate separately for each of the six upcoming half-yearly periods the maximum interest rate HIC would be willing to pay to borrow additional money elsewhere during that period only.

(c) It turns out that HIC has other possibilities for taking out loans at the bank. In addition to the existing half-yearly loans, the bank offers the investment company HIC the further possibility to borrow money for 1 year at a time at an interest rate of 7.5% per year. However, the condition that, at all times, the outstanding loans with the bank may not exceed 2.5 million euros still applies. Adapt the LP formulation and solve the revised LP problem. Can you explain why this additional possibility for a loan may be interesting despite the higher interest rate?

References

1. S. P. Bradley, A. C. Hax, and T. L. Magnanti, *Applied Mathematical Programming*, Addison-Wesley, Reading, Massachusetts, 1977.
2. V. Chvátal, *Linear Programming*, W. H. Freeman, New York, 1983.
3. P. A. Jensen and J. F. Bard, *Operations Research, Models and Methods*, John Wiley & Sons, Inc., Hoboken, New Jersey, 2003.

Chapter 2

Integer Programming

2.1 Introduction

The LP model we discussed in Chapter 1 implicitly assumes that the decision variables may take on fractional values. However, many real-world problems require some or all of the decision variables to be integers. An *integer LP (ILP) problem* is a linear programming problem in which some or all of the decision variables are required to be integers. A surprisingly large class of real-world problems can be formulated as integer programming problems, although this formulation sometimes requires a certain amount of ingenuity. For example, "if-then" constraints are often added to LP models. An example of an "if-then" constraint is the requirement that if product A is made, then at least a certain quantity of it is made, or that if product A is made, then product B is also made. Such requirements can be modeled as linear constraints using integer variables that can only take on the values 0 and 1 (*binary variables*). Many combinatorial optimization problems can be formulated as integer programming problems with 0-1 variables. Binary variables play a key role in integer programming.

The fact that integer programming is extremely useful for modeling many decision problems is the "good" news. The "bad" news is that the computation time required to solve an integer programming problem is many times greater than the computation time for a continuous LP problem of the same magnitude. Continuous LP models with thousands of constraints and variables can be solved on the computer within a reasonable amount of time. Unfortunately, this is not the case for integer programming models. The computational complexity of a *discrete* optimization problem is generally much greater than that of a *continuous* optimization problem. Discrete (combinatorial) optimization problems are usually *NP-complete*; that is, there is a polynomial-time algorithm for the problem, but the computation time increases *exponentially* in the problem size. See Appendix A for a description of complexity. In an integer programming problem, the computation time strongly depends on the following two factors:

- the number of integer variables,
- the chosen formulation.

The second point deserves some clarification. In specific applications, it is often possible to give more than one formulation as an integer programming problem, particularly as far as the linear constraints are concerned. In general, for integer programming, the more severe the restrictions on the variables, the less computation time solving the problem takes. More constraints do not mean more computation time. The importance of the chosen formulation is also reflected in the fact that an integer programming problem can be solved using (continuous) linear programming if the coefficient matrix of the constraints is *totally unimodular*.[1] In that case, any optimal basic solution to the LP relaxation of the integer programming model automatically has *integer* values. The *LP relaxation* of an integer programming problem is generally defined as the LP problem that arises by dropping the integer constraints (for example, the constraint $x_j \in \{0, 1\}$ is weakened to $0 \le x_j \le 1$). The total unimodularity property is the reason why in specific applications, one should always try to find an integer formulation for which the coefficients of the variables in the constraints are equal to 0, 1, or -1, in the hope that the resulting coefficient matrix is (almost) totally unimodular. Of course, this is only possible in special cases, but it is a good rule of thumb to follow.

The computation time of an integer programming model strongly depends on the number of variables declared as integers. This will also be evident later on from the nature of the so-called *branch-and-bound algorithm*, which is often used to solve integer programming problems. In specific applications, it is always worth checking whether certain integer variables can be declared as continuous variables in order to save computation time. The structure of the specific application sometimes ensures that in the optimal solution, the variables have integer values even though they were declared as continuous variables. To limit the number of integer variables, it is not uncommon, in practice, to declare integer variables that one knows in advance will assume values of, for example, 20 or more as continuous variables.

The LP relaxation of an ILP problem is extremely useful for solving the original ILP problem. Two simple basic principles play a crucial role in this. We formulate these basic principles in the case where the integer problem concerns a maximization.

Principle 2.1. *The maximum objective value of the original ILP problem is always less than or equal to the optimal objective value of the LP relaxation.*

This purely follows from the simple fact that the feasible region for the LP relaxation is larger than the feasible region for the original integer problem. The larger the feasible region, the higher the value the objective function can take on ("the more choice you have, the better off you are"). The following terminology is often used

---

[1] A matrix is called totally unimodular if it has the following properties: (**a**) every entry of the matrix is 0, 1, or -1; (**b**) every column of the matrix has at most two nonzero elements; and (**c**) the rows of the matrix can be partitioned into two subsets such that if a column contains two nonzero elements with the same sign (respectively, different signs), then the two rows containing these elements are not (respectively, are) in the same subset.

for the first principle: the optimal solution to the LP relaxation leads to an *upper bound* for the maximum objective value in the original ILP problem. A *lower bound* for this maximum objective value is the subject of Principle 2.

Principle 2.2. *The objective value of a feasible solution to the ILP problem gives a lower bound for the maximum objective value of the ILP problem.*

In some cases, rounding off the optimal solution to the LP relaxation leads to a feasible solution to the original ILP problem and therefore to a lower bound for the maximum objective value of the ILP problem. It is important to note that if the lower bound produced by the feasible solution is sufficiently close to the upper bound for the maximum objective value of the ILP problem, the feasible solution is a good suboptimal solution to the original ILP problem. In practice, people are often satisfied with such a solution. The upper and lower bounds for the maximum objective value of the original ILP problem play a leading role in the branch-and-bound algorithms for integer linear programming problems. Most software packages with a branch-and-bound code allow the user to stop the calculations prematurely based on the calculated upper and lower bounds. Namely, the user may want to stop the calculations if these bounds are sufficiently close to each other. The software package then gives the best feasible solution found up to that point, and for this solution, we know that the corresponding objective value can only differ slightly from the optimal objective value. Finally, we note that unlike in the case of linear programming, there is no satisfactory method of sensitivity analysis for integer programming. A partial explanation for this is that the duality theory for linear programming does not apply to integer programming.

2.2 Applications of Integer Programming

In this section, we present important optimization problems that can be formulated within the framework of integer programming models. In formulating the integer programming models, we will come across generally useful tricks for formulating "almost-linear" constraints mathematically using linear constraints.

2.2.1 *An Investment Problem*

Problem statement. An investor considers seven possible investment projects for the next five years. Each selected project requires a certain amount of annual investment. Each year, only a limited amount of capital is available for investment. Table 2.1 shows the investment requirements and the amount of capital available each year. The table also gives the net present value (NPV) of a selected project. All amounts are expressed in units of one thousand euros. Which projects should be selected in order to maximize the overall net present value?

Table 2.1   Data for the investment problem.

		Investment Requirements for Projects						
	1	2	3	4	5	6	7	Available
Year 1	40	20	25	80	20	90	50	250
Year 2	10	30	30	40	20	25	10	125
Year 3	25	0	20	30	20	0	0	75
Year 4	25	0	10	0	10	10	30	50
Year 5	10	35	0	15	10	20	0	50
NPV	250	180	225	300	150	275	200	

Formulation. Binary variables are used to model "yes-no" decisions. Define a binary variable x_j for each project j $(= 1, \ldots, 7)$ to indicate whether or not a project is selected:

$$x_j = \begin{cases} 1 & \text{if project } j \text{ is selected,} \\ 0 & \text{otherwise.} \end{cases}$$

This leads to the following integer programming model for the investment problem:

$$\begin{aligned}
\text{Maximize} \quad & 250x_1 + 180x_2 + 225x_3 + 300x_4 + 150x_5 + 275x_6 + 200x_7 \\
\text{subject to} \quad & 40x_1 + 20x_2 + 25x_3 + 80x_4 + 20x_5 + 90x_6 + 50x_7 \leq 250 \\
& 10x_1 + 30x_2 + 30x_3 + 40x_4 + 20x_5 + 25x_6 + 10x_7 \leq 125 \\
& 25x_1 \qquad\quad + 20x_3 + 30x_4 + 20x_5 \qquad\qquad\qquad \leq 75 \\
& 25x_1 \qquad\quad + 10x_3 \qquad\quad + 10x_5 + 10x_6 + 30x_7 \leq 50 \\
& 10x_1 + 35x_2 \qquad\quad + 15x_4 + 10x_5 + 20x_6 \qquad\quad \leq 50
\end{aligned}$$

and $x_1, x_2, \ldots, x_7 = 0$ or 1.

The solution for this model is given in Figure 2.1. The optimal solution is to invest in projects 1, 3, 4, and 6. The total net present value of this investment is 1 million 50 thousand euros.

Fig. 2.1   Computer output for the investment problem.

```
INTOPT Version 2.1                          IEOR   VU Amsterdam

The following model was read:

Objective Function :
  MAX 250.0000 X1 +180.0000 X2 +225.0000 X3 +300.0000 X4 +
      150.0000 X5 +275.0000 X6 +200.0000 X7
Subject to :
  1. 40.0000 X1 +20.0000 X2 +25.0000 X3 +80.0000 X4 +
     20.0000 X5 +90.0000 X6 +50.0000 X7 <= 250.0000
  2. 10.0000 X1 +30.0000 X2 +30.0000 X3 +40.0000 X4 +
     20.0000 X5 +25.0000 X6 +10.0000 X7 <= 125.0000
  3. 25.0000 X1 +20.0000 X3 +30.0000 X4 +20.0000 X5 <=
     75.0000
  4. 25.0000 X1 +10.0000 X3 +10.0000 X5 +10.0000 X6 +
```

```
    30.0000 X7 <= 50.0000
 5. 10.0000 X1 +35.0000 X2 +15.0000 X4 +10.0000 X5 +
    20.0000 X6 <= 50.0000
BINARY  X1, X2, X3, X4, X5, X6, X7
```

Summary of Results

Value Objective Function :   1050.0000

Variable		Activity Level
X1	:	1.0000
X2	:	0.0000
X3	:	1.0000
X4	:	1.0000
X5	:	0.0000
X6	:	1.0000
X7	:	0.0000

In real-world investment problems, there are often additional "if-then"-type re-quirements. As stated earlier, such requirements can be modeled using 0-1 variables. Suppose, for example, that project 5 must be selected if project 3 is selected. This requirement can be modeled by adding the constraint

$$x_5 \geq x_3.$$

Since x_5 is a 0-1 variable, this constraint ensures that $x_5 = 1$ if $x_3 = 1$. The constraint is not unnecessarily restrictive for x_5 if $x_3 = 0$. Likewise, the requirement that project 4 not be selected without also selecting project 2 can be modeled using the constraint $x_4 \leq x_2$. This constraint ensures that project 4 is not chosen (that is, $x_4 = 0$) if $x_2 = 0$. If $x_2 = 1$, the constraint $x_4 \leq x_2$ leaves one free to choose project 4 or not. Another requirement that can easily be translated to a linear constraint using 0-1 variables is the following. Suppose that at most two of the projects 1, 3, 4, and 6 can be selected. This requirement can be modeled by adding the constraint

$$x_1 + x_3 + x_4 + x_6 \leq 2.$$

Since the variables are 0 or 1, this constraint ensures that at most two of the variables x_1, x_3, x_4, and x_6 take on the value 1. As a last example, consider the requirement that project 6 not be selected if both project 2 and project 5 are selected. This requirement can be modeled using the constraint

$$x_6 \leq 2 - x_2 - x_5.$$

Verify that this constraint ensures that the stated requirement is met.

2.2.2   A Location Problem

Problem statement.   A typical real-world problem is finding the least number of fire stations in a region such that every city in the region is within a prespecified distance from a fire station. Suppose that fire stations are to be built in a region with five cities. Each city is a potential location for a fire station, and every city is required to be within 10 minutes travel time of a fire station. Table 2.2 gives the travel times in minutes between any two cities. What is the least number of fire stations required, and where should they be built?

Table 2.2   Travel times.

From\To	City 1	City 2	City 3	City 4	City 5
City 1	0	11	12	9	15
City 2	11	0	14	12	10
City 3	9	8	0	15	12
City 4	10	12	15	0	13
City 5	17	12	10	13	0

Formulation.   For each city $j = 1, \ldots, 5$, define the decision variable

$$x_j = \begin{cases} 1 & \text{if a fire station is built in city } j, \\ 0 & \text{otherwise.} \end{cases}$$

The goal is to minimize the sum $x_1 + x_2 + x_3 + x_4 + x_5$. For a clear formulation of the constraints, it is helpful to use the following notation. For each city i, we denote by N_i the set of cities with a travel time of 10 minutes or less to city i. For example, $N_1 = \{1, 3, 4\}$. The constraint

$$\sum_{j \in N_i} x_j \geq 1$$

ensures that city i is within 10 minutes of a fire station. So we find the integer programming model

$$\begin{aligned}
\text{Minimize} \quad & x_1 + x_2 + x_3 + x_4 + x_5 \\
\text{subject to} \quad & x_1 + x_3 + x_4 \geq 1 \\
& x_2 + x_3 \geq 1 \\
& x_3 + x_5 \geq 1 \\
& x_4 + x_1 \geq 1 \\
& x_5 + x_2 \geq 1 \\
\text{and} \quad & x_1, x_2, x_3, x_4, x_5 = 0 \text{ or } 1.
\end{aligned}$$

The computer output for the solution to this integer programming model is given in Figure 2.2. The optimal solution is to build three fire stations, in cities 3, 4, and 5.

Fig. 2.2   Computer output for the location problem.

```
INTOPT Version 2.1                          IEOR  VU Amsterdam

The following model was read:

Objective Function :
   MIN  1.0000 X1 +1.0000 X2 +1.0000 X3 +1.0000 X4 +1.0000 X5
Subject to :
   1.  1.0000 X1 +1.0000 X3 +1.0000 X4 >=  1.0000
   2.  1.0000 X2 +1.0000 X3 >=  1.0000
   3.  1.0000 X3 +1.0000 X5 >=  1.0000
   4.  1.0000 X4 +1.0000 X1 >=  1.0000
   5.  1.0000 X5 +1.0000 X2 >=  1.0000
BINARY  X1, X2, X3, X4, X5


   Summary of Results

   Value Objective Function :          3.0000

   Variable             Activity Level
   --------             --------------

   X1        :              0.0000
   X2        :              0.0000
   X3        :              1.0000
   X4        :              1.0000
   X5        :              1.0000
```

Set Covering

The integer programming problem above is a particular case of a so-called "set-covering" problem. The set-covering problem can be formulated in general terms as follows. Suppose given a number of elements and a number of groups, each group covering some of those elements. Which groups should one choose in order to cover every element by at least one of the chosen groups and minimize the total cost, where the cost of choosing group j is c_j? In the location problem, each group corresponds to a fire station in that city, and the elements covered by a group are the cities that are no more than 10 minutes away. The set-covering problem has many practical applications. For example, suppose that someone needs to deliver goods to a number of customers and needs to choose between a number of possible routes. For each route, the customers on that route and the cost of following the route are known. The question which routes should be chosen can be answered with a set-covering model.

Experience has shown that, in practice, the set-covering model in which all costs c_j equal 1 can be solved using two "inexpensive" LP problems instead of an "expensive" integer programming problem. Let us explain this using the location problem for fire stations introduced above:

1. First, solve the LP relaxation of the integer programming problem obtained by replacing the 0-1 constraints on the variables x_j with the constraints $0 \leq x_j \leq 1$ for $j = 1, \ldots, 5$. The optimal solution to the LP relaxation turns out to be fractional, with objective value 2.5. Based on the form of Principle 2.1 (for a minimization problem) from Section 2.1 and on the form of the objective function, we can directly conclude that the optimal objective value of the integer problem must be at least 3. In other words, we can restrict ourselves to feasible solutions of which the sum of the values of the decision variables is at least 3.
2. Solve a second LP problem obtained by adding the constraint
$$x_1 + x_2 + x_3 + x_4 + x_5 \geq 3$$
to the LP relaxation of the integer problem. If the optimal solution to this LP problem turns out to be integer valued, then this solution gives the optimal solution to the original integer problem.

In applications of set-covering problems with costs $c_j = 1$ for all j, it is recommended to first try the trick above before using time-consuming computer codes for integer programming.

2.2.3   A Distribution Problem

Problem statement.   For the distribution problem in Section 1.2.3, we now consider the situation that extra storage capacity is needed. The first possibility is to open a new depot D_4 with a capacity of 50 thousand tons. It is also possible to expand the capacity of depot D_2 from 40 thousand tons to 65 thousand tons. The monthly depreciation amounts to 10 thousand euros for the new depot and 3 thousand euros for the capacity expansion of depot D_2. The distribution cost from the factory to the new depot D_4 is 5 euros per ton. The distribution costs from the new depot D_4 to the customers C_1, C_2, C_3, and C_4 are, respectively, 7, 5, 8, and 10 euros per ton. How should the depot capacity be increased, and what is the optimal distribution schedule in the new situation to minimize the total cost?

Formulation.   In order to answer the questions, the LP model from Section 1.2.2 needs to be extended to an integer programming model. This extension requires the following binary variables:
$$\delta_1 = \begin{cases} 1 & \text{if the new depot } D_4 \text{ is built,} \\ 0 & \text{otherwise,} \end{cases}$$
and
$$\delta_2 = \begin{cases} 1 & \text{if the capacity of depot } D_2 \text{ is expanded,} \\ 0 & \text{otherwise.} \end{cases}$$

In addition to the decision variables FD_i, FC_j, and DC_{ij} already introduced in Section 1.2.2 ($1 \le i \le 3$ and $1 \le j \le 4$), we also define the variables

$$FD_4 = \text{number of tons to send from the factory to depot } D_4,$$
$$DC_{4j} = \text{number of tons to send from depot } D_4 \text{ to customer } C_j.$$

The following cost term must be added to the original objective function in the LP formulation:

$$10\,000\delta_1 + 3000\delta_2 + 5FD_4 + 7DC_{41} + 5DC_{42} + 8DC_{43} + 10DC_{44}.$$

With regard to the new depot D_4, the following constraints must be added:

$$FD_4 \le 50\,000\delta_1$$
$$DC_{41} + DC_{42} + DC_{43} + DC_{44} = FD_4.$$

In case depot D_4 is not opened ($\delta_1 = 0$), the first constraint ensures that nothing is sent from the factory to depot D_4, and the second constraint ensures that nothing is sent from depot D_4 to the customers. The first constraint also ensures that $\delta_1 > 0$ if $FD_4 > 0$. In other words, there cannot be a solution with $FD_4 > 0$ while depot D_4 has not been opened. In the event that depot D_4 is opened ($\delta_1 = 1$), the constraints ensure that the flow into depot D_4 is not greater than the capacity and that the outflow is equal to the inflow.

Because of the possibility to expand the capacity of depot D_2 from 40 thousand tons to 65 thousand tons, in the original formulation, the constraint $FD_2 \le 40\,000$ must be replaced by

$$FD_2 \le 40\,000 + 25\,000\delta_2.$$

These are the main changes to the original formulation. The further changes in the LP formulation speak for themselves. The complete formulation of the integer programming model and the optimal solution are given in Figure 2.3. In the optimal solution, the new depot D_4 is opened, and the capacity of depot D_2 is expanded. The total monthly costs of the new solution are 2 million 210.500 thousand euros.

Fig. 2.3 Computer output for the distribution problem.

```
INTOPT Version 2.1                          IEOR  VU Amsterdam


The following model was read:


Objective Function :
  MIN 10000.0000 DELTA1 +3000.0000 DELTA2 +4.5000 FD1
       +4.0000 FD2 +7.0000 FD3 +5.0000 FD4 +12.0000 FC1
       +10.0000 FC2 +15.0000 FC3 +18.0000 FC4 +10.0000 DC11
       +5.0000 DC12 +10.0000 DC13 +12.0000 DC14 +7.0000 DC21
       +5.0000 DC22 +8.0000 DC23 +12.0000 DC24 +7.0000 DC31
       +8.0000 DC32 +5.0000 DC33 +10.0000 DC34 +7.0000 DC41
       +5.0000 DC42 +8.0000 DC43 +10.0000 DC44
```

Subject to :
   1.   1.0000 FD1 +1.0000 FD2 +1.0000 FD3 +1.0000 FD4 +
        1.0000 FC1 +1.0000 FC2 +1.0000 FC3 +1.0000 FC4 <=
        250000.0000
   2.   1.0000 FD1 <= 80000.0000
   3.   1.0000 FD2 <= 40000.0000 +25000.0000 DELTA2
   4.   1.0000 FD3 <= 75000.0000
   5.   1.0000 FD4 <= 50000.0000 DELTA1
   6.   1.0000 FC1 +1.0000 DC11 +1.0000 DC21 +1.0000 DC31
        +1.0000 DC41 = 50000.0000
   7.   1.0000 FC2 +1.0000 DC12 +1.0000 DC22 +1.0000 DC32
        +1.0000 DC42 = 70000.0000
   8.   1.0000 FC3 +1.0000 DC13 +1.0000 DC23 +1.0000 DC33
        +1.0000 DC43 = 45000.0000
   9.   1.0000 FC4 +1.0000 DC14 +1.0000 DC24 +1.0000 DC34
        +1.0000 DC44 = 30000.0000
  10.   1.0000 DC11 +1.0000 DC12 +1.0000 DC13 +1.0000 DC14 <=
        1.0000 FD1
  11.   1.0000 DC21 +1.0000 DC22 +1.0000 DC23 +1.0000 DC24 <=
        1.0000 FD2
  12.   1.0000 DC31 +1.0000 DC32 +1.0000 DC33 +1.0000 DC34 <=
        1.0000 FD3
  13.   1.0000 DC41 +1.0000 DC42 +1.0000 DC43 +1.0000 DC44 <=
        1.0000 FD4
BINARY DELTA1, DELTA2

Summary of Results

Value Objective Function :   2210500.0000

Variable	Activity Level
DELTA1 :	1.0000
DELTA2 :	1.0000
FD1 :	55000.0000
FD2 :	65000.0000
FD3 :	45000.0000
FD4 :	30000.0000
FC1 :	0.0000
FC2 :	0.0000
FC3 :	0.0000
FC4 :	0.0000

DC11	:	0.0000
DC12	:	55000.0000
DC13	:	0.0000
DC14	:	0.0000
DC21	:	50000.0000
DC22	:	15000.0000
DC23	:	0.0000
DC24	:	0.0000
DC31	:	0.0000
DC32	:	0.0000
DC33	:	45000.0000
DC34	:	0.0000
DC41	:	0.0000
DC42	:	0.0000
DC43	:	0.0000
DC44	:	30000.0000

2.2.4 A Timetabling Problem

Problem statement. A bus company needs the following numbers of drivers every week:

Monday	Tuesday	Wednesday	Thursday	Friday	Saturday	Sunday
25	27	23	21	25	20	15

Each bus driver works five consecutive days a week and then has two days off. The salary is 100 euros for each weekday, 115 euros for Saturdays, and 125 euros for Sundays. What work schedule minimizes the weekly salary costs?

Formulation. The seven days of the week are numbered $1, \ldots, 7$, where day 1 = Monday, day 2 = Tuesday, and so on. We use the following decision variables:

$$x_i = \text{number of bus drivers beginning work on day } i$$

for $i = 1, \ldots, 7$. We do not take the number of bus drivers working on day i as a decision variable because that variable does not give us any information about the remaining number of days the bus drivers still have to work. The chosen decision variables x_i immediately allow us to specify the objective function and constraints. The number of bus drivers working on day 1 is equal to $x_1 + x_7 + x_6 + x_5 + x_4$. This number must be at least 25, which results in the inequality $x_1 + x_7 + x_6 + x_5 + x_4 \geq 25$. The constraints for the other days are found likewise. The aim is to minimize the weekly salary costs. The total salary of a bus driver who starts working on day 7, for example, is $125 + 4 \times 100 = 525$ euros, which contributes $525x_7$ to the objective function. The complete formulation of the integer programming model and the computer solution are given in Figure 2.4.

Fig. 2.4   Computer output for the timetabling problem.

```
INTOPT Version 2.1                      IEOR  VU Amsterdam

The following model was read:

Objective Function :
   MIN 500.0000 X1 +515.0000 X2 +540.0000 X3 +540.0000 X4 +
       540.0000 X5 +540.0000 X6 +525.0000 X7
Subject to :
   1.  1.0000 X1 +1.0000 X7 +1.0000 X6 +1.0000 X5 +1.0000 X4 >=
       25.0000
   2.  1.0000 X2 +1.0000 X1 +1.0000 X7 +1.0000 X6 +1.0000 X5 >=
       27.0000
   3.  1.0000 X3 +1.0000 X2 +1.0000 X1 +1.0000 X7 +1.0000 X6 >=
       23.0000
   4.  1.0000 X4 +1.0000 X3 +1.0000 X2 +1.0000 X1 +1.0000 X7 >=
       21.0000
   5.  1.0000 X5 +1.0000 X4 +1.0000 X3 +1.0000 X2 +1.0000 X1 >=
       25.0000
   6.  1.0000 X6 +1.0000 X5 +1.0000 X4 +1.0000 X3 +1.0000 X2 >=
       20.0000
   7.  1.0000 X7 +1.0000 X6 +1.0000 X5 +1.0000 X4 +1.0000 X3 >=
       15.0000
INTEGER   X1, X2, X3, X4, X5, X6, X7


Summary of Results

Value Objective Function :      16675.0000


Variable          Activity Level
--------          --------------

X1        :          10.0000
X2        :           7.0000
X3        :           0.0000
X4        :           4.0000
X5        :           4.0000
X6        :           5.0000
X7        :           2.0000
```

An LP Approach

This specific timetabling problem can also be solved with linear programming. An optimal solution to the continuous LP relaxation of the integer model above is given by $x_1^* = 10.6$, $x_2^* = 5.6$, $x_3^* = 0.6$, $x_4^* = 3.6$, $x_5^* = 4.6$, $x_6^* = 5.6$, and $x_7^* = 0.6$, with objective value 16.275 euros. In this case, one can always obtain a feasible integer solution by rounding up each fractional value (why?). The rounded solution is given by $x_1 = 11$, $x_2 = 6$, $x_3 = 1$, $x_4 = 4$, $x_5 = 5$, $x_6 = 6$, and $x_7 = 1$, with objective value 17.775 euros. The rounded solution is not optimal. In general, one cannot expect that an optimal integer solution will result from rounding off the optimal LP solution. Nevertheless, this specific timetabling problem and several other timetabling problems can be solved with linear programming. To this end, three LP problems need to be solved. The procedure, for which we do not give a proof (see the paper J. J. Bartholdi et al., Cyclic scheduling via integer programs with circular ones, *Operations Research*, 21 (1989), 1074–1084), goes as follows:

Step 1. Solve the LP relaxation of the timetabling problem. If the optimal LP solution (x_i^*) has integer values, then this solution gives the optimum for the timetabling problem. Otherwise, proceed to step 2.

Step 2. Calculate V^* as the sum of the decision variables in the optimal LP solution (x_i^*). Let V_1 be the greatest integer that is less than or equal to V^*, and let V_2 be the smallest integer that is greater than or equal to V^*. Form the two LP problems LP1 and LP2 that arise by adding the extra constraints

$$\sum_{i=1}^{7} x_i = V_1 \quad \text{and} \quad \sum_{i=1}^{7} x_i = V_2,$$

respectively, to the LP relaxation. Solve the two LP problems LP1 and LP2. At least one of these two problems has a feasible solution. The optimal LP solution is automatically integer valued. The best solution to the two LP problems is the optimal solution to the timetabling problem.

In the example above, the sum of the decision variables in the optimal LP solution is equal to $V^* = 31.2$. The LP problems LP1 and LP2 are obtained by adding the respective equations $\sum_{i=1}^{7} x_i = 31$ and $\sum_{i=1}^{7} x_i = 32$ to the LP relaxation. If we solve these two LP problems, we find that LP1 does not have a feasible solution and that LP2 has optimal solution $x_1 = 12$, $x_2 = 5$, $x_3 = 2$, $x_4 = 2$, $x_5 = 4$, $x_6 = 7$, and $x_7 = 0$, with objective value 16.675 euros. This is the optimal solution to the timetabling problem. Since solving integer problems can be quite time consuming, it is easier to solve three LP problems rather than one integer problem.

2.2.5   *A Partitioning Problem*

Problem statement. Suppose that n goods are given, with values $a_1, \ldots, a_n$. How should these goods be split up into two disjoint groups S_1 and S_2 so that

$$\left| \sum_{i \in S_1} a_i - \sum_{i \in S_2} a_i \right|$$

is as small as possible? In other words, the value of the goods must be divided as evenly as possible.

Formulation. The partitioning problem can be formulated as an integer linear programming problem by defining the 0-1 variables

$$x_i = \begin{cases} 1 & \text{if product } i \text{ is assigned to group } S_1, \\ 0 & \text{if product } i \text{ is assigned to group } S_2 \end{cases}$$

for $i = 1, \ldots, n$. There are various ways to give a formulation as an integer programming problem. The most direct one is to choose the variables x_i in a way that minimizes

$$\left| \sum_{i=1}^{n} a_i x_i - \sum_{i=1}^{n} a_i (1 - x_i) \right|.$$

By using deviation variables as in Section 1.3.5, we can translate this minimization problem into an integer linear problem: minimize $d^+ + d^-$ subject to $\sum_{i=1}^{n} a_i x_i - \sum_{i=1}^{n} a_i (1 - x_i) = d^+ - d^-$ with $x_i \in \{0,1\}$ for $i = 1, \ldots, n$ and $d^+, d^- \geq 0$. A more subtle way to formulate the partitioning problem as an integer problem is as follows. Suppose that the goods have to be split up over two persons. For reasons of symmetry, it is not restrictive to assume that the total value received by the first person is greater than or equal to the total value acquired by the second person. However, the amount $\sum_{i=1}^{n} a_i x_i$ that person 1 receives should be as small as possible subject to the constraint $\sum_{i=1}^{n} a_i x_i \geq \sum_{i=1}^{n} a_i (1 - x_i)$. This leads to the integer programming formulation

$$\text{Minimize} \quad \sum_{i=1}^{n} a_i x_i$$

$$\text{subject to} \quad \sum_{i=1}^{n} a_i x_i \geq \frac{1}{2} \sum_{i=1}^{n} a_i$$

$$\text{and} \qquad x_i \in \{0, 1\} \quad \text{for } i = 1, \ldots, n.$$

As an illustration, consider the following numerical data:

$$a_1 = 85, \ a_2 = 81, \ a_3 = 71, \ a_4 = 70, \ a_5 = 67, \ a_6 = 65, \ a_7 = 59,$$
$$a_8 = 52, \ a_9 = 43, \ a_{10} = 40, a_{11} = 36, a_{12} = 33, a_{13} = 33, a_{14} = 32,$$
$$a_{15} = 32, a_{16} = 29, a_{17} = 26, a_{18} = 25, a_{19} = 21, a_{20} = 11.$$

The optimal solution to the corresponding integer programming problem is shown in Figure 2.5. It follows from the computer output that the optimal partition of the goods $1, \ldots, 20$ is given by

$$S_1 = \{5, 7, 8, 9, 10, 11, 12, 13, 14, 15, 16\}, \quad S_2 = \{1, 2, 3, 4, 6, 17, 18, 19, 20\}.$$

The sets S_1 and S_2 have the respective values 456 and 455.

Fig. 2.5   Computer output for the partitioning problem.

```
INTOPT Version 2.1                          IEOR   VU Amsterdam

The following model was read:

Objective Function :
   MIN 85.0000 X1 +81.0000 X2 +71.0000 X3 +70.0000 X4 +
       67.0000 X5 +65.0000 X6 +59.0000 X7 +52.0000 X8 +
       43.0000 X9 +40.0000 X10 +36.0000 X11 +33.0000 X12 +
       33.0000 X13 +32.0000 X14 +32.0000 X15 +29.0000 X16 +
       26.0000 X17 +25.0000 X18 +21.0000 X19 +11.0000 X20
Subject to :
   1. 85.0000 X1 +81.0000 X2 +71.0000 X3 +70.0000 X4 +
      67.0000 X5 +65.0000 X6 +59.0000 X7 +52.0000 X8 +
      43.0000 X9 +40.0000 X10 +36.0000 X11 +33.0000 X12 +
      33.0000 X13 +32.0000 X14 +32.0000 X15 +29.0000 X16 +
      26.0000 X17 +25.0000 X18 +21.0000 X19 +11.0000 X20 >=
      455.5000
BINARY     X1, X2, X3, X4, X5, X6, X7, X8, X9, X10, X11, X12,
           X13, X14, X15, X16, X17, X18, X19, X20


   Summary of Results

   Value Objective Function :        456.0000

   Variable          Activity Level
   --------          --------------
   X5, X7-X16 :          1.0000
   X1-X4, X17-X20 :      0.0000
```

2.2.6   *A Production-Stock Problem*

Problem statement. The company Jones Chemical has entered into a contract to deliver a special type of sulfuric acid over the next six months. The agreement provides for the supply of 100 tons of sulfuric acid on July 1st, 75 tons on August 1st, 90 tons on September 1st, 60 tons on October 1st, 40 tons on November 1st, and 85 tons on December 1st. The production of sulfuric acid requires a few special measures. The production manager has therefore decided that the sulfuric acid can only be produced on the first day of the month. The production takes a negligible amount of time. The fixed setup cost of the production process is 500 euros. Each time, any desired amount of sulfuric acid can be produced. The company has sufficient storage capacity for the sulfuric acid; the storage cost is 4 euros per ton of sulfuric acid per month. At the moment, there is no stock of the product. What production plan has the lowest total cost?

Formulation. We number the six months July through December as periods $1, \ldots, 6$. As decision variables, we take

$$x_i = \text{amount of sulfuric acid to make at the beginning of period } i,$$
$$v_i = \text{amount of sulfuric acid in stock at the end of period } i.$$

The variables $v_1, \ldots, v_6$ are auxiliary variables that simplify the formulation and lead to the "input = output" equations $v_{i-1} + x_i = a_i + v_i$, where a_i is the given demand in period i. In addition to the continuous variables x_i and v_i, we need the 0-1 variables $\delta_1, \ldots, \delta_6$, where

$$\delta_i = \begin{cases} 1 & \text{if production takes place at the beginning of period } i, \\ 0 & \text{otherwise.} \end{cases}$$

These variables are needed to model the fixed setup cost in the objective function using the additional constraints $x_i \leq M_i \delta_i$, which ensure that $\delta_i = 1$ if $x_i > 0$. Here, M_i is a sufficiently large number such that x_i is always less than or equal to M_i (take, for example, M_i equal to the total demand in periods i through 6). The production-stock problem is modeled by the following (mixed) ILP problem:

$$\text{Minimize} \quad 500 \sum_{i=1}^{6} \delta_i + 4 \sum_{i=1}^{6} v_i$$
$$\text{subject to} \quad v_{i-1} + x_i = a_i + v_i \qquad \text{for } i = 1, \ldots, 6$$
$$x_i \leq M_i \delta_i \qquad \text{for } i = 1, \ldots, 6$$
$$\text{and} \qquad x_i, v_i \geq 0, \ \delta_i \in \{0, 1\} \text{ for } i = 1, \ldots, 6,$$

with constants $v_0 = 0$, $a_1 = 100$, $a_2 = 75$, $a_3 = 90$, $a_4 = 60$, $a_5 = 40$, $a_6 = 85$, and $M_i = a_i + \cdots + a_6$ for $i = 1, \ldots, 6$. The optimal solution to the problem is given by

$$x_1 = 175, \ x_2 = 0, \ x_3 = 190, \ x_4 = 0, \ x_5 = 0, x_6 = 85,$$
$$v_1 = 75, \ v_2 = 0, \ v_3 = 100, \ v_4 = 40, \ v_5 = 0, \ v_6 = 0,$$
$$\delta_1 = 1, \ \delta_2 = 0, \ \delta_3 = 1, \ \delta_4 = 0, \ \delta_5 = 0, \ \delta_6 = 1.$$

The minimum total cost is 2360 euros. The optimal production plan is to produce 175 tons on July 1st for the demand in the months of July and August, 190 tons on September 1st for the demand in the months of September, October, and November, and 85 tons on December 1st for the demand in December. In the optimal solution, production only takes place when there is no stock. This property holds in general for the considered production-stock problem and will be used for an alternative solution method in Section 5.5.2.

2.3   General Modeling Tricks

In this section, we discuss a number of generally applicable modeling tricks to model "almost-linear" programming problems as (mixed) ILP problems. Some of these tricks have already been used in the examples in Section 2.2. Other tricks will come back in the exercises.

Almost-Linear Objective Function

Suppose that a production-stock problem is given, with aim to minimize the costs, and the objective function contains an almost-linear term $P(x)$, where $P(x)$ represents the production costs of x units. The following three cases are of interest for applications:

(a) variable production costs plus fixed setup cost:

$$P(x) = \begin{cases} K + cx & \text{for } x > 0, \\ 0 & \text{for } x = 0, \end{cases} \tag{2.1}$$

where K and c are constants with $K > 0$ and x is a variable;

(b) piecewise-linear production costs:

$$P(x) = \begin{cases} c_1 x & \text{for } 0 \le x \le a_1, \\ c_1 a_1 + c_2(x - a_1) & \text{for } x > a_1, \end{cases} \tag{2.2}$$

where c_1 and c_2 are constants with $0 < c_2 < c_1$ and x is a variable;

(c) linear production costs with bulk discount:

$$P(x) = \begin{cases} cx & \text{for } 0 \le x \le a, \\ c(1 - d)x & \text{for } x > a, \end{cases} \tag{2.3}$$

where c, d, and a are constants with $c > 0$ and $0 < d < 1$ and x is an integer variable.

Case (a).   The idea is to introduce a 0-1 variable δ and use suitably selected linear constraints to ensure that

$$\delta_1 = \begin{cases} 1 & \text{if } x > 0, \\ 0 & \text{if } x = 0. \end{cases}$$

In that case, $P(x)$ can be replaced by the linear function $K\delta + cx$. To linearize the constraint on δ, choose a sufficiently large number M such that $x \leq M$ always holds. Add the following linear constraints to the optimization problem:

$$x \leq M\delta$$
$$\delta \in \{0, 1\}.$$

Then we have $\delta = 1$ if $x > 0$. The condition that $\delta = 0$ if $x = 0$ is met by minimizing the objective function in which $P(x)$ has been replaced with $K\delta + cx$. Because $K > 0$, we have $\delta = 0$ in the optimal solution if $x = 0$.

Case (b). The first step is to write the variable x as the sum of two variables:

$$x = y_1 + y_2 \quad \text{with} \quad 0 \leq y_1 \leq a_1 \quad \text{and} \quad y_2 \geq 0.$$

Then

$$P(x) = c_1 y_1 + c_2 y_2$$

provided that we have $y_1 = a_1$ if $y_2 > 0$. This can be achieved by introducing a 0-1 variable δ and imposing the following linear constraints:

$$a_1 \delta \leq y_1 \leq a_1$$
$$y_2 \leq M\delta$$
$$\delta \in \{0, 1\}$$
$$y_2 \geq 0,$$

where M is a sufficiently large number such that $x - a_1 \leq M$ always holds. Verify that subject to these linear constraints, $P(x)$ may be replaced by the linear term $c_1 y_1 + c_2 y_2$.

Case (c). For this case, x is assumed to be an integer variable, so that the number a can also be assumed to be integer valued. The inequality $x > a$ is then the same as $x \geq a + 1$. To linearize the function $P(x)$, we introduce a 0-1 variable δ and a nonnegative variable z with

$$\delta = \begin{cases} 1 \text{ if } x \geq a + 1, \\ 0 \text{ if } x \leq a \end{cases} \quad \text{and} \quad z = \begin{cases} x \text{ if } \delta = 1, \\ 0 \text{ if } \delta = 0. \end{cases}$$

If these requirements on δ and z are satisfied, then for every x, the term $P(x)$ can be written as the linear function

$$P(x) = cx - cdz.$$

The following linear constraints ensure that the requirements on δ and z are met (verify!):

$$x \leq a + M\delta, \quad x \geq (a + 1)\delta,$$
$$z \leq x, \quad z \geq x - M(1 - \delta), \quad z \leq M\delta,$$
$$\delta \in \{0, 1\}, \quad z \geq 0,$$

where M is a sufficiently large number such that $x \leq M$ always holds.

Almost-Linear Constraints

We consider five cases:

(a) The variable x can take on only one of the values $\alpha_1, \ldots, \alpha_r$.
(b) Only one of the two nonnegative variables x_1 and x_2 can be positive.
(c) The variable x must satisfy either $x = 0$ or $x \geq \alpha$, with $\alpha > 0$ a given constant.
(d) The variables $x_1, \ldots, x_n$ must satisfy the condition that at least k of a given m-tuple of constraints $\sum_{j=1}^{n} a_{ij} x_j \leq b_i$, for $i = 1, \ldots, m$, are satisfied.
(e) The variables x_1 and x_2 must satisfy the condition that if $x_1 > 0$, then $x_2 \leq \beta$, where $\beta \geq 0$ is a given constant.

Case (a). Introduce the 0-1 variables $\delta_1, \ldots, \delta_r$. Replace x with $x = \sum_{i=1}^{r} \alpha_i \delta_i$ and add the linear constraints $\sum_{i=1}^{r} \delta_i = 1$ and $\delta_i \in \{0, 1\}$ for $i = 1, \ldots, r$.

Case (b). Introduce two 0-1 variables δ_1 and δ_2, and let M_1 and M_2 be sufficiently large numbers such that $x_1 \leq M_1$ and $x_2 \leq M_2$ always hold. Add the following linear constraints:

$$x_i \leq M_i \delta_i \quad \text{for } i = 1, 2$$
$$\delta_1 + \delta_2 \leq 1$$
$$\delta_1, \delta_2 \in \{0, 1\}.$$

Case (c). Introduce a 0-1 variable δ and add the linear constraints

$$x \leq M\delta$$
$$x \geq \alpha\delta$$
$$\delta \in \{0, 1\},$$

where M is a sufficiently large number such that $x \leq M$ always holds.

Case (d). Introduce the 0-1 variables $\delta_1, \ldots, \delta_m$ and add the linear constraints

$$\sum_{j=1}^{n} a_{ij} x_j \leq b_i + (1 - \delta_i) M_i$$
$$\sum_{i=1}^{m} \delta_i \geq k$$
$$\delta_i \in \{0, 1\} \qquad \text{for } i = 1, \ldots, m,$$

where M_i is a sufficiently large number such that $\sum_{j=1}^{n} a_{ij} x_j \leq b_i + M_i$ always holds.

Case (e). Introduce a 0-1 variable δ and add the linear constraints

$$x_1 \leq M\delta$$
$$x_2 \leq \beta + M(1 - \delta)$$
$$\delta \in \{0, 1\},$$

where M is a sufficiently large number such that $x_1 \leq M$ and $x_2 \leq \beta + M$ always hold.

Product Term

In many applications, a product term occurs that is of the form $x_1 x_2$, where x_1 and x_2 are both 0-1 variables. This is easy to linearize. Namely, note that $x_1 x_2 = 1$ if $x_1 = x_2 = 1$ and $x_1 x_2 = 0$ otherwise. Replace $x_1 x_2$ with the variable y and add the linear constraints

$$y \leq x_1$$
$$y \leq x_2$$
$$y \geq x_1 + x_2 - 1$$
$$y \geq 0.$$

Since x_1 and x_2 are both 0-1 variables, these constraints imply that $y = 1$ if $x_1 = x_2 = 1$ and $y = 0$ otherwise. The variable y is therefore automatically an integer variable and can therefore be declared as a continuous variable, which is favorable from a computational point of view.

2.4   The Branch-and-Bound Method

Branch-and-bound is a flexible approach that is generally applicable to solving discrete optimization problems, including integer programming problems. The branch-and-bound approach is not a method with a set procedure; rather, it consists of a few simple basic ideas, the details of which depend on the problem and leave room for personal input. The basic idea is to split the total collection of feasible solutions into ever smaller subsets and, during this process, try to eliminate certain subsets that do not need to be searched. The elimination is carried out using upper and lower bounds. To further explain this, we assume, for the sake of simplicity, that the optimization problem is a *maximization problem*. The following holds for the upper and lower bounds:

- An upper bound is associated with every subset of solutions created during the splitting (branching) process. The upper bound for a given subset of solutions is a number such that the objective value of every feasible solution in the subset is less than or equal to that number.
- The lower bound is the objective value of the best feasible solution known so far.

Every subset therefore has its own upper bound. The lower bound, however, is not directly linked to a particular subset. The initial value of the lower bound is often the objective value of a feasible solution found using a heuristic method. What is the point of the upper and lower bounds? Suppose that for a subset of solutions, the upper bound for that subset is less than the best lower bound known so far. Then this subset can be *eliminated*. After all, the objective value of every feasible solution in the subset is less than the objective value of the best feasible solution known so far.

What do the *branching rule* and the *calculation of the upper bound* look like? This depends strongly on the application in question. If the discrete optimization problem is an ILP problem, then one can find an upper bound for a given subset of solutions by solving the LP relaxation of the ILP problem associated with the subset. If a subset is not eliminated, then it is split further by choosing one variable that should have been integer valued but assumes a fractional value in the solution to the LP relaxation. Suppose, for example, that the variable x_5 with value 7.47 is chosen; then the subset under consideration is split into two smaller subsets. One smaller subset consists of all solutions with $x_5 \leq 7$, while the other subset consists of all solutions with $x_5 \geq 8$.

A typical phenomenon in the branch-and-bound method is that *during* the branching process, other feasible solutions to the original maximization problem are found. The *lower bound* is then tightened if the objective value of a newly found feasible solution is better than the best objective value known so far. An improved lower bound can then lead to the elimination of certain subsets. It should also be clear that a tight upper bound for a subset is extremely important. The tighter the upper bound, the sooner a subset will be eliminated. In practice, however, the calculation of a tighter upper bound requires more computation time. In the implementation of a branch-and-bound method, these two aspects must therefore be weighed against each other.

Finally, a branch-and-bound method must specify a *search strategy*. A search strategy indicates which of the subsets that have not been eliminated yet is next in line for further splitting. The search strategy is often based on the "last-in-first-out" rule, in which the last-generated subset is chosen as the next one to be split. This strategy not only limits the amount of data that has to be stored in the computer's memory, but also has the great advantage that it often quickly leads to a feasible solution. A feasible solution is important for calculating the lower bound.

As an illustration of the above, consider the following ILP problem:

$$\text{Maximize} \quad 60x_1 + 60x_2 + 40x_3 + 10x_4 + 20x_5 + 10x_6 + 3x_7$$
$$\text{subject to} \quad 3x_1 + 5x_2 + 4x_3 + 1.4x_4 + 3x_5 + 3x_6 + x_7 \leq 11$$
$$\text{and} \qquad x_j \in \{0,1\}, \quad j = 1, \ldots, 7.$$

This is an example of the so-called *knapsack problem*, whose general formulation is as follows:

$$\text{Maximize} \quad c_1 x_1 + c_2 x_2 + \cdots + c_n x_n$$
$$\text{subject to} \quad a_1 x_1 + a_2 x_2 + \cdots + a_n x_n \leq b$$
$$0 \leq x_j \leq u_j \quad \text{for } j = 1, \ldots, n$$
$$\text{and} \qquad x_j \text{ integer} \quad \text{for } j = 1, \ldots, n,$$

where $c_j > 0$, $a_j > 0$, $b > 0$, and $u_j > 0$ (with u_j integer) are given constants. The name knapsack problem comes from the interpretation that refers to n articles $j = 1, \ldots, n$ that can be carried in a backpack, where each specimen of article j has value c_j and takes up a_j units of the total capacity b of the backpack. The solution to the LP relaxation of the knapsack problem is straightforward. Without loss of generality, we may assume that the variables $x_1, x_2, \ldots, x_n$ are ordered in such a way that

$$\frac{c_1}{a_1} \geq \frac{c_2}{a_2} \geq \cdots \geq \frac{c_n}{a_n};$$

that is, the variables are ordered by decreasing value of profit per unit of capacity use. It is not difficult to see that the optimal solution to the LP relaxation of the knapsack problem is then given by

$$\bar{x}_1 = \min\left(u_1, \frac{b}{a_1}\right), \quad \bar{x}_2 = \min\left(u_2, \frac{b - a_1\bar{x}_1}{a_2}\right), \quad \text{etc.}$$

In the numerical example above, the variables are already in the desired order. The optimal continuous solution to the LP relaxation of the specific example is

$$\bar{x}_1 = 1, \ \bar{x}_2 = 1, \ \bar{x}_3 = \frac{3}{4}, \ \text{remaining } \bar{x}_i = 0, \ \text{with LP objective value } = 150.$$

The numerical value 150 is an upper bound on the total set of feasible solutions to the ILP problem. A feasible initial solution to the integer problem can be found using the following *heuristic*: following the sequence of the ordered ratios c_j/a_j, set as many of the variables x_j as possible equal to 1 without violating the capacity restriction. This gives the initial solution (verify!)

$$x_1^{(0)} = 1, \ x_2^{(0)} = 1, \ x_3^{(0)} = 0, \ x_4^{(0)} = 1, \ x_5^{(0)} = x_6^{(0)} = 0, \ x_7^{(0)} = 1.$$

This feasible solution to the integer problem has objective value 133. The number 133 is therefore the initial value for the lower bound. The next step is to branch the total set of feasible solutions to the integer problem on the variable x_3 that has a fractional value in the optimal solution to the LP relaxation. One subset consists of all feasible solutions $(x_1, \ldots, x_7)$ with x_3 equal to 0, while the other subset consists of all feasible solutions with x_3 equal to 1 (the solution to the LP problem at the root node **1** was only needed to find the upper bound 150 and to find the branching variable x_3). It should be clear that the integer problem corresponding to a given subset of feasible solutions is still a knapsack problem. For example, the following integer problem corresponds to node **7**: maximize $60x_1 + 60x_2 + 10x_4 + 20x_5 + 10x_6 + 3x_7 + 40$ subject to $3x_1 + 5x_2 + 1.4x_4 + 3x_5 + 3x_6 + x_7 \leq 11 - 4$ and $x_j \in \{0, 1\}$. The LP relaxation of the integer programming problem can therefore easily be solved at each node. The calculations in the branch-and-bound approach for the specific knapsack problem given above are summarized in Figure 2.6, where the symbol UB indicates the upper bound and the symbol LB indicates the lower bound.

In the search strategy, the "last-in-first-out" rule has been used, where the depth of the tree increases when splitting the subsets. The nodes in Figure 2.6 are numbered according to this search strategy and have therefore been gone through in the order given by the numbering. Note that the lower bound LB remains unchanged up to node **6**, where an improved feasible solution with objective value 140 is found.

After seven steps, the branch-and-bound method leads to the conclusion that

$$x_1^* = x_2^* = x_5^* = 1, \qquad x_3^* = x_4^* = x_6^* = x_7^* = 0$$

is the optimal solution to the original integer problem, with maximum objective value 140.

Fig. 2.6   The branch-and-bound steps.

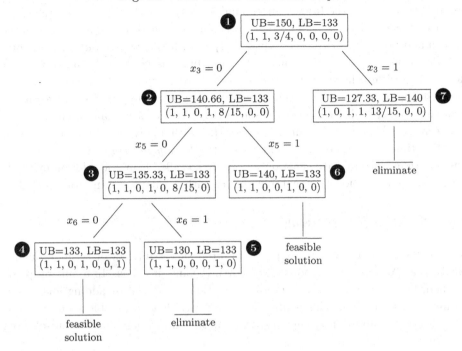

Stopping Rule

In applications of the branch-and-bound algorithm, it often happens that a feasible solution is found right at the beginning, but that many more calculations are necessary to conclude, at the very end of the elimination process, that this solution is optimal. This is the reason why most branch-and-bound codes give the user the possibility to stop the calculations before completion when the upper and lower bounds found up to that point are sufficiently close to each other. The objective value of the feasible solution at the termination point can then only differ slightly from the optimal objective value. For example, suppose that in a maximization problem, at a certain point, the highest upper bound over a not-yet-eliminated subset is equal to 500, while a feasible solution has been found that gives a lower bound of 493. Then the objective value of this feasible solution is known to differ by no more than $\frac{500-493}{500} \times 100\% = 1.4\%$ from the maximum objective value.

Nonpolynomial Computation Time

The branch-and-bound method for integer programming problems is an algorithm whose computation time generally explodes when the number of integer variables becomes big. This can easily be seen with the help of the knapsack problem discussed above. If the number of 0-1 variables in the knapsack problem is n, then in the worst case, one needs to consider 2^n subsets in the branching process. The computation time of the branch-and-bound algorithm is then 2^n and therefore increases *exponentially* with n. This non–polynomially bound computation time of the branch-and-bound algorithm contrasts sharply with the *polynomially bound* computation time of, for example, network algorithms discussed in Chapter 3. Most network algorithms have a computation time of order n^2 or n^3 if n is the number of nodes in the network. For the general integer programming problem and for many combinatorial optimization problems such as the traveling salesman problem, it does not seem possible to develop an exact algorithm with a polynomially bounded computation time. This type of problem is called *NP-complete*. For large NP-complete optimization problems, one generally resorts to so-called heuristic solution methods that produce a solution of acceptable quality within a limited computation time.

2.5   Cutting Plane Methods

Integer linear programming problems are often successfully dealt with through a combination of branch-and-bound and cutting plane methods. In a cutting plane method, the feasible region for the LP relaxation is reduced by adding linear constraints, so-called cuts. This is done in such a way that no feasible solution is cut off. Ideally, the addition of cuts ensures that the corner points of the cut-off area are integer points.

The most common cutting plane method is that of Gomory cuts. In this section, we restrict ourselves to this classical method. Gomory cuts date back to the 1960s and were discarded for a long time because they did not work fast enough in real-world applications. New developments in the theory of polyhedrons have led to significant improvements in the cutting plane methods.[2] The idea of Gomory cuts is very simple. For the sake of convenience, we assume that we have an integer problem in which *every* variable must be integer valued. Suppose that for the original integer problem, the continuous LP problem

$$\text{Maximize } \mathbf{cx} \text{ subject to } \mathbf{Ax} = \mathbf{b}, \ \mathbf{x} \geq \mathbf{0}$$

has been solved. Let the matrix $\mathbf{B}$ be the optimal basis (consisting of column vectors of $\mathbf{A}$), and let the matrix $\mathbf{R}$ consist of the remaining column vectors of $\mathbf{A}$. In the final tableau of the LP solution, the system $\mathbf{Ax} = \mathbf{b}$ has been transformed into the equivalent system

$$\mathbf{x_B} + \mathbf{B}^{-1}\mathbf{R}\mathbf{x_R} = \mathbf{B}^{-1}\mathbf{b};$$

cf. Section 1.7. The optimal basic solution to the continuous LP problem is $(\mathbf{x_B^*} = \mathbf{B}^{-1}\mathbf{b}, \ \mathbf{x_R^*} = \mathbf{0})$. Now, choose an index i for which the numerical value

---

[2]The interested reader is referred to the paper J. E. Mitchell, Branch-and-cut algorithms for combinatorial optimization problems, in *Handbook of Applied Optimization*, Oxford University Press, 2000.

of $(\mathbf{B}^{-1}\mathbf{b})_i$ is not an integer while the value of the variable x_{B_i} must be an integer. If we denote the number a rounded down by $\lfloor a \rfloor$, then we can rewrite the equation

$$x_i + \sum_{j \in \mathbf{R}} (\mathbf{B}^{-1}\mathbf{R})_{ij} x_j = (\mathbf{B}^{-1}\mathbf{b})_i$$

as

$$x_i + \sum_{j \in \mathbf{R}} (\lfloor (\mathbf{B}^{-1}\mathbf{R})_{ij} \rfloor + f_{ij}) x_j = \lfloor (\mathbf{B}^{-1}\mathbf{b})_i \rfloor + g_i,$$

where f_{ij} is the fractional part of the number $(\mathbf{B}^{-1}\mathbf{R})_{ij}$ and g_i is the fractional part of the number $(\mathbf{B}^{-1}\mathbf{b})_i$ (for example, the fractional part of the number $7\frac{1}{3}$ is $\frac{1}{3}$; the fractional part of an integer is, of course, 0). We then rewrite this equation as

$$x_i + \sum_{j \in \mathbf{R}} \lfloor (\mathbf{B}^{-1}\mathbf{R})_{ij} \rfloor x_j - \lfloor (\mathbf{B}^{-1}\mathbf{b})_i \rfloor = g_i - \sum_{j \in \mathbf{R}} f_{ij} x_j.$$

The right-hand side of the equality is strictly less than 1, while the left-hand side is an integer for every feasible integer solution to the original problem. If we add the constraint (cut)

$$x_i + \sum_{j \in \mathbf{R}} \lfloor (\mathbf{B}^{-1}\mathbf{R})_{ij} \rfloor x_j - \lfloor (\mathbf{B}^{-1}\mathbf{b})_i \rfloor \leq 0$$

to the LP problem, then we reduce the feasible region without cutting off a feasible integer solution to the original problem.

2.6 Lagrangian Relaxation[3]

Lagrangian relaxation is a tool that has been used increasingly in recent years to tackle large-scale integer programming problems. This approach is based on the fact that many real-world ILP problems can be seen as combinations of a relatively simple basic problem and a number of difficult constraints. The idea of this approach is to include these difficult constraints in the objective function with a penalty term and then to solve a sequence of relatively simple optimization problems. Instead of dropping the integer requirements as in the LP relaxation, the Lagrangian relaxation follows a completely different approach and drops a number of constraints. The precise implementation of the basic idea strongly depends on the problem. In this section, we only roughly outline the Lagrangian approach; for more details, see the paper M. L. Fisher, An applications oriented guide to Lagrangian relaxation, *Interfaces*, 15 (1985), 10–21.

Consider, in matrix notation, an integer programming problem of the following form:

$$\begin{aligned} \text{Maximize} \quad & \mathbf{c}^T \mathbf{x} \\ \text{subject to} \quad & \mathbf{A}\mathbf{x} \leq \mathbf{b} \\ & \mathbf{D}\mathbf{x} \leq \mathbf{e} \\ \text{and} \quad & \mathbf{x} \geq \mathbf{0}, \ \mathbf{x} \text{ integer,} \end{aligned} \qquad (IP)$$

---

[3]This section contains more advanced material and can be skipped at first reading.

where $\mathbf{c}$ and $\mathbf{b}$ are given vectors and $\mathbf{A}$ and $\mathbf{D}$ are given matrices. We assume that the constraints of (IP) have been split up into the two groups $\mathbf{Ax} \leq \mathbf{b}$ and $\mathbf{Dx} \leq \mathbf{e}$ in such a way that the problem (IP) is relatively easy to solve if the constraints $\mathbf{Ax} \leq \mathbf{b}$ are dropped. To define the Lagrangian problem, we choose a vector $\mathbf{u}$ of nonnegative numbers and add the term $\mathbf{u}^T(\mathbf{b} - \mathbf{Ax})$ to the objective function of (IP) and therefore remove the difficult constraints $\mathbf{Ax} \leq \mathbf{b}$. So for a fixed vector $\mathbf{u} \geq \mathbf{0}$, the relaxed ILP problem (LR) is defined by

$$
\begin{array}{ll}
\text{Maximize} & \mathbf{c}^T\mathbf{x} + \mathbf{u}^T(\mathbf{b} - \mathbf{Ax}) \\
\text{subject to} & \mathbf{Dx} \leq \mathbf{e} \qquad\qquad\qquad\qquad\qquad (LR) \\
\text{and} & \mathbf{x} \geq \mathbf{0}, \ \mathbf{x} \text{ integer.}
\end{array}
$$

The assumption is that this Lagrangian problem is relatively easy to solve. If we define

$$Z_{\text{IP}}^* = \text{maximum objective value of the original problem (IP)}$$
$$Z_{\text{LR}}^*(\mathbf{u}) = \text{maximum objective value of the Lagrangian relaxation (LR)}$$
$$\text{for the vector } \mathbf{u},$$

then we have the following result.

Theorem 2.1. *For every* $\mathbf{u} \geq \mathbf{0}$, *it holds that* $Z_{\text{IP}}^* \leq Z_{\text{LR}}^*(\mathbf{u})$.

Proof. Fix $\mathbf{u} \geq \mathbf{0}$ and observe that $\mathbf{u}^T(\mathbf{b} - \mathbf{Ax}) \geq \mathbf{0}$ for every $\mathbf{x}$ with $\mathbf{Ax} \leq \mathbf{b}$. This implies

$$
\begin{aligned}
Z_{\text{IP}}^* &= \max\left\{\mathbf{c}^T\mathbf{x} \mid \mathbf{Ax} \leq \mathbf{b}, \mathbf{Dx} \leq \mathbf{e}, \mathbf{x} \geq \mathbf{0}, \ \mathbf{x} \text{ integer}\right\} \\
&\leq \max\left\{\mathbf{c}^T\mathbf{x} + \mathbf{u}^T(\mathbf{b} - \mathbf{Ax}) \mid \mathbf{Ax} \leq \mathbf{b}, \mathbf{Dx} \leq \mathbf{e}, \mathbf{x} \geq \mathbf{0}, \mathbf{x} \text{ integer}\right\} \\
&\leq \max\left\{\mathbf{c}^T\mathbf{x} + \mathbf{u}^T(\mathbf{b} - \mathbf{Ax}) \mid \mathbf{Dx} \leq \mathbf{e}, \mathbf{x} \geq \mathbf{0}, \mathbf{x} \text{ integer}\right\} \\
&= Z_{\text{LR}}^*(\mathbf{u}),
\end{aligned}
$$

where the second inequality follows by considering that the maximum can only increase if the feasible region becomes larger. This completes the proof. □

In the next theorem, we show that the number Z_{LD} defined by

$$Z_{\text{LD}} = \min_{\mathbf{u} \geq \mathbf{0}} Z_{\text{LR}}^*(\mathbf{u})$$

gives a tighter upper bound for the maximum of the integer programming problem (IP) than the maximum of the LP relaxation of the problem (IP). Later, we will discuss how the upper bound Z_{LD} can be calculated efficiently. Define

$$Z_{\text{LP}}^* = \text{maximum objective value of the LP relaxation of the}$$
$$\text{integer problem (IP).}$$

Theorem 2.2. *It holds that* $Z_{\text{IP}}^* \leq Z_{\text{LD}} \leq Z_{\text{LP}}^*$.

Proof. The inequality $Z_{IP}^* \leq Z_{LD}$ is a direct consequence of Theorem 2.1. The proof that $Z_{LD} \leq Z_{LP}^*$ is based on the duality result that the optimal objective value of an LP problem is equal to the optimal objective value of its dual when both problems have a finite optimum. We now find

$$
\begin{aligned}
Z_{LD} &= \min_{\mathbf{u} \geq 0} \left[\max \left\{ \mathbf{c}^T \mathbf{x} + \mathbf{u}^T (\mathbf{b} - \mathbf{Ax}) \mid \mathbf{Dx} \leq \mathbf{e}, \mathbf{x} \geq \mathbf{0}, \mathbf{x} \text{ integer} \right\} \right] \\
&\leq \min_{\mathbf{u} \geq 0} \left[\max \left\{ \mathbf{c}^T \mathbf{x} + \mathbf{u}^T (\mathbf{b} - \mathbf{Ax}) \mid \mathbf{Dx} \leq \mathbf{e}, \mathbf{x} \geq \mathbf{0} \right\} \right] \\
&= \min_{\mathbf{u} \geq 0} \left[\mathbf{u}^T \mathbf{b} + \max \left\{ \left(\mathbf{c}^T - \mathbf{u}^T \mathbf{A} \right) \mathbf{x} \mid \mathbf{Dx} \leq \mathbf{e}, \mathbf{x} \geq \mathbf{0} \right\} \right] \\
&\overset{(dual)}{=} \min_{\mathbf{u} \geq 0} \left[\mathbf{u}^T \mathbf{b} + \min \left\{ \mathbf{v}^T \mathbf{e} \mid \mathbf{v}^T \mathbf{D} \geq \mathbf{c}^T - \mathbf{u}^T \mathbf{A}, \mathbf{v} \geq \mathbf{0} \right\} \right] \\
&= \min \left\{ \mathbf{u}^T \mathbf{b} + \mathbf{v}^T \mathbf{e} \mid \mathbf{u}^T \mathbf{A} + \mathbf{v}^T \mathbf{D} \geq \mathbf{c}^T, \ \mathbf{u}, \mathbf{v} \geq \mathbf{0} \right\} \\
&\overset{(dual)}{=} \max \left\{ \mathbf{c}^T \mathbf{x} \mid \mathbf{Ax} \leq \mathbf{b}, \mathbf{Dx} \leq \mathbf{e}, \mathbf{x} \geq \mathbf{0} \right\} \\
&= Z_{LP}^*.
\end{aligned}
$$

This completes the proof. $\qquad\square$

Next, we discuss how to minimize the function $Z_{LR}^*(\mathbf{u})$ for $\mathbf{u} \geq 0$. Note that, in general, $Z_{LR}^*(\mathbf{u})$ is a function of more than one variable, where the number of variables is equal to the number of constraints in $\mathbf{Ax} \leq \mathbf{b}$.

Subgradient Method for the Minimization of $Z_{LR}^*(\mathbf{u})$

It should be clear, intuitively, that an optimal solution to the Lagrangian relaxation problem ($LR_{\mathbf{u}}$) for a given vector $\mathbf{u}$ remains optimal in a sufficiently small neighborhood of the point $\mathbf{u}$. This means that $Z_D(\mathbf{u})$ is a piecewise-linear function of $\mathbf{u} \geq 0$. It can also be proved that $Z_D(\mathbf{u})$ is a convex function of $\mathbf{u} \geq 0$. The function $Z_D(\mathbf{u})$ is differentiable at almost every point $\mathbf{u}$, except for the points $\mathbf{u}$ where the Lagrangian problem ($LR_{\mathbf{u}}$) has several optimal solutions $\mathbf{x}$ and a transition takes place from a region with a specific optimal solution to a region with another optimal solution. At every point $\mathbf{u}$ where the function $Z_D(\mathbf{u})$ is differentiable, the gradient of $Z_D(\mathbf{u})$ is given by $\mathbf{b} - \mathbf{Ax}$ with $\mathbf{x}$ the optimal solution associated with the point $\mathbf{u}$. The fact that the gradient is readily available means that we can use a gradient method to minimize $Z_D(\mathbf{u})$. An efficient procedure, called the subgradient method, has been developed. At the points $\mathbf{u}$ where $Z_D(\mathbf{u})$ is not differentiable, the subgradient method randomly chooses one of the optimal solutions $\mathbf{x}$ to the problem ($LR_{\mathbf{u}}$) and pretends that $\mathbf{b} - \mathbf{Ax}$ is the gradient at the point $\mathbf{u}$. Starting out with an arbitrary point $\mathbf{u}^{(0)}$, the subgradient method generates a sequence of $\mathbf{u}$-points according to the formula

$$
\mathbf{u}^{(k+1)} = \max \left\{ 0, \mathbf{u}^{(k)} - t_k \left(\mathbf{b} - \mathbf{Ax}^{(k)} \right) \right\},
$$

where the scalar t_k is a cleverly chosen step size and $\mathbf{x}^{(k)} \in \Omega$ is an optimal solution to the Lagrangian problem with $\mathbf{u} = \mathbf{u}^{(k)}$. The step size t_k must be selected according to a specific rule, in part because the function $Z^*_{\mathrm{LR}}(\mathbf{u})$ is not everywhere differentiable. The step size t_k must converge to zero, but not too quickly. We state the following result without proof. If t_k is chosen in such a way that

$$t_k \to 0 \text{ and } \sum_{i=1}^{k} t_i \to \infty \text{ as } k \to \infty,$$

then $Z^*_{\mathrm{LR}}(\mathbf{u}^{(k)})$ converges to the minimum value Z_{LD}. Experiments have shown that the following technical formula for t_k works remarkably well in practice:

$$t_k = \frac{\lambda_k \left\{ Z^*_{\mathrm{LR}}(\mathbf{u}^{(k)}) - \overline{Z} \right\}}{\displaystyle\sum_{i=1}^{m} \left(b_i - \sum_{j=1}^{n} a_{ij} x_j^{(k)} \right)^2},$$

where $\overline{Z}$ is the objective value of the best feasible solution to the problem (IP) known so far and λ_k is a scalar between 0 and 2. In solving the Lagrangian relaxation problem, one regularly obtains a solution $\mathbf{x}^{(k)}$ that is also feasible for the original problem (IP) and that can be used to adjust $\overline{Z}$. It is common to start the sequence λ_k with $\lambda_k = 2$ and divide λ_k by 2 when $Z^*_{\mathrm{LR}}(\mathbf{u}^{(k)})$ has not decreased for a set number of iterations. We leave out the details; these can be found in the article Held, Wolfe, and Crowder, Validation of subgradient optimization, *Mathematical Programming*, 6 (1974), 62–88. This paper also gives details for the criterion for terminating the subgradient method. The algorithm obviously stops intermediately if it turns out that $Z^*_{\mathrm{LR}}(\mathbf{u}^k) = \overline{Z}$. After all, since $\overline{Z} \leq Z^*_{\mathrm{IP}} \leq Z^*_{\mathrm{LR}}(\mathbf{u}^{(k)})$, it then follows that the feasible solution with objective value $\overline{Z}$ is optimal for the problem (IP).

Before giving an illustration of the Lagrangian relaxation method, let us note that this method can also be used within the branch-and-bound algorithm. As we have shown in Theorem 2.2, the Lagrangian relaxation method gives a tighter upper bound than the LP relaxation. Moreover, with the Lagrangian relaxation method, a good feasible interim solution is often found that can serve as the lower bound for the elimination process in the branch-and-bound algorithm. A successful application of the Lagrangian relaxation method requires that the optimization problem has a special structure. It is only by using that structure that we can hope that large-scale integer problems can become solvable in practice. How exactly to use the structure still strongly depends on the problem.

Illustration

Consider the ILP problem

$$\begin{aligned}
\text{Maximize} \quad & 16x_1 + 10x_2 + 4x_4 \\
\text{subject to} \quad & 8x_1 + 2x_2 + x_3 + 4x_4 \leq 10 \\
& x_1 + x_2 \leq 1 \\
& x_3 + x_4 \leq 1 \\
\text{and} \quad & x_j \in \{0, 1\}, \; j = 1, \ldots, 4.
\end{aligned}$$

In this example, different relaxations are possible. A relaxation that leads to an extremely easily solvable Lagrangian problem is that where the first constraint is included in the objective function. This gives the Lagrangian relaxation problem

$$\text{Maximize} \quad (16 - 8u)\, x_1 + (10 - 2u)\, x_2 + (0 - u)\, x_3 + (4 - 4u)\, x_4 + 10u$$
$$\text{subject to} \quad x_1 + x_2 \le 1$$
$$x_3 + x_4 \le 1$$
$$\text{and} \quad x_j \in \{0, 1\}, \; j = 1, \ldots 4.$$

For fixed u, this problem is trivially solvable. Every variable with a negative coefficient is given the value 0, while in each of the groups $\{x_1, x_2\}$ and $\{x_3, x_4\}$, the variable with the highest positive coefficient is given the value 1.

An alternative relaxation is to include the second and third constraints in the objective function. The Lagrangian relaxation problem then becomes

$$\text{Maximize} \quad (16 - u_1)\, x_1 + (10 - u_1)\, x_2 + (0 - u_2)\, x_3 + (4 - u_2)\, x_4 + u_1 + u_2$$
$$\text{subject to} \quad 8x_1 + 2x_2 + x_3 + 4x_4 \le 10$$
$$\text{and} \quad x_j \in \{0, 1\} \text{ for } j = 1, \ldots, 4.$$

This relaxation problem is an example of the 1-dimensional knapsack problem. This problem is not trivially solvable, but can, for example, be solved efficiently using dynamic programming.

The question is, of course, which of the two relaxation options is preferable. Experience with the Lagrangian relaxation method has shown that the relaxation must simplify the original problem considerably, but not too much. This means that the second relaxation given above is preferable. In Table 2.3, we give the results for this relaxation when we start with $u_1 = u_2 = 0$ and apply the subgradient method mentioned above with initial value $\overline{Z} = 0$ and $\lambda_k = 1$ in the formula for the step size. In the table, we write a $(*)$ if a feasible interim solution is found. After four iterations, we find a feasible solution whose objective value is equal to the upper bound $Z_{\text{LR}}^*(u_1, u_2)$, so that not only has the algorithm converged to $Z_{\text{LD}} (= 16)$, but an optimal solution to the original problem has also been found.

Table 2.3 The Lagrangian Iterations.

u_1	u_2	λ_k	x_1	x_2	x_3	x_4	$Z_{\text{LR}}^*(u_1, u_2)$	$\overline{Z}$
0	0	1	1	1	0	0	26	0
13	0	1	0	0	0	1(*)	17	4
0	0	1	1	1	0	0	26	4
11	0	1	1	0	0	0(*)	16	16

If we had done the calculations for the first relaxation, we would have found a looser upper bound of $Z_{\text{LD}} = 18$ (associated with $u = 1$). This upper bound is, by the way, the same as the upper bound Z_{LP}^* we would have found using the LP relaxation.

2.7   Heuristics

The computational complexity of the integer programming model associated with the distribution problem rapidly surpasses any practical limit when n (= the number of goods) is large. The problem cannot be solved in a time that is a polynomial function of the problem size; rather, the computation time grows exponentially in that size. As in most combinatorial optimization problems, it is practically impossible to find an optimal integer solution when the problem becomes very large. In such situations, we fall back on methods that restrict themselves to finding a reasonably good solution within an acceptable amount of computation time. Such methods are called *heuristics*. Often, one can develop different heuristics for a certain class of problems. When designing a heuristic, one must always weight the computation time and the quality of the calculated solution against each other. Ideally, a heuristic provides a near-optimal solution within a relatively short computation time. A good heuristic should have the following properties:

- intuitively appealing,
- easy to apply,
- good solution in a relatively short computation time.

In Section 2.4, we already gave an example of a heuristic method for the knapsack problem. In this section and Sections 2.8 and 2.9, we give other examples of heuristics. Designing a heuristic is to a large extent an art that requires, in particular, creativity from the designer. The exact form of a heuristic strongly depends on the specific problem.

2.7.1   *A Partitioning Problem*

A simple heuristic for the partitioning problem from Section 2.2.5 is as follows:

1. Order the goods in decreasing order of value.
2. Successively assign the goods to one of the two persons, where a product is assigned to the person who has the lowest total value of assets at that time.

Let us apply the heuristic to the numerical example from Section 2.2.5, where the numbers a_i have already been ordered in the desired way. The heuristic provides the two groups

$$S_1 = \{1, 4, 6, 8, 9, 11, 13, 16, 17, 19\}, \quad S_2 = \{2, 3, 5, 7, 10, 12, 14, 15, 18, 20\}.$$

In the heuristic solution, persons 1 and 2 receive goods worth 460 and 451, with a difference of 9, while in the optimal solution, the assigned goods have values 456 and 455, with a difference of 1.

2.7.2   *A Container Loading Problem*

Suppose that we have to put n goods in containers with the same capacity C. Product i requires c_i units of container capacity. How should we put the goods in the containers in order to minimize the number of containers we need? A simple heuristic works as follows:

1. Renumber the goods so that $c_i \geq c_{i+1}$ for $1 \leq i < n$. The order in which the goods are put into the containers corresponds to the new numbering $1, \ldots, n$.
2. Place product i in the container with the smallest remaining capacity in which it fits well.

The intuitive idea behind this heuristic is to first pack the largest goods and to pack them in such a way that as little space as possible is left. This is exactly what most of us would do when packing up for a move. Here, too, we have a heuristic that is easy to apply and provides reasonably good solutions.

The heuristics from Sections 2.7.1 and 2.7.2 are so-called *constructive heuristics*, which build a feasible solution step by step. There also exist so-called *improvement heuristics*. Unlike constructive heuristics, an improvement heuristic does not create new solutions from scratch; rather, it adapts existing solutions. By gradually adapting an existing solution in a clever way, an attempt is made to improve it. This is often done using an *exchange procedure*; for example, in a timetabling problem, one can try to improve a given solution by exchanging two jobs. The exchange heuristic has proved to be extremely successful for the traveling salesman problem; see Section 2.8. The exchange heuristic is a local search procedure. A local search procedure that only accepts improvements to the current solution has the disadvantage of being able to get stuck in a local optimum. A search method such as simulated annealing accepts not only improvements of the objective function, but also deteriorations. Consequently, a local optimum can be left and a larger part of the solution space can be searched. If multiple feasible initial solutions can easily be generated, then, in practice, applying a local search method to each of these initial solutions and then taking the best resulting solution leads to excellent results.

Designing a heuristic is a creative process that depends strongly on the problem. This is also illustrated in the following two subsections.

2.7.3   *The Set-Covering Problem*

We have already come across this problem in Section 2.2.2. The integer formulation of the set-covering problem is

$$\text{Minimize} \quad \sum_{j=1}^{n} c_j x_j$$
$$\text{subject to} \quad \sum_{j=1}^{n} a_{ij} x_j \geq 1 \text{ for } i = 1, \ldots, m$$
$$\text{and} \quad x_j \in \{0, 1\} \quad \text{for } j = 1, \ldots, n,$$

where (a_{ij}) is an $m \times n$ matrix consisting of zeros and ones. So at a cost of c_j for choosing column j, we are looking for columns that cover every row, at the lowest total cost: row i is covered by a column j if $a_{ij} = 1$. This problem has many applications, including routing problems. Of particular interest is the set-covering problem with $c_j = 1$ for all j.

Let us discuss two basic heuristics. We begin with the case

$$c_j = 1 \text{ for } j = 1, \ldots, n;$$

that is, we are looking for the least number of columns such that every row is covered. Let us first introduce some notation. Let $I = \{1, \ldots, m\}$ be the index set of the rows and $J = \{1, \ldots, n\}$ that of the columns. For each column j, let M_j be the set consisting of the rows that are covered by column j, and for each row i, let N_i be the set consisting of the columns that cover row i; that is,

$$M_j = \{i \in I \mid a_{ij} = 1\} \quad \text{and} \quad N_i = \{j \in J \mid a_{ij} = 1\}.$$

In the heuristics below, we adjust the sets R and S step by step, where R consists of the rows that have not been covered so far and S consists of the columns that are being used for the cover so far.

Heuristic 1

In the first heuristic, in each iteration, we add a column that covers the greatest number of still uncovered rows.

Algorithm 2.1.

Step 0. Set $R := I$ and $S := \emptyset$.
Step 1. If $R = \emptyset$, then stop. Otherwise, determine the column $j \notin S$ that covers the greatest number of still uncovered rows, that is, the j for which $|R \cap M_j|$ is the greatest. Adjust R and S as follows: $R := R \backslash M_j$ and $S := S \cup \{j\}$. Repeat Step 1.

In this form, the heuristic is not yet very useful. The reason is that this heuristic does not look ahead enough; that is, it could initially include a column in S and later come to regret this. This is why the heuristic is extended with an exchange procedure. An exchange procedure is frequently incorporated into a heuristic. The idea is to replace an existing column in the solution with a yet unused column. The column that is best qualified to be removed from the solution is the one for which the number of rows only covered by that column is the least. This column is compared to the column that is not yet in the solution and covers the greatest number of rows that are not yet covered. Removing the former in favor of the latter only takes place if it results in an improved number of covered rows. So, the algorithm is adjusted to

Algorithm 2.2.

Step 0. Set $R := I$ and $S := \emptyset$.

Step 1. If $R = \emptyset$, then stop. Otherwise, determine the column $j \notin S$ that covers the greatest number of still uncovered rows, that is, the j for which $|R \cap M_j|$ is the greatest. Adjust R and S as follows: $R := R \backslash M_j$ and $S := S \cup \{j\}$.

Step 2. (Exchange step) Determine the column $\ell \in S$ for which $|X_\ell|$ with

$$X_\ell = \{i \in I \mid a_{i\ell} = 1, \ a_{ij} = 0 \text{ for } j \in S, \ j \neq \ell\}$$

is the least. Determine the column $j \notin S$ for which $|R \cap M_j|$ is the greatest. If $|R \cap M_j| > |X_\ell|$, then replace column ℓ in S with column j.
Go to Step 1.

Calculating the numbers $|R \cap M_j|$ and $|X_\ell|$ in Steps 1 and 2 must, of course, be done in a "smart" way, so that the number of calculations is not unnecessarily high. Much also depends on the chosen implementation. Ensuring that when adjusting the numbers and sets one checks only the rows and columns that really need to be updated can save an enormous amount of time. We will not go into further detail on updating these numbers.

Heuristic 2

The next heuristic is based on probabilistic principles. The columns added to the solution are randomly selected from the unused columns. The probability that a column is chosen depends on its "frequency," that is, the number of uncovered rows the column covers. The probability of a column is taken to be the frequency of the column divided by the sum of the frequencies of the unused columns. These probabilities are, of course, adjusted after each iteration.

Algorithm 2.3.

Step 0. Set $R := I$ and $S := \emptyset$.

Step 1. If $R = \emptyset$, then stop. For every column $j \notin S$, calculate

$$p_j := \frac{|R \cap M_j|}{\sum_{k \notin S} |R \cap M_k|}.$$

Randomly select a column $j \notin S$ according to the probability distribution $\{p_j \mid j \notin S\}$. Adjust R and S as follows: $R := R \backslash M_j$ and $S := S \cup \{j\}$.
Repeat Step 1.

One could carry out Heuristic 2 a great number of times and then take the best solution that has been found.

General c_j

The heuristics mentioned above need to be adjusted slightly if the c_j are general. The selection criterion for a column to be admitted to the solution is now based on the number of uncovered rows the column covers per unit of cost: instead of $|R \cap M_j|$, consider $|R \cap M_j| / c_j$.

2.7.4   The Facility Location Problem

Consider a directed or undirected network with n points (for example, cities) in which a facility (for example, a hospital or nursing home) needs to be placed in some of the points. For every pair of points i and j, a number a_{ij} is given that represents the travel distance from i to j; these numbers may be weighted to take into account the population or the importance of the cities. We may assume, without loss of generality, that all the numbers a_{ij} are nonnegative.

Suppose that we can place a facility in (at most) p of the n cities, with p given, and that every other city is assigned to the nearest facility. The aim is to keep the total sum of the travel distances as small as possible. Below, we discuss a constructive heuristic for this. The constructive heuristic is endowed with an additional procedure to correct for unfavorable choices in the first steps of the heuristic.

Heuristic

In order to motivate the steps of the constructive heuristic, assume that a number of facilities have already been assigned to cities. Let V be the current set of cities that have (provisionally) been allocated a facility. The total travel distance associated with this allocation is

$$\sum_{i=1}^{n} a_i^* \quad \text{with} \quad a_i^* = \min_{j \in V} a_{ij}, \ i = 1, \ldots, n.$$

Now, suppose that we expand the current V with city $t \notin V$ by building a facility in city t; then the total travel distance decreases by

$$\Delta_t = \sum_{i=1}^{n} \max \left(a_i^* - a_{it}, 0 \right).$$

Obviously, we add the city t for which Δ_t is the greatest, provided that we do not have $\Delta_j = 0$ for all $j \notin V$, in which case we do not expand V. This *allocation procedure* adds one city at a time to the set of cities with a facility until either the costs no longer decrease, or a facility has been built in p cities. It should be clear that it may be useful to consider an *elimination procedure* in addition to this allocation procedure that can reverse the previous heuristic allocation of a facility to a city. Suppose that we remove city t from V. Then the total travel distance of this new allocation is equal to

$$\sum_{i=1}^{n} \min_{j \in V, j \neq t} a_{ij},$$

so that the effect of eliminating city t is an increase in the total travel distance by

$$\overline{\Delta}_t = \sum_{i=1}^{n} \left[\min_{j \in V, j \neq t} a_{ij} - \min_{j \in V} a_{ij} \right], \ t \in V.$$

The elimination procedure now removes the city t from V for which $\overline{\Delta}_t$ is the least, provided that this city was not the last one added by the allocation procedure (otherwise, city t will be added again in the next step).

In the manner described above, the allocation procedure and elimination procedure are now alternated until the total travel distance no longer changes or p facilities have been allocated.[4] The algorithm can be initialized as follows. Calculate $c_j = \sum_{i=1}^{n} a_{ij}$ for all $j = 1, \ldots, n$ and set $V = \{t\}$ with t the j for which c_j is the smallest.

2.8   The Traveling Salesman Problem

The traveling salesman problem is a classic problem from combinatorial optimization. Although the problem is easy to formulate, it is challenging to solve. A traveling salesman must visit each city in his sales area exactly once and then return to his starting point. For every pair of cities i and j, the distance from city i to city j is given. We speak of a *symmetric* traveling salesman problem if for every pair of cities i and j, the distance c_{ij} from city i to city j is equal to the distance c_{ji} from city j to city i. What is the shortest tour? The problem is not only interesting; it also has many practical applications outside of the context of a traveling salesman who wants to minimize his travel time. An important application is a routing problem, where a truck must deliver goods to a number of customers and a route must be set out that begins and ends at a central depot and has the lowest total cost. Other applications include collecting mail from mailboxes, making the optimal planning for the production sequence of different products on a single machine with change-over times, determining the optimal sequence in which a laser drills holes into an integrated circuit from a fixed starting position, and so on. These applications can be modeled as traveling salesman problems through "creative" choices of the cities and distances between the cities. The following problem is also a nice application of the traveling salesman problem. Is it possible to visit all 64 squares on a chessboard with a knight and be back at the start after 64 steps? The answer is "yes." The problem is a traveling salesman problem with the chessboard's squares as cities and the number of knight jumps needed to go from one square

---

[4]The elimination procedure can sometimes be skipped, for example if otherwise the heuristic algorithm does not converge.

to the other as the distance between two cities. The knight's tour over the chessboard is given in Figure 2.7. This chess problem had already been studied in Arab literature in the ninth century. In Chapter 1 of the fascinating book David Applegate et al., The Traveling Salesman Problem, Princeton University Press, 2007, interesting historical applications of the traveling salesman problem are discussed.[5]

Fig. 2.7   A knight's tour over the chessboard.

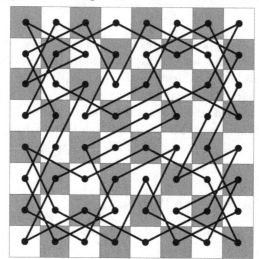

The computational difficulty of the traveling salesman problem is immense. Suppose that there are n cities. The number of possible tours is then equal to $(n-1) \times (n-2) \times \cdots \times 1 = (n-1)!$ (the mathematical notation for the product $1 \times 2 \times \cdots \times k$ is $k!$). For the symmetric traveling salesman problem with n cities, the number of possible tours is, of course, equal to $\frac{1}{2}(n-1)!$. In the nonsymmetric case, for $n = 15$, this is approximately 87×10^9, and for $n = 25$, it is approximately 620×10^{21}. So, the number of possible tours shoots up like a rocket as the number of cities increases. One can see this using Stirling's famous approximation formula that states that

$$k! \approx \sqrt{2\pi k}\ (k/e)^k$$

when k is sufficiently large (say, $k \geq 10$), where $e \approx 2.7183$. The term $(k/e)^k$ explains the explosive rise of $k!$ as k increases. So, it is clear that it quickly becomes impossible in practice to find the optimal tour by enumeration (that is, by examining

---

[5]On http://press.princeton.edu/chapters/s8451.pdf, the first chapter of this book can be downloaded. A brilliant historical note in that chapter concerns the contest that Procter & Gamble held for the public in 1962 to find the shortest possible tour through 33 cities in the United States for representatives of the company. This competition, with a main prize of $10.000, also attracted the attention of applied mathematicians and led to the first solution methods for the traveling salesman problem, including the "branch-and-bound" principle.

all the different tours individually). Mathematicians have done extensive research into fast algorithms for the traveling salesman problem. However, it is almost certain that a "fast" algorithm that always leads to the shortest tour does not exist.[6] This means that in practice, one often uses calculation methods that lead to a good but not necessarily optimal tour.

Figure 2.8 shows a huge traveling salesman problem for which the optimal tour was found in 1998. This figure shows the shortest tour through 13509 cities in the United States. An even larger traveling salesman problem that was solved exactly in 2004 concerns the shortest tour through all 24978 cities in Sweden. The largest traveling salesman problem solved by 2006 concerns 85900 cities. One might wonder about the practical use of solving traveling salesman problems with an ever increasing number of cities. The answer is that the associated challenge often leads to new ideas and methods that can also be applied to entirely different real-world large-scale optimization problems.

Fig. 2.8   A shortest tour through 13509 cities in the US.

Two essentially different methods are available for solving the traveling salesman problem.

---

[6]In theoretical sense, in the literature, one speaks of a "good" algorithm if for constants c and K, the number of calculations does not increase more quickly than Kn^c as the problem size n increases. For many combinatorial optimization problems, including the traveling salesman problem, there is almost certainly no "good" algorithm in this theoretical sense. However, this does not prevent the development of practical algorithms that work satisfactorily "on average."

1. Use a sophisticated search technique that drastically reduces the computation time of an enumeration and guarantees an optimal solution. An example of such a sophisticated search technique is the branch-and-bound approach. However, the computation time increases rapidly when the problem becomes large.
2. Use a heuristic to produce a "good" suboptimal solution. Heuristic methods do not guarantee optimal solutions, but the emphasis is on finding good solutions without using too much computation time.

Below, we discuss a branch-and-bound method and some heuristics for the traveling salesman problem. The heuristics are generally applicable, except for the greatest angle insertion heuristic, which can only be used for the *Euclidean* traveling salesman problem. A traveling salesman problem is called Euclidean if the distance c_{ij} between any two cities i and j is measured as the length of the line segment between the two points in the Euclidean plane.

Little's Algorithm

A simple branch-and-bound algorithm for the traveling salesman problem is Little's algorithm. The reason for developing this algorithm was a contest held at the end of the 1950s by a large American soap company to find the shortest tours in a number of states for its representatives. Many university researchers focused on this problem, and so the soap company obtained good solutions in a relatively inexpensive way. We illustrate the broad lines of Little's branch-and-bound algorithm using an example with $n = 5$ cities and the following distance matrix $C = (c_{ij})$:

$$
\begin{array}{c c c c c c}
 & 1 & 2 & 3 & 4 & 5 \\
1 & - & 11 & 15 & 28 & 18 \\
2 & 11 & - & 7 & 18 & 10 \\
3 & 15 & 7 & - & 21 & 16 \\
4 & 28 & 18 & 21 & - & 11 \\
5 & 18 & 10 & 16 & 11 & -
\end{array}
$$

The traveling salesman problem is a minimization problem. This means that for every subset of feasible solutions, a lower bound is calculated and not an upper bound. The length of each tour in the subset is then greater than or equal to the calculated lower bound. Moreover, in the elimination process of the branch-and-bound method, an upper bound is needed. The upper bound is the length of the best tour found so far. The initial value of the upper bound is usually determined by a tour that is determined using a heuristic method. What does the calculation of the lower bound look like for a given subset of tours? Let us explain this calculation for the set of all tours. How to calculate the lower bound for a subset then immediately follows. Take the set of all tours. Every city, for example city 3, is exited to go to another city. If the travel distance from city 3 to every other city is reduced by the same amount, then the length of every tour is reduced by that same amount. Every city, for example city 5, is entered via another city. If the travel distance from every other city to city 5 is reduced by the same amount, then the length of every tour is reduced by that same amount. The idea should now be clear; choose a city i and reduce the travel distance from city i to every other city by the smallest number in the ith row of the distance matrix. This means that from city i, the travel distances are reduced to

$$c'_{ij} = c_{ij} - e_i \quad \text{for } j = 1,\ldots,n$$

with $e_i = \min_{j \neq i} c_{ij}$. Do this for every city i. In a tour, every city i is exited exactly once. This means that the length of the tour subject to the original distances c_{ij} is equal to the length of the tour subject to the reduced distances c'_{ij} plus $\sum_{i=1}^{n} e_i$. Next, choose a city j and reduce the travel distance from every other city to city j by the smallest number in the jth column of the reduced distance matrix. This means that the travel distances to city j are reduced to

$$c''_{ij} = c'_{ij} - f_j \quad \text{for } i = 1,\ldots,n$$

with $f_j = \min_{i \neq j} c'_{ij}$. In a tour, every city j is entered exactly once. This means that the length of the tour subject to the distance matrix (c'_{ij}) is the length of the tour subject to the distance matrix c''_{ij} plus $\sum_{j=1}^{n} f_j$. It follows that the length of a tour subject to the original distance matrix (c_{ij}) is equal to the length of the tour subject to the reduced distance matrix c''_{ij} plus $\sum_{i=1}^{n} e_i + \sum_{j=1}^{n} f_j$. Based on the choices of the e_i and f_j, all c''_{ij} are nonnegative, so that the length of a tour subject to the distance matrix (c''_{ij}) is nonnegative. This means that $\sum_{i=1}^{n} e_i + \sum_{j=1}^{n} f_j$ gives a lower bound for the length of every tour in the set of all tours.

We apply the reduction method described above to the example. If we subtract 11, 7, 7, 11, and 10, respectively, from rows 1, 2, 3, 4, and 5, and then subtract 4 and 1, respectively, from columns 1 and 4, we obtain the following reduced distance matrix $C'' = (c''_{ij})$:

	1	2	3	4	5
1	–	0	4	16	7
2	0	–	0	10	3
3	4	0	–	13	9
4	13	7	10	–	0
5	4	0	6	0	–

Since a total of 51 has been subtracted and all elements of the reduced distance matrix are nonnegative, the length of every tour must be at least 51. So the number 51 is a lower bound for the lengths of the tours in the set of all tours.

How does one split up the set of all possible tours? The set is split up into two subsets. The first one contains all tours including a specific edge (that is, segment of the tour) (s,t), and the second subset contains all tours without that edge. The edge (s,t) between the cities s and t is chosen as the basis for the following reasoning. Let (s,t) be a pair of cities with reduced distance $c''_{st} = 0$. If the edge (s,t) is excluded, then someone leaving city s must go to a city j with $j \neq t$, and city t

can only be entered from a city i with $i \neq s$. This results in an "additional" travel
distance of at least

$$p_{st} = \min_{j \neq t} c''_{sj} + \min_{i \neq s} c''_{it}.$$

A good choice is to branch on the edge (s, t) for which the "penalty" p_{st} is the
greatest. The underlying idea is that a large value of p_{st} is an indication that (s, t)
could be a crucial edge. In the example, we find $p_{12} = p_{21} = 4$, $p_{23} = p_{32} = 4$,
$p_{52} = 0$, $p_{45} = 10$, and $p_{54} = 10$, and therefore branch on the edge $(4, 5)$. For
the subset of tours where the edge $(4, 5)$ is excluded, the reduced distance matrix is
adjusted by setting c''_{45} equal to ∞. For the subset of tours that do contain the edge
$(4, 5)$, the reduced distance matrix can be reduced by deleting row 4 and column 5.
Then, c''_{54} is set to ∞ so that the partial tour $4 \to 5 \to 4$ does not appear. After
the branching, a new lower bound for the next subset to be studied is calculated
with the previously described reduction method. The exclusion of partial tours is
somewhat more subtle after the first step.

Next, we discuss different heuristics for the traveling salesman problem. These
heuristics are easy to use and require relatively little computation time.

Nearest Neighbor Heuristic

This heuristic is a "short-sighted" procedure that constructs a tour step by step.
Once the starting point is chosen, the city nearest to the initial city is the next one
on the route. Similarly, at each step, the route is expanded by adding the city that
is nearest to the most recently added one. It should not come as a surprise that this
heuristic can lead to somewhat unsatisfactory results. In particular, the last edge in
the tour can be very long because of the "short-sightedness" at the beginning. Note
that the tour constructed by the heuristic depends on the chosen starting point.
Different starting points will generally result in different tours. Choosing the best
of those tours for a number of randomly chosen starting points often leads to a good
tour.

Greatest Angle Insertion Heuristic

This heuristic supposes that the traveling salesman problem is Euclidean. The
Euclidean distance between two points (x_1, y_1) and (x_2, y_2) in the Euclidean plane
is given by

$$\sqrt{(x_1 - x_2)^2 + (y_1 - y_2)^2}.$$

The greatest angle insertion algorithm specifically uses the special geometric properties of the Euclidean traveling salesman problem. If the cities do not lie on one
line, it can be proved that an optimal tour does not intersect itself. This is the
reason why the first step of the heuristic looks as follows:

Step 1. First, construct the convex hull of the cities, which is the smallest polygon such that every city lies either on or inside the polygon; see Figure 2.9.[7]

The convex hull gives the first partial tour, which is then expanded step by step. The expansion step is based on a geometric argument. Suppose that only one more city can be added to the current partial tour. Which city should that be? The answer is that it should, of course, be the city that is nearest one of the edges in the current tour. Then, bear in mind that the nearer a point lies to an edge, the greater the angle that point makes with the two endpoints of that edge. This leads to Step 2 of the heuristic.

Step 2. For a given partial tour, measure the angles between every interior point and every pair of directly connected points on the current partial tour. The interior point with the greatest angle is inserted between the two relevant points of the partial tour; cf. Figure 2.9. Repeat Step 2 for the partial tour until all cities have been inserted.

Fig. 2.9 The convex hull and an insertion.

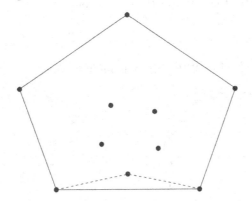

The greatest angle insertion heuristic usually gives a surprisingly good result. It is an excellent example of a heuristic that is intuitively appealing, easy to apply, and gives a good solution in a relatively short computation time.

Insertion Heuristic

The previous heuristic is an example of an insertion heuristic. Another insertion heuristic is the so-called farthest insertion heuristic. Start with a partial tour (in the Euclidean traveling salesman problem, take the convex hull of all of the cities).

---

[7]An algorithm for determining the convex hull is given in the paper R. L. Graham, An efficient algorithm for determining the convex hull of a finite planar set, *Information Processing Letters*, 1 (1972), 131–132.

In every step, add the city that has the greatest minimum distance to the current partial tour. The idea behind this heuristic is to fix the broad lines of the tour right from the start.

We conclude by discussing a so-called exchange heuristic, which can be used to construct an improved tour from a given one. The idea that underlies this heuristic can be applied in general.

Exchange Heuristic (2-Opt Heuristic)

This heuristic removes a number of edges from a given tour and then combines the resulting fragments into a new tour by adding other edges. Of course, the aim is to construct a new tour that is shorter than the old one. We will elaborate on this idea in the case where two edges are broken open (the *2-opt heuristic*). Suppose that for a given tour, we break open the two edges (r, s) and (v, w); see the first part of Figure 2.10. There is only one way to combine the fragments into a new tour. To this end, the new edges (r, v) and (s, w) are made, and the direction of travel in the section between s and v is reversed; see the second part of Figure 2.10. Assuming that the distances between the cities are symmetrical, it follows that the length of the new tour is shorter than that of the old one only if

$$c_{rv} + c_{sw} < c_{rs} + c_{vw}.$$

Fig. 2.10   Exchanging two edges.

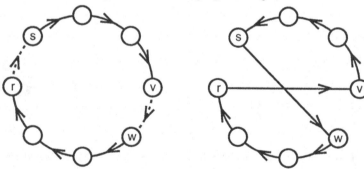

Roughly speaking, the exchange heuristic works as follows:

Step 1. Choose two edges (r, s) and (v, w) in the current tour.
Step 2. If $c_{rv} + c_{sw} < c_{rs} + c_{vw}$, construct a new tour as in Figure 2.10.

Steps 1 and 2 are repeated until all pairs of edges have the property that exchanging does not lead to an improvement. The exchange heuristic will generally stop at a local optimum. The computation time of the heuristic is of order n^2 if n is the number of cities. After all, in a tour that consists of n edges, there are $\binom{n}{2} = \frac{1}{2}n(n-1)$ ways to choose two edges.

An extension of the 2-opt heuristic is of course the 3-opt heuristic, where three edges can be replaced. By far the best exchange heuristic is that of Lin and Kernighan; see their paper An effective heuristic approach for the traveling salesman problem, *Operations Research*, 21 (1973), 498–516. This heuristic also tries to improve a given tour. However, at each iteration, the heuristic replaces not a fixed number of edges, but rather a variable number. We will not go into the details of this heuristic, which can be applied to a broad class of combinatorial optimization problems. The Lin–Kernighan heuristic gives astonishingly good results, even for problems with (tens of) thousands of cities, when it is applied to a number of randomly chosen tours (for example, 100 initial tours) and the best of the improved tours is taken. This approach often also works very well when the simple 2-opt heuristic is used instead of the rather complicated Lin–Kernighan heuristic. In general, the strength of an exchange heuristic, of whatever problem, is that it can be applied repeatedly to different initial solutions, allowing a choice between different "good" solutions.

In the next section, we discuss the famous Clarke and Wright heuristic for routing problems.

2.9   A Routing Problem

Suppose that n customers $i = 1, \ldots, n$ must be supplied with stock from a central depot. Sufficiently many trucks are present at the depot for this. Each customer may be supplied by one truck only. The number of units to be delivered to customer i is a given quantity q_i. Every truck has a limited capacity. For the sake of convenience, we assume that the trucks have the same capacity, namely Q units. Further, the travel times between the customers and the depot and between the customers themselves are given, where it is assumed that for any two points i and j, the travel time from i to j is equal to the travel time from j to i. Define

$$c_{0i} = \text{travel time between the depot and customer } i,$$
$$c_{ij} = \text{travel time between customer } i \text{ and customer } j.$$

An additional requirement is that the total travel time of a truck should not exceed a given value T. Every truck begins and ends the route at the depot.

The question is how many trucks must be used and what the route should be for each of the trucks to minimize the sum of the travel times of these trucks.

The above problem, which is in fact a traveling salesman problem with multiple salesmen, occurs in real-world situations in many different forms. In those situations, the scale of the problem is almost always such that an exact solution method is not appropriate because the computation time becomes too high. Therefore, in such cases, a heuristic approach is sought that leads to a best-possible solution. A commonly used heuristic is Clarke and Wright's savings heuristic. This heuristic is not only easy to apply but also flexible enough to take into account all kinds of other restrictions that may occur in practice; for example, that certain customers may not be supplied by the same truck or that a particular customer can only be supplied during a limited number of hours.

Clarke and Wright's Heuristic

The basic idea of the heuristic is straightforward; we explain it using the situation shown in Figure 2.11. In this figure, two routes are given, where one truck supplies customers A, B, and C with goods and the other truck supplies customers D, E, F, and G.

Fig. 2.11   Two routes.

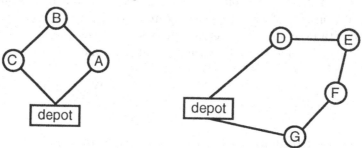

Now, on each of the two routes, consider a customer who is supplied *directly* from the depot, for example the customers A and D. Then the routes can easily be merged to one route by connecting customers A and D; see Figure 2.12. In this new route, the same truck supplies customers C, B, A, D, E, F, and G. Merging to one route is, of course, only possible if the truck has sufficient capacity to supply all customers on the route and the travel time restriction on the truck is not violated.

Fig. 2.12   A combined route.

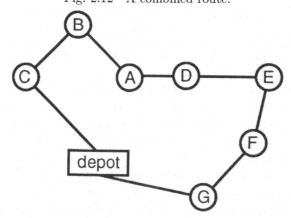

If merging to one route is possible, then in comparison to the original two routes, the savings in travel time are equal to

$$c_{0A} + c_{0D} - c_{AD}.$$

Of course, it only makes sense to actually merge to one route if the savings are greater than or equal to zero. This describes the principle of the heuristic: try to obtain savings in travel time by merging routes via customers who are supplied directly from the depot on the routes in question.

Clarke and Wright's savings heuristic looks as follows:

Step 1. Calculate the savings

$$s_{ij} = c_{0i} + c_{0j} - c_{ij}, \quad i, j = 1, \ldots, n.$$

Consider only the s_{ij} that are greater than or equal to zero, and order these by decreasing size on a list of possible savings.

Step 2. As an initial solution, assign an individual truck to every customer.

Step 3. Take the first number s_{ij} on the savings list, and check whether the customers i and j lie on different routes and are supplied directly from the depot. If this is the case, merge the two routes on which i and j are located to one route provided that the capacity of the truck allows this and the travel time constraint is not violated. Repeat Step 3 for the next number on the savings list until the list has been exhausted.

In real-world situations, this heuristic often provides very satisfactory solutions. The obtained solution can possibly be improved further by optimizing the order of the customers to be supplied by each of the individual routes using an exchange heuristic as described in Section 2.7.

Numerical Example

Suppose that we have $n = 6$ customers, where for each customer i, the number q_i of units to be delivered is equal to 1. Every truck has a capacity of $Q = 4$ units and is subject to a travel time constraint of $T = 72$. The travel times c_{0i} and c_{ij} are given in Table 2.4.

Table 2.4   Travel times.

	1	2	3	4	5	6
0	11	21	26	25	21	11
1		12	21	22	25	22
2			11	12	23	33
3				10	9	26
4					14	20
5						14

We first determine the savings list:

$$s_{34} = 41, \ s_{35} = 38, \ s_{23} = 36, \ s_{24} = 34, \ s_{45} = 32, \ s_{12} = 20,$$
$$s_{25} = 19, \ s_{56} = 18, \ s_{13} = 16, \ s_{46} = 16, \ s_{14} = 14, \ s_{36} = 11,$$
$$s_{15} = 7, \ s_{16} = 0.$$

Fig. 2.13   The iteration steps.

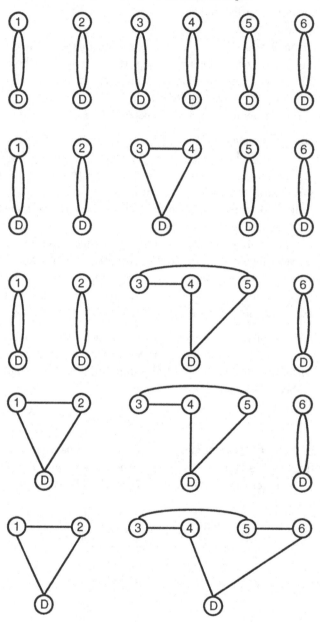

We then apply the iterative heuristic. The initial routes are shown in the first part of Figure 2.13. The first savings value is s_{34}. The two routes on which customers 3 and 4 are located can be merged. This gives the second part of Figure 2.13. The next savings value is s_{35}. The two routes on which customers 3 and 5 are located can be merged. The new situation is shown in the third part of Figure 2.13. The next savings value s_{23} does not lead to a merging of routes because customer 3 is not supplied directly from the depot. The savings value s_{24} also does not lead to a merging of routes because merging would lead to a violation of the travel time constraint. The savings value s_{45} does not change the situation either. Next, the savings value s_{12} does lead to a merging of routes, whose result is given in the fourth part of Figure 2.13.

The savings value s_{25} does not lead to a merging of routes because merging would lead to a violation of both the travel time constraint and the capacity constraint. The two routes on which customers 5 and 6 are located can be merged. The new situation is shown in the fifth part of Figure 2.13. The savings values s_{13} and s_{46} do not change the situation. The savings value s_{14} does not lead to a merging of routes because of both the travel time constraint and the capacity constraint. The savings values s_{36} and s_{15} also leave the situation unchanged. Finally, the savings value s_{16} does not lead to a merging of routes because of both the travel time constraint and the capacity constraint.

Clarke and Wright's heuristic therefore gives a solution that requires two trucks to drive the route depot – customer 1 – customer 2 – depot and the route depot – customer 6 – customer 5 – customer 3 – customer 4 – depot. The total travel time is $44 + 69 = 113$.

2.10   Exercises

2.1 During the Easter weekend, ten soccer matches that can be regarded as risky are scheduled in the West Friesland region. The mayors of the various municipalities in the region therefore have major objections. An analysis of the objections gives the following conclusions:

- Match 7 cannot be played if match 3 is played.
- If match 4 is not played, then match 5 must take place.
- If both match 2 and match 5 are played, then match 8 cannot take place.
- Playing match 8 means that no more than one of the four matches 1, 7, 9 and 10 can take place.
- If match 7 takes place, then neither match 4 nor match 6 can be played.

Which matches should take place in order to maximize the number of matches that are played? Formulate this as an ILP problem.

2.2 Seven different types of cargo are ready to be loaded onto an aircraft. The aircraft has weight and volume capacities of 25 tons and 4000 cubic meters, respectively. The table below gives an overview of the weight, volume, and profit per unit for each of the seven types of cargo. For each type of cargo, an unlimited number of units is available to be transported.

- **(a)** How should the aircraft be loaded in order to maximize the total profit on the freight?
- **(b)** Adjust the integer formulation for item (a) to include the additional constraint that no more than two different types of cargo can be loaded.

128   *Operations Research: Introduction to Models and Methods*

Cargo	1	2	3	4	5	6	7
Weight (Tons)	1.8	2.8	3.0	3.6	3.8	4.6	5.0
Volume (Cubic Meters)	250	300	500	400	550	800	750
Profit (Euros)	310	400	500	500	650	800	850

2.3 Four trucks are available to deliver goods to seven customers. The weight capacities of the four trucks 1, 2, 3, and 4 are 9, 12, 15, and 11 tons, respectively. The respective weights of the goods for customers 1 through 7 are 6, 3, 5, 2, 4, 5, and 3 tons. Each customer can be supplied by one truck only, but the same truck can supply several customers. In addition, the truck that supplies a customer may not spread this out over multiple trips. The operational costs of trucks 1, 2, 3, and 4 are 125, 150, 170, and 135 euros, respectively. What supply schedule minimizes the total cost?

2.4 After the performance of the satirical play "Yup and Yul" by the theater group Sater at the municipal theater, violent unrest broke out on the Amsterdam canals. The unrest is concentrated at the house of the cabaret artist Youp de Hacker on the Prinsengracht, at the university institute on the Keizersgracht where the cultural sociologist Bram Zwetsloot works, and at writers' café De Rode Neus on the Spui. The officers present at these three locations have asked the police department for backup: 7 extra police officers for the Prinsengracht, 4 extra for the Keizersgracht, and 12 extra for the Spui. Police chief Baantjer must immediately decide how to send reinforcement from the offices on the Marnixstraat, Nieuwe Zijds, and Warmoesstraat, where 10, 5, and 8 police officers, respectively, are still available. The chief decides to send 3 officers from the Marnixstraat and 4 from the Warmoesstraat to the Prinsengracht, 2 officers from the Marnixstraat and 2 from the Warmoesstraat to the Keizersgracht, and 5 officers from the Marnixstraat, 5 from the Nieuwe Zijds, and 2 from the Warmoesstraat to the Spui. The following day, the police chief's disastrous decision is evaluated at the city hall. The chief defends himself by saying that, in accordance with the rules in force, he opted for a distribution that minimizes the total time needed to get the officers in place (that is, the sum of all officers' travel times). The table below shows the travel times in minutes from the various police stations to the places of unrest (the travel time does not depend on the number of officers sent):

	Prinsengracht	Keizersgracht	Spui
Marnixstraat	2	3	4
Nieuwe Zijds	3	4	2
Warmoesstraat	5	3	3

(a) Draw up an integer programming problem the mayor can use to calculate whether chief Baantjer's solution is optimal subject to the constraints mentioned above.

(b) The police chief challenges his dismissal in court. His lawyer argues that, under the stress of the evaluation at the town hall, the commissioner made a mistake in stating that the goal was to minimize the sum of the travel times, whereas the goal was to minimize the earliest moment when all officers were on the scene. Indeed, this second criterion appears to be stated (in more formal language) in the municipal regulations. Verify for the lawyer whether the chief did act optimally subject to this criterion.

2.5 Every day, the manager of a power plant must decide which generators to use. The power plant has two different electricity generators. The required amount of electricity fluctuates during the day, which is divided into six periods. The number of megawatts requested is equal to 2000 in period 1, 4000 in period 2, 6000 in period 3, 9000 in period 4, 7000 in period 5, and 3000 in period 6. At the beginning of each period, generators can be switched on or off. A generator that is not used in the coming period will be switched off at the beginning of that period. The maximum capacities of generators 1 and 2 are 4000 mW and 6000 mW, respectively, for each period. The following costs are incurred. Every time generator j is started up after it has been turned off, a setup cost of C_j euros is made, where $C_1 = 500$ and $C_2 = 300$. In addition, there is an operational cost of a_j euros for each period that generator j is used, where $a_1 = 1000$ and $a_2 = 1100$. Both generators are turned off for testing at the end of the day. Use integer programming to determine for what schedule the total daily cost is the lowest.

2.6 A pharmaceutical company wants to launch a new product within the next 12 months. This process requires four consecutive steps, each of which can be performed at a different speed. The necessary time in months and the costs in thousands of euros for the different steps are shown in the table below, with the costs in parentheses. What is the least expensive way to get the product released within 12 months?

Execution	Theoretical Research	Laboratory Experiments	Government Approval	Marketing
Slow	5 (40)	3 (50)	5 (7)	5 (65)
Normal	4 (56)	2 (65)	3 (7)	4 (80)
Fast	2 (75)	1 (95)	2 (15)	3 (120)

2.7 Suppose that we want to give someone a hundred Dutch guilders and only have access to quarters, guilders, and rijksdaalders (Dutch coins from before the introduction of the euro: a quarter is $\frac{1}{4}$ of a guilder and a rijksdaalder is $2\frac{1}{2}$ guilders). How should we do this if we want to give, as much as possible, the same number of quarters, guilders, and rijksdaalders? Formulate an ILP model and solve it. What will the solution be if we add the constraint that we may not use more than 25 of one of the three types of coins?

2.8 Twenty gold coins with the respective values $\sqrt{1}, \sqrt{2}, \ldots, \sqrt{20}$ are given. How can these coins be split up as fairly as possible between two persons?

2.9 Consider the production-stock problem from Section 2.2.6. Solve this problem again with the additional constraint that no more than two rounds of production are allowed and see how the optimal solution changes.

2.10 Consider the production-stock problem from Section 2.2.6 once more. Adapt the ILP formulation to each of the following situations and then solve the problem again:

    (a) If the decision is made to produce in September, then at least 200 tons must be produced that month.

    (b) For a production run, the variable production cost per ton is equal to 18 euros per ton for the first 200 tons that are produced and 10 euros per ton for each additional ton.

    (c) The variable production cost per ton is equal to 18 euros per ton if fewer than 100 tons are produced and is 14 euros per ton if the production size is 100 tons or more.

2.11 A bus company needs the following numbers of drivers every week:

Mon	Tue	Wed	Thu	Fri	Sat	Sun
21	20	24	17	21	18	14

Each bus driver works five consecutive days a week and can, in addition, be asked to work one day overtime. The usual salary is 100 euros for each weekday and 125 euros for both Saturdays and Sundays. The salary for overtime work is 40% higher than the regular salary. Calculate an optimal work schedule.

2.12 A large shopping center employs 30 guards. Each guard works 5 days a week. The guard duties vary according to the day of the week. The number of guards needed is 18 on Monday, 24 on Tuesday, 25 on Wednesday, 20 on Thursday, 20 on Friday, 28 on Saturday, and 15 on Sunday. Determine a duty roster that minimizes the number of guards that are not free on two consecutive days.

2.13 A refinery runs 24 hours a day, where a_i operators are needed from hour i to hour $i+1$ for $i = 0, 1, \ldots, 23$ (hour 0 is at midnight). Every 4 hours exactly, from midnight on, a new team of operators can start work. Every team of operators must take a 1-hour break after working 4 or 5 hours. Every team works 8 hours (excluding the hour of the break). For each operator from a team starting at hour i ($= 0, 4, 8, 12, 16, 20$), the cost is equal to c_i for the shift in question. What work schedule has minimum total cost? Solve this problem using ILP with the data $c_i = 1$ for all i, $a_0 = 12$, $a_1 = 15$, $a_2 = 11$, $a_3 = 17$, $a_4 = 17$, $a_5 = 18$, $a_6 = 22$, $a_7 = 24$, $a_8 = 19$, $a_9 = 21$, $a_{10} = 24$, $a_{11} = 27$, $a_{12} = 28$, $a_{13} = 31$, $a_{14} = 36$, $a_{15} = 29$, $a_{16} = 28$, $a_{17} = 25$, $a_{18} = 23$, $a_{19} = 20$, $a_{20} = 20$, $a_{21} = 20$, $a_{22} = 16$, $a_{23} = 12$. Then, solve the problem subject to the additional constraint that the percentage of operators that take a break after 5 hours is not more than 15%.

2.14 A contractor must decide which of the projects she has been offered she should carry out herself in the spring of next year and which she should outsource to subcontractors. She has been offered five projects numbered $1, \ldots, 5$, and these projects require 855, 1340, 900, 1260, and 1070 hours of labor, respectively. In all, the contractor can do 4320 hours of labor. The subcontractors can provide sufficiently many hours of labor. The contractor makes a profit of 11 000 euros, 22 000 euros, 28 600 euros, 19 800 euros, and 33 000 euros, respectively when carrying out the projects herself, and respective profits of 2200 euros, 4950 euros, 8800 euros, 9900 euros, and 6600 euros when outsourcing. Use the branch-and-bound algorithm to calculate which projects the contractor should carry out herself and which she should outsource to maximize the total profit.

2.15 Consider the cutting stock problem from Section 1.3.3 in Chapter 1. Suppose that there is an additional constraint that the length of every roll that is supplied is at least 75 m. Solve the problem again as an integer programming problem.

2.16 Consider again the location problem from Section 2.2.2. Now, assume that no more than two fire stations may be built. In which cities must these be located to minimize the greatest value of the travel time to a city from the nearest fire station?

2.17 A company can make three different products with one and the same raw material. Each unit of product j needs a_j units of raw material and delivers a net profit of p_j euros, where $a_1 = 1.5$, $a_2 = 1.25$, $a_3 = 2$, $p_1 = 30$, $p_2 = 25$, and $p_3 = 35$. The available quantity of raw material is 1500 units. If the company decides to make product 1, it must make at least 200 units of this product. Furthermore, only one of the two available machines A and B may be used. Products 1, 2, and 3 require 0.23, 0.20, and 0.27 hours per unit, respectively, for machine A and 0.30, 0.25, and 0.28 hours per unit for machine B. The respective capacities of machines A and B are 200 and 250 hours. Determine an optimal product mix with integer programming.

2.18 Use the branch-and-bound algorithm to determine an optimal solution to the following three integer problems:

(a) Maximize $8x_1 + 11x_2 + 6x_3 + 4x_4 + 2.5x_5$
subject to $10x_1 + 14x_2 + 8x_3 + 6x_4 + 4x_5 \leq 28$
and $\quad x_1, x_2, x_3, x_4, x_5 \in \{0, 1\}$.

(b) Minimize $3.5x_1 + 6x_2 + 2.5x_3 + 7x_4$
subject to $6x_1 + 12x_2 + 10x_3 + 15x_4 \geq 14$
and $\quad x_1, x_2, x_3, x_4 \in \{0, 1\}$.

(c) Maximize $21x_1 + 11x_2$
subject to $7x_1 + 4x_2 \leq 13$,
and $\quad x_1, x_2 \geq 0$ and integer.

Begin the calculations with the lower bound LB $= 0$.

2.19 Develop and apply an exchange procedure to improve the heuristic solution from the constructive heuristic in Section 2.7.1.

2.20 Consider the traveling salesman problem with the seven cities $A = (10, 23)$, $B = (0, 13)$, $C = (4, 0)$, $D = (21, 2)$, $E = (10, 10)$, $F = (11, 6)$, and $G = (13, 4)$ and Euclidean distances between the cities. Draw the seven cities in a coordinate plane on graph paper, and use the greatest angle insertion heuristic to determine a tour. Then, try to improve this tour using the 2-opt exchange heuristic. *Remark*: The length of the shortest tour is 73.07.

2.21 (a) Suppose that a company has to make n different types of paints on a machine. The machine can make only one type of paint at a time. A switching cost of C_{ij} is incurred when paint type j is produced directly after paint type i. In what order should the paints be made to minimize the total switching costs? Formulate this as a traveling salesman problem. *Hint*: Introduce an auxiliary city for the beginning and end of the production process.

(b) Suppose that a company wants to use four vans to supply customers in cities $1, \ldots, n$ from a central depot, where every city can be visited exactly once and every van must return to the depot. The travel distances between the depot and the cities and between the cities themselves are given. How can the company reduce the problem of determining the four routes with minimum total travel distance to the standard traveling salesman problem?

2.22 A shipping company has been instructed to deliver goods from a central depot O to customers in seven cities A through G. The quantities to be delivered and the travel times are given in the table below. The company's trucks each have a capacity of 45 units. There are no restrictions on travel time. Use Clarke and Wright's heuristic to determine a delivery schedule.

From/To	A	B	C	D	E	F	G	Delivery
O	52	56	47	30	56	12	63	
A		60	94	38	22	62	87	9
B			69	32	44	60	34	12
C				62	92	38	52	18
D					32	38	52	6
E						65	74	3
F							62	12
G								6

2.23 A gasoline company must temporarily store three different types of fuel. Three tanks are available on site for this purpose. Only one type of fuel can be stored at the same time in each tank. Fuel that cannot be stored on site will be stored elsewhere. The table below shows the quantities of fuel and the capacities of the tanks (in tons). The table also shows the storage costs per ton of fuel. For storage elsewhere, the storage costs are 330, 360, and 330 euros per ton for fuel 1, fuel 2, and fuel 3, respectively. The aim is to store the fuel at the lowest possible cost. Develop a heuristic for this problem. Apply it and compare the solution to the optimal solution.

	Tank 1	Tank 2	Tank 3	Quantity
Fuel 1	270 euros	270 euros	225 euros	14
Fuel 2	300 euros	210 euros	285 euros	2
Fuel 3	225 euros	195 euros	300 euros	16
Capacity	6	14	8	

2.24 Members for an important government committee must be chosen from a group of 10 persons. The members must represent seven policy areas. The sets of representatives for policy areas 1, 2, 3, 4, 5, 6, and 7 are $\{1,5,6,7\}$, $\{6,7,8,9,10\}$, $\{6,7,8,9,10\}$, $\{1,2,3,4,5\}$, $\{2,3,8,9,10\}$, $\{3,4,6,7,9\}$, and $\{1,2,8,10\}$, respectively. The aim is to set up the committee in such a way that it is as small as possible and that every policy area is represented. Calculate a heuristic solution and compare this to the optimal solution.

2.25 Apply the constructive heuristic from Section 2.7.4 to the following numerical example with $n = 7$ and $p = 3$:

	$j = 1$	2	3	4	5	6	7
$i = 1$	750	1806	4644	2353	2222	4962	3744
2	6402	690	10902	7158	9090	10626	6534
3	3022	1454	798	5214	7374	4654	4382
4	3750	7440	8910	750	3500	9440	6990
5	3505	3799	7587	2079	599	8127	6159
6	3175	739	3861	5096	7151	907	2573
7	2130	1496	872	1966	2018	1816	678

References

1. G. L. Nemhauser and L. A. Wolsey, *Integer and Combinatorial Optimization*, John Wiley & Sons, Inc., Hoboken, New Jersey, 1988.
2. L. A. Wolsey, *Integer Programming*, John Wiley & Sons, Inc., Hoboken, New Jersey, 1988.

Chapter 3

Network Analysis

Many real-world problems in various fields can be formulated as optimization problems on networks. A network consists of nodes and edges that connect two nodes of the network to each other. A network is called *directed* if all edges can only be traversed in one direction. In an undirected network, every edge may be traversed in either direction. An undirected network can always be reduced to a directed network (why?). Network optimization problems can generally be solved with (continuous) linear programming. However, using the special structure of such problems, one can develop algorithms that reach the optimal solution much more quickly than linear programming algorithms. In this chapter, we discuss some of the most important models from network theory. In Section 3.1, we consider the shortest-path problem. We discuss Dijkstra's algorithm for finding the shortest path between nodes in a network with nonnegative distances. Section 3.2 deals with the maximum-flow problem, in which the maximum amount of flow (for example, liquid or traffic) that can go from one node in the network to another node is sought, taking into account capacity restrictions on the edges between the nodes. In Section 3.3, we discuss the minimum spanning tree problem. We discuss Prim's algorithm, which can be used to calculate how to connect a given number of nodes by a spanning tree at the lowest cost. We also briefly touch upon the related Steiner tree problem. Section 3.4 deals with the Euler circuit problem, where in an undirected graph, a path is sought that passes through every edge of the graph exactly once and ends at the starting point of the path. We then discuss the Chinese postman problem, in which costs are attached to the edges of the graph and a circuit is sought with the lowest total cost. Finally, in Section 3.5, we briefly discuss the general minimum-cost flow model that can be used for a large number of network optimization problems; in particular, we mention the assignment problem and the associated Hungarian solution method.

3.1   Shortest-Path Problems

Shortest-path problems are at the heart of network problems. They occur frequently in real-world situations. In various applications, the aim is to send material as quickly as possible, as cheaply as possible, or as reliably as possible between two points in a network. Moreover, solving a shortest-path problem is often part of the solution process for more complex network problems. Shortest-path problems can be solved particularly efficiently, even problems with hundreds of thousands or even millions of nodes.

We assume given a directed network with nonnegative distances (costs), as shown in Figure 3.1. This network consists of the nodes $N_1, \ldots, N_7$. Certain pairs of these nodes are connected by directed edges (called *arcs*). The number associated with an arc (N_i, N_j) represents the distance from node N_i to node N_j. Suppose that we want to find the shortest path from node N_1 to node N_7. A path is a sequence of connecting edges. Let us discuss an extremely efficient algorithm that calculates the shortest path, namely the famous Edsger Dijkstra algorithm that, among other things, forms the basis for calculating the fastest routes in navigation devices.

Fig. 3.1   Example of a shortest path.

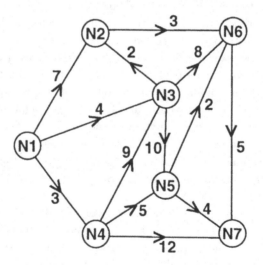

LP formulation

Define the variables

$$x_{ij} = \begin{cases} 1 \text{ if the path uses arc } (N_i, N_j), \\ 0 \text{ otherwise.} \end{cases}$$

The ILP formulation is

$$\text{Minimize} \quad 7x_{12} + 4x_{13} + 3x_{14} + 3x_{26} + 2x_{32} + 10x_{35} + 8x_{36}$$
$$+9x_{43} + 5x_{45} + 12x_{47} + 2x_{56} + 4x_{57} + 5x_{67}$$
$$\text{subject to} \quad x_{12} + x_{13} + x_{14} = 1$$
$$x_{26} - x_{12} - x_{32} = 0$$
$$x_{32} + x_{35} + x_{36} - x_{13} - x_{43} = 0$$
$$x_{43} + x_{45} + x_{47} - x_{14} = 0$$
$$x_{56} + x_{57} - x_{35} - x_{45} = 0$$
$$x_{67} - x_{26} - x_{36} - x_{56} = 0$$
$$-x_{47} - x_{57} - x_{67} = -1$$
$$\text{and} \quad x_{ij} \in \{0,1\}.$$

The constraints describe the "inflow = outflow" principle for the intermediate nodes, where "outflow − inflow" must have value +1 at the initial node N_1 and value −1 at the terminal node N_7. The coefficient matrix of the equality constraints is totally unimodular (see Section 2.1). This means that the optimal basic solution to the LP relaxation of the integer problem is automatically integer valued (furthermore, $x_{ij} \geq 0$ suffices for the relaxation $0 \leq x_{ij} \leq 1$). The optimal LP solution turns out to be $x_{14} = x_{45} = x_{57} = 1$ and $x_{ij} = 0$ for the other variables, with optimal objective value 12. The optimal path is therefore $N_1 \to N_4 \to N_5 \to N_7$.

Dijkstra's Algorithm

The shortest-path problem occurs not only as an independent problem but sometimes also as a subproblem in more complex optimizations. In some situations, large-scale shortest-path problems with hundreds of thousands or even millions of nodes must be solved, while in other situations, a shortest-path problem has to be solved a great number of times in a row. No matter how good LP codes are, in these situations, the computation time becomes too high if special codes for the shortest-path problem are not used. An efficient algorithm to determine the shortest path from a given initial node (say, N_1) to another node is Dijkstra's algorithm. This algorithm requires that *every* arc have *nonnegative* length and not only calculates the shortest path from N_1 to a given endpoint but at the same time also gives the shortest path from N_1 to every other node. Dijkstra's algorithm is an iterative procedure, in which "temporary" paths are improved step by step and each step gives the definite shortest path from N_1 to one of the nodes. The algorithm uses a smart labeling procedure. A node's label gives an upper bound for the length of the shortest path from node N_1 to the node in question, namely the length of a *temporarily* constructed path from the initial node N_1 to the node in question. The labels are adjusted iteratively. After each iteration, one of the labels becomes *permanent.* We will shortly see that if a node gets a permanent label, the (definite) shortest path to that point has been found. The algorithm can best be explained using an example. Table 3.1 shows the iterations that correspond to the shortest-path problem of Figure 3.1. In this figure, not every pair of nodes is connected by an edge. Where there is no edge between two nodes, we take a dummy edge with an extremely high length so that it will never be used in the optimal route. We denote such a length by the mathematical symbol ∞ for infinitely large.

Table 3.1   Iterations for the shortest-path problem.

	It. 1	It. 2	It. 3	It. 4	It. 5	It. 6
N_1	–	–	–	–	–	–
N_2	$(7, N_1)$	$(7, N_1)$	$(6, N_3)^*$	–	–	–
N_3	$(4, N_1)$	$(4, N_1)^*$	–	–	–	–
N_4	$(3, N_1)^*$	–	–	–	–	–
N_5	(∞, N_1)	$(3 + 5, N_4)$	$(8, N_4)$	$(8, N_4)^*$	–	–
N_6	(∞, N_1)	(∞, N_1)	$(4 + 8, N_3)$	$(6 + 3, N_2)$	$(9, N_2)^*$	–
N_7	(∞, N_1)	$(3 + 12, N_4)$	$(15, N_4)$	$(15, N_4)$	$(8 + 4, N_5)$	$(12, N_5)^*$

To explain the calculations in the table, we first note that the shortest path from node N_1 to itself is the path $N_1 \to N_1$ and therefore has length 0. After all, according to the assumption, every arc has positive length, so that every other path from N_1 to N_1 will have length greater than 0 and therefore cannot be optimal. Because the optimal path from N_1 to N_1 is known in advance, the algorithm can be initiated by giving N_1 a permanent label. Let us now explain how the first two columns of the table are created. The other columns then speak for themselves.

- For every other node N_i, the first column of the table gives an upper bound for the length of the shortest path from N_1 to N_i by taking the length of the arc (N_1, N_i) (if such an arc exists). The smallest upper bound in column 1 is 3 and determines the definite shortest path to the point corresponding to the upper bound. The smallest upper bound corresponds to node N_4 and gives the length of the shortest path from N_1 to N_4. Every other path from N_1 to N_4 has length greater than 3: such a path must go through another point, but a path from N_1 to another point than N_4 already has length greater than 3, and a subsequent path from the other point to N_4 has nonnegative length. Node N_4 is therefore given a permanent label, indicated by a star *.
- To obtain the second column, we consider all nodes N_i with a temporary label for which an arc (N_4, N_i) exists. These are N_3, N_5, and N_7. We can now calculate the length of the path that consists of the shortest path from N_1 to N_4 plus the arc (N_4, N_i). The length of this path is $3 + 9 = 12$ for $N_i = N_3$, $3 + 5 = 8$ for $N_i = N_5$, and $3 + 12 = 15$ for $N_i = N_7$. For nodes N_5 and N_7, the length of this path is less than the current upper bound for the length of the shortest path. We therefore change the upper bound for these two nodes. The second component of the labels of nodes N_5 and N_7 is changed to N_4. This second component indicates the direct predecessor of the node on the best path from N_1 to the node in question constructed so far. Node N_3 now has the smallest upper bound in column 2. This upper bound of 4 gives the length of the shortest path from N_1 to N_3.
- The fact that the shortest path from N_1 to N_3 has been found is explained as

follows. Suppose that there is another path $\mathcal{P}$ from N_1 to N_3 with length less than 4. Then this path must have a node with a temporary label as *intermediate* node (in this case, one of the nodes N_2, N_5, N_6, or N_7). The part of the path $\mathcal{P}$ from N_1 to the *first* node with a temporary label has length at least 4 because the smallest upper bound for the nodes with temporary labels is 4. This means that the total length of the path $\mathcal{P}$ is greater than or equal to 4 because the arc lengths are nonnegative. This contradicts the assumption that the path $\mathcal{P}$ has length less than 4. So the second iteration gives the shortest path from N_1 to N_3.

It should now be clear how to continue the algorithm. In the third iteration, we try to modify the labels of all nodes N_i with temporary labels that are directly accessible from node N_3 that was just given a permanent label, and so on. In each iteration, the definite shortest path is found for the point that is given a permanent label in that iteration. To prove the correctness of this claim, apply the same argument as given above for the second iteration (using the fact that on each "temporary" path, the intermediate nodes have permanent labels). The reader should verify the calculations of Table 3.1 to see whether he has understood the essence of the algorithm correctly: at each iteration, for each node with a temporary label, compare the current estimate for the shortest distance to that node to the sum of the length of the optimal path to the node that was last to get a permanent label and the direct distance from that node to the one being considered.

Each optimal path can be easily deduced from Table 3.1 because in the calculations, we have kept track of the direct predecessors of the nodes. Suppose that we want to know the optimal path from N_1 to N_7. For this, look up the iteration in the table in which node N_7 is given a permanent label. The permanent label of N_7 is $(12, N_5)$. This means that the shortest path from the initial node N_1 to N_7 is the shortest path from N_1 to N_5 plus the arc (N_5, N_7). Next, look up the permanent label of N_5, and so on. On this optimal path from N_1 to N_7, one finds that N_5 is the predecessor of node N_7, node N_4 is the predecessor of node N_5, and node N_1 is the predecessor of node N_4. The optimal path is therefore $N_1 \to N_4 \to N_5 \to N_7$, with total length 12. This is the same solution as the LP solution.

General Formulation

It is useful to give the algorithm for the general case. We consider a directed network with n nodes $N_1, \ldots N_n$, where the length of an existing arc (N_i, N_j) is given by c_{ij}. The algorithm can also be applied to an undirected network. Simply replace every edge between two nodes with two arcs in opposite directions that have the same length. We assume

$$c_{ij} \geq 0 \quad \text{for all arcs } (N_i, N_j).$$

This assumption of *nonnegative* distances is essential for Dijkstra's algorithm. Suppose that we want to determine the shortest path from a specific initial node

N_s to one or more other nodes. In the algorithm, we use a label $(\alpha_i,\ \beta_i)$ for each node N_i, where α_i is the current best estimate for the length of the shortest path from N_s to N_i and β_i indicates the *direct* predecessor of N_i in the current best path from N_s to N_i.

Algorithm 3.1 (Dijkstra's Algorithm).

Initialization. The node s is given a permanent label. Give the nodes N_i with
$i \neq s$ the temporary labels

$$\alpha_i := c_{si} \qquad \text{and} \qquad \beta_i := N_s,$$

with c_{si} a very large number if the arc $(N_s,\ N_i)$ does not exist.

Step 1. Among the nodes N_i with temporary labels, find the one for which α_i is
the smallest.[1] Suppose that this holds for node N_k. Mark the label of node N_k
as permanent, and let $p := k$.

Step 2. For every node N_i with a temporary label for which the arc (N_p, N_i) exists
and $\alpha_p + c_{pi} < \alpha_i$, adjust the labels α_i and β_i as follows:

$$\alpha_i := \alpha_p + c_{pi} \quad \text{and} \quad \beta_i := N_p.$$

Step 3. Repeat Steps 1 and 2 until all nodes have a permanent label.

Computation Time for the Algorithm

The algorithm requires the most computation time if all n nodes are connected to one another. Let us determine, for that case, the number of computations in case there is a direct connection between any two nodes. In the first iteration of the algorithm, $n - 1$ additions must be carried out, and the minimum of two numbers must be determined $2(n - 1)$ times. In the kth iteration, these numbers of operations are equal to $n - k$ and $2(n - k)$, respectively. So the algorithm described above requires a total of $\sum_{k=1}^{n}(n - k) = \frac{1}{2}n(n - 1)$ additions and $n(n - 1)$ determinations of the minimum of two numbers. In other words, the total computation time for the algorithm is proportional to n^2, where n is the number of nodes. Modifications to Dijkstra's basic algorithm have been developed for which the number of computations is much lower, so that the algorithm is much faster (for example, the so-called A^*-algorithm).

Mathematical Proof of Correctness

Above, we have used an abundance of words to argue that Dijkstra's algorithm leads to the shortest paths. In the beautiful language of mathematics, the correctness of Dijkstra's algorithm can be proved in a much shorter and more elegant way. For this, we need some mathematical notation. For notational clarity, we denote the

---

[1] If α_i is the smallest for more than one node, choose one randomly (in each iteration, only one node may receive a permanent label).

nodes by letters such as s, x, and y instead of N_s, N_i, and N_j, where s denotes the initial node. Let L_{pq}^* be the length of the shortest path from node p to node q. Denote the set of nodes with a permanent label by S: in the first step of the algorithm, we begin with S consisting of only the initial node s, and in every subsequent step, S is expanded by exactly one node. Denote by $\alpha(p)$ the most recent estimate from the algorithm for the length of the shortest path from the initial node s to node p, and by $\pi(p)$ the length of the (definite) shortest path from s to p. Armed with this notation, we can give an elegant mathematical proof. This proof is based on the mathematical method of induction. For the first step of the algorithm, the correctness of the algorithm is trivial (the shortest path from s to itself has length 0). Now, suppose that for a number of consecutive steps of the algorithm, the correctness has been proved; we then prove the correctness for the next step as follows:

Fig. 3.2   Iteration of Dijkstra's algorithm.

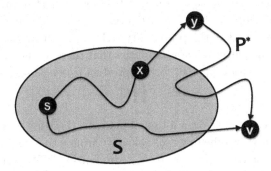

- Suppose that in this step, Dijkstra's algorithm wants to add node v to the current set S.
- Now, suppose that the shortest path P^* from s to v has a smaller length than $\alpha(v)$. The path P^* must contain an arc that leaves the set S, say the arc (x, y) with x in S and y outside S; see Figure 3.2.
- Under the assumption above and the fact that the distances are nonnegative, we find

$$\alpha(v) > L_{sv}^* = L_{sx}^* + c_{xy} + L_{yv}^*$$
$$\geq L_{sx}^* + c_{xy} = \pi(x) + c_{xy}$$
$$= \alpha(y).$$

- The conclusion $\alpha(v) > \alpha(y)$ is in contradiction with the fact that in the current step of Dijkstra's algorithm, the smallest value of $\alpha(w)$ for the nodes w outside S is for $w = v$. The assumption $\alpha(v) > L_{sv}^*$ therefore cannot be correct. This means that $\alpha(v) = L_{sv}^*$, which shows the correctness of Dijkstra's algorithm in every step following the first trivial step.

Most Reliable Path

Until now, we have assumed that the numbers associated with the arcs are lengths (or costs) and that the length of a path is equal to the sum of the lengths of the arcs that form the path. Now, suppose that we have a directed network in which a number $p_{ij} > 0$ is associated with each arc (N_i, N_j) and that this number p_{ij} represents the probability that a message is lost when it is sent through arc (N_i, N_j). We assume that disturbances in the different arcs occur independently of one another. What is the most reliable path from a given initial node N_s to another node N_t?

This problem can easily be transformed into a standard shortest-path problem. To this end, we work with the so-called complementary probabilities $1 - p_{ij}$, where $1 - p_{ij}$ is the probability that a message is not lost while passing through arc (N_i, N_j). A message sent along a path $\mathbf{P}$ is *not* lost somewhere along this path only if the message passes through *every* arc of the path undisturbed. Denote by $\alpha(\mathbf{P})$ the probability that the message is not lost along path $\mathbf{P}$ from node N_s to node N_t. Under the assumption that disturbances in the arcs occur independently of one another, it follows that the probability $\alpha(\mathbf{P})$ is given by the product over the individual arcs in the path of the probabilities that there are no disturbances. Using the mathematical notation $\prod_{i=1}^{n} a_i$ for the product $a_1 \times a_2 \times \cdots \times a_n$, it follows that

$$\alpha(\mathbf{P}) = \prod_{(i,j)} (1 - p_{ij}),$$

where the product is taken over all combinations (i, j) for which (N_i, N_j) is an element of the path $\mathbf{P}$. So the most reliable path from s to t is found by maximizing $\alpha(\mathbf{P})$ over all possible paths from N_s to N_t. Maximizing $\alpha(\mathbf{P})$ over $\mathbf{P}$ gives the same maximizing $\mathbf{P}$ as maximizing $\log \alpha(\mathbf{P})$ over $\mathbf{P}$: after all, $\alpha(\mathbf{P}_1) \leq \alpha(\mathbf{P}_2)$ if and only if $\log \alpha(\mathbf{P}_1) \leq \log \alpha(\mathbf{P}_2)$. Based on the fact that

$$\log(ab) = \log(a) + \log(b) \quad \text{for all } a, b > 0,$$

the original problem can therefore be reduced to the problem of maximizing

$$\log \alpha(\mathbf{P}) = \sum_{(i,j)} \log(1 - p_{ij}).$$

The solution to this problem, in turn, can be found by minimizing

$$-\sum_{(i,j)} \log(1 - p_{ij})$$

over all possible paths $\mathbf{P}$ (observe that the point where the function $f(x)$ has a maximum is the same as the point where the function $h(x) = -f(x)$ has a minimum). The problem of finding the most reliable path has therefore been reduced to a standard shortest-path problem in which the lengths of the arcs are given by

$$c_{ij} = -\log(1 - p_{ij}).$$

The numbers p_{ij} are between 0 and 1, and consequently so are the numbers $1 - p_{ij}$. Since the function $\log(x)$ is negative for $0 < x < 1$, it follows that the numerical values $c_{ij} = -\log(1 - p_{ij})$ are nonnegative. This means that one can apply Dijkstra's algorithm to find the most reliable path.

Illustration

As an illustration, consider a network with five nodes in which a message must be sent from node N_1 to node N_5 in a way that maximizes the probability of the message reaching node N_5 safely.

Table 3.2 Loss probabilities p_{ij}.

From\To	N_1	N_2	N_3	N_4	N_5
N_1	-	0.03	0.05	0.03	*
N_2	*	-	0.01	0.02	0.035
N_3	*	0.01	-	0.04	0.05
N_4	*	*	0.02	-	0.04

For every arc (N_i, N_j), Table 3.2 gives the probability p_{ij} that the message is lost when sent over this arc (a star indicates that there is no such arc). To apply Dijkstra's algorithm, we first give the "distances" $c_{ij} = -\log(1 - p_{ij})$ in Table 3.3, where we have taken the natural logarithm with base the number $e \approx 2.7183$.

Table 3.3 The distances c_{ij}.

From\To	N_1	N_2	N_3	N_4	N_5
N_1	-	0.0305	0.0513	0.0305	*
N_2	*	-	0.0101	0.0202	0.0356
N_3	*	0.0101	-	0.0408	0.0513
N_4	*	*	0.0202	-	0.0408

The application of Dijkstra's algorithm to the distances c_{ij} from Table 3.3 is summarized in Table 3.4. Note that in iteration 1, the estimate α_i for the shortest distance is the least for both $i = 2$ and $i = 4$; we have randomly chosen the node N_4 (recall that in each iteration, only one node may receive a permanent label).

Table 3.4 shows that subject to the distances c_{ij}, the shortest path from N_1 to N_5 is given by

$$N_1 \to N_2 \to N_5, \quad \text{with length } 0.0661.$$

This path also is the most reliable one for sending a message from N_1 to N_5 subject to the information loss probabilities in Table 3.2. The probability that the message arrives safely when sent along this path can be calculated as $0.97 \times 0.965 = 0.936$ (it can also be calculated as $e^{-0.0661} = 0.936$).

Table 3.4 The iterations of Dijkstra's algorithm.

	Iteration 1	Iteration 2	Iteration 3	Iteration 4
N_1	–	–	–	–
N_2	$(0.0305, N_1)$	$(0.0305, N_1)^*$	–	–
N_3	$(0.0513, N_1)$	$(0.0305 + 0.0202, N_4)$	$(0.0305 + 0.0101, N_2)^*$	–
N_4	$(0.0305, N_1)^*$	–	–	–
N_5	(∞, N_1)	$(0.0305 + 0.0408, N_4)$	$(0.0305 + 0.0356, N_2)$	$(0.0661, N_2)^*$

Arcs with Negative Costs

Dijkstra's algorithm can only be applied if the costs (lengths) of all arcs are non-negative. In case the costs of some of the arcs are negative, it may happen that an optimal path does not exist. Consider Figure 3.3 as an illustration of this. In this figure, the circuit $N_2 \to N_3 \to N_4 \to N_2$ has the negative cost of -2. By walking along this circuit infinitely many times, we find a path from node N_1 to node N_5 whose cost is equal to $-\infty$.

Fig. 3.3   A cycle with a negative cost.

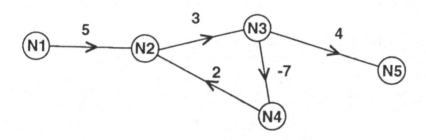

In real-world applications, some arcs may have positive costs while others have negative costs. In those cases, Dijkstra's algorithm cannot be used. The shortest-path problem with general costs cannot be reduced to a shortest-path problem with nonnegative costs by adding the same large number to the costs of all arcs (why not?). For the general case, the shortest-path algorithms are based on dynamic programming; see Section 5.7.

We conclude this section by giving an example of a shortest-path problem in which some of the arcs have negative costs. Suppose that we want to convert American dollars to Swiss francs. The conversion may take place via several other currencies. We number the currencies $i = 1, \ldots, n$, with currency 1 corresponding to US dollars and currency n corresponding to Swiss francs. Let $g(i, j)$ be the exchange rate for converting currency i to currency j. How should we convert dollars to other currencies in order to ultimately obtain a maximum amount of Swiss francs? To simplify, we ignore the transaction costs. We can formulate this problem as a shortest-path problem. Let node i correspond to currency i, and let arc (i, j) correspond to converting currency i to currency j. Suppose that we convert currency $i_0 = 1$ to currency i_1, then currency i_1 to i_2, and so on, and finally currency i_{k-1} to currency i_k with $i_k = n$ for some k with $1 \le k \le n$. We then obtain

$$g(i_0, i_1) \times g(i_1, i_2) \times \cdots \times g(i_{k-1}, i_k)$$

Swiss francs for every dollar. We want to find a path $\mathbf{P} = (i_0, i_1, \ldots, i_k)$ for which this product is at its maximum. Maximizing the product is equivalent to maximizing the logarithm of the product. The latter is equal to the sum of the logarithms of the individual terms of the product. We are therefore looking for a path $P = (i_0, i_1, \ldots, i_k)$ that maximizes

$$\sum_{m=1}^{k} \log g(i_{m-1}, i_m).$$

This, in turn, amounts to minimizing $-\sum_{m=1}^{k} \log g(i_{m-1}, i_m)$ over all possible paths **P**. The currency problem can therefore be translated into a standard shortest-path problem in which the cost

$$c_{ij} = -\log g(i, j)$$

is associated with the arc (i, j). These costs c_{ij} satisfy

$$c_{ij} = \begin{cases} < 0 \text{ if } g(i, j) > 1, \\ > 0 \text{ if } g(i, j) < 1. \end{cases}$$

So some arcs have negative costs, and other arcs have positive costs. It is interesting to note that the existence of a cycle with negative costs indicates the existence of a "gap" in the exchange rate market.

3.2 Maximum-Flow Problems

Another important problem in network optimization is the maximum-flow problem. In this problem, the maximum amount of "flow" is sought that can be transported from a given node (the source) to another one (the sink). The flow can consist of messages, vehicles, liquids, and so on. The size of the flow is restricted by capacity constraints on the arcs of the network. The maximum-flow problem has many real-world applications. An example is finding the highest traffic flow between two locations in a road network. The solution to this problem also provides insight into which roads are congested and can therefore form a bottleneck for traffic. The maximum-flow problem can be formulated as an LP problem. However, as for the shortest-path problem, it is possible to formulate a more efficient algorithm based on the special structure of the network.

The best way to explain the solution method is to use an example. Consider the directed network from Figure 3.4. Suppose that the nodes represent pumping stations and that the arcs represent pipelines. Oil must be pumped from an oil field to a refinery. Node N_1 (source s) represents the oil field, and node N_7 (sink t) represents the refinery. The integer associated with each arc is an upper bound for the amount of oil that can flow through the pipeline in question per unit of time. We assume that the flow is measured in thousands of liters per minute. What is the maximum amount of oil that can be pumped per minute from the oil field to the refinery, and how does this flow pass through the network?

Fig. 3.4　Example of a flow.

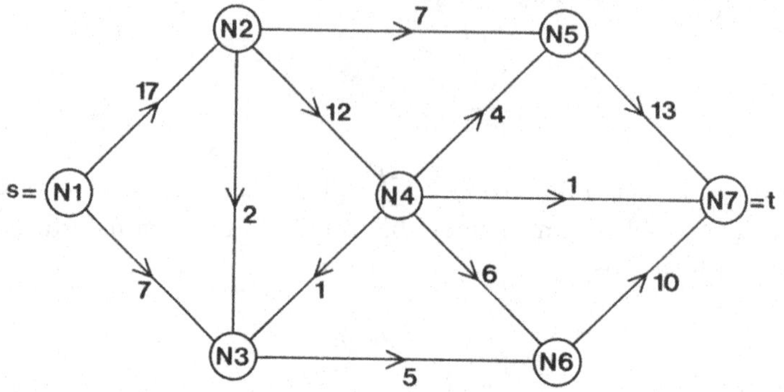

If we denote the capacity of arc (N_i, N_j) by q_{ij}, then the LP formulation is

$$\text{Maximize} \quad v$$

$$\text{subject to} \quad x_{12} + x_{13} = v$$

$$x_{23} + x_{24} + x_{25} = x_{12}$$

$$x_{36} = x_{13} + x_{23} + x_{43}$$

$$x_{43} + x_{45} + x_{46} + x_{47} = x_{24}$$

$$x_{57} = x_{25} + x_{45}$$

$$x_{67} = x_{36} + x_{46}$$

$$x_{47} + x_{57} + x_{67} = v$$

$$\text{and} \quad v \geq 0, \qquad 0 \leq x_{ij} \leq q_{ij} \quad \text{for all } i, j.$$

The variable x_{ij} represents the amount of flow that passes through arc (N_i, N_j), while the variable v represents the net amount of flow from the source to the sink. For the intermediate nodes, the constraints are that the flow out of a node must equal the flow into it. The total unimodularity of the coefficient matrix of the constraints ensures that the LP solution is integer valued, so that no integer constraints are necessary for the x_{ij}. The optimal LP solution is $x_{12} = 17$, $x_{13} = 5$, $x_{23} = 0$, $x_{24} = 10$, $x_{25} = 7$, $x_{36} = 5$, $x_{45} = 4$, $x_{46} = 5$, $x_{47} = 1$, $x_{57} = 11$, and $x_{67} = 10$, with objective value 22. The maximum amount of oil per minute that can be pumped from the oil field to the refinery is therefore equal to 22 thousand liters. A more efficient method to determine the optimal solution is the maximum-flow algorithm, which we discuss below using this example.

Basic Idea of the Maximum-Flow Algorithm

Before explaining the basic idea of the algorithm, let us introduce some notation. A *feasible flow* x is defined as a set of nonnegative integer numbers x_{ij} such that

1. at every intermediate node, the flow out of the node is equal to the flow into it;
2. for every arc (N_i, N_j), the flow x_{ij} passing through the arc is not greater than the arc capacity q_{ij}.

The *value* $v(x)$ of a feasible flow x is defined as the net amount of flow that leaves the source s (= the net flow that enters the sink t). As an illustration, we give a feasible flow $x^{(0)}$ in Figure 3.5. The underlined numbers next to the arcs indicate the flows $x_{ij}^{(0)}$ (the numbers in parentheses indicate the capacities of the arcs). The value of the flow $x^{(0)}$ is equal to 19.

Fig. 3.5   A feasible flow $x^{(0)}$.

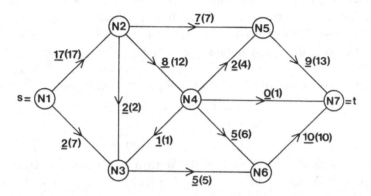

An essential concept is that of an augmenting chain. Such a chain is defined with respect to a *given* flow. An *augmenting path* for a given flow $x^{(0)}$ is a chain of arcs from the source s to the sink t such that

1. $x_{ij}^{(0)} < q_{ij}$ for every forward arc (N_i, N_j) in the chain,
2. $x_{ij}^{(0)} > 0$ for every backward arc (N_i, N_j) in the chain,

where a forward (respectively, backward) arc is an arc toward the sink t (respectively, source s). An example of an augmenting path for the flow $x^{(0)}$ is given in Figure 3.6. We see in this figure that the augmenting path contains the forward arcs (N_1, N_3) and (N_4, N_7) and the backward arc (N_4, N_3).

Fig. 3.6   An augmenting path for $x^{(0)}$.

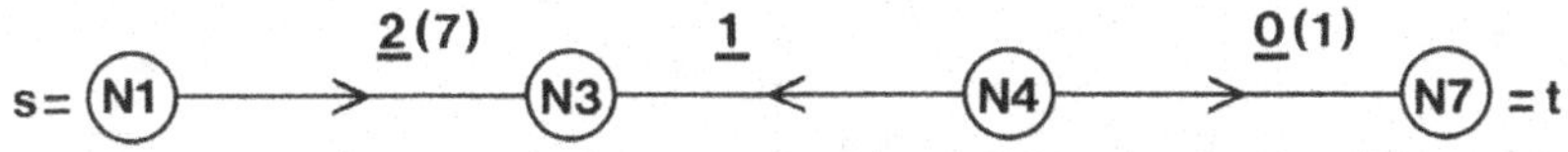

$$\underline{2}\,(7) \qquad \underline{1} \qquad \underline{0}\,(1)$$

$$s = \boxed{N1} \longrightarrow \boxed{N3} \longleftarrow \boxed{N4} \longrightarrow \boxed{N7} = t$$

For a given augmenting path for a flow $x^{(0)}$, we define the slack numbers Δ_f and Δ_b as

$$\Delta_f = \min_{i,j}\{q_{ij} - x_{ij}^{(0)} \mid (N_i, N_j) \text{ forward arc}\}$$
$$\Delta_b = \min_{i,j}\{x_{ij}^{(0)} \mid (N_i, N_j) \text{ backward arc}\}.$$

We denote the least of the two slack numbers by

$$\Delta = \min\{\Delta_f, \Delta_b\}.$$

We can now construct a new feasible flow $x^{(1)}$ from the flow $x^{(0)}$ by adding Δ to each forward arc in the augmenting path and subtracting Δ from each backward arc in the path. The value of the new flow $x^{(1)}$ is Δ more than the value of the previous flow $x^{(0)}$. To see that the new flow is feasible, we observe that when we add $\Delta(\leq \Delta_f)$ to the flow in the forward arcs, the capacities of those arcs are not exceeded. Likewise, subtracting $\Delta(\leq \Delta_b)$ from the backward arcs does not cause the flows in those arcs to become negative. Since the "inflow = outflow" principle still holds for every intermediate node, the new flow $x^{(1)}$ is indeed feasible. Let us now carry out the construction given above for the current flow $x^{(0)}$ from Figure 3.5 and the corresponding augmenting path in Figure 3.6. We find

$$\Delta_f = \min\{7 - 2, 1 - 0\} = 1,$$
$$\Delta_b = \min\{1\} = 1,$$

and therefore $\Delta = 1$. This results in the new feasible flow $x^{(1)}$ given in Figure 3.7. The value of the flow $x^{(1)}$ is $19 + 1 = 20$. We then try to find an augmenting path for the flow $x^{(1)}$. When we have found such a path, we can improve the flow $x^{(1)}$. This procedure is repeated until there is no augmenting path for the current flow. We have then found a flow for which the value is as high as possible.

Fig. 3.7   The improved flow $x^{(1)}$.

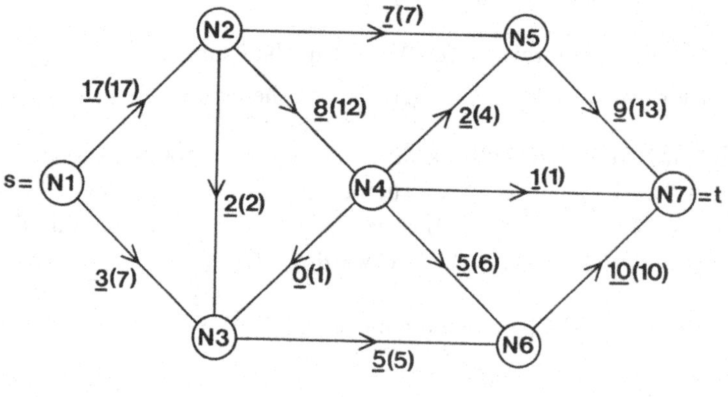

Constructing a series of improved flows using augmenting paths is at the heart of the maximum-flow algorithm. However, we still need a procedure that systematically finds an augmenting path. Such a procedure is discussed below.

The Labeling Procedure

A node is labeled if additional flow can be made available in the node using an augmenting path from the source. This can be either because the node receives more flow from an adjacent node, or because the node sends less flow to an adjacent node. First, all nodes are labeled that are directly connected to the source and can receive more flow. Next, we choose a labeled node and consider the unlabeled nodes that are directly connected to that node. An unlabeled node is then labeled if more flow can be made available to this node by either having it receive more flow from the chosen labeled node, or having it send less flow to that node. If we continue in this way, we will find an augmenting path from the source to the sink, provided that such a path exists.

How do we implement this labeling technique? Suppose that the current flow is given by $\mathbf{x} = (x_{ij})$. At the beginning, none of the nodes are labeled. We initially label the source s with $(s, +\infty)$ to indicated that an unlimited amount of flow is available at node s. In the general step of the labeling procedure, we take a labeled node N_i that has not yet been used. We consider all unlabeled nodes N_j that are directly connected to the labeled node N_i:

1. In the case of a (forward) arc (N_i, N_j), we label node N_j with $(+, N_i)$ if $x_{ij} < q_{ij}$, where q_{ij} is the capacity of the arc (N_i, N_j). The label $(+, N_i)$ indicates that additional flow can be sent to node N_j by increasing the flow through the arc (N_i, N_j).
2. In the case of a (backward) arc (N_j, N_i), we label node N_j with $(-, N_i)$ if $x_{ji} > 0$. The label $(-, N_i)$ indicates that additional flow can be made available in node N_j by reducing the flow through the arc (N_j, N_i).

Illustration

As an illustration, we consider the feasible flow $x^{(1)}$ in Figure 3.7. We first label the source s with $(s, +\infty)$. From the labeled node s, we label node N_3 with $(+, s)$. No other nodes can be labeled from node s. Next, we take the labeled node N_3 to continue labeling from that node. Node N_2 is the only node we can label using N_3. This node is given the label $(-, N_2)$. Using node N_2, we label N_4 with $(+, N_2)$. No other nodes can be labeled from node N_2. We then continue the labeling procedure at node N_4. From that node, we label each of the nodes N_5 and N_6 with $(+, N_4)$. Further labeling is not possible from the node N_4. From node N_5, we can label the sink t with $(+, N_5)$. Now that we have labeled the sink, we can stop the labeling procedure. We have found an augmenting path from the source to the sink with respect to the flow $x^{(1)}$. The labels are used to trace the augmenting path. The label $(+, N_5)$ from node t indicates that the forward arc (N_5, t) belongs to the augmenting path. It follows from the label $(+, N_4)$ from node N_5 that the augmenting path uses the forward arc (N_4, N_5), and so on. The augmenting path is given in Figure 3.8.

This path allows us to improve the current flow $x^{(1)}$. To this end, we calculate

$$\Delta_f = \min\{7 - 3, 12 - 8, 4 - 2, 13 - 9\} = 2,$$
$$\Delta_b = \min\{2\} = 2,$$

Fig. 3.8   An augmenting path for the flow $x^{(1)}$.

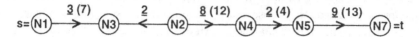

which gives $\Delta = 2$. To find the new flow $x^{(2)}$, we increase the flow in all forward arcs of the path by 2 and decrease the flow in the backward arc by 2. The new flow $x^{(2)}$ is given in Figure 3.9, where the underlined numbers next to the arcs indicate the amount of flow through the arc. The value of the improved flow $x^{(2)}$ is equal to $20 + 2 = 22$.

Fig. 3.9   The improved flow $x^{(2)}$.

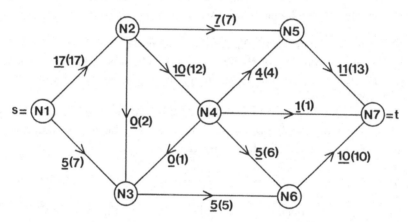

Next, we try to improve the flow $x^{(2)}$. The calculations are as follows. Label the source s with $(s, +\infty)$. From node s, we can only label node N_3, which we give the label $(+, s)$. No other nodes can be labeled from node N_3. This means that no augmenting path exists from the source to the sink for the flow $x^{(2)}$. The flow $x^{(2)}$ in Figure 3.9 is therefore the maximum flow.

Maximum Flow/Minimum Cut

An interesting question regarding the oil example in Figure 3.4 is the following. Which pipelines should be blocked in case of an emergency so that no oil flow is possible from the oil field to the refinery?[2] This question can easily be answered if the cost of blocking a pipeline is proportionate to its capacity (where the proportionality constant is the same for all pipelines). To do this, we need the concept

---

[2]An interesting application concerning the Russian rail network from the time of the Cold War can be found in the article A. Schrijver, Flows in railway optimization, Nieuw Archief voor Wiskunde, June 2008, 126–131.

of cuts. Let S be a set of nodes such that the source s is in S and the sink t is not. The set of arcs (N_i, N_j) with $N_i \in S$ and $N_j \notin S$ is then called a *cut* for the network. Removing the arcs of a cut from the network makes it impossible to travel from the source to the sink. We denote the capacity of the cut corresponding to S by $c(S)$; it is defined as the sum of the capacities of the arcs from the nodes in S to the nodes in $\overline{S}$, where $\overline{S}$ consists of the nodes outside S. The least expensive way of blocking the flow is found by determining the set S for which $c(S)$ is the lowest. We can prove that the function $c(S)$ takes its minimum for

$S^* = $ set of nodes that have been labeled in the last iteration of the maximum-flow algorithm.

The proof is as follows. Suppose that the maximum-flow algorithm gives the maximum flow x^*. In the last iteration of the algorithm, the labeling procedure is stopped as soon as the sink can no longer be labeled. This happens because

1. every forward arc (N_i, N_j) between a labeled node N_i and an unlabeled node N_j is saturated (that is, $x^*_{ij} = q_{ij}$),
2. every backward arc (N_j, N_i) between a labeled node N_i and an unlabeled node N_j has flow zero (that is, $x^*_{ji} = 0$).

The following holds for the value $v(x)$ of an arbitrary flow x:

$$v(x) = \text{(total amount of flow from } S \text{ to } \overline{S})$$
$$- \text{(total amount of flow from } \overline{S} \text{ to } S)$$

for every set S with $s \in S$ and $t \notin S$, where $\overline{S}$ consists of the nodes that do not belong to S. Since $x_{ij} \leq q_{ij}$ and $x_{ji} \geq 0$, this gives

$$v(x) \leq c(S)$$

for the flow x and the cut corresponding to S. For $x = x^*$ and $S = S^*$, we have $x^*_{ij} = q_{ij}$ for the arcs from S^* to $\overline{S}^*$ and $x^*_{ji} = 0$ for the arcs from $\overline{S}^*$ to S^*; that is,

$$v(x^*) = c(S^*).$$

The inequality $v(x) \leq c(S)$ holds for every x and S. Hence, max $v(x) \leq$ min $c(S)$, and consequently

$$v(x^*) \leq \max_x v(x) \leq \min_S c(S) \leq c(S^*).$$

Since $v(x^*) = c(S^*)$, equality must hold in each of these inequalities. Not only does this prove that S^* is the minimum cut, we also find the interesting result

value of the maximum flow = capacity of the minimum cut.

This result is, in fact, a duality result and could also have been proved using the duality theorem from linear programming.

In the oil example, we have $S^* = \{s, N_3\}$. By blocking the pipelines (s, N_2) and (N_3, N_6), oil is prevented from flowing from the oil field to the refinery at minimum cost.

3.3   Minimum Spanning Trees

Suppose that we have a given set of nodes and that for each pair of nodes, the cost of establishing an edge between the two nodes is known. How can we construct, at minimum cost, a network of edges such that any two nodes are connected by a path? This is the minimum spanning tree problem for an undirected network.

The minimum spanning tree problem arises in the construction of telephone and pipeline networks to connect a given number of places. As an illustration, consider the example of an oil company that wants to construct a network of pipelines to connect five oil extraction sites to a receiving terminal. Every extraction site must be connected to the terminal, either directly or indirectly. Table 3.5 shows the distances between the extraction sites and the terminal. The five extraction sites are denoted by $N_1, \ldots, N_5$, and the terminal is denoted by N_0. How should the oil company connect the extraction sites and the terminal to minimize the overall length of the pipelines?

Table 3.5   Distances.

	N_0	N_1	N_2	N_3	N_4	N_5
N_0	–	21	13	45	47	53
N_1		–	7	53	69	71
N_2			–	43	57	67
N_3				–	9	39
N_4					–	37

We need the concept of a *spanning tree* to solve the problem. A spanning tree is a network in which any two nodes are connected by a path and there are no *cycles* (a cycle is a path in which the starting point and endpoint coincide). In a spanning tree, there is a *unique* path between any two nodes (verify this result using a proof by contradiction: assume that there are two nodes between which two different paths exist and then deduce a contradiction). Also verify that a spanning tree has $n - 1$ arcs if there are n nodes. Figure 3.10 gives two spanning trees for the oil example. These spanning trees have respective lengths 171 and 143.

Fig. 3.10   Two spanning trees.

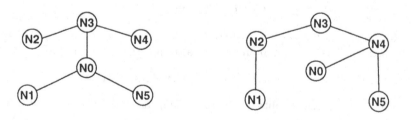

A *minimum spanning tree* is a spanning tree with minimum cost (length). Prim's algorithm constructs a minimum spanning tree by gradually expanding a subtree.

Algorithm 3.2 (Prim's Algorithm).

Step 1. Begin with the trivial subtree consisting of a randomly chosen node.

Step 2. Determine the unconnected node for which the cost of connecting it directly to the current subtree is the least. Extend the subtree with this node.

Step 3. Stop when all nodes are connected to one another; otherwise, go back to Step 2 with the new subtree.

We call this algorithm "greedy" because, at each stage, it acts as if it is the last possibility to add an edge. The minimum spanning tree problem is one of the exceptions for which a greedy algorithm provides an optimal solution.

Let us apply the algorithm to the data in Table 3.5.

Step 1. Start with the trivial subtree consisting of only node N_0.

Step 2. Node N_2 is closest to node N_0; connect these two nodes.

Step 3. The unconnected node that is closest to node N_0 or N_2 is node N_1. It is closest to node N_2; connect these two nodes.

Step 4. The unconnected node that is closest to node N_0, N_1, or N_2 is node N_3. It is closest to node N_2; connect these two nodes.

Step 5. The unconnected node that is closest to node N_0, N_1, N_2, or N_3 is node N_4. It is closest to node N_3; connect these two nodes.

Step 6. The only remaining unconnected node is node N_5. It is closest to node N_4; connect these two nodes.

The minimum spanning tree for the oil example is given in Figure 3.11. The overall length of the pipelines is 109.

Fig. 3.11 The minimum spanning tree.

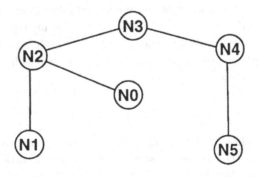

It is important to have an efficient implementation of the procedure for finding the unconnected node that is closest to the current set of connected nodes. Let us now formulate, in general terms, Prim's algorithm for finding a minimum spanning tree for an undirected network of n nodes $N_1, \ldots, N_n$, where the length (cost) of an edge between the nodes N_i and N_j is c_{ij}. The algorithm keeps track of the set S_v of connected nodes and adjusts the labels of the unconnected nodes in each step. Every unconnected node N_j has two labels α_j and β_j, where α_j is the shortest distance between node N_j and the current subtree and β_j is the node in the current subtree that is closest to N_j.

Algorithm 3.3 (Prim's Algorithm for an Undirected Network).

Initialization. Randomly select a node N_s. Initially, let the set S_v of nodes contain
   only N_s, and let the set T_v of edges be empty. For every node N_j unequal to
   N_s, set $\alpha_j := c_{sj}$ and $\beta_j := N_s$.

Step 1. Determine the node N_j outside of S_v for which α_j is the least. Suppose
   that this is node N_k.

Step 2. Extend S_v by adding node N_k, and extend T_v by adding the edge (β_k, N_k).
   For every node N_j that does not yet belong to S_v and for which $c_{kj} < \alpha_j$, the
   labels are changed to

$$\alpha_j := c_{kj} \quad \text{and} \quad \beta_j := N_k.$$

Step 3. Repeat Steps 1 and 2 until all nodes are connected.

The computational complexity of the algorithm is low. Verify that the total number of operations in the algorithm is of order n^2 if n is the number of nodes and an edge can be added between any pair of nodes. It is important to have a fast algorithm for finding a minimum spanning tree. For many complex combinatorial optimization problems, the solution method requires repeatedly solving minimum spanning tree problems as subproblems.

In addition to the previously mentioned application for telephone and pipeline networks, the minimum spanning tree can also be used for so-called *cluster analysis*. Suppose, for example, that we want to bundle a number of points in the plane into four groups. We could then determine a minimum spanning tree (with costs equal to the Euclidean distances between the points, that is, the distances measured along a straight line). If we then leave out the three longest edges in the minimum spanning tree, we obtain four clusters of points.

Optimality Proof

It is not difficult to show that Prim's greedy algorithm generates a minimum spanning tree. Let T_k be the subtree constructed in the kth iteration of the algorithm $(k = 1, \ldots, n)$. In particular, $T_1 = \{N_s\}$ and T_n is the spanning tree for the network. Now, suppose that T_n is not a minimum spanning tree. Then there exists a smallest index k (≥ 2) such that the subtree T_k is not part of a minimum spanning tree. Let T^* be a minimum spanning tree that contains the subtree T_{k-1}. Suppose that T_k was constructed by connecting N_w to $N_v \in T_{k-1}$. In the spanning tree T^*, the nodes N_v and N_w are connected by a path. This path must use an edge

$\{N_g, N_h\}$ with $N_g \in T_{k-1}$ and $N_h \notin T_{k-1}$; see Figure 3.12 (in this figure, the solid edges belong to T_{k-1} and T^*, while the dashed edges belong to T^* but not to T_{k-1}). Because of the choice $\{N_v, N_w\}$ in the kth iteration, we have $c_{vw} \leq c_{gh}$. Removing the edge $\{N_g, N_h\}$ from T^* and replacing it with the edge $\{N_v, N_w\}$ gives a spanning tree of length less than or equal to that of the minimum spanning tree T^*. This contradicts the assumption that the subtree T_k is not part of the minimum spanning tree. Prim's algorithm therefore provides a minimum spanning tree.

Fig. 3.12   The optimal tree T^* and the subtree T_{k-1}.

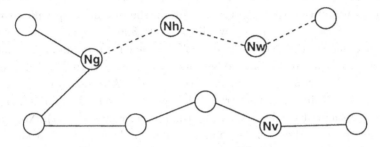

Incidentally, Prim's algorithm was also developed independently by Dijkstra. The story goes that Dijkstra came up with the idea at the end of the 1950s while sitting on a terrace on the Damrak in Amsterdam; someone knocked over a cup of coffee, and Dijkstra saw how the puddle of coffee slowly spread over the table. This gives a striking image of how the algorithm works (private communication with J. K. Lenstra). Another algorithm for constructing a minimum spanning tree is *Kruskal's algorithm*. This algorithm first sorts the edges by increasing cost and then builds up a set of edges step by step. The algorithm begins with the lowest-cost edge and then, at each step, adds the first not-yet-considered edge of the sorted list that does not form a cycle with the already selected edges. The algorithm stops as soon as the constructed set consists of $n-1$ edges. This set then forms a minimum spanning tree.

The Minimax Path Problem

Suppose given an undirected network with n nodes $N_1, .., N_n$, where an edge between node N_i and node N_j costs c_{ij}. How can one find a path between two given nodes such that the maximum of the individual costs of the edges in the path is as small as possible? This is called the minimax path problem. As an application, think of determining a route from one city to another in a desolate region such that the longest march distance per day is minimized. The minimax problem is solved by the following algorithm:

Algorithm 3.4 (Minimax Problem).

Step 1. Determine the minimum spanning tree for the network.
Step 2. For every given pair of nodes N_k and N_ℓ, the *unique* path between N_k and N_ℓ in the minimum spanning tree is the path between N_k and N_ℓ for which the maximum of the individual costs of the edges in the path is as low as possible.

Proving that this algorithm works is not difficult. Let P^* be the unique path between N_k and N_ℓ in the minimum spanning tree T^*, and suppose that in this path, the edge (N_a, N_b) has the highest cost; that is, the path P^* has objective value c_{ab}. If we remove the edge (N_a, N_b) from the minimum spanning tree, then the set of all nodes is divided into two disjoint sets S and $\bar{S}$, where S consists of the nodes that remain connected to the node N_k in the cut-up minimum spanning tree and $\bar{S}$ consists of the remaining nodes that stay connected to N_ℓ. For every edge (N_i, N_j) with $N_i \in S$ and $N_j \in \bar{S}$, we have $c_{ab} \leq c_{ij}$. After all, if this were not the case, we would obtain a spanning tree with lower costs than the minimum spanning tree T^* by replacing the edge (N_a, N_b) with another edge (N_i, N_j). Now, consider an arbitrary path P between N_k and N_ℓ. This path must contain an edge (N_i, N_j) with $N_i \in S$ and $N_j \in \bar{S}$. This means that the objective value of the path is at least c_{ij} and therefore at least as great as the objective value c_{ab} of the path P^*. This proves that P^* is a path between N_k and N_ℓ with minimum objective value.

The following is a nice example. A desert traveler wants to make a trip from the starting point A to the endpoint G. The possible day trips are $A-B$, $A-C$, $A-D$, with respective distances 26, 21, 32; $B-E$, $B-F$, $C-E$, $C-F$, $D-E$, $D-F$, with respective distances 10, 12, 17, 21, 7, 8; and $E-G$, $F-G$ with respective distances 20, 22. For which route is the greatest distance to be traveled on one day the least? The minimum spanning tree for the network is given in Figure 3.13 below (verify). In this spanning tree, the unique path from A to G is given by $A-C-E-G$. This is the desired path for which the greatest distance traveled on one day is the least; it has value 21.

Fig. 3.13   Minimax solution.

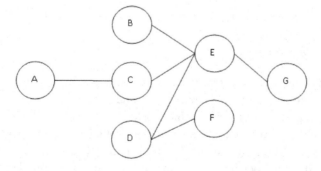

The Steiner Tree Problem

This fascinating problem is closely related to the minimum spanning tree problem but is much more difficult to solve. The Steiner tree problem also concerns searching for the shortest (cheapest) network that connects a given number of nodes, but the problem allows the use of auxiliary points, called Steiner points. These Steiner points serve as additional support points in the network of edges. This problem has important real-world applications, for example the construction of a cable network to connect villages in an area with mountains and rivers. In such a situation, it can be very cost effective to determine not a minimum spanning tree with only the villages as nodes, but rather a minimum spanning tree with the villages and carefully selected support points outside the villages as nodes. Telephone companies in the United States have been able to save millions of dollars with Steiner tree constructions when establishing a network of connections in remote areas. The difficulty of the Steiner tree problem is how many Steiner points to choose and where to place them. Once the Steiner points have been fixed, the matter is reduced to the simple problem of finding a minimum spanning tree for the original nodes and the Steiner points.

The Steiner tree problem goes back to the 19th-century mathematician Jacob Steiner, who asked how a number of points in the plane should be connected by a system of line segments to minimize the total length of the segments. This problem is called the Euclidean Steiner tree problem if the distance between two points is the Euclidean distance, that is, the distance measured along a straight line. For three nodes that form an equilateral triangle, we give, in Figure 3.14, both the minimum spanning tree for the three points and the Steiner tree, where it should be clear that the Steiner point is the geometric center of the triangle.

Fig. 3.14   A three-point Steiner tree problem.

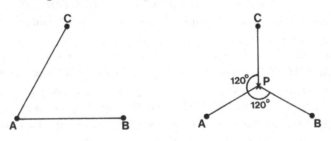

Verify that the total required line length is reduced by $\left(1 - \frac{1}{2}\sqrt{3}\right) \times 100\% = 13.4\%$. In Figure 3.15, we give the optimal Steiner tree for a simple four-point problem with original points A, B, C, and D, where we now use two Steiner points P_1 and P_2.

Fig. 3.15   A four-point Steiner tree problem.

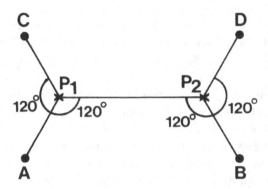

It can be proved that for the Euclidean Steiner tree problem, the optimal Steiner tree has the following properties:

- The angle between two adjacent edges at a Steiner point is equal to 120 degrees, and exactly three edges meet at a Steiner point.
- The number of Steiner points is at most $n - 2$ if n is the original number of nodes.
- For the length L_{Stein} of the best Steiner tree and the length L_{tree} of the minimum spanning tree (without auxiliary points), we have

$$0 \le L_{tree} - L_{Stein} \le \frac{1}{2}\sqrt{3}\, L_{tree}.$$

These properties hold in general for the Euclidean Steiner tree problem. The last property tells us that the length of a minimum spanning tree can be reduced by at most $\left(1 - \frac{1}{2}\sqrt{3}\right) \times 100\% = 13.4\%$ by adding Steiner points. The maximum reduction of 13.4% is achieved in the case of three given points forming an equilateral triangle; see Figure 3.14. In real-world problems, the reduction is much smaller and typically in the order of 3 to 4%.

To conclude the section, let us discuss an interesting geometric procedure for finding the shortest Steiner tree for the special case of a three-point Euclidean Steiner tree problem. Consider three given points A, B, and C. Assume that triangle ABC has its greatest angle at point C, and suppose that angle C is less than 120 degrees (otherwise, one can see that the shortest Steiner tree is given by the minimum spanning tree). The procedure is as follows:

1. Draw an equilateral triangle ABZ on side AB such that Z lies across from point C on the other side of AB.
2. Construct a circle through the three points A, B, and Z of the equilateral triangle. Draw a line from point C to point Z. The Steiner point P is the intersection point of this line and the circle.

Fig. 3.16 The construction of a Steiner point.

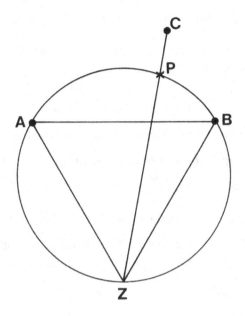

This geometric procedure is illustrated in Figure 3.16. The shortest Steiner tree is given by the line segments AP, BP, and CP. An elegant, geometric construction also exists for the four-point Euclidean Steiner tree problem. This solution is the basis for a good heuristic for the Euclidean Steiner tree problem with more than four points; see the article J. E. Beasley, A heuristic for Euclidean and rectilinear Steiner problems, *European Journal of Operational Research*, Vol. 58 (1992), 284–292. For the Steiner tree problem, heuristic solutions must suffice in general because exact solution methods are practically impossible to implement for larger problems. The number of calculations of an exact algorithm increases exponentially in the number of nodes.

3.4 Minimum-Cost Flow Problems

Many optimization problems in networks can be formulated as minimum-cost flow problems. This problem in fact comes down to the determination of a distribution plan to send a certain product through a network at minimum cost in order to satisfy demand for the product in certain nodes using the supply in other nodes. For the general formulation of the minimum-cost flow problem, we assume that we have a *directed* network $G = (X, A)$, where $X = \{N_1, \ldots, N_m\}$ is the set of nodes and A is the set of arcs (N_i, N_j) between the nodes. The cost c_{ij} is given for every arc (N_i, N_j), where

$$c_{ij} = \text{cost per unit of product passing through arc } (N_i, N_j).$$

Integers $l_{ij} \geq 0$ and $u_{ij} \geq 0$ are associated with every arc (N_i, N_j), where

$$l_{ij} = \text{minimum quantity of product to be transported}$$
$$\text{through arc } (N_i, N_j),$$
$$u_{ij} = \text{maximum quantity of product that can be transported}$$
$$\text{through arc } (N_i, N_j).$$

Of course, we must have $0 \leq l_{ij} \leq u_{ij}$ for the lower and upper bounds l_{ij} and u_{ij}. In addition, for every node, an integer b_i is given, where

$$b_i = \text{supply of or demand for the product at node } N_i.$$

Node N_i is called a supply node if $b_i > 0$, a demand node if $b_i < 0$, and a transshipment node if $b_i = 0$. We assume $\sum_{i=1}^{m} b_i = 0$ (why?). The decision variables x_{ij} are defined by

$$x_{ij} = \text{amount of product passing through arc } (N_i, N_j).$$

The minimum-cost flow problem can be formulated as the following optimization model:

$$\text{Minimize} \quad \sum_{(i,j)} c_{ij} x_{ij}$$

$$\text{subject to} \quad \sum_{j} x_{ij} - \sum_{k} x_{ki} = b_i \qquad \text{for every node } N_i,$$

$$\text{and} \qquad l_{ij} \leq x_{ij} \leq u_{ij}, \quad x_{ij} \text{ integer} \text{ for every arc } (N_i, N_j).$$

For the ILP problem above, the special structure of the coefficient matrix of the constraints can be used to prove that every optimal basic solution of the LP relaxation is automatically integer valued. This means that the minimum-cost flow problem can be solved with linear programming. However, using the special structure of the network problem, one can develop a network algorithm that is much more efficient than the simplex method. We will not discuss this network algorithm here, not even for the particular case of the minimum-cost flow problem. Despite the existence of such an efficient network algorithm, in practice, an LP code is often used for the simple reason that fast LP codes with good input and output features are readily available. Only in situations where large-scale minimum-cost flow problems must be solved repeatedly is it worthwhile to use special software for network problems.

We conclude this section by discussing two special cases of the minimum-cost flow problem.

3.4.1 *A Transportation Problem*

In a transportation problem, the set of nodes consists of two disjoint sets $\{S_1, \ldots, S_r\}$ and $\{D_1, \ldots, D_s\}$, where every node S_i is a supply node ($b_i > 0$) and every node D_j is a demand node ($b_j < 0$). Moreover, only arcs of the form (S_i, D_j) exist. The lower bounds l_{ij} are equal to 0, and the upper bounds u_{ij} are equal to ∞. The classic example of a transportation problem is the distribution of goods from factories to distribution centers. If a quantity b_i is present at supply node S_i and the demand at node D_j is d_j, with $\sum_{i=1}^{r} b_i = \sum_{j=1}^{s} d_j$, then the minimum-cost flow problem reduces to the LP problem

$$\text{Minimize} \quad \sum_{i=1}^{r} \sum_{j=1}^{s} c_{ij} x_{ij}$$

$$\text{subject to} \quad \sum_{j=1}^{s} x_{ij} = b_i \quad \text{for } i = 1, \ldots, r$$

$$\sum_{i=1}^{r} x_{ij} = d_j \quad \text{for } j = 1, \ldots, s,$$

$$\text{and} \quad x_{ij} \geq 0 \text{ for all } i, j.$$

In some applications of the transportation problem, the total supply S may be higher than the total demand D. This situation can be reduced to the standard form of the transportation problem by introducing an auxiliary node D_{s+1} for which the demand is equal to the difference $S - D$ and the transportation cost $c_{i,s+1}$ is equal to zero for every supply point S_i. A well-known algorithm for the transportation problem is the so-called "stepping-stone" algorithm. We will not discuss it.

3.4.2 An Assignment Problem

In the assignment problem, the set of nodes is made up of two disjoint sets $\{P_1, \ldots, P_r\}$ and $\{K_1, \ldots, K_s\}$ with $r = s$, and all arcs are of the form (P_i, K_j). For every node P_i, we have $b_i = 1$, while for every node K_j, we have $b_j = -1$. The lower bounds l_{ij} are equal to 0, and the upper bounds u_{ij} are equal to 1. The assignment problem arises, for example, when a company wants to outsource s tasks to s persons at minimum cost, subject to the constraint that every person carry out only one task and that every task be carried out by only one person, where the cost of having task i carried out by person j is c_{ij}. If we define the decision variables x_{ij} to equal 1 if task i is carried out by person j and to equal 0 otherwise, the formulation of the minimum-cost flow problem reduces to the LP formulation

$$\text{Minimize} \quad \sum_{i=1}^{s} \sum_{j=1}^{s} c_{ij} x_{ij}$$

$$\text{subject to} \quad \sum_{j=1}^{s} x_{ij} = 1 \quad \text{for } i = 1, .., s,$$

$$\sum_{i=1}^{s} x_{ij} = 1 \quad \text{for } j = 1, .., s,$$

$$\text{and} \quad x_{ij} \geq 0 \text{ for all } i, j.$$

For every optimal solution to this LP problem obtained using the simplex method, every variable automatically has value 0 or 1 because the coefficient matrix of the constraints is totally unimodular. The assignment problem in which the number of persons is greater than the number of tasks can be reduced to the standard assignment problem by introducing phantom tasks and setting the cost of carrying out a phantom task to 0. Several specialized algorithms for the assignment problem exist that are much faster than the general minimum-cost flow algorithm.

The Hungarian Method

This is a classic method—especially for manual calculations—to determine an optimal assignment. In this method, we assume $c_{ij} \geq 0$ for all $i, j = 1, 2, \ldots, s$. This is not a limitation (why not?). The method is as follows:

Step 1. Created a square s by s cost matrix with elements the costs of assigning persons to tasks.

Step 2. Begin by subtracting the minimum element of each column from every element in that column. Then, subtract the minimum element in each row from every element in that row.

Step 3. In the reduced matrix, determine a *minimum* number of rows and columns needed to cover all zeros in the rows and columns. Once the number of covering lines is equal to the dimension s of the cost matrix, go to Step 5. Otherwise, go to Step 4.

Step 4. Reduce the matrix further by subtracting the smallest *uncovered* element from every uncovered element and adding it to every element at the *intersection* of two covering lines. Go back to Step 3.

Step 5. An optimal assignment results by assigning a column to each row in such a way that the column has a zero in the row in question and every column is assigned only once.

As an illustration, consider the numerical example

Person\Task	1	2	3
1	150	150	75
2	80	115	90
3	125	140	80

Carrying out Steps 2 and 3 leads to

Person\Task	1	2	3	Person\Task	1	2	3
1	70	35	0	1	70	35	**0**
2	0	0	15	2	**0**	**0**	15
3	45	25	5	3	40	20	**0**

Carrying out Steps 4 and 3 leads to

Person\Task	1	2	3	Person\Task	1	2	3
1	50	15	0	1	50	15	**0**
2	0	0	35	2	**0**	**0**	35
3	20	0	0	3	20	**0**	**0**

The covered rows and columns are indicated by numbers in bold. The optimal assignment is to assign person 1 to task 3, person 2 to task 1, and person 3 to task 2. The total cost is 295.

The proof of the optimality of the Hungarian method is simple and is based on the principle

> *the optimal solution remains the same if in the matrix of cost coefficients, the same constant is subtracted from every element in a specific row or column.*

This principle is applied in Step 2 of the algorithm and repeatedly applied in Step 4: if $a > 0$ is the smallest uncovered number, then the result of Step 4 is the same as when in every non–fully covered column, the number a is subtracted from every element in that column, and then in every fully covered row, the number a is added to every element in that row. Furthermore, the following holds. If before applying Step 4, the number of covered rows is equal to h and the number of covered columns is equal to v with $h + v < s$ (where s is the dimension of the cost matrix), then applying Step 4 reduces the sum of the elements in the reduced cost matrix by a positive amount while all elements remain nonnegative (this positive amount is $as(s-h-v)$ since there are hv elements that are covered twice and $s^2 - s(h+v) + hv$ uncovered elements, so that the decrease in the matrix is $a(s^2 - s(h+v) + hv) - a(hv) = as(s - (h+v))$). This means that after a finite number of steps, the algorithm converges, where the number of covering lines is equal to the dimension s of the cost matrix.

The Nonbipartite Matching Problem

The assignment problem is also called the "bipartite matching problem." A related problem that is much more difficult to solve is the "nonbipartite matching problem" (minimum-matching problem). This is a network problem with an *even* number of nodes in which you want to create a perfect matching of the nodes in disjoint pairs. The cost of pairing the two nodes N_i and N_j is c_{ij}. An interesting application of the nonbipartite matching problem is the following: Suppose given $2m$ points (x_i, y_i) in the plane. How should one pair these points to minimize the average distance between the two points in a pair? In this problem, the constants c_{ij} are given by $1/m$ times the Euclidean distance:

$$c_{ij} = \frac{1}{m}\sqrt{(x_i - x_j)^2 + (y_i - y_j)^2}.$$

This problem arose when, in 1995, paleontologists discovered a nearly intact nest of 22 dinosaur eggs in Montana. The positioning of the eggs raised the question whether, 80 million years ago, dinosaurs laid two eggs at a time. To answer this question, the researchers carried out a computer simulation in which 22 points were repeatedly randomly chosen in a two-dimensional area in the form of the nest. For each simulation, the average pairwise distance was calculated for the randomly chosen points. By carrying out a significant number of simulations, the researchers could determine the empirical probability distribution of the average pairwise distance when choosing 22 points randomly. It turned out that for the nest with 22 eggs, the average distance fell so far in the left tail of the probability distribution that the hypothesis of eggs being laid in pairs was supported.

The nonbipartite matching problem can be formulated as a 0–1 ILP problem, but for this network, it does not hold that the LP relaxation automatically has an integer solution. For this problem, a graph-theoretical solution method exists that gives an optimal solution in polynomial time. We will not discuss the exact algorithm.

A simple heuristic is to generate a number of feasible matchings and then improve each of these matchings with the 2-opt exchange heuristic we discussed with the traveling salesman problem. Then, take the best of the matchings that are found. The more complex exchange heuristic of Lin and Kernighan (see Section 2.8) can also be applied to the matching problem and, of course, generally gives better results than the 2-opt exchange heuristic. A "smart" way to construct two acceptable matchings is to first solve a traveling salesman problem for the $2n$ points that need to be matched, where in the traveling salesman problem, the distance between two points is given by the cost of connecting them. As a heuristic solution for the minimum-matching problem, take alternating segments from the tour found for the traveling salesman problem. This gives two possible matchings. For example, the tour $P_1 \rightarrow P_3 \rightarrow P_5 \rightarrow P_8 \rightarrow P_6 \rightarrow P_2 \rightarrow P_4 \rightarrow P_7 \rightarrow P_1$ leads to the two possible matchings $\{(P_1, P_3), (P_5, P_8), (P_6, P_2), (P_4, P_7)\}$ and $\{(P_3, P_5), (P_8, P_6), (P_2, P_4), (P_7, P_1)\}$.

3.5   Euler Circuit and the Chinese Postman Problem

The traveling salesman problem corresponds to finding a minimum-cost cycle in a graph such that every point of the graph is visited exactly once. A kind of dual of this problem is the graph problem of looking for a cycle that uses every edge of the graph exactly once. The latter is known as the Euler problem. This problem and its variants have important practical applications.

3.5.1   *Euler Circuit*

The concept of an Euler circuit has its foundation in the famous Königsberg bridge problem (1736). The city of Königsberg (now Kaliningrad) is built on the two banks of the Pregel River and on two islands in the river. The banks and the islands are connected by seven bridges, as shown in Figure 3.17.

The question asked in 1736 was whether it is possible to devise a walk during which you cross every bridge exactly once and return to the starting point of the walk. Basing himself on a graph representation of the problem, Euler (1707–1783) concluded that such a walk is not possible.

By representing each riverbank and island by a node and each bridge by an edge, the map in Figure 3.17 can be depicted by an equivalent graph; see Figure 3.18. This graph is undirected; that is, every edge of the graph can be traversed in either

Fig. 3.17   A map of Königsberg.

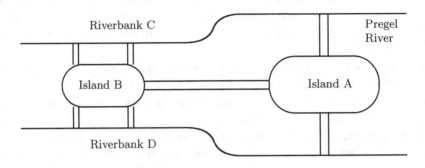

direction. Using this graph, we introduce a few notions we will need further on. The graph is *connected*, that is, any two vertices are connected by a path in the graph. The *degree* of a vertex is defined as the number of edges that connect to it. For example, the degree of vertex A is equal to 3, and points B, C, and D have degrees 5, 3, and 3, respectively.

Fig. 3.18   Equivalent graph.

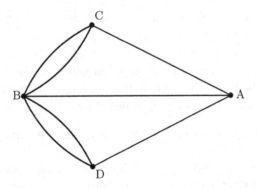

An *Euler circuit* is defined as a circuit that passes through every edge of the graph exactly once (a circuit, or cycle, is a path whose starting point and endpoint coincide). Similarly, an *Euler path* is defined as a path that uses every edge of the graph exactly once. For a path, the starting point and endpoint may differ.

Below, we restrict ourselves to the case of an undirected graph. The basic setup for the Euler problem is as follows

Theorem 3.1. *A connected undirected graph contains an Euler circuit (respectively, path) if and only if the number of vertices of odd degree is 0 (respectively, 0 or 2).*

Proof. Let us prove this property for the case of a circuit. The proof for a path is completely analogous.

(**a**) *Necessity.* Suppose that the graph contains an Euler circuit. Since the Euler circuit uses every edge exactly once, every vertex has the property that one edge is used to enter the vertex and *another* edge is used to leave it. So the degree of the vertex is twice the number of times the Euler circuit enters the vertex and is therefore even.

(**b**) *Sufficiency.* Suppose that the degree of every vertex is even. Randomly select a vertex x_0. Construct a circuit Φ_0 from x_0 to x_0 such that every edge in the circuit is used exactly once. The construction of such a circuit is possible because of the assumption that every vertex has even degree and the fact that the number of vertices is finite. If the circuit Φ_0 contains all of the graph's edges, we are done. Otherwise, proceed as follows. Since the graph is connected, Φ_0 must contain a vertex x_1 at which an edge arrives that is not in Φ_0. Remove all edges that are in Φ_0 from the graph. This gives a subgraph (that may consist of a number of disjoint connected components). The degree of every vertex in the subgraph remains even. Now, in the subgraph, construct a circuit Φ_1 from x_1 to x_1 such that every edge in Φ_1 is used exactly once. If Φ_1 contains all of the subgraph's edges, we are done. In this case, the desired Euler circuit is found by following the part of Φ_0 from x_0 to x_1, then following Φ_1, and then following the remaining part of Φ_0 from x_1 to x_0. If Φ_1 does not contain all of the subgraph's edges, then there exists a vertex x_2 on Φ_0 or Φ_1 at which an unused edge arrives. Remove all edges that are in Φ_1 from the subgraph and repeat the procedure above with x_2, and so on. This method leads to a number of circuits Φ_0, Φ_1, ... that can be linked together to form an Euler circuit.   □

The proof given above is a constructive proof on which an algorithm to determine an Euler circuit can be based. An entirely different algorithm that leads to an Euler circuit, if such a circuit exists, is Fleury's algorithm.

Algorithm 3.5 (Fleury's Algorithm for an Euler Circuit).

Initialization. Randomly select a vertex N_s as the starting point.

Step 1. From the current vertex, choose an unused edge while observing the rule that you may not choose a "bridge" in the reduced subgraph. The reduced subgraph is the graph in which the edges that have already been used and the vertices that have become isolated have been left out, while a "bridge" is an edge whose deletion causes the subgraph to split into two separate pieces.

Step 2. Pass through the chosen edge to the next vertex.

Step 3. Repeat Steps 1 and 2 until all edges have been passed through and you are back at the starting vertex.[3]

---

[3] If one carries out this algorithm by hand, after the removal of a used edge, the reduced graph and "bridges" are immediately obvious to the eye. A computer code requires a nontrivial procedure to test whether an edge is a "bridge."

Fig. 3.19   A scimitar network.

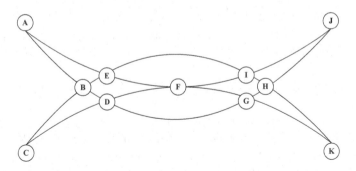

A well-known puzzle is how, in a scimitar-like network as in Figure 3.19, one can trace all edges with a pencil in a continuous movement without lifting the pencil from the paper, in such a way that every edge is traced exactly once and one finishes at the starting point. This corresponds to finding an Euler circuit for the network. Use Fleury's algorithm to check that an Euler circuit is given by $A, B, D, G, H, J, I, H, K, G, F, D, C, B, E, I, F, E, A$.

3.5.2   *The Chinese Postman Problem*

Consider, again, a connected, undirected graph. Suppose that no Euler circuit exists. Then a natural question is how to construct the smallest cycle that passes through every edge of the graph at least once. More generally, suppose that positive costs of $c(a)$ are made every time edge a is traversed. The question is then how to construct, at the lowest possible cost, a circuit such that every edge of the graph is passed through at least once. This problem is known as the Chinese postman problem, named after the Chinese scholar M. K. Kwan. The Chinese postman problem has many real-world applications:

1. *Trash pickup.* Suppose that a garbage truck has to pick up the trash in a specific district. The edges of the graph represent the streets, and the nodes correspond to street intersections. The cost assigned to an edge could be the street length.
2. *Mail or newspaper delivery.* This is also typically an example of a problem where a route is sought such that every street is visited at least once and the total distance traveled is as small as possible.

If the graph contains an Euler circuit, then the Euler circuit provides an optimal solution to the Chinese postman problem. Suppose that the graph does not contain an Euler circuit because at least one of the vertices has odd degree. To prepare for how we then proceed, we define

$$X^- = \text{set of vertices in the graph with odd degree.}$$

Theorem 3.2. *The number of vertices in X^- is even.*

Proof. For any point x_i of the graph, define the number g_i as its degree. If m is the number of edges in the graph, then the sum of the degrees g_i over all vertices x_i must equal $2m$. Hence,

$$\sum_{x_i \in X^+} g_i + \sum_{x_i \in X^-} g_i = \text{even},$$

where X^+ is the set of vertices of even degree. For every $x_i \in X^+$, the number g_i is even. The sum of the even numbers is again even. Consequently, the sum of the g_i over $x_i \in X^-$ must also be an even number. Since odd numbers only add up to an even number when the number of terms in the sum is even, this means that the number of points in X^- is even.

How does the algorithm for the Chinese postman problem work? The first step is to construct a shortest path in the original graph between any two vertices in X^-. Based on the lengths of these shortest paths, we then match every point of X^- to exactly one other point of X^- (this is where we use that the number of points in X^- is even). □

Once a matching M has been fixed between pairs of points in X^-, the next step is to expand the original graph using the shortest paths that underlie this matching. This is done by adding the edges in these paths as auxiliary edges to the original graph, with the provision that an edge is added as an auxiliary edge as many times as it is used by the paths in the matching M. This way, we construct an extended graph $G(M)$ that has the property that the degree of every vertex is *even* (verify). The extended graph $G(M)$ therefore has an Euler circuit. This Euler circuit is a "general" circuit for the original graph in which every edge of the original graph is passed through at least once. This circuit gives the solution to the Chinese postman problem with minimum cost $\sum_a n(a) \times c(a)$, where $n(a)$ is the number of times the edge a from the original graph is used as an auxiliary edge in the extended graph $G(M)$.

Algorithm 3.6 (The Chinese Postman Problem).

Step 1. For the original graph, determine the shortest path P_{ij}^* between each pair of points x_i and x_j from X^-, where the length of an edge a is taken to be the cost $c(a)$ of using that edge. Let d_{ij} be the length of the shortest path P_{ij}^*.

Step 2. Match every point in X^- to exactly one other point in X^- at minimum cost, where d_{ij} is the cost of matching $x_i \in X^-$ and $x_j \in X^-$. Let M^* be the resulting matching.

Step 3. Form an extended graph $G(M^*)$ by adding the edges from the paths P_{ij}^* between the matched pairs of points (x_i, x_j) as auxiliary edges to the original graph, where every edge is added as often as it is used in the paths of the matching M^*.

Step 4. Construct an Euler circuit in the graph $G(M^*)$. This is the desired minimum-cost circuit for the original problem.

As one can see, the solution to the Chinese postman problem is based on an interesting combination of standard algorithms from network theory.

Fig. 3.20 Nonoptimal matching.

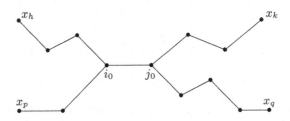

Remark. In the optimal solution of the Chinese postman problem, no edge is passed through more than twice. The explanation for this is simple. If, in the optimal matching, the pairs of vertices (x_h, x_k) and (x_p, x_q) are formed, then the corresponding paths P_{hk} and P_{pq} do not have any edges in common. Suppose that the paths P_{hk} and P_{pq} share an edge $a_0 = (i_0, j_0)$. Then we can save the amount of $2c(a_0)$ in connection costs by matching x_h to x_p and x_k to x_q; see Figure 3.20.

Example

Consider the network in Figure 3.21, where the number along an edge indicates its length. The aim is to determine a cycle such that every edge is used at least once and the total length is as low as possible.

Not all nodes have even degree, so the graph in Figure 3.21 does not have an Euler circuit in which every edge is passed through exactly once. We therefore need the Chinese postman problem algorithm. The four nodes B, D, E, and F have odd degrees. We first determine the shortest paths between the nodes of odd degree for the original graph:

- The shortest path between B and D is $B - C - D$, with length 9.
- The shortest path between E and F is $E - D - G - F$, with length 12.
- The shortest path between B and E is $B - A - E$, with length 12.
- The shortest path between D and F is $D - G - F$, with length 7.
- The shortest path between B and F is $B - C - D - G - F$, with length 16.
- The shortest path between D and E is $D - E$, with length 5.

Fig. 3.21   Example of a Chinese postman problem.

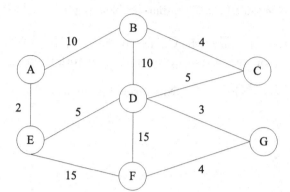

There are only three pairings possible to match the nodes of odd degree, so that we can find the optimal matching through inspection and therefore do not need an algorithm. For the matching (BD, EF), the total "cost" is equal to $9 + 12 = 21$; for the matching (BE, DF), the total "cost" is equal to $12 + 7 = 19$, and for the matching (BF, DE), the total "cost" is equal to $16 + 5 = 21$. The optimal matching is therefore (BE, DF), with corresponding shortest paths $B - A - E$ and $D - G - F$. Figure 3.22 shows the extended graph in which the edges in these two paths have been included as additional edges.

Fig. 3.22   Extended graph.

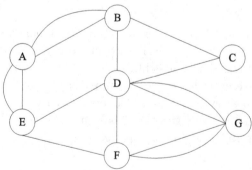

In this graph, every node has even degree, so that we can apply Fleury's algorithm to find an Euler circuit. This Euler circuit is $A - B - C - D - G - F - D - E - A - B - D - G - F - E - A$ (verify!). This is also the solution to the original problem, and the total distance traveled using this circuit is 92.

3.6 Exercises

3.1 In a neighborhood in a large city, there is a fire station, from which seven places (nodes) must be reached quickly. We number the nodes 1 through 8, where node 1 is the fire station. The nodes are connected by streets, of which some are one-way streets and others allow two-way traffic. In the two-way streets, the driving time is not the same in the two directions. The table below gives the possible routes and driving times. Calculate the shortest path from the fire station to every other node.

From\To	2	3	4	5	6	7	8
1	5		2				
2		9	1		7	10	
3				10	5		
4	2				3		
5		7				4	7
6						2	9
7			7				6
8				1	10		

3.2 A vast oil supply has been found in a remote area. An oil company wants to build a pipeline from the site to a refinery in a neighboring country. The pipeline must go through one or more of a number of given nodes. The nodes are numbered 1 through 7, where node 1 represents the location of the oil and node 7 is the refinery. The table below gives the distances in tens of kilometers between the nodes and the costs (in parentheses) in thousands of euros per ten kilometers for the possible routes. Moreover, in every node that is used (except for node 7), a pumping station must be built. The cost of building a pumping station in any of the nodes 2, 4, and 6 is 25 000 euros, and the cost in any of the nodes 1, 3, and 5 is 45 000 euros. What pipeline network has the lowest total cost?

From\To	2	3	4	5	6	7
1	20(5)	15(4)				
2			7(8)	2(5)	15(6)	
3			7(5)	6(7)		
4					10(5)	14(9)
5					11(3)	
6						12(4)

3.3 A car rental agency has a surplus of two cars in both city 1 and city 2 and a shortage of one car in each of the cities 5, 6, 7, and 8. How should the cars be brought from the cities with a surplus to the cities with a shortage to minimize the total number of kilometers driven? From city 1, only the cities 2, 3, and 5 can be reached directly, with travel distances 5, 2, and 11; from city 2, the cities 4, 5, and 6 can be reached directly, with travel distances 10, 7, and 10; from city 3, the cities 2, 4, 5, and 6 can be reached directly, with travel distances 2, 9, 6, and 11; from city 4, the cities 6, 7, and 8 can be reached directly, with travel distance 2, 4, and 5; from city 5, the cities 4 and 6 can be reached directly, with travel distances 5 and 2; from city 6, the cities 7 and 8 can be reached directly, with travel distances 2 and 4; and from city 7, the cities 6 and 8 can be reached directly, with travel distances 1 and 2.

3.4 Five cities are candidates for the construction of a hospital. The table below gives the direct distance between any two of these cities. Where should the hospital be built to minimize the total sum of the travel distances between the cities and the hospital? Where should it be built to minimize the greatest distance from a city to the hospital?

From\To	City 1	City 2	City 3	City 4	City 5
City 1	–	25	17		15
City 2	25	–	12	9	
City 3	17	12	–	19	12
City 4		9	19	–	10
City 5	15		12	10	–

3.5 **(a)** The student bar "Reintje de Vos" has a special offer for the coming Friday evening. Anyone can drink as much beer as he wants for free provided that he drinks it in portions of 0.4 liters. The bar provides a full beer mug with 0.8 liters of beer and two empty beer mugs that can hold 0.3 liters and 0.5 liters. These are the only tools given to divide the beer into two equal portions of 0.4 liters each. Is this possible and if it is, how does one do this in the least number of steps? Model this as a shortest-path problem and give the meaning of the nodes and arcs. However, do not use Dijkstra's algorithm to find the optimal solution, but creative thinking (Dijkstra's algorithm requires many calculations and is impractical to do by hand).

(b) A group of four persons is standing on the southern bank of a river. The group must cross the bridge to get to the other side. However, it is pitch dark, and the group carries only one flashlight. The flashlight is indispensable to cross the rather rickety bridge. As no more than two persons can cross the bridge at the same time, several crossings are required. Not all of the four persons can cross the bridge at the same speed: person 1 needs 1 minute, person 2 needs 2 minutes, person 3 needs 5 minutes, and person 4 needs 10 minutes. Of course, the crossing time of a group consisting of two persons is equal to the time of the slowest of the two. How should the river crossing be carried out to minimize the total time needed until everyone is on the northern bank of the river? (Use the obvious fact that after a crossing, the fastest person from the group on the northern bank returns to the other bank if there are still people there.) Model this as a shortest-path problem and give the meaning of the nodes and arcs. However, do not use Dijkstra's algorithm to find the optimal solution, but creative thinking (Dijkstra's algorithm requires many calculations and is impractical to do by hand).

3.6 Using an existing network of pipelines, two oil fields must supply three refineries. The maximum quantities the oil fields can deliver are 10 000 liters per hour for oil field 1 and 15 000 liters per hour for oil field 2. The following quantities of oil are needed at the refineries: 9000 liters per hour for refinery 1, 7000 liters per hour for refinery 2, and 7500 liters per hour for refinery 3. The transportation cost is €0.01 per liter per kilometer. The pipelines in the network and their lengths (in kilometers) are given in the table below, where O_1 and O_2 represent the oil fields, R_1, R_2, and R_3 the refineries, and J_1, J_2, and J_3 the intersection points of the network. The capacities of the pipelines are not limiting. Calculate how the refineries should be supplied by the oil fields to minimize the total cost.

From\To	O_1	O_2	J_1	J_2	J_3	R_1	R_2	R_3
O_1	$-$	$*$	30	70	$*$	$*$	$*$	$*$
O_2	$*$	$-$	50	35	25	$*$	$*$	55
J_1	$*$	$*$	$-$	$*$	$*$	$*$	$*$	70
J_2	$*$	$*$	$*$	$-$	35	45	$*$	$*$
J_3	$*$	$*$	$*$	$*$	$-$	30	40	25
R_1	$*$	$*$	$*$	$*$	$*$	$-$	50	$*$
R_2	$*$	$*$	$*$	$*$	$*$	$*$	$-$	$*$
R_3	$*$	$*$	$*$	$*$	25	$*$	$*$	$-$

3.7 Consider a directed network with five nodes $N_1, \ldots, N_5$. A message must be sent from node N_1 to node N_5. The table below gives, for each arc, the probability that the message is lost when it passes through that arc (a star means that there is no arc between the two nodes in question). How should the message be sent to maximize the probability that it arrives safely at node N_5?

From\To	N_1	N_2	N_3	N_4	N_5
N_1	$-$	0.06	0.10	0.06	$*$
N_2	$*$	$-$	0.02	0.04	0.07
N_3	$*$	0.02	$-$	0.08	0.10
N_4	$*$	$*$	0.02	$-$	0.08

3.8 Liquid has to be pumped from tank A to tank D via one or more tanks. When the liquid is transferred from one tank to another, some is lost. When transferring from A to B, C, or D, the respective loss rates are 1%, 3%, and 4%; when transferring from B to A, C, or D, the loss rates are 2%, 2%, and 3%; and when transferring from C to A, B, or D, the loss rates are 4%, 3%, and 2%. How should the liquid be transferred from tank A to tank D to minimize the loss?

3.9 A confidential message must be read by five persons. The message is intercepted with probability p_{ik} if it is passed directly by person i to person k. The same probability holds if the message is passed directly by person k to person i; that is, $p_{ik} = p_{ki}$. These probabilities are given in the table below. How should the message be circulated among the five persons to minimize the probability of it being intercepted?

	1	2	3	4	5
1	$-$	0.05	0.07	0.03	0.08
2		$-$	0.09	0.04	0.05
3			$-$	0.07	0.04
4				$-$	0.06

3.10 After the minimum spanning tree has been calculated for a network with costs $c_{ij} \geq 0$, it turns out that the costs should have been $c_{ij}\sqrt{c_{ij}}$ rather than c_{ij}. Explain why the minimum spanning tree remains the same and there is no need to redo the calculations.

3.11 An inhabitant of Rotterdam has severe back pain and can therefore not sit very long. She needs to go for a short walk frequently. Yet, this person needs to travel to Enschede. She has decided to do so by train and wants to minimize the maximum time she sits in one train. Based on the travel times (in minutes) from city to city in the table, what route should she take?

From\To	Am.	Ar.	Br.	En.	Go.	He.	Ro.	Ut.	Zu.	Zw.
Amersfoort	–	*	*	82	38	73	57	14	*	36
Arnhem		–	77	*	*	*	*	35	21	60
Breda			–	*	*	*	24	*	102	141
Enschede				–	122	7	*	98	*	50
Gouda					–	113	18	18	*	77
Hengelo						–	*	90	37	42
Rotterdam							–	37	*	96
Utrecht								–	*	53
Zutphen									–	37
Zwolle										–

3.12 As part of a temporary detour, highway traffic must be routed through a city. The highway can handle 2000 cars per hour during peak hours. A set of possible routes through the city has been proposed. The streets of the road network and the corresponding capacities are given in the table below, where E is the point where the cars enter the city, J_1, J_2, J_3, J_4, and J_5 are the intersection points of the road network, and L is the point where the cars leave the city. The flow capacities are given in units of 100 cars per hour. Most streets are one-way streets. Is it possible to accommodate 2000 cars per hour with this road network?

From\To	E	J_1	J_2	J_3	J_4	J_5	L
E	–	7	9	8	*	*	*
J_1	*	–	4	*	7	*	*
J_2	*	4	–	5	3	7	*
J_3	*	*	5	–	*	8	*
J_4	*	*	*	*	–	5	10
J_5	*	*	*	*	*	–	12

3.13 There are several routes between city A in Texaco and city F in Utopia, which pass through one or more of the cities B, C, D, and E. The border guard must set up enough checkpoints along the existing roads so that everyone has to pass through at least one checkpoint on a trip from city A to city F. The existing connecting roads are (A, B), (A, C), (A, D), (B, E), (C, F), (D, E), (D, F), (E, C), and (E, F), and the respective costs of checkpoints along these roads are 30, 20, 40, 24, 20, 30, 30, 10, and 40. What is the least expensive solution for setting up checkpoints?

3.14 Five instructors are available to teach four subjects. One instructor can only teach one of the four subjects at a time. Every instructor has already taught every subject several times in the past. Based on the results of student surveys, a rating out of 10 has been given to every combination of an instructor and a subject. These ratings are given in the table. To which four instructors should the subjects be assigned to maximize the average rating? Formulate this problem as an assignment problem and solve it using the Hungarian method.

Teacher\Subject	1	2	3	4
1	6.5	6.0	7.8	7.0
2	8.0	7.5	8.2	8.2
3	8.5	6.5	8.4	7.6
4	8.1	7.3	7.0	6.9
5	8.2	7.1	8.2	7.3

3.15 For the Dutch club championships, a swimming coach has to form a relay team for the 4 × 100-meter medley relay. After preselection, five swimmers are eligible for the relay team. The best times of the five swimmers in the different swimming styles are given in the table below. Solve the problem using the Hungarian method.

Swimmer	1	2	3	4	5
Backstroke	59.2	55.3	57.9	59.8	55.7
Breaststroke	64.9	64.5	66.8	63.4	65.9
Butterfly Stroke	57.3	54.2	55.7	56.2	55.8
Freestyle	51.3	49.3	50.2	51.9	50.8

3.16 Consider the maximization version of the assignment problem, where a profit w_{ij} is made if person i is assigned to task j and the goal is to maximize the total gain. Explain that this maximization assignment problem can be reduced to the standard assignment problem with minimization by taking c_{ij} equal to $-w_{ij} + M$, where M is the greatest value of all the w_{ij}.

3.17 An international contracting company has asked three different airlines for quotations for flights to five cities. In the coming quarter, the company employees must take 30, 20, 35, 40, and 25 flights to the respective cities of Rio de Janeiro, Sydney, Bangkok, Singapore, and Nairobi. The table below gives the amounts charged by the three airlines A, B, and C per flight to each of these cities. The contracting company wishes to allocate 30% of the flights to company A, 40% to company B, and 30% to company C. How should the flights be allocated to minimize the total cost? Formulate this problem as a transportation problem and solve it using an LP code.

	Rio	Sydney	Bangkok	Singapore	Nairobi
A	€1800	€2250	€1225	€1425	€1250
B	€1750	€2350	€1300	€1450	€1275
C	€1775	€2100	€1250	€1390	€1290

3.18 Determine a tour for the Königsberg bridge problem such that every bridge is crossed at least once and the total number of times a bridge is crossed is minimized.

3.19 Solve the Chinese postman problem in the figure below.

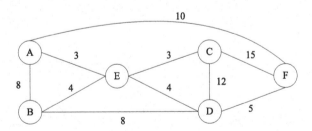

References

1. R. K. Ahuja, T. L. Magnanti, and J. B. Orlin, *Network Flows: Theory, Algorithms and Applications*, Prentice–Hall, Inc., Upper Saddle River, New Jersey, 1993.

2. N. Christofides, *Graph Theory: An Algorithmic Approach*, Academic Press, Inc., London, UK, 1975.

3. P. A. Jensen and J. F. Bard, *Operations Research, Models and Methods*, John Wiley & Sons, Inc., Hoboken, New Jersey, 2003.

Chapter 4

Decision Trees

Decision trees are an extremely useful tool for analyzing decision problems which, in a situation of uncertainty, require one to take a series of related decisions. This tool enables the decision maker to assess the consequences of the various decisions and determine the most desirable one given the available information. Decision trees can be interpreted as a formalization of common sense for more complex decision-making situations under uncertainty. We explain the basic idea of the approach with a few examples. First, we introduce the notion of a (probability) tree diagram, after which we give an example of the use of decision trees.

4.1 Tree Diagrams

A (probability) tree diagram provides a clear way to map out uncertainties for situations where there is a process that unfolds over several points in time and is surrounded by uncertainties.

4.1.1 *The Monty Hall Problem*

Rarely has a probability problem aroused emotions as much as the Monty Hall problem. This problem, named after the quizmaster of a then-famous TV show, became known worldwide when the American columnist Marilyn vos Savant raised the problem in *Parade* magazine in 1990. The finalist of a television quiz has to choose one of three doors. Behind one door is an expensive car, and behind the other two are goats. The finalist randomly chooses one of the doors. As promised beforehand, the quizmaster then opens one of the two remaining doors hiding a goat. He then asks the candidate whether she wants to switch doors. The candidate faces a dilemma. What should she do? Vos Savant's advice was to switch and take the only remaining door. In this way, the candidate would increase the probability of winning the car to $\frac{2}{3}$. Vos Savant was inundated by thousands of letters from readers who for the most part did not agree with her solution. The challenges to her solution were sometimes worded quite fiercely. Ninety percent of the letter writers, including mathematicians, said that there was no point in switching doors.

They argued that the two unopened doors that remained were each hiding the car with probability $\frac{1}{2}$. In other countries, too, the problem received the necessary attention and provoked many emotional reactions. In the Netherlands, a reader wrote to his newspaper: "I am aware that it is of great brutality, as someone who has always failed mathematics, to dispute the conclusion that the probability of winning becomes $\frac{2}{3}$ when switching doors. Let me use an analogous example to show that the columnist is wrong. Suppose that there are 100 doors and that the candidate stands in front of door 1. She then has probability 1% of standing in front of the correct door, while the probability that the car is behind one of the other doors is 99%. The quizmaster then opens doors 2 through 99. The car is behind none of them. Then the car must be behind door 1 or door 100. If I were to follow the columnist's reasoning, the entire 99% probability would transfer to door 100. Needless to say, this is complete nonsense. Here, too, a new situation has arisen with only two possibilities, each with an equal probability." This reader is not only completely wrong but inadvertently contributes a strong argument for switching doors. Another reader wrote: "As a professional mathematician, I am increasingly concerned about the lack of mathematical understanding among the general public. Just realize that the probability must be $\frac{1}{2}$ and, in the future, refrain from commenting on matters you do not understand." Martin Gardner, the spiritual father of the Monty Hall problem, wrote a long time ago: "In no other branch of mathematics is it so easy for experts to blunder as in probability theory."

The fact that there was so much disagreement about the correct answer can be explained by the psychological phenomenon that many people naturally tend to give equal probabilities to the remaining possibilities. Some readers may have thought that the quizmaster opened one of the two remaining doors arbitrarily. If that were the case, there would indeed be no advantage in switching doors. However, the quizmaster has promised to open a door that does not hide the car. This changes the situation. The candidate received additional relevant information. At the moment of the candidate's first choice, the probability that the car is behind the chosen door is $\frac{1}{3}$, and the probability that it is not is $\frac{2}{3}$. The quizmaster then opens a door. Beforehand, the quizmaster has promised to always open a door without a car, regardless of the candidate's choice (this is an essential fact). In the situation where the car is not behind the door chosen by the candidate, it is certain that the car is behind the door that remains after the quizmaster has opened a door. In other words, switching doors will lead to winning the car with probability $\frac{2}{3}$ and to not winning the car with probability $\frac{1}{3}$. It is, of course, harsh, if the candidate switches doors and her first choice subsequently turns out to be the right one.

Tree Diagram

The reasoning that leads to the correct answer of $\frac{2}{3}$ is simple, but the thought does have to occur to one. How can one reach the right answer more systematically without running the risk of getting caught up in wrong intuitive reasoning? One way is to use computer simulation (or mimic the game with three beer mats, each time placing a ten-euro bill under one of them). A systematic approach that only requires pen and paper is also possible. This is based on the concept of a "tree diagram." A tree diagram clearly shows how the probabilities depend on the available information. Figure 4.1 shows the tree diagram for the Monty Hall problem. This diagram shows all possible events and their associated probabilities. For the sake of convenience, we have numbered the candidate's first chosen door in the diagram as door 1. The diagram also shows the quizmaster's promise to always open a door that does not hide the car. For example, the quizmaster opens door 2 if the car is behind door 3 and vice versa, while the quiz master opens either of the two doors 2 and 3 if the car is behind door 1. The numbers on the line segments coming from a chance node give the probabilities of the events that can occur at that chance node. The probabilities of the possible paths are calculated by multiplying the probabilities of the line segments. The tree diagram shows that the last two paths lead to winning the car when switching doors. The probability of winning the car is given by the sum of the probabilities of these two paths and is therefore equal to $\frac{1}{3} + \frac{1}{3} = \frac{2}{3}$. So the correct answer is indeed $\frac{2}{3}$.

Fig. 4.1 The tree diagram for the Monty Hall problem.

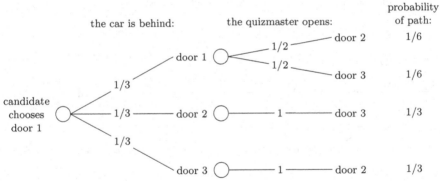

The Monty Hall problem shows that in some probability problems, one can easily come to an intuitive argument in which the error is difficult to pinpoint. This fact is further confirmed by the following closely related problem.

The Three Prisoners Problem

Each of three prisoners A, B, and C qualifies for early release for good behavior. The prison warden has decided to release one of the three prisoners and to let the choice be determined by chance. The three prisoners know that one of them is being released, but they do not know who the lucky one is. The correctional officer does know. Arguing that it makes no difference to his probability of being released, prisoner A asks the correctional officer to reveal to him the identity of one of his fellow prisoners who will not be released. The correctional officer refuses because providing the requested information would increase prisoner A's probability of being released to $\frac{1}{2}$. Who is right, the correctional officer or prisoner A? The correct answer is that prisoner A is right, but only under the assumption that the correctional officer chooses one of the two prisoners B and C randomly if neither of them is released (if the correctional officer hates B and only mentions C if he has no choice and prisoner A knows this, then the situation is entirely different!). If the correctional officer provides the requested information, the probability of A being released does not change and remains equal to $\frac{1}{3}$. This immediately follows from the tree diagram in Figure 4.2. Another way to see that the probability must be $\frac{1}{3}$ is to reduce the three-prisoner problem to the Monty Hall problem. Identify the prisoner who is to be released with the door hiding the car. The essential difference with the Monty Hall problem is that there is no question of switching doors. If one does not switch doors, the probability of winning the car remains $\frac{1}{3}$ after the quizmaster has opened a door without a car.

Fig. 4.2   The tree diagram for the prisoner problem.

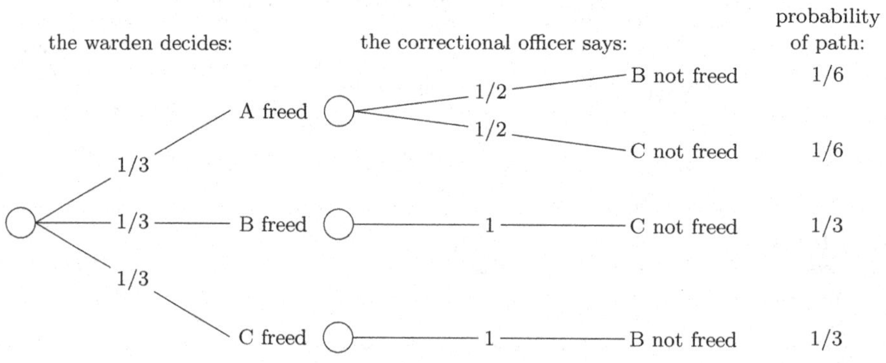

4.1.2   *The Test Paradox*

An inexpensive diagnostic test is available for a particular disease. Although the test is very reliable, it does not provide 100% certainty. If the test result is positive, a more extensive screening follows to determine with absolute certainty whether the disease is present. In individuals who have the disease, the test reacts to the disease in 99% of the cases by giving a positive result. If someone does not have the disease, the probability of a false positive result is 2%. In the current situation, the test is used only on a group of outpatients of whom it is known that, on average, 1 in 2 persons has the disease. What is the probability, for someone in this group, that in case of a positive test result, the person in question has the disease? To determine this, set up a tree diagram as in Figure 4.3. The probability associated with each path in the tree is the product of the probabilities along the edges of the path. In the diagram, the first and third paths lead to a positive test result, with respective probabilities 0.495 and 0.01. The probability that a person has the disease given that the test result is positive is equal to the probability of the occurrence of the first path given that either the first or the third path has occurred. The definition of conditional probability gives

$$\mathbb{P}(\text{path 1} \mid \text{path 1 or path 3}) = \frac{\mathbb{P}(\text{path 1})}{\mathbb{P}(\text{path 1}) + \mathbb{P}(\text{path 3})}.$$

The desired probability is therefore equal to

$$\frac{0.495}{0.495 + 0.01} = 0.9802.$$

In other words, in the test group, on average, 98.0% of the positive results are correct.

Fig. 4.3   The tree diagram for the test group.

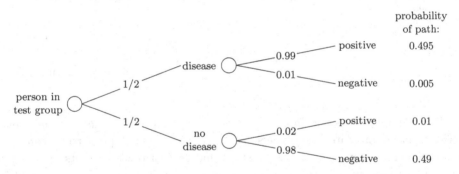

Now, suppose that based on the success of the test, it is proposed that the test be applied annually to everyone. On average, 1 in every 1000 persons in the total population suffers from the disease. Is the proposal above a good idea? To answer this question, we calculate, for an arbitrary person, the probability that she has the disease if tested positive. To this end, we use the tree diagram in Figure 4.4. It follows from that diagram that for a randomly selected person, the probability of having the disease given a positive test result is equal to

$$\frac{0.00099}{0.00099 + 0.01998} = 0.0472.$$

This leads to the seemingly paradoxical answer that, on average, in more than 95% of the cases with a positive test result, the disease is not present. Many people would therefore be unnecessarily worried if the test were applied to everyone. The reason why a test that is, in itself, reliable is so inadequate for the whole population lies in the fact that the vast majority of the population does not have the disease.

As a result, although the probability of a positive test result in the absence of the disease is small, the persons without the disease, due to their overwhelming majority, still produce a much larger number of positive reactions than the small group of actual sick persons. In other words, the number of false positive results far exceeds the number of correct diagnoses if everyone is subjected to the test. The following reasoning supports this. Suppose that 10 000 randomly selected persons are tested. On average, 9990 persons will not have the disease, while 10 will have it. This means, on average, $0.02 \times 9990 = 199.8$ positive test results that come from healthy individuals and $0.99 \times 10 = 9.9$ positive test results that come from sick individuals.

Fig. 4.4   The tree diagram for the entire population.

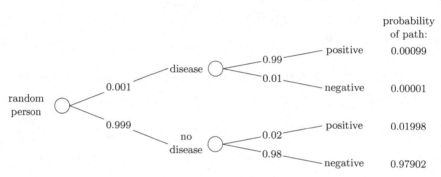

This example shows how important it is to keep an eye on the *basic proportions* between the various categories of persons. Ignoring these basic proportions can lead to strange statements like "Statistics show that 10% of road accidents are caused by drunk drivers, meaning that sober drivers cause the other 90%..., so would it not be better to allow only drunk drivers on the road?"

Bayes' Rule

A different method of calculating the probability of having the disease given a positive test result is based on Bayes' famous rule. The concept of conditional probability is essential in Bayesian probability theory. This is an intuitive concept that is immediately clear to most people without any theoretical consideration. For example, almost everyone argues as follows when asked about the probability of two aces when randomly drawing two cards from a thoroughly shuffled deck of 52 cards: The probability of an ace with the first card is $\frac{4}{52}$. Given that one ace has been removed from the deck, the probability of an ace with the second card is equal to $\frac{3}{51}$. The requested probability is therefore $\frac{4}{52} \times \frac{3}{51}$. This is an application of the basic rule

$$\mathbb{P}(A \text{ and } B) = \mathbb{P}(A)\mathbb{P}(B \mid A).$$

In words, the probability that event A and event B both happen is calculated by multiplying the probability that event A happens with the conditional probability that event B will happen given that event A has happened.

We now calculate the conditional probability $\mathbb{P}(\text{disease} \mid \text{positive})$ using Bayes' rule for the situation of applying the test to the entire population. First, note that the starting data are

$$\mathbb{P}(\text{disease}) = 0.001, \quad \mathbb{P}(\text{no disease}) = 0.999,$$

$$\mathbb{P}(\text{positive} \mid \text{disease}) = 0.99, \quad \mathbb{P}(\text{negative} \mid \text{disease}) = 0.01,$$

$$\mathbb{P}(\text{positive} \mid \text{no disease}) = 0.02, \quad \mathbb{P}(\text{negative} \mid \text{no disease}) = 0.98.$$

Applying first the formula $\mathbb{P}(B \mid A) = \mathbb{P}(A \text{ and } B)/\mathbb{P}(A)$ for the conditional probability and then the law of total probability leads to

$$\mathbb{P}(\text{disease} \mid \text{positive}) = \frac{\mathbb{P}(\text{positive and disease})}{\mathbb{P}(\text{positive})}$$

$$= \frac{\mathbb{P}(\text{positive and disease})}{\mathbb{P}(\text{positive and disease}) + \mathbb{P}(\text{positive and no disease})}.$$

Repeatedly applying the basic rule $\mathbb{P}(A \text{ and } B) = \mathbb{P}(A)\mathbb{P}(B \mid A)$ gives Bayes' rule for the desired probability $\mathbb{P}(\text{disease} \mid \text{positive})$:

$$\mathbb{P}(\text{disease} \mid \text{positive}) =$$

$$\frac{\mathbb{P}(\text{positive} \mid \text{disease})\mathbb{P}(\text{disease})}{\mathbb{P}(\text{positive} \mid \text{disease})\mathbb{P}(\text{disease}) + \mathbb{P}(\text{positive} \mid \text{no disease})\mathbb{P}(\text{no disease})}.$$

Substituting the data gives $\mathbb{P}(\text{disease} \mid \text{positive}) = (0.99 \times 0.001)/(0.99 \times 0.001 + 0.02 \times 0.999) = 0.0472$, the same value as that found using the tree diagram.

Doctors should be more familiar with the concept of a probability tree diagram and Bayes' rule. Consider the following situation: "0.8% of all 40-year-old women who undergo routine screening have breast cancer, 90% of the women with breast cancer test positive, while 7.5% of the women without breast cancer test positive. What is the probability that a 40-year-old woman who tests positive in a routine screening has breast cancer?" In an American study, many doctors gave a wrong answer and estimated the probability at around 75%; apparently, they thought that the probability of having breast cancer given a positive result should be close to the probability of a positive result given the occurrence of breast cancer. The correct answer is 8.8%. The percentage of wrong answers was much lower when the problem was presented in terms of "natural frequencies": "On average, 80 out of 10 thousand 40-year-old women who undergo routine screening have breast cancer. Of any 80 women with breast cancer, 72 have a positive test result, while of the 9920 women without breast cancer, 744 have a positive test result. If 10 thousand 40-year-old women undergo routine screening, what percentage of the women with a positive test result have breast cancer?" The calculation $\frac{72}{72+744} \times 100\% = 8.8\%$ was carried out correctly by most doctors.

4.1.3   *Bayes' Rule in Odds Form*

In practice, the most convenient representation of Bayes' rule is the one in odds form. This is illustrated with the following example. Suppose that a murder has been committed in a large city. The police have found traces of the killer's DNA at the crime scene, and a suspect has been arrested. The only evidence against the suspect is that his DNA profile matches the DNA traces that were found. An expert estimates that, on average, 1 in a million people have a DNA profile that matches the found DNA traces. This is a very small probability, and the prosecutor concludes that there is no doubt that the suspect is guilty. This is a classic error in reasoning known as the *prosecutor's fallacy*: the probability that the suspect is innocent should not be confused with the probability 10^{-6} that a randomly selected person's DNA profile matches the DNA traces. Suppose that the suspect lives in a densely populated region with 4 million people, each of whom could theoretically be the killer. The suspect's lawyer will then say: "There are about 4 million people who could each theoretically be the killer. So, in addition to my client, the expected number of people whose DNA profile matches the DNA traces that were found is approximately equal to 4. There is no convincing evidence that my client is guilty." This conclusion, which is correct as long as no other direct or indirect evidence has been found, also follows naturally from Bayes' rule in odds form. This is a more insightful representation of Bayes' rule. The concept of "odds" is another way of representing probabilities. The odds of a hypothesis are given by the probability that the hypothesis is true divided by the probability that it is false. A hypothesis that is true with probability p therefore has odds $p/(1-p)$. Conversely, odds a mean a probability of $a/(1+a)$.

The Bayesian model enables you to give a judgment about the probability that a hypothesis is true. For example, in a court case, the suspect is guilty or innocent. Let H be the event that the hypothesis is true (the suspect is guilty) and $\bar{H}$ the event that the hypothesis is false (the suspect is innocent). The Bayesian model assumes that, *beforehand*, (subjective) estimates are made for the probabilities of $\mathbb{P}(H)$ and $\mathbb{P}(\bar{H}) = 1 - \mathbb{P}(H)$. These are called the *prior* probabilities. It should be emphasized that these probabilities concern the situation *before* there is any further evidence. If, in the example above, we assume that the killer belongs to the population of 4 million people from the densely populated area, then the prior probability $\mathbb{P}(H)$ that the suspect is guilty is equal to $\mathbb{P}(H) = \frac{1}{4} \times 10^{-6}$. How does the prior probability change when the information becomes available that a relevant event E has occurred? In the example above, event E is given by the observation that the DNA profile of the suspect corresponds to the DNA traces. The judge is now interested in the revised value of the probability that the suspect is guilty given the occurrence of event E. This *posterior* probability is given by $\mathbb{P}(H \mid E)$ and is calculated using the probabilities $\mathbb{P}(E \mid H)$ and $\mathbb{P}(E \mid \bar{H})$. In the murder case, the latter probabilities are given by $\mathbb{P}(E \mid H) = 1$ and $\mathbb{P}(E \mid \bar{H}) = 10^{-6}$ based on an expert's statement that, on average, the DNA profile of 1 in a million people matches the found DNA traces. The desired probability $\mathbb{P}(H \mid E)$ is calculated using *Bayes' rule in odds form*:

$$\frac{\mathbb{P}(H \mid E)}{\mathbb{P}(\bar{H} \mid E)} = \frac{\mathbb{P}(H)}{\mathbb{P}(\bar{H})} \times \frac{\mathbb{P}(E \mid H)}{\mathbb{P}(E \mid \bar{H})},$$

where $\mathbb{P}(\bar{H} \mid E) = 1 - \mathbb{P}(H \mid E)$. In words,

$$posterior\ odds = prior\ odds \times likelihood\ ratio.$$

The derivation is based on the basic formula $\mathbb{P}(A \text{ and } B) = \mathbb{P}(A \mid B)\mathbb{P}(B)$. The latter leads to $\mathbb{P}(A \mid B) = \mathbb{P}(A \text{ and } B)/\mathbb{P}(B) = \mathbb{P}(B \mid A)\mathbb{P}(A)/\mathbb{P}(B)$. This gives

$$\mathbb{P}(H \mid E) = \mathbb{P}(E \mid H)\frac{\mathbb{P}(H)}{\mathbb{P}(E)} \quad \text{and} \quad \mathbb{P}(\bar{H} \mid E) = \mathbb{P}(E \mid \bar{H})\frac{\mathbb{P}(\bar{H})}{\mathbb{P}(E)},$$

from which Bayes' rule in odds form immediately follows.

The ratio of $\mathbb{P}(H)$ to $\mathbb{P}(\bar{H})$ is called the *prior odds* of the hypothesis H, and the ratio of $\mathbb{P}(H \mid E)$ to $\mathbb{P}(\bar{H} \mid E)$ is called the *posterior odds* of the hypothesis. The ratio of $\mathbb{P}(H \mid E)$ to $\mathbb{P}(\bar{H} \mid E)$ is called the *likelihood ratio*. This ratio is a measure of the diagnostic value of the evidence. If the likelihood ratio is greater than 1, then the evidence speaks in favor of the hypothesis, and the posterior odds are higher than the prior odds. It is important that one does not need to know the exact values of the numerator and denominator in the likelihood ratio, as only the ratio matters. In a court case where the evidence is a DNA trace, it is the forensic expert who assigns a value to the likelihood ratio, but it is the court that assigns a value to the prior odds. Bayes' insightful rule in odds form has the pleasant property that new (independent) information can simply be added to the analysis. To calculate the new posterior odds, take the old posterior odds as new prior odds.

Applying the above to the murder case gives

$$\frac{\mathbb{P}(H \mid E)}{\mathbb{P}(\bar{H} \mid E)} = \frac{1/4\,000\,000}{1 - 1/4\,000\,000} \times \frac{1}{10^{-6}} = \frac{1}{4}.$$

So the posterior probability of guilt is $\mathbb{P}(H \mid E) = \frac{1/4}{1+1/4} = \frac{1}{5}$, that is, 20%.

Prosecutor's Fallacy

The prosecutor's fallacy is a common misconception that had dramatic consequences in the Sally Clark case in England. Within slightly more than a year, Sally Clark and her husband had two newborn children die of sudden infant death syndrome (SIDS, or cot death). In 1999, Sally Clark was convicted of double murder and sentenced to life in prison. In the absence of any direct evidence, the jury was decisively influenced by a renowned pediatrician's statement that the estimated probability of both infants dying of SIDS was 1 in 73 million, an estimate that, incidentally, was completely wrong. It took four years for the miscarriage of justice to be overturned and for Sally Clark to be released. Bayesian analysis could have prevented much misery: a very small probability of two subsequent occurrences of SIDS in the same family should have been weighed against the tiny prior probability that a mother would kill her newborn child. The internet has much information on the Sally Clark case.

Insight from Bayes' Rule

In the media, we regularly hear about wrongful convictions that were, in fact, based solely on the suspect's confession. Bayes' rule in odds form provides insight into the underlying error. Let H be the hypothesis that the suspect is guilty, and let E be the event that the suspect confesses. For the confession to contribute to the conviction, the posterior probability $\mathbb{P}(H \mid E)$ must be greater than the prior probability $\mathbb{P}(H)$. Using simple algebra, it follows from Bayes' rule that $\mathbb{P}(H \mid E) > \mathbb{P}(H)$ holds only if $\mathbb{P}(E \mid H)$ is greater than $\mathbb{P}(E \mid \bar{H})$ (for $0 < p, q < 1$, we have $p/(1-p) > q/(1-q)$ only if $p > q$). In other words, $\mathbb{P}(H \mid E)$ is greater than $\mathbb{P}(H)$ only if the probability that a confession comes from a guilty party is greater than the probability that it comes from someone who is innocent. In real-world situations, it is very questionable whether this is always the case. Bayes' rule underlines that a conviction should never be based on a confession without further evidence.

Examples of Bayes' Rule in Odds Form

Bayes' rule in odds form is an intuitively appealing formula. In the application of this rule, the focus is on translating the problem situation into the hypothesis H and the evidence E. Let us illustrate this with two simple examples. The first example is the mammogram problem at the end of the previous subsection. For H, take the hypothesis that the tumor is malignant, with $\mathbb{P}(H) = 0.01$ and $\mathbb{P}(\bar{H}) = 0.99$. The evidence E is a positive test result, with $\mathbb{P}(E \mid H) = 0.90$ and $\mathbb{P}(E \mid \bar{H}) = 0.10$. This gives

$$\frac{\mathbb{P}(H \mid E)}{\mathbb{P}(\bar{H} \mid E)} = \frac{0.01}{0.99} \times \frac{0.90}{0.10} = \frac{1}{11}.$$

This means that the probability of having a malignant tumor given a positive test result is equal to $\frac{1/11}{1+1/11} = \frac{1}{12}$, that is 8.3%.

The second example is as follows. In a certain region, it rains on average once every ten days. On the days it rains, this is correctly predicted 85% of the cases, while on the days it does not rain, rain is predicted in 25% of the cases. The prediction for tomorrow is rain. What is the probability that it will rain tomorrow? For H, we take the hypothesis that it rains tomorrow, with $\mathbb{P}(H) = 0.10$ and $\mathbb{P}(\bar{H}) = 0.90$. The evidence E is that rain has been predicted for tomorrow; we have $\mathbb{P}(E \mid H) = 0.85$ and $\mathbb{P}(E \mid \bar{H}) = 0.25$. This gives

$$\frac{\mathbb{P}(H \mid E)}{\mathbb{P}(\bar{H} \mid E)} = \frac{0.10}{0.90} \times \frac{0.85}{0.25} = \frac{17}{45}.$$

The probability that it rains tomorrow when this has been predicted is $\frac{17/45}{1+17/45} = \frac{17}{62}$, that is, 27.4%.

4.2   Decision Trees

The main purpose of decision trees is scenario analysis. Each path in the tree represents a scenario. Scenarios are necessary when the outcome of a decision depends not only on the decision itself but also on uncertain outcomes of external events that cannot be influenced by the decision maker. We give two illustrative examples to show how a decision tree is built and analyzed.

4.2.1 *A Choice Problem*

The Dutch electronics group NViews has been offered the opportunity to deliver a large number of night vision goggles to Libya in six months' time. However, that country is currently subject to an arms embargo. If the group decides to accept the offer now, it will make a profit of 10 million euros if the embargo is lifted by the time of delivery, while a loss of 2.5 million euros will be incurred if the embargo is not lifted and delivery has to be stopped. In two months' time, the European Union meets to discuss the lifting of the arms embargo. The electronics group believes that there is a 50% probability that the EU lifts the arms embargo. Instead of taking an immediate decision, the group can also wait until the EU has taken a decision. If the group decides to wait, there is a 20% probability that a competitor takes over the order in the meantime. What should the electronics group do to maximize the expected final yield?

Fig. 4.5   The decision tree for the choice problem.

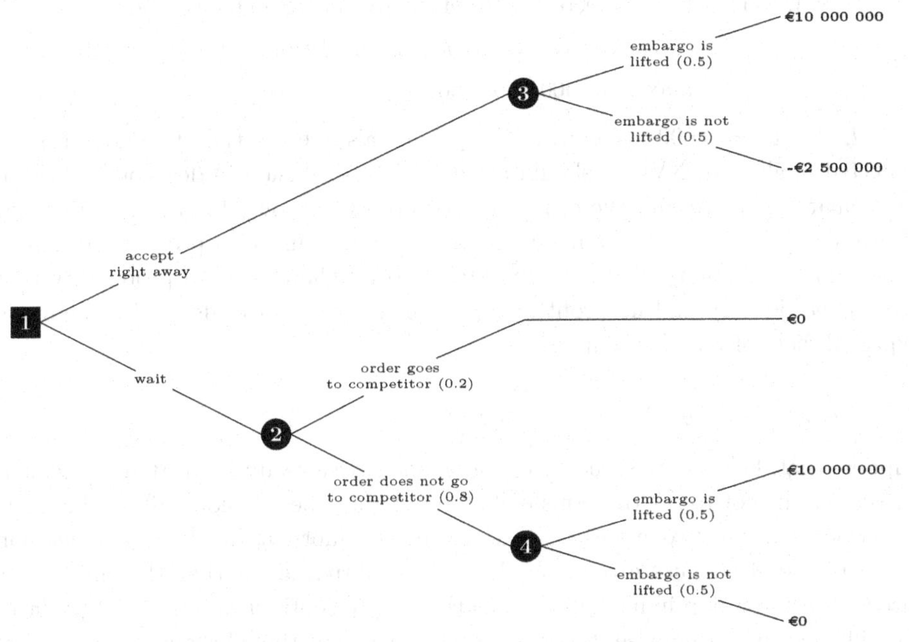

To construct the decision tree for this problem, we use decision nodes and chance nodes. A decision node is indicated by a small square and a chance node by a small circle. Decision branches originate from a square and chance branches from a circle.

The decision tree is built up chronologically over time from left to right. Figure 4.5 shows the decision tree for the choice problem. The numbers in parentheses on the chance branches give the probabilities of the different events. At the end of each path, the revenue in euros for the electronics group is given.

In order to find the maximum expected final yield, we work our way back through the decision tree from right to left. First, we compute

$$\text{expected value of final yield from chance node 4}$$
$$= 0.5 \times 10\,000\,000 + 0.5 \times 0 = 5\,000\,000 \text{ euros}$$

and   expected value of final yield from chance node 3
$$= 0.5 \times 10\,000\,000 - 0.5 \times 2\,500\,000 = 3\,750\,000 \text{ euros.}$$

Next,   expected value of final yield from chance node 2
$$= 0.2 \times 0 + 0.8 \times (\text{expected value of final yield from chance node 4})$$
$$= 0.2 \times 0 + 0.8 \times 5\,000\,000 = 4\,000\,000 \text{ euros.}$$

Finally, we calculate the maximum expected final yield from decision node 1. We denote it by $EMV(1)$, where EMV stands for "expected monetary value." It follows from the relation

$$EMV(1) = \max[\text{expected value of final yield from chance node 3,}$$
$$\text{expected value of final yield from chance node 2}]$$
$$= \max[3\,750\,000, 4\,000\,000]$$

that $EMV(1) = 4\,000\,000$ euros. The above also shows that to maximize the expected final yield, NViews should await the EU's decision in decision node 1. If the embargo is lifted and the order has not been taken over by a competitor, the electronics group delivers the night vision goggles and makes a profit of 10 million euros. The probability of this is $0.8 \times 0.5 = 0.4$. There is a 0.6 probability that there is no delivery and no profit or loss is made. This all leads to the maximum expected yield of 4 million euros.

The Expected Value of Perfect Information

Suppose that, behind the scenes, the politicians have already reached an agreement on whether or not to lift the arms embargo but that the decision will not be made public for another two months. What is inside information about the decision worth to the electronics group? If the inside information is that the embargo is lifted, then the group immediately signs the supply contract and makes a profit of 10 million euros. Otherwise, the group does not accept the offer and has a yield of 0 euros. Inside information gives an expected return of $0.5 \times 10\,000\,000 + 0.5 \times 0 = 5\,000\,000$ euros. This means that the company is prepared to pay a maximum of $5\,000\,000 - 4\,000\,000 = 1\,000\,000$ euros in bribes for inside information. This difference of 1 million euros is called the *expected value of perfect information*.

4.2.2 *An Oil Drilling Problem*

The Wildcat Company of the business duo Jacobse & Van Es has acquired an option in Texas to search for oil on a specific piece of land. The option expires if drilling does not start within the next two weeks. The company must therefore decide soon; the business partners have three possible choices: 1. drill immediately, 2. get a seismic test before deciding whether or not to drill, and 3. let the option expire. At the moment, the company's working capital is 1 275 000 euros. Moreover, the following data are relevant to the decision-making process. Based on the currently available information on the condition of the site, the estimated probability that the relevant piece of land contains oil is equal to 0.50. In order to obtain more information, it is also possible to have a seismic test carried out within a few days by an experienced geologist. In cases where oil drilling took place and the geologist had given prior advice, the geologist had advised to drill in 85% of the cases where oil was found and in 25% of the cases where no oil was found. The seismic test costs 100 000 euros. If, after the test, the company decides to drill for oil, sufficient funds will still be available to finance the drilling cost of 1 000 000 euros. A large oil company has agreed with the duo Jacobse & Van Es to take over the rights to the area for 3 million euros if oil is found. What should the business partners do?

   This decision problem under uncertainty can be structured clearly in the form of a decision tree; see Figure 4.6. As usual, the tree is built up chronologically over time from left to right, where a decision node is indicated by a square and a chance node by a circle. The numbers in parentheses along the decision branches give the direct revenue of the decisions. At the end of each path in the tree, the end capital obtained by following the path is given. The end capital for a path is obtained by taking the sum of the starting capital of 1 275 000 euros and the amounts along the decision branches of the path. The numbers in parentheses along the chance branches give the probabilities of the different events that can occur. These probabilities require some explanation. The probability of oil being present changes when information from the seismic test becomes available. The numerical values $\frac{17}{22}$ and $\frac{5}{22}$ along the chance branches from chance node 4a give the probability of there being oil given the advice to drill and the probability of there not being oil given the advice to drill. The numerical values $\frac{1}{6}$ and $\frac{5}{6}$ along the chance branches from chance node 4b give the probability of there being oil given the advice not to drill and the probability of there not being oil given the advice not to drill. These probabilities can easily be deduced from Bayes' rule in odds form. To calculate the posterior probability of finding oil given the advice to drill, apply this rule with H the hypothesis that oil is present and E the evidence that the geologist has advised to drill. This gives

$$\frac{\mathbb{P}(H \mid E)}{\mathbb{P}(\bar{H} \mid E)} = \frac{\mathbb{P}(H)}{\mathbb{P}(\bar{H})} \times \frac{\mathbb{P}(E \mid H)}{\mathbb{P}(E \mid \bar{H})} = \frac{0.5}{0.5} \times \frac{0.85}{0.25} = \frac{17}{5},$$

that is, $\mathbb{P}(H \mid E) = \frac{17}{5}/(1 + \frac{17}{5}) = \frac{17}{22}$. The probability of oil given the advice not to drill is calculated in the same way as $\frac{1}{6}$. For chance node 2, one needs

Fig. 4.6   The decision tree for the oil drilling problem.

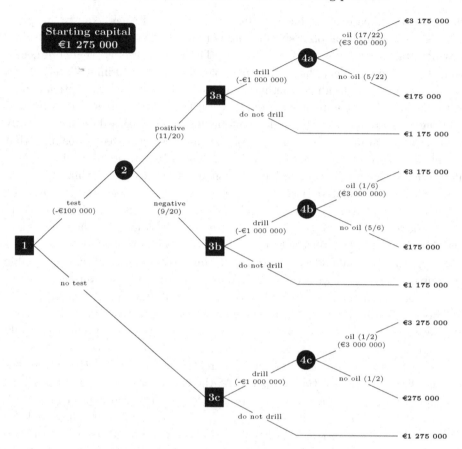

the probability that the test is positive. This probability can be calculated as
$\mathbb{P}(\text{positive} \mid \text{oil})\mathbb{P}(\text{oil}) + \mathbb{P}(\text{positive} \mid \text{no oil})\mathbb{P}(\text{no oil}) = 0.85 \times 0.5 + 0.25 \times 0.5 = \frac{11}{20}$.

How is the best strategy calculated when the expected value of the final capital is taken as the optimization objective? For each of the decision nodes 1, 3*a*, 3*b*, and 3*c*, we calculate the maximum expected final capital that can be obtained from the decision node. The calculations are carried out backward, that is, first for the final decision node and then for the preceding one. We denote the maximum expected final capital from decision node or chance node P by $EMV(P)$. First, consider the decision node 3*a*. The decision "drill" leads to a chance node from which the expected value of the final capital is equal to

$$\frac{17}{22} \times 3\,175\,000 + \frac{5}{22} \times 175\,000 = 2\,493\,182 \text{ euros,}$$

while the decision "do not drill" leads to a final capital of 1 175 000 euros. This means that

$$EMV(3a) = \max[2\,493\,182, 1\,175\,000] = 2\,493\,182 \text{ euros.}$$

So the best decision in node $3a$ is to drill for oil. For node $3b$, we find

$$EMV(3b) = \max[\tfrac{1}{6} \times 3\,175\,000 + \tfrac{5}{6} \times 175\,000, \ 1\,175\,000]$$
$$= \max[675\,000, 1\,175\,000] = 1\,175\,000 \text{ euros.}$$

The best decision in node $3b$ is not to drill. For node $3c$, we have

$$EMV(3c) = \max[0.5 \times 3\,275\,000 + 0.5 \times 275\,000, \ 1\,275\,000]$$
$$= \max[1\,775\,000, 1\,275\,000] = 1\,775\,000 \text{ euros.}$$

The best decision in node $3c$ is to drill. Next, we calculate the EMV for decision node 1. The expected value of the final capital from node 2 is

$$EMV(2) = \tfrac{11}{20} \times EMV(3a) + \tfrac{9}{20} \times EMV(3b) = 1\,900\,000 \text{ euros.}$$

This gives

$$EMV(1) = \max[EMV(2), EMV(3c)]$$
$$= \max[1\,900\,000, 1\,775\,000] = 1\,900\,000 \text{ euros.}$$

The best decision in node 1 is to request a seismic test.

In summary: if the objective is to maximize the expected value of the final capital, then the best strategy is as follows: 1) first, carry out a seismic test, and 2) if the test leads to the advice drill for oil, follow that advice; otherwise, do not drill and let the option expire. Following this strategy leads to an expected value of the final capital equal to €1 900 000. For the management of the Wildcat Company, it is also important to know the probability distribution of the final capital for this strategy. Suppose that the probability of substantial losses is nonnegligible; the management will think twice before accepting the strategy. It immediately follows from the decision tree in Figure 4.6 that the final capital has probability $\tfrac{11}{20} \times \tfrac{17}{22} = 0.425$ of being equal to 3 175 000 euros, probability $\tfrac{11}{20} \times \tfrac{5}{22} = 0.125$ of being equal to 175 000 euros, and probability 0.45 of being equal to 1 175 000 euros.

Utility Criterion

The major advantage of a decision tree is that it maps out all possible scenarios, including worst-case scenarios in which large losses could be possible. In situations of nonrecurring uncertainty, people often do not base their decision on the economic criterion of the expected value. For example, suppose that a gambler is offered the choice between the following two lotteries: (1) winning €100 000 with absolute certainty, (2) winning €2 000 000 with probability $\tfrac{1}{10}$ and winning nothing with probability $\tfrac{9}{10}$. Unlike the Emir of Kuwait, the gambler probability prefers the first lottery even though the second lottery has the higher *expected* revenue of €200 000. Different persons may have different utility functions. A utility function transforms "money" into "utility." In situations where large losses are possible, the expected value criterion is often an inappropriate criterion on which to base decisions.

4.3   Exercises

4.1 Consider the following variant of the Monty Hall problem. There are now four doors, with a car behind one of them. First, the contestant points to a door. The quiz master then opens another door that does not hide the car and gives the player the opportunity to switch doors. Regardless of whether the contestant switches, the quiz master opens another door, different from the contestant's current choice, which does not hide the car and gives the contestant one last opportunity to switch doors. What is the contestant's best strategy?

4.2 In the Monty Hall show there are 10 cups and under one of the cups is a prize. The cups are in random order. Monty knows under which cup the prize is located. A contestant chooses two cups and marks them. Then Monty removes 6 of the unmarked 8 cups but not the cup covering the prize. So there are four cups left and the price is under one of these four cups. Then, the contestant gets the option to choose one of the remaining four cups and exchange it for the two previously chosen cups. What should the contestant do in order to maximize the chance of winning the prize?

4.3 On the sidewalk, passers-by are invited to take part in the following bet. Three cards go into a hat. One card is red on both sides, one is black on both sides, and the third has one red side and one black side. A player is asked to draw a card from the hat in such a way that only one side of the card is visible. No matter which color the player draws, the hat's owner bets that the other side of the card has the same color; the owner wins if the two sides of the card have the same color, while the player wins if they do not. Is this bet fair to the player? Determine this both with a tree diagram and with Bayes' rule in odds form.

4.4 In a particular region, alcohol tests are regularly carried out. A driver who has been stopped first undergoes a breath alcohol test. If this test is positive, the driver is taken in for a blood alcohol test. The breath test gives a positive reaction in 90% of the cases of too high blood alcohol content (BAC), while it gives a positive reaction in 5% of the cases where the BAC is not too high. Currently, a driver is stopped for a breath alcohol test only in the case of suspicious driving. It has been proposed to randomly stop road users for breath tests. Of all road users in the region in question, on average, 1 in 25 drives under the influence of alcohol. What is the probability that a randomly stopped driver with a positive breath test is unnecessarily subjected to a blood alcohol test? Determine this probability both with a tree diagram and with Bayes' rule in odds form.

4.5 In a city, there are two taxi companies, Yellow Riders and White Riders; 85% of the taxis in the city belong to the first company and 15% to the second. On a rainy evening, a taxi is involved in a hit-and-run accident. A witness to the accident believes that the taxi belongs to the White Riders. When the court tests the witness's reliability under similar weather conditions, it turns out that in 80% of the cases, the witness identifies the taxi's color correctly, and in 20% of the cases, the witness is wrong. What is the probability that the taxi that drove off was indeed a White Riders taxi? Determine this probability both with a tree diagram and with Bayes' rule in odds form.

4.6 A shipwreck that has been sought for a long time has been located, with a 40% probability, in a specific area. Further investigation would lead to the wreck with a probability of 90% if it indeed is in the area. The investigation is carried out but does not unearth the wreck. What is the probability that the wreck is nevertheless located in the area in question? Determine this probability both with a tree diagram and with Bayes' rule in odds form.

4.7 Of twins, 30% are identical (and necessarily of the same sex), and 70% are fraternal. Elvis Presley (the "King of Rock and Roll") had a twin brother who died at birth. Determine the probability that Elvis was an identical twin both with a tree diagram and with Bayes' rule in odds form.

4.8 A murder has been committed. The killer is one of two persons X and Y, both of whom have fled. After an initial investigation, they are equally likely to be the perpetrator. Testing shows that the killer has blood group A. This blood group occurs in 10% of the population. Further investigation shows that person X has blood group A, but does not give any indication of person Y's blood group. In the light of this new information, what is the probability that person Y is the killer? Apply Bayes' rule in odds form.

4.9 The well-known thriller writer Stephanie Queen has offered the exclusive film rights for her new book, which will be released in a year, to film studio Van Gogh for the sum of 1 million dollars. The film studio must now decide whether or not to accept this offer. The decision whether or not to turn the book into a movie will only be taken after the book has been published. The film studio's expert estimates that there is a 60% probability that the upcoming book will be a success and a 40% probability that it will not. Furthermore, a film adaptation of the book is expected to generate a net profit of 30 million dollars if the movie is a blockbuster and a loss of 10 million dollars if it is a flop. The probabilities that the movie is a blockbuster or flops are 75% and 25%, respectively, if the book is a success, and 20% or 80%, respectively, if the book is not a success. Use a decision tree to determine what the studio should do to maximize the expected net profit.

4.10 A contestant on a TV quiz can win ten thousand euros by correctly guessing the ratio of red marbles to white marbles in an opaque vase. The vase contains a great number of marbles. The contestant must guess whether the vase contains twice as many red marbles as white marbles or, conversely, twice as many white marbles as red marbles. Both options have the same prior probability. To help the contestant guess, he is given the opportunity to pull one, two, or three marbles from the vase, just once. This does, however, reduce the prize money of ten thousand euros by 750 euros if the contestant draws one marble, by 1000 euros if he draws two, and by 1500 euros if he draws three. Use a decision tree to decide which strategy maximizes the expected yield.

4.11 A new element must be built into an electronic system. This element is crucial for the system's functioning. Experience shows that, on average, 3 out of 10 elements require an adjustment for the system to work. Before the element is installed, it may be subjected to a thorough test that determines with 100% accuracy whether or not the element needs to be modified. This test costs €1500. If the element is installed without thorough testing and it turns out that the system does not work, the element must be taken out of the system and adjusted. This costs €4500. It is also possible to carry out a quick test beforehand and use the results of this test to decide whether or not to carry out the more elaborate test. The quick test costs €500 but is not completely accurate. This test gives the indication "safe" or "unsafe." If the element is good, the quick test gives the "safe" signal with probability 0.8 and the "unsafe" signal with probability 0.2. If the element needs adjustment, the signals "unsafe" and "safe" are given with respective probabilities 0.9 and 0.1. If the signal is "unsafe," the thorough test is carried out. For which approach is the expected cost the lowest?

4.12 The insurance company Hollandia has reached a verbal agreement to take over the specialized insurance company Calisure. However, before the official papers can be signed, the management of Calisure sells the company, against the agreement, to the insurance company Imperia. Hollandia does not accept the course of events and engages the Insurance Disciplinary Board. This board stipulates that Imperia must pay Hollandia compensation of 20 million euros. The company Imperia refuses to pay this amount and offers 5 million euros as compensation. Hollandia's lawyers analyze the situation and recommend offering a 10 million euros settlement before considering a lawsuit. The lawyers foresee two possible reactions by Imperia to the settlement proposal. One response is to make a final counterproposal of $7\frac{1}{2}$ million euros, and the other is to simply let it go to trial. The two responses are considered equally likely. If Hollandia decides not to accept any of Imperia's counterproposals and files a lawsuit, the lawyers foresee three possibilities: the court awards the original indemnity of 20 million euros, or the proposed settlement of 10 million euros, or nothing at all. The experts estimate the probabilities of these three options at 10%, 50%, and 40%, respectively. What is the best strategy for Hollandia to maximize the expected value of the settlement?

4.13 A TV production company has acquired the rights to a soap opera. The company can sell the rights directly to a competitor at a profit of 150 thousand euros, but can also decide to make a pilot episode for a national TV network. Beforehand, the company estimates that the probability that the TV network buys the entire series based on the pilot episode is 70%, while the probability that it does not is 30%. In the latter case, the production company will have incurred a loss of 300 thousand euros. A profit of 500 thousand euros is expected if the TV network does purchase the series. Before making a decision, the production company can seek advice from an independent TV consultant. In the past, the consultant has given correct advice in 80% of the cases in which the TV network made the purchase and in 85% of the cases in which the TV network did not make the purchase. The advice costs 25 thousand euros. What strategy leads to the highest expected return?

Reference

1. H. C. Tijms, *Understanding Probability*, 3rd ed., Cambridge University Press, Cambridge, UK, 2012.

Chapter 5

Dynamic Programming

5.1 Introduction

Dynamic programming is a solution method for problems that require a series of related decisions to be taken in succession. These problems are often nonlinear and therefore do not allow the application of linear programming. Sequential decision problems occur in a variety of fields, such as production planning, maintenance and replacement, investment, and raw material requirements. Dynamic programming is not a method with a fixed procedure like linear programming. Every dynamic optimization problem requires its own formulation, often involving a certain degree of creativity. The underlying concept that comes up in every formulation is *recursion*. Recursion is an essential concept in applied mathematics and computer science. Recursion relations are of great importance for both theoretical and computational purposes.

We first discuss sequential decision problems of a deterministic nature. Once one understands the basic principle of dynamic programming for deterministic applications, it is clear how a similar approach works for stochastic sequential decision problems.

Fig. 5.1   What is the shortest path from A to B?

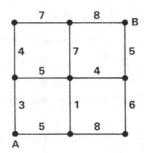

Shortest-path problems are particularly suited for introducing the basic principle of dynamic programming. Let us begin with the elementary network shown in Figure 5.1. The number written along an edge is the travel distance between the two nodes on that edge. The network can be seen as a map of city districts, where the edges correspond to streets and the nodes to intersections. Suppose that one wishes to determine the shortest path from the starting point A to the endpoint B. At each intersection, one can only go up or go right. Detours are not allowed. For the simple example in Figure 5.1, the shortest path can easily be found by enumerating all possibilities. There are six possible paths from A to B. Calculating the lengths of all of these paths shows that the shortest path from A to B has length 15 and is obtained by successively making the following decisions: go right – go up – go right – go up.

Complexity Analysis

For larger networks, it is practically impossible to determine all possible paths and their lengths. To show the fundamental difficulty of the method of complete enumeration, we consider the problem in a more general form. Consider an $n \times m$ Manhattan network as shown in Figure 5.2. It is easy to order the intersections (x, y) in a naturally way, where $x = 0, 1, \ldots, n$ and $y = 0, 1, \ldots, m$. We call the point $(0, 0)$ at the bottom left A and the point (n, m) at the top right B. Only one-way traffic is allowed in the network. The intersection (x, y) only has direct connections to the intersections $(x + 1, y)$ and $(x, y + 1)$, via a line segment to the right and a line segment going up. Suppose that each segment has a given travel distance. The problem is finding the shortest path from the starting point A to the endpoint B. To show the numerical complexity of the method in which all possible paths are enumerated, we determine the total number of possible paths from A to B. Combinatorics tells us that

$$\text{total number of paths from } A \text{ to } B = \binom{n + m}{n} = \frac{(n + m)!}{n!\, m!}.$$

The argument is simple. To go from A to B, one needs to go right n steps and up m steps. The total number of paths from A to B is then the same as the number of different ways of placing n elements of one kind and m elements of another kind in a sequence. In particular, for an $n \times m$ network with $m = n$, we have

$$\text{total number of paths from } A \text{ to } B = \frac{(2n)!}{n!\, n!}.$$

To estimate the order of magnitude of this number, we use Stirling's approximation

$$n! \approx \sqrt{2\pi}\, n^{n+\frac{1}{2}}\, e^{-n} \quad \text{for large } n.$$

In practice, this formula can already be used for $n \geq 10$. The relative error in the approximation is about $(100/12n)\%$. By applying Stirling's approximation, we find for the $n \times n$ network that

$$\text{total number of paths from } A \text{ to } B \approx \frac{1}{\sqrt{\pi n}}\, 2^{2n}.$$

Every path requires $2n - 1$ additions. So for the method of total enumeration, we have

$$\text{total number of operations} \approx \sqrt{(n/\pi)}\, 2^{2n+1}.$$

Fig. 5.2 An $n \times m$ Manhattan network.

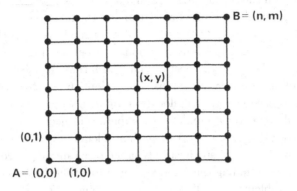

Suppose that we have a high-speed computer that can carry out 10^{10} operations per second. Then for the modest value of $n = 29$, we already need

$$\frac{1.75 \times 10^{18}}{365 \times 24 \times 60 \times 60 \times 10^{10}} \approx 5.5 \text{ years}$$

of computation time to determine all possible paths and their lengths. For a 35×35 network, the computation time grows to approximately 25 000 years. This method therefore cannot be used in practice. A faster supercomputer will not help much. What we need is a modest amount of mathematics. Without mathematical insight, brute computational force does not help. Dynamic programming is a mathematical method that is particularly effective and makes it possible to reduce an exponentially exploding computation time to one that is polynomially bound in the problem size.

Recursion plays a key role in dynamic programming. The basic principle of dynamic programming is essentially simple. A complex problem is divided into a series of smaller nested subproblems with the same structure. These subproblems are related by a recursive relation. Applying the basic principle of dynamic programming on specific problems is partly an art. The basic idea is powerful and generally applicable to a wide variety of sequential decision problems. The best way to understand the problem is through examples. The shortest-path problem is particularly suited for learning the principles of dynamic programming. As already stated, dynamic programming is not a standard solution method, but a general principle for dealing with specific problems. How this approach is to be worked out depends on the specific application, but the general reasoning is essentially the same for every application. Just like the branch-and-bound approach in integer programming, dynamic programming solves the problem by cleverly excluding solutions that cannot be optimal. For calculating the solution of many integer programming problems, the method of dynamic programming is an alternative for the branch-and-bound method.

In Section 5.2, we explain the principle of dynamic programming using the shortest-path problem in Figure 5.2. This problem is particularly well suited to this purpose. The recursive approach of dynamic programming enables us to find a solution very quickly on the computer. Section 5.3 shows us that the versatile technique of dynamic programming can also be used to find the most reliable path if the number along each edge represents the probability of safely passing through the edge in question. In Section 5.4, we apply dynamic programming to the shortest-path problem in an acyclic network. Section 5.5 shows that this shortest-path problem has a number of important applications. In Section 5.6, we show how dynamic programming can be used to solve an allocation problem where a given budget must be optimally distributed over a number of competing projects. A dynamic programming algorithm for the general shortest-path problem is given in Section 5.7. Finally, in Section 5.8, we introduce stochastic dynamic programming. We do this through several dice games, including the game Pig, by presenting the famous Kelly strategy for investment problems, as well as optimal stopping. We will see that, conceptually, the application of dynamic programming to stochastic sequential problems is not more difficult than the application to deterministic problems.

5.2   Shortest Path in a Manhattan Network

An efficient algorithm for the shortest-path problem in Figure 5.2 is the *recursive approach* of dynamic programming. The basic principle of this approach is to divide the original problem into a series of related and easily solvable subproblems. The main observation of the recursive approach is that a shortest path from the starting point $A = (0,0)$ to the endpoint B would be easy to calculate if a shortest path from each of the points $(1,0)$ and $(0,1)$ to B were known. In general, one can observe that a shortest path from point (x,y) to the endpoint B could easily be calculated if the shortest path to B from each of the points $(x+1,y)$ and $(x,y+1)$ were known. The original problem can therefore be divided into a series of nested subproblems of decreasing size. The smallest subproblem is the problem that calculates the shortest path to the endpoint B from each of the points $(n-1,m)$ and $(n,m-1)$. The solution to this subproblem is trivial. To concretize the ideas, we define

$$f(x,y) = \text{ minimum travel distance from } (x,y) \text{ to the endpoint } B.$$

This function is called the *value function* and is crucial in dynamic programming. Note that this function is defined for every point (x,y) even though the goal is to find $f(0,0)$. However, by defining $f(x,y)$ for every point (x,y), it is possible to create a recursive algorithm for $f(x,y)$ that will eventually lead to the desired value $f(0,0)$ for the starting point $A = (0,0)$. The data of the problem are

$$R(x,y) = \text{travel distance from point } (x,y) \text{ to point } (x+1,y)$$
$$U(x,y) = \text{travel distance from point } (x,y) \text{ to point } (x,y+1).$$

The algorithm is initiated with

$$f(n-1,m) = R(n-1,m) \quad \text{and} \quad f(n,m-1) = U(n,m-1).$$

The general recursion step of the dynamic programming algorithm is as follows:

$$f(x,y) = \min \{R(x,y) + f(x+1,y), \ U(x,y) + f(x,y+1)\}.$$

The argument behind this recursive relation is simple and generally applicable. Suppose that one knows the shortest path to the endpoint B from each of the points $(x+1,y)$ and $(x,y+1)$. Then one finds the shortest path from point (x,y) to B by considering the following two paths:

(a) Go right to $(x+1,y)$ and then follow the shortest path from point $(x+1,y)$ to the endpoint B.
(b) Go up to $(x,y+1)$ and then follow the shortest path from point $(x,y+1)$ to the endpoint B.

The total length of path (a) is equal to $R(x,y) + f(x+1,y)$, while the total length of path (b) is equal to $U(x,y) + f(x,y+1)$. It should be clear that the smallest of these two path lengths gives $f(x,y)$. If this were not the case, we would obtain a contradiction, namely that one of the subpaths from $(x+1,y)$ or $(x,y+1)$ is not optimal. Moreover, this calculation of $f(x,y)$ indicates which of the two decisions "go right" and "go up" is optimal at point (x,y). It is clear that the recursive relation requires a minor adjustment for the boundary points (x,m) and (n,y):

$$f(x,m) = R(x,m) + f(x+1,m) \quad \text{and} \quad f(n,y) = U(n,y) + f(n,y+1).$$

The only possible decision in point (x,m) is to go right, and in point (n,y), the only possible decision is to go up.

The recursive relation allows one to determine the function $f(x,y)$ recursively "from back to front." One first calculates $f(n-1,m)$ and $f(n,m-1)$, then $f(n-2,m)$, $f(n-1,m-1)$, and $f(n,m-2)$, etc., and finally $f(0,0)$. During the calculations, one remembers, for every point (x,y), the decision for which the minimum in the recursive equation for $f(x,y)$ is reached. Once the value function has been calculated, the shortest path from A to B can be deduced from these data, working backward. In fact, the dynamic programming algorithm gives the shortest path from every point (x,y) to the endpoint B.

The dynamic programming algorithm significantly reduces the number of calculations compared to the method where all possible paths are considered. It is easy to verify that for the Manhattan network with $m = n$, the total number of operations required by the recursive algorithm is of order $3(n+1)^2$ as opposed to $\sqrt{(n/\pi)}\, 2^{2n+1}$ for the complete enumeration method. This means that even large-scale shortest-path problems can easily be solved on a computer if the recursive method is used.

At first glance, the recursion in dynamic programming is surprisingly simple. Still, some time may be required to fully understand the essence. Most people find the concept behind dynamic programming difficult when they encounter it for the first time, in particular when complex notation is used for the value function. In the example above, the notation could be kept simple. Most applications require more complex notation. However, if one fully understands the idea behind the dynamic programming algorithm for the Manhattan network, then one masters the general principle of dynamic programming. Dynamic programming was designed to solve multistep decision problems. The shortest-path problem from Figure 5.2 is in fact a multistep decision problem. There are $n + m - 1$ decision stages. At the first decision stage, the state is $(0,0)$; at the second stage, it is $(0,1)$ or $(1,0)$; at the third stage, it is one of $(0,2), (1,1)$, and $(2,0)$; and so on.

Numerical Example

Let us illustrate the algorithm with the numerical example from Figure 5.1. The calculations are as follows. First, calculate

$$f(1,2) = 8, \qquad\qquad \text{``go right,''}$$
$$f(2,1) = 5, \qquad\qquad \text{``go up.''}$$

Then, calculate

$$f(0,2) = 7 + f(1,2) = 15, \qquad\qquad \text{``go right,''}$$
$$f(1,1) = \min\{4 + f(2,1),\ 7 + f(1,2)\},$$
$$= \min\{4 + 5, 7 + 8\} = 9 \qquad \text{``go right,''}$$
$$f(2,0) = 6 + f(2,1),$$
$$= 6 + 5 = 11 \qquad\qquad \text{``go up.''}$$

Next, calculate

$$f(0,1) = \min\{5 + f(1,1),\ 4 + f(0,2)\}$$
$$= \min\{5 + 9, 4 + 15\} = 14 \qquad \text{``go right,''}$$
$$f(1,0) = \min\{8 + f(2,0),\ 1 + f(1,1)\}$$
$$= \min\{8 + 11, 1 + 9\} = 10 \qquad \text{``go up.''}$$

Finally, calculate

$$f(0,0) = \min\{5 + f(1,0),\ 3 + f(0,1)\}$$
$$= \min\{5 + 10, 3 + 14\} = 15 \qquad \text{``go right.''}$$

This means that the minimum travel distance from A to B is equal to $f(0,0) = 15$. One finds the shortest path by going backward through the calculations. The minimizing decision at $f(0,0)$ indicates that we must go from $(0,0)$ to $(1,0)$; the minimizing decision at $f(1,0)$ indicates that we must go from $(1,0)$ to $(1,1)$; and so on. In this way, we find that the shortest path from A to B is given by the path $(0,0) \to (1,0) \to (1,1) \to (2,1) \to (2,2)$. This path can, of course, also be found with Dijkstra's algorithm from Section 3.1.

Longest Path

Suppose that we wish to find the longest path from A to B (imagine that the numbers along the edges represent revenue). In contrast to Dijkstra's shortest-path algorithm, the dynamic programming algorithm can still be applied. The only change is replacing the min by max in the recursion relation for the value function. This shows how flexible the dynamic programming approach is.

5.3 Flexibility of Dynamic Programming

In this section, we first describe the general structure of dynamic programming problems. Then, we show how flexible the dynamic programming approach is by considering the Manhattan network for two other optimization criteria. First, we consider the determination of the safest path from A to B when risks are associated with passing through edges in the network. Then, we consider the determination of the path from A to B for which the greatest distance covered in one step is the least.

5.3.1 *General Structure of Dynamic Programming Problems*

Every dynamic programming problem consists of several key components. The problem can be divided into *stages* n, with a decision required at each stage. Stages are also called *decision epochs*: the moments at which a decision must be made. Each stage has a number of *states* associated with it. The *state space* S_n is the set of possible states i which can occur at stage n. The state contains all the information that is needed to make an optimal *decision*. Decisions are also called *actions*. The *decision space* $D_n(i)$ is the set of decisions d which are feasible in state i at stage n. As a consequence of a decision, two things happen: the decision maker receives an *immediate reward*, and there is a *transition* to another state in the next stage. We define $r_n(i, d)$ as the immediate reward during stage n as a consequence of decision d in state i. Naturally, these are rewards in a maximization setting and costs in a minimization setting. Next to the immediate reward, the decision d in state i at stage n causes a transition to state j in stage $n + 1$. In deterministic dynamic programming problems, which we are considering right now, the decision chosen at any stage fully characterizes how the state at the current stage is transformed into the state at the next stage. The fact that a decision causes an immediate reward as well as a transition to another state is at the heart of optimization in dynamic programming problems: a decision is optimal if it achieves the maximum value of the sum of the immediate reward and the rewards that can be earned from the next stage onward.

More formally, when optimizing in a dynamic programming problem, the objective is to maximize the total reward over all stages:

$$\max\left\{\sum_{n=0}^{N} r_n(i,d)\right\}.$$

This is done recursively, using the *optimal value function* $f_n(i)$, which is the maximum total reward that can be obtained from stages n through N if the system is in state i at stage n. The optimal value function is characterized by a *recursion relation* which has the following general structure:

$$f_n(i) = \max_{d \in D_n(i)} \{r_n(i,d) + f_{n+1}(d)\}.$$

At the final stage, N, there is no transition to another state in the next stage, so only the immediate reward plays a role. Hence, $f_N(i)$ can be found easily for all states i and is therefore a natural starting point for the recursive calculations. $f_N(i)$ is called the *salvage value*. An important principle in dynamic programming is the *principle of optimality*: given the current state, the optimal decision for each of the remaining stages must not depend on previously reached states or previously chosen decisions.

In summary, a dynamic programming problem consists of stages, states, decisions, and immediate rewards, which come together in an optimal value function.

5.3.2   *The Safest Path*

Suppose that the numbers $R(x,y)$ and $U(x,y)$ along the edges have the following interpretations:

$$R(x,y) = \text{probability of going safely from point } (x,y) \text{ to } (x+1,y),$$
$$U(x,y) = \text{probability of going safely from point } (x,y) \text{ to } (x,y+1).$$

Think of a communication network with unreliable transmission lines along which a message must be sent from source A to destination B. For every transmission line, there is a given probability that the message will be lost if it passes through that line. The different transmission lines are assumed to behave independently of one another. The safest-path problem is the problem of finding a path from the source A to the destination B that maximizes the probability that the message arrives safely at B. The dynamic programming algorithm requires only a small adjustment. The value function $f(x,y)$ is now defined as

$$f(x,y) = \text{maximum probability of getting the message safely from}$$
$$\text{point } (x,y) \text{ to the destination } B.$$

This function is again defined for every point (x,y). To obtain a recursive relation for $f(x,y)$, we follow the same type of reasoning as before. Suppose that we know a safe path to the destination B from each of the points $(x+1,y)$ and $(x,y+1)$. A safest path from point (x,y) to point B can be obtained by considering the following two paths:

(a) Go right to $(x+1, y)$ and then follow the safest path from point $(x+1, y)$ to point B.
(b) Go up to $(x, y+1)$ and then follow the safest path from point $(x, y+1)$ to point B.

Since the different transmission lines of the network fail independently of one another, the probability of safe transmission along a given path is equal to the product of the probability of safe transmission along the first edge of the path and the probability of safe transmission along the remainder of the path. The probability of safe transmission for path (a) is given by $R(x, y) \times f(x+1, y)$, while the probability of safe transmission for path (b) is equal to $U(x, y) \times f(x, y+1)$. We therefore obtain the following recursive equation:

$$f(x, y) = \max \left\{ R(x, y) \times f(x+1, y), \ U(x, y) \times f(x, y+1) \right\}.$$

The dynamic programming algorithm is initiated with

$$f(n-1, m) = R(n-1, m) \quad \text{and} \quad f(n, m-1) = U(n, m-1).$$

Next, calculate $f(n-2, m)$, $f(n-1, m-1)$, and $f(n, m-2)$, etc., and finally $f(0, 0)$. For every point (x, y), remember the decision for which the maximum of the recursive relation for $f(x, y)$ is reached.

Numerical Example

As a numerical example, consider the network in Figure 5.3. The number along an edge gives the probability as a percentage of safe transmission along the edge. The numbers must therefore be divided by 100.

First, calculate

$f(1, 2) = 0.97$,   "go right,"
$f(2, 1) = 0.94$,   "go up."

Then, calculate

$f(0, 2) = 0.93 \times f(1, 2) = 0.9021$,   "go right,"
$f(1, 1) = \max \{0.96 \times f(2, 1), \ 0.92 \times f(1, 2)\}$,
$\quad = \max \{0.9024, 0.8924\} = 0.9024$ "go right,"
$f(2, 0) = 0.93 \times f(2, 1) = 0.8724$,   "go up."

Next, calculate

$f(0, 1) = \max \{0.98 \times f(1, 1), \ 0.96 \times f(0, 2)\}$,
$\quad = \max \{0.8844, 0.8660\} = 0.8844$,
$f(1, 0) = \max \{0.95 \times f(2, 0), \ 0.93 \times f(1, 1)\}$,   "go right,"

Fig. 5.3   What is the safest path from A to B?

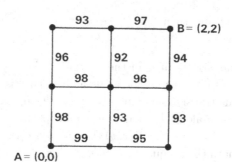

$$= \max\{0.8288, 0.8392\} = 0.8392, \qquad \text{"go up."}$$

Finally, calculate

$$f(0,0) = \max\{0.99 \times f(1,0),\ 0.98 \times f(0,1)\},$$
$$= \max\{0.8308, 0.8667\} = 0.8667, \qquad \text{"go up."}$$

The maximum probability of safe transmission from the source A to the destination B is therefore 0.8667. Going through the calculations backward shows that the safest path is given by $(0,0) \to (0,1) \to (1,1) \to (2,1) \to (2,2)$. The safest path could also have been calculated using Dijkstra's algorithm from Section 3.1. In contrast to the dynamic programming algorithm, in Dijkstra's algorithm, it is necessary to transform the products into summations by taking the logarithm.

5.3.3   The Minimax Optimal Path

Now, suppose that in the Manhattan network in Figure 5.2, we are looking for the path from A to B in which the greatest distance covered in one step is as small as possible. For this minimax criterion, we define the value function $f(x,y)$ by

$f(x,y) =$ smallest possible value of the greatest one-step distance in a path from point (x,y) to the endpoint B.

Of course, we have the constraints

$$f(n-1,m) = R(n-1,m) \text{ and } f(n,m-1) = U(n,m-1).$$

To define the recursion relation for $f(x,y)$, we use the same reasoning as for the shortest path and the safest path. If in point (x,y), one decides to first go to $(x+1,y)$ and then continue optimally from there to the endpoint B, then the greatest one-step distance in this path from (x,y) to B is equal to

$$\max\{R(x,y), f(x+1,y)\}.$$

After all, the greatest travel distance from point $(x + 1, y)$ to the endpoint B is, by definition, equal to $f(x + 1, y)$. If one goes from (x, y) to $(x, y + 1)$ and then continues optimally to endpoint B, then the greatest one-step distance in this path from (x, y) to B is equal to

$$\max \{U(x, y), f(x, y + 1)\}.$$

This leads to the recursion relation

$$f(x, y) = \min \{\max \{R(x, y), f(x + 1, y)\}, \max \{U(x, y), f(x, y + 1)\} \}.$$

Numerical Illustration

As a numerical example, take the network in Figure 5.1. First, calculate

$$f(1, 2) = 8 \qquad \qquad \text{"go right,"}$$
$$f(2, 1) = 5, \qquad \qquad \text{"go up."}$$

Then, $\quad f(0, 2) = \max \{7, f(1, 2)\},$
$$\qquad \qquad = \max \{7, 8\} = 8, \qquad \qquad \text{"go right,"}$$
$$f(1, 1) = \min \{\max \{4, f(2, 1)\}, \ \max \{7, f(1, 2)\} \},$$
$$\qquad \qquad = \min \{\max \{4, 5\}, \ \max \{7, 8\} \},$$
$$\qquad \qquad = \min \{5, 8\} = 5, \qquad \qquad \text{"go right,"}$$
$$f(2, 0) = \max \{6, f(2, 1)\},$$
$$\qquad \qquad = \max \{6, 5\} = 6, \qquad \qquad \text{"go up."}$$

Next, $\quad f(0, 1) = \min \{\max \{5, f(1, 1)\}, \ \max \{4, f(0, 2)\} \},$
$$\qquad \qquad = \min \{\max \{5, 5\}, \ \max \{4, 8\} \},$$
$$\qquad \qquad = \min \{5, 8\} = 5, \qquad \qquad \text{"go right,"}$$
$$f(1, 0) = \min \{\max \{8, f(2, 0)\}, \ \max \{1, f(1, 1)\} \},$$
$$\qquad \qquad = \min \{\max \{8, 6\}, \ \max \{1, 5\} \},$$
$$\qquad \qquad = \min \{8, 5\} = 5, \qquad \qquad \text{"go up."}$$

Finally, $\ f(0, 0) = \min \{\max \{5, f(1, 0)\}, \ \max \{3, f(0, 1)\} \},$
$$\qquad \qquad = \min \{\max \{5, 5\}, \ \max \{3, 5\} \},$$
$$\qquad \qquad = \min \{5, 5\} = 5, \qquad \qquad \text{"go right or go up."}$$

The smallest possible value of the greatest one-step travel distance from A to B is therefore 5 and is attained by both the path $(0, 0) \to (1, 0) \to (1, 1) \to (2, 1) \to (2, 2)$ and the path $(0, 0) \to (0, 1) \to (1, 1) \to (2, 1) \to (2, 2)$.

Value Function and State Variables

The value function is always a function of so-called *state variables*. In specific dynamic programming applications, one should always first decide what the state variables are before defining the value function. In the Manhattan network problem, it was evident that for each of the three "cost structures" considered, the current node determines the state of the system. In general, however, it is not always that simple. We can illustrate this using the Manhattan network. Assume that additional costs are made every time there is a bend in the path; see Exercise 5.1. In this situation, it does not suffice to take the current node as the only state variable; the description of the state must also include the direction from which one has arrived at that node. In order to correctly choose the state variables in specific applications, it helps to put oneself in the decision maker's shoes and ask what information is needed to make the decisions and allocate the associated costs.

5.4 Shortest Path in an Acyclic Network

Shortest-path problems are the prototypes of dynamic programming problems. As a further illustration of dynamic programming, consider the shortest-path problem in Figure 5.4. In the acyclic network of Figure 5.4, the nodes are numbered $M, M - 1, \ldots, 0$, and from every node i, there is an arc to node j only if $j < i$. This network is called acyclic because all arcs point forward and therefore no cycle is possible. A cost of c_{ij} is incurred when arc $i \to j$ is used. The problem is how to go from node M to node 0 at minimum cost.

Fig. 5.4   A minimum-cost problem for an acyclic network.

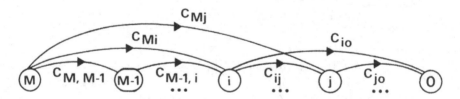

The acyclic shortest-path problem has many applications; see Section 5.5 and the exercises. A nice application is the following. The thickness of steel plates must be reduced from D cm to 1 cm through the process of rolling. The cost of rolling the plates once to reduce the thickness by d cm is equal to $a(d)$. How should the steel plates by rolled to minimize the costs? This can be translated into a shortest-path problem in an acyclic network with the following specifications for the nodes, arcs, and costs associated with the arcs:

- nodes $i = 1, \ldots, D$, where i indicates that the current thickness of the steel plate is i cm;
- an arc from node i to node j means that there is a rolling process that reduces the thickness of the steel plate by $i - j$ cm;
- the cost $c_{ij} = a(i - j)$ is associated with the arc from i to j.

We now give the dynamic programming algorithm for the shortest-path problem of the acyclic network in Figure 5.4. To determine an optimal path from node M to node 0, we define the value function as follows:

$$f(i) = \text{minimum cost to go from node } i \text{ to node } 0$$

for $i = 0, 1, \ldots, M$. The original problem is again divided into a series of simpler nested subproblems. The subproblems can be solved recursively. To determine the recursive relation for $f(i)$, we use the same type of reasoning as before. Fix node i. Suppose that an optimal path has been found from node j to node 0 for every $j < i$. Now, consider the following path from node i to node 0. First, go from node i to node j, and then continue along the optimal path from node j to node 0. The total cost of this path is equal to $c_{ij} + f(j)$. By minimizing this expression over all $j < i$, we find the recursive relation

$$f(i) = \min_{0 \leq j < i} \{c_{ij} + f(j)\} \quad \text{for } i = 1, \ldots, M.$$

Starting with $f(0) = 0$, we can determine $f(1), f(2), \ldots, f(M)$ recursively. In order to later find the optimal path from node M to node 0, for every i, during the calculation, the node $j = j(i)$ for which the minimum is reached in the recursive relation for $f(i)$ must be remembered.

Fig. 5.5   What is the minimum-cost path from 5 to 0?

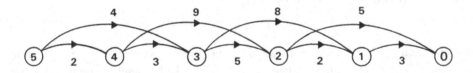

Numerical Example

We illustrate the algorithm with the numerical example from Figure 5.5. The recursive calculations are as follows:

$f(0) = 0,$
$f(1) = 3,$ "go to node 0"
$f(2) = \min\{2 + f(1),\ 5 + f(0)\}$
     $= \min\{5, 5\} = 5,$ "go to node 1 or node 0"
$f(3) = \min\{5 + f(2),\ 8 + f(1)\}$
     $= \min\{10, 11\} = 10,$ "go to node 2"
$f(4) = \min\{3 + f(3),\ 9 + f(2)\}$
     $= \min\{13, 14\} = 13,$ "go to node 3"
$f(5) = \min\{2 + f(4),\ 4 + f(3)\}$
     $= \min\{15, 14\} = 14,$ "go to node 3."

The minimum cost to go from node 5 to node 0 is therefore $f(5) = 14$. Verify that both of the paths $5 \to 3 \to 2 \to 1 \to 0$ and $5 \to 3 \to 2 \to 0$ are optimal. These paths can be deduced from the calculations above by going through them backward.

Extension

Suppose that the additional requirement is set that a path from the starting point M to the endpoint 0 may consist of a maximum of R arcs for a given value of R. How can one find the minimum-cost path subject to this constraint? In this case, the state variable that indicates the current node does not suffice for the value function. One must also know how many more arcs may be used for the remainder of the path to the endpoint 0. Therefore, define the value function $f_r(i)$ as

$$f_r(i) = \text{minimum cost to go from node } i \text{ to node } 0$$
$$\text{when no more than } r \text{ arcs may be used}$$

for $i = 0, \ldots, M$ and $r = 1, \ldots, R$, where $f_r(0) = 0$ for all r. To calculate $f_R(M)$ and find an optimal path from node M to node 0, the original problem is again divided into a series of nested subproblems. The subproblems can be solved recursively. Initiate the algorithm with $f_1(i) = c_{i0}$ for $i = 1, \ldots, M$. Let us now define the recursion relation for $f_r(i)$. Suppose that one first uses the arc from node i to node j and then continues optimally from node j to node 0 along a path with at most $r - 1$ arcs. The total cost of this path is equal to $c_{ij} + f_{r-1}(j)$. Minimizing this expression over all $j < i$ leads to the recursion relation

$$f_r(i) = \min_{0 \le j < i} \{c_{ij} + f_{r-1}(j)\}.$$

Starting with $f_1(i) = c_{i0}$ for all i, calculate successively for $r = 2, \ldots, R$ the value function $f_r(i)$ for all i, where $f_k(0) = 0$ for all $k \ge 1$. Remember the decision for which the minimum for $f_r(i)$ is reached in the recursive relation. This way, the sequence of optimal decisions can be deduced as soon as $f_R(M)$ is calculated.

5.5   Applications of the Acyclic Network

This section gives some applications of the shortest-path problem in an acyclic network. The first two applications concern planning problems in maintenance and replacement and in production and inventory control. Then, we discuss the important knapsack problem. Modeling these problems as shortest-path problems in acyclic networks requires some creativity.

5.5.1   *A Replacement Problem*

A car wash is planning the purchase of a new wash system. Such a system can be used for a maximum of three years but can be traded in sooner for a new one. A new wash system costs 50 thousand euros. The residual value of a one-year-old machine is 40 thousand euros, that of a two-year-old machine is 30 thousand euros, and that of a three-year-old machine is 24.5 thousand euros. The maintenance cost of a newly purchased machine is 3 thousand euros the first year of operation, 4 thousand euros the second year of operation, and 6 thousand euros the third year. Due to stricter environmental requirements, the car wash will have to close down in five years' time. What is the optimal replacement schedule for the next five years?

This problem can be formulated as a shortest-path problem in an acyclic network as follows:

- node i for $i = 1, \ldots, 6$ means that we are at the beginning of year i and have just gotten rid of a machine (node 6 is an auxiliary node);
- an arc from node i to node j with $j > i$ means that at the beginning of year i, a new wash system is purchased that will be sold at the end of year $j - 1$;
- an arc from i to j is assigned a total cost of c_{ij} that consists of the purchase cost at the beginning of year i plus the maintenance costs for years i through $j - 1$ minus the residual value at the end of year $j - 1$.

If we take a thousand euros as unit, we find

$$c_{12} = 50 + 3 - 40 = 13 = c_{23} = c_{34} = c_{45} = c_{56}.$$

Likewise, we have $c_{13} = c_{24} = c_{35} = c_{46} = 50 + 3 + 4 - 30 = 27$ and $c_{14} = c_{25} = c_{36} = 50 + 3 + 4 + 6 - 24.5 = 38.5$. The value function $f(i)$ is defined by

$$f(i) = \text{minimum costs from the beginning of year } i \text{ to the end of year } 5$$
$$\text{if a new wash system is necessary at the beginning of year } i.$$

We are looking for $f(1)$ and an optimal replacement schedule. The recursion is

$$f(i) = \min_{j=i+1,\ldots,6} \{c_{ij} + f(j)\} \quad \text{for } i = 5, \ldots, 1,$$

with $f(6) = 0$. The calculations are as follows:

$$f(5) = c_{56} = 13, \quad \text{go to } 6$$

$$f(4) = \min_{j=5,6} \{c_{45} + f(5), \, c_{46}\}$$
$$= \min_{j=5,6} \{13 + 13, \, 27\} = 26, \quad \text{go to } 5$$

$$f(3) = \min_{j=4,5,6} \{c_{34} + f(4), \, c_{35} + f(5), \, c_{36}\}$$
$$= \min_{j=4,5,6} \{13 + 26, \, 27 + 13, \, 38.5\} = 38.5, \quad \text{go to } 6$$

$$f(2) = \min_{j=3,4,5} \{c_{23} + f(3), \, c_{24} + f(4), \, c_{25} + f(5)\}$$
$$= \min_{j=3,4,5} \{13 + 38.5, \, 27 + 26, \, 38.5 + 13\} = 51.5, \quad \text{go to } 3 \text{ or } 5$$

$$f(1) = \min_{j=2,3,4} \{c_{12} + f(2), \, c_{13} + f(3), \, c_{14} + f(4)\}$$
$$= \min_{j=2,3,4} \{13 + 51.5, \, 27 + 38.5, \, 38.5 + 26\} = 64.5, \quad \text{go to } 2 \text{ or } 4.$$

The minimum total cost over the next five years is therefore 64.5 thousand euros and is achieved for several optimal replacement schedules: trade in at the end of year 3 (from $f(1)$, go to 4), trade in at the end of year 4 (from $f(4)$, go to 5), and trade in at the end of year 5 (from $f(5)$, go to 6). The other two optimal schedules are: (a) trade in at the end of years 1, 2, and 5; (b) trade in at the end of years 1, 4, and 5.

5.5.2   A Production-Stock Problem

The company Jones Chemical has entered into a contract to deliver a special type of sulfuric acid over the next six months. The agreement provides for the supply of 100 tons of sulfuric acid on July 1st, 75 tons on August 1st, 90 tons on September 1st, 60 tons on October 1st, 40 tons on November 1st, and 85 tons on December 1st. The production of sulfuric acid requires a few special measures. The production manager has therefore decided that the sulfuric acid can only be produced on the first day of the month. The production takes a negligible amount of time. The fixed setup cost of the production process is 500 euros. Each time, any desired amount of sulfuric acid can be produced. The company has sufficient storage capacity for the sulfuric acid; the storage cost is 4 euros per ton of sulfuric acid per month. At the moment, there is no stock of the product. What production plan has the lowest total cost?

This problem was solved in Section 2.2.6 as an ILP problem but can also be formulated as a shortest-path problem in an acyclic network. This is possible because an optimal production plan has the property that production only takes place if there is no stock and then has a size that exactly covers the demand in a yet to be determined number of periods. This property is easy to see. Suppose that production were to take place when there is a positive inventory level V; then if the fixed production cost remains the same, the storage costs could be reduced by producing a quantity V less in the previous production run and V more than originally planned in the next production run. If we number the months July through December as periods 1 through 6, then we can formulate the problem as follows as a shortest-path problem in an acyclic network:

- node i with $i = 1, \dots, 7$ means that we are at the beginning of period i with no stock (node 7 is an auxiliary node);
- an arc from node i to node j with $j > i$ means that at the beginning of period i, a quantity is produced that is exactly enough to cover the demand in periods i through $j - 1$;
- an arc from i to j is assigned a total cost c_{ij} that consists of the fixed setup cost at the beginning of period i and the storage costs for periods i through $j - 1$.

The calculation of the costs c_{ij} is simple:

$$c_{12} = 500, \quad c_{13} = 500 + 1 \times 4 \times 75 = 800,$$
$$c_{14} = 500 + 1 \times 4 \times 75 + 2 \times 4 \times 90 = 1520.$$

Likewise, we find $c_{15} = 2240$, $c_{16} = 2880$, $c_{17} = 4580$, $c_{23} = 500$, $c_{24} = 860$, $c_{25} = 1340$, $c_{26} = 1820$, $c_{27} = 3180$, $c_{34} = 500$, $c_{35} = 740$, $c_{36} = 1060$, $c_{37} = 2080$, $c_{45} = 500$, $c_{46} = 660$, $c_{47} = 1340$, $c_{56} = 500$, $c_{57} = 840$, $c_{67} = 500$. The value function $f(i)$ is defined by

$$f(i) = \text{minimum total cost to cover the demand in the periods}$$
$$i \text{ through 6 if there is no stock at the beginning of period } i.$$

We are looking for $f(1)$ and an optimal production schedule. The recursion is

$$f(i) = \min_{j=i+1,\dots,7} \{c_{ij} + f(j)\} \quad \text{for } i = 6, \dots, 1,$$

beginning with $f(7) = 0$. We leave out the details of the numerical calculations and give only the results: $f(6) = 500$ (go to node 7), $f(5) = 840$ (go to node 7), $f(4) = 1160$ (go to node 6), $f(3) = 1560$ (go to node 6), $f(2) = 2020$ (go to node 4), and $f(1) = 2360$ (go to node 3). The optimal path therefore goes from node 1 to node 3, from node 3 to node 6, and from node 6 to node 7. In other words, the optimal production schedule is to produce 175 tons for periods 1 and 2 at the beginning of period 1, 190 tons for periods 3, 4, and 5 at the beginning of period 3, and 85 tons for period 6 at the beginning of period 6. The total cost for this schedule is 2360 euros.

5.5.3 The Knapsack Problem

The knapsack problem is given by the ILP problem

$$\text{Maximize} \quad \sum_{k=1}^{n} r_k x_k$$
$$\text{subject to} \quad \sum_{k=1}^{n} b_k x_k \leq B$$
$$\text{and} \quad x_k \geq 0,\ x_k \text{ integer for } k = 1, \dots, n,$$

where the r_k, the b_k, and B are given positive numbers with the b_k and B integers. In certain algorithms for large-scale optimization problems, the knapsack problem must be solved a great number of times as a subproblem. It is therefore important to have access to an extremely fast algorithm for the knapsack problem. The shortest-path algorithm in an acyclic network is such an algorithm. To formulate the knapsack problem as a shortest-path algorithm in an acyclic network, the following physical interpretation of the knapsack problem is very useful:

- in a backpack, one can put one or more units of each of the articles $k = 1, 2, \dots, n$;
- there are sufficiently many units of every article;
- the backpack has capacity B and every unit of article k that is put inside takes up b_k units and gives a revenue of r_k for $k = 1, 2, \dots, n$;
- how many units of every article should one put in the backpack to maximize the revenue without exceeding the backpack's capacity?

This interpretation of the knapsack problem is the key to the formulation of the problem as a shortest-path algorithm in an acyclic network with nodes B, $B - 1, \ldots, 1, 0$. Node i corresponds to the situation where the remaining capacity of the backpack is i. An arc from node i to node j with $j < i$ means that one unit of article k is added to the backpack with capacity use b_k equal to $i - j$. To this arc is assigned a cost equal to $c_{ij} = -r_k$. After thus gaining the insight that the knapsack problem can be interpreted as a shortest-path problem in an acyclic network, one can also directly formulate a dynamic programming algorithm. For $i = 0, \ldots, B$, define the value function $f(i)$ by

$f(i)$ = maximum revenue from a remaining capacity of i in the backpack.

We are looking for the maximum revenue $f(B)$ and the optimal backpack filling. A now familiar argument leads to the recursion relation

$$f(i) = \max_{k: b_k \le i} \{r_k + f(i - b_k)\}.$$

Initiate the algorithm with $f(i) = 0$ for every $0 \le i < \min_k b_k$. Then, determine $f(i)$ for increasing values of i. As an illustration, consider the numerical example

$$\text{Maximize } 11x_1 + 7x_2 + 12x_3$$
$$\text{subject to } 4x_1 + 3x_2 + 5x_3 \le 10$$
$$\text{and} \qquad x_1, x_2, x_3 \ge 0 \text{ and integer.}$$

To find the maximum revenue $f(10)$ and the optimal allocation, begin with

$$f(0) = f(1) = f(2) = 0.$$

Then, we successively calculate the function values of $f(i)$ for the remaining i. We denote the decision k in the recursion that maximizes $f(i)$ by k_i. Verify the following calculations:

$$f(3) = \max_{k=2}\{7 + f(3-3)\} = 7, \ k_3 = 2$$

$$f(4) = \max_{k=1,2}\{11 + f(4-4), \ 7 + f(4-3)\}$$

$$= \max_{k=1,2}\{11 + 0, \ 7 + 0\} = 11, \ k_4 = 1$$

$$f(5) = \max_{k=1,2,3}\{11 + f(1), 7 + f(2), 12 + f(0)\}$$

$$= \max_{k=1,2,3}\{11 + 0, \ 7 + 0, \ 12 + 0\} = 12, \ k_5 = 3$$

$$f(6) = \max_{k=1,2,3}\{11 + f(2), 7 + f(3), 12 + f(1)\}$$

$$= \max_{k=1,2,3}\{11 + 0, \ 7 + 7, \ 12 + 0\} = 14, \ k_6 = 2$$

$$f(7) = \max_{k=1,2,3}\{11 + f(3), 7 + f(4), 12 + f(2)\}$$

$$= \max_{k=1,2,3}\{11 + 7, \ 7 + 11, \ 12 + 0\} = 18, \ k_7 = 1 \text{ or } 2$$

$$f(8) = \max_{k=1,2,3}(11 + f(4), 7 + f(5), 12 + f(3)\}$$

$$= \max_{k=1,2,3} \{11 + 11, \ 7 + 12, \ 12 + 7\} = 22, \ k_8 = 1$$

$$f(9) = \max_{k=1,2,3} \{11 + f(5), 7 + f(6), 12 + f(4)\}$$

$$= \max_{k=1,2,3} \{11 + 12, \ 7 + 14, \ 12 + 11\} = 23, \ k_9 = 1 \text{ or } 3$$

$$f(10) = \max_{k=1,2,3} \{11 + f(6), 7 + f(7), 12 + f(5)\}$$

$$= \max_{k=1,2,3} \{11 + 14, \ 7 + 18, \ 12 + 12\} = 25, \ k_{10} = 1 \text{ or } 2.$$

The maximum profit is therefore 25. An optimal allocation can be deduced from the calculations. Taking $k_{10} = 1$ at $f(10)$ leads to $k_6 = 2$ at $f(6)$ and $k_3 = 2$ at $f(3)$. In other words, the optimal allocation is to take along 1 unit of article 1 and 2 units of article 2. The same optimal solution is found by taking $k_{10} = 2$ at $f(10)$.

5.6 An Allocation Problem

Before we discuss the allocation problem, we first look at shortest-path problems from a different angle. The shortest-path problems we have dealt with so far can be seen as typical examples of decision problems with a series of decisions. In sequential decision problems, a variable (phase variable) is needed to indicate how many decisions have been taken so far or how many still need to be taken. The phase variable is often a time variable and increases (or decreases) by one after every decision. Apart from the concept of phases, other essential features of sequential decision problems are the concepts of state and state transition. State variables are needed to describe the current situation. It is not always clear how to specify state variables; this sometimes requires some ingenuity. In order to find a suitable description of the state in a specific application, it is advisable to put oneself in the decision maker's shoes and ask what information is needed to make a decision. Once the phase variables, state variables, and effects of the decisions on the states have been identified, the value function can be defined, and a recursive relation can be established.

Dynamic programming can sometimes also be applied to static decision problems that do not require a series of decisions. As an illustration, consider the following allocation problem:

$$\text{Maximize } \sum_{k=1}^{n} r_k(x_k)$$

$$\text{subject to } \sum_{k=1}^{n} b_k x_k \leq B$$

$$\text{and} \qquad 0 \leq x_k \leq m_k, \quad x_k \text{ integer for } k = 1, \ldots, n,$$

where the b_k, the m_k, and B are given positive numbers. This optimization problem occurs when a budget of B euros is available for investing in n projects. Allocating x_k euros to project k leads to a revenue of $r_k(x_k)$ and reduces the budget by $b_k x_k$

euros. No more than m_k euros may be allocated to project k. The problem is finding an allocation that maximizes the total gain.

The allocation problem can be solved with dynamic programming since the objective function and the constraints are expressed as functions of only one variable. The problem can be divided into a series of smaller problems that are linked using a recursive relation. Define, for $k = 1, \dots, n$ and $b = 0, \dots, B$, the value function

$$f_k(b) = \text{maximum attainable revenue with the remaining budget}$$
$$\text{of } b \text{ euros when this budget may be allocated only to}$$
$$\text{the projects } k, \dots, n.$$

The recursion relationship for $f_k(b)$ is found through the usual reasoning. Suppose that one allocates x_k euros to project k and then allocates the remaining budget of $b - b_k x_k$ euros optimally to the projects $k + 1, \dots, n$. The total revenue of this allocation is $r_k(x_k) + f_{k+1}(b - b_k x_k)$. With the notation $A_k(b) = \{x \mid 0 \le x \le \min(b/b_k, m_k), x \text{ integer}\}$, the recursion becomes

$$f_k(b) = \max_{x \in A_k(b)} \{r_k(x) + f_{k+1}(b - b_k x)\}.$$

Starting with $f_{n+1}(b) = 0$, the functions $f_n(b), f_{n-1}(b), \dots, f_1(b)$ are calculated recursively for $b = 0, 1, \dots, B$. This way, we find the desired value $f_1(B)$. In order to find the optimal allocation, we need to remember, for every combination (b, k), the value of x_k for which the right-hand side of the recursive relation above takes on its maximum. The knapsack problem from Section 5.5.3 is, of course, a special case of the allocation problem above. However, for the knapsack problem, we could give a much more efficient dynamic programming algorithm because the revenue functions $r_k(x_k)$ are linear in x_k and no upper bounds were imposed on the values of the variables x_k. For the so-called 0-1 knapsack problem in which the x_k may only be 0 or 1, the efficient shortest-path algorithm is not applicable, and the more generally applicable dynamic programming approach given above must be followed.

A Static Problem Viewed as a Multistep Decision Problem

Although the allocation problem is a static decision problem, it can also be seen as a sequential decision problem. In the first phase, allocate an amount of x_1 to project 1. Then, in the second phase, allocate an amount of x_2 from the remaining budget of $B - b_1 x_1$ to project 2, and so on. In this context, the decision problem is sequential, and one can interpret the value function $f_k(b)$ defined above as the maximum revenue that can be achieved over the decision times $k, \dots, n$ if at time k, the state (= budget) equals b. In the setup of a sequential decision problem, there is a repeating sequence of events, in which the decision maker

- monitors the situation at the current decision time,
- makes a decision,
- receives a direct return or incurs direct costs,
- observes a new situation at the next decision time,

and so on. In dynamic programming, there are in fact two strongly related approaches: the "creative" approach, in which one defines a value function by splitting the problem into smaller problems of the same structure, and the more "mechanical" approach, in which one first identifies decision times and states. For some, one approach has didactic advantages, for others the other approach does.

As an illustration, consider the separable optimization problem

$$\text{Minimize} \quad 0.6x_1^2 + 2.5x_2^2 + 5x_3^2$$
$$\text{subject to } 2x_1 + 3x_2 + 5x_3 \geq 15$$
$$\text{and} \qquad x_1, x_2, x_3 \geq 0 \text{ and integer.}$$

To see this static optimization problem as a multistep decision problem, imagine that a waste processing plant is given three days to remove at least 15 m^3 of waste from a large garbage heap. If on the first day, x_1 laborers work, then on that day, $2x_1$ m^3 of garbage is removed at a direct cost of $0.6x_1^2$. Likewise, $3x_2$ m^3 of garbage is removed the second day at a cost of $2.5x_2^2$ if x_2 laborers work that day, and $5x_3$ m^3 of garbage is removed on day 3 at a cost of $5x_3^2$ if x_3 laborers work. The optimization problem can therefore be seen as a three-step decision problem in which the state on a given day indicates how many m^3 of garbage still need to be removed to reach the target level of 15 m^3. For $t = 1, 2, 3$, define the value function $g_t(a)$ by

$$g_t(a) = \text{minimum cost of removing at least a total of } a \text{ m}^3$$
$$\text{of garbage on days } t \text{ through } 3.$$

For $t = 3$, we calculate the function $g_3(a)$ as follows:

$$g_3(a) = 5 \, (d_3(a))^2 \text{ with } d_3(a) = \left\lceil \frac{a}{5} \right\rceil,$$

where $\lceil y \rceil$ is the integer that results from rounding y up to the nearest integer. Of course, we take $g_3(a) = 0$ for $a < 0$. For $t = 2$, we calculate the function $g_2(a)$ as follows:

$$g_2(a) = \min_{d_2=0,1,\ldots,\lceil \frac{a}{3} \rceil} \{2.5d_2^2 + g_3(a - 3d_2)\}.$$

We denote the decision d_2 that minimizes $g_2(a)$ by $d_2(a)$. Finally, for $t = 1$, we calculate

$$g_1(15) = \min_{d_1=0,1,\ldots,8} \{0.6d_1^2 + g_2(15 - 2d_1)\}.$$

The calculations give $g_1(15) = 17.1$ with $d_1(15) = 4$. From $g_1(15)$, $g_2(15 - 8)$, and $g_3(7 - 3)$, we derive that $x_1^* = 4$, $x_2^* = 1$, and $x_3^* = 1$ are the optimal values.

5.7 The General Shortest-Path Problem

The basic principle of dynamic programming can also be used to develop an efficient algorithm for calculating the shortest paths in networks in general form. Consider a directed network with N nodes. We use the symbols x and y to indicate nodes. For each existing arc (x, y) in the network, we set

$$c(x, y) = \text{cost for passing through arc } (x, y).$$

Suppose that the objective is to calculate minimum-cost paths from a given node s to every other node x. If all costs $c(x, y)$ are nonnegative, we can use Dijkstra's algorithm, discussed in Chapter 3. If some costs $c(x, y)$ are positive and some are negative, then the problem is complicated by the fact that a cycle (that is, a path from a node to itself) can exist with negative total cost. In this case, a path with cost $-\infty$ can be constructed by passing through the cycle an infinite number of times. How can we design an algorithm that uncovers the existence of a negative cycle and gives the optimal paths if such a cycle does not exist? The following observations give the key to the algorithm:

1. An optimal path from node s to any other node x exists only if there is no cycle with negative cost.
2. If an optimal path from node s to node x exists, then there exists an optimal path from s to x that consists of at most $N - 1$ arcs, where N is the number of nodes.

The second observation follows from the first one by taking into account that a path with N or more arcs must contain a cycle. The algorithm applies the basic principle of dynamic programming: split the original problem into a series of subproblems of the same structure but smaller in size. As a subproblem, we seek to determine the best paths within the following subclass $\mathcal{P}_m$ of paths:

$$\mathcal{P}_m = \text{class of paths that consist of at most } m \text{ arcs.}$$

For $m = 1, \ldots, N$, we define the value function

$$\pi_m(x) = \text{minimum of the costs of the paths from } s \text{ to } x \text{ within the}$$
$$\text{subclass } \mathcal{P}_m.$$

Based on the observations above, we can state that we have $\pi_N(x) = \pi_{N-1}(x)$ for every node x only if no negative cycle exists. In this case, the number $\pi_N(x)$ gives the minimum cost over the paths from s to x within the class of all possible paths. If $\pi_N(x_0) < \pi_{N-1}(x_0)$ for some node x_0, then there is a negative cycle that contains node x_0. We can calculate the value functions $\pi_m(x)$ recursively. The recursion is initiated with

$$\pi_1(x) = c(s, x) \quad \text{for every node } x,$$

where we set $c(s, x) = \infty$ if arc (s, x) does not exist. Now, suppose that we have calculated the value functions $\pi_k(x)$ for $k = 1, \ldots, m-1$. To find $\pi_m(x)$, we consider both the best path from s to x within the class $\mathcal{P}_{m-1}$ and the paths that consist of the best path from s to y within the class $\mathcal{P}_{m-1}$ plus the direct connection (y, x) for $y \neq x$. These are the only paths consisting of at most m arcs that qualify. So

$\pi_m(x)$ is the least of the two numbers $\pi_{m-1}(x)$ and $\min_{y \neq x}\{\pi_{m-1}(y) + c(y, x)\}$. If we set $c(x, x) = 0$ for ease of notation, we find the recursive relation

$$\pi_m(x) = \min_y \{\pi_{m-1}(y) + c(y, x)\}.$$

Starting with $\pi_1(x)$, we can successively calculate the value functions $\pi_i(x)$ for $i = 2, \ldots, N$. In case $\pi_N(x) = \pi_{N-1}(x)$ for all x, then the numbers $\pi_N(x)$ give the costs of the desired optimal paths from node s to every other node x. The optimal paths themselves are found by keeping track of the predecessor of x on the current best path from s to x in the recursive calculations for each node x. If $\pi_N(x_0) < \pi_{N-1}(x_0)$ for some node x_0, then there is a negative cycle that contains x_0. The algorithm described above is known as the Bellman–Ford algorithm.

5.8 Stochastic Dynamic Programming

In the previous sections, we discussed sequential decision problems that are deterministic in nature. The application of dynamic programming to stochastic sequential decision problems is not conceptually more difficult. In the case of problems with deterministic state transitions, one can determine beforehand what sequence of decisions to make. In stochastic dynamic programming problems, the future states are uncertain, and it is therefore impossible to say beforehand what sequence of decisions will be taken. What we need is a collection of conditional decisions of the form "If the realized state is equal to ..., make decision ..." Such a policy can be obtained by applying the backward recursion of dynamic programming. The reasoning behind the recursive relation for stochastic dynamic programming problems is essentially the same as that for deterministic problems. In the stochastic context, writing down the recursive relation in fact always comes down to applying the law of total expectation. In this section, we give the general structure of stochastic dynamic programming problems followed by a few characteristic examples.

5.8.1 *General Structure of Stochastic Dynamic Programming*

Similar to deterministic dynamic programming problems (see Section 5.3.1), stochastic dynamic programming problems are characterized by stages, states, and decisions. The immediate rewards become *expected* immediate rewards as a consequence of stochasticity. And the big difference is that the transition to the next state is now stochastic. Hence, *transition probabilities* are added to the model: $p_n(j|i, d)$ is the probability to transition to state j at stage $n + 1$ as a consequence of decision d in state i at stage n. When optimizing in a stochastic dynamic programming problem, the objective is to maximize the expected total reward over all stages:

$$\max \mathbb{E}\left[\sum_{n=0}^{N} r_n(i, d)\right].$$

This is done recursively, using the *optimal value function* $f_n(i)$, which is now the maximum *expected* total reward that can be obtained from stages n through N if the system is in state i at stage n. The recursion relation of the optimal value function for stochastic dynamic programming problems has the following general structure:

$$f_n(i) = \max_{d \in D_n(i)} \left\{ r_n(i,d) + \sum_{j \in S_{n+1}} p_n(j|i,d) f_{n+1}(j) \right\}.$$

This recursion relation does indeed contain the expected reward from stage $n+1$ onward. The immediate reward may in this case also depend on the next state j, in which case we get $r_n(i,j,d)$, and the reward should be included in the sum to obtain the expected immediate reward:

$$f_n(i) = \max_{d \in D_n(i)} \left\{ \sum_{j \in S_{n+1}} p_n(j|i,d) \Big(r_n(i,j,d) + f_{n+1}(j) \Big) \right\}.$$

5.8.2   *A Dice Game and Optimal Stopping*

An interesting game is the following. A fair die will be rolled up to six times. After each throw, one decides whether to continue or stop. The payout of the game is the number of dots rolled the last time. What policy maximizes the expected payout?

   The problem is a typical example of a stochastic sequential decision problem. The state of a dynamic process is observed at discrete time points. After each observation of a state, a decision is taken. A direct payout is received that is based only on the state at that moment and the decision. Then, the process moves on to the next state following a given probability distribution. What policy must be followed to maximize the expected value of the total yield over a given period of finite length? Dynamic programming allows one to solve such problems using a recursive algorithm. In the same way as was done in the previous sections for deterministic sequential decision problems, the original problem is divided into a series of nested subproblems that are linked recursively. To do this, we need the basic concept of a value function. For the problem in question, for $k = 0, 1, \ldots, 6$, we define the function $f_k(i)$ by

$f_k(i)$ = maximum expected number of dots if there are still k throws to go
           and the number of dots of the last throw of the die is equal to i.

The goal is to determine $f_6(0)$ and the optimal policy. To determine the recursive relation for $f_k(i)$, reason as follows. There are two possible actions, "stop" and "continue," in the decision state where there are still k throws to go and the number of dots of the last throw is i. If the decision to stop is made, an immediate reward of i is paid out and the game is over. If the game continues, then there is no payout and the next state of the process will be equal to j with probability $1/6$ for $j = 1, \ldots, 6$. If the choice is made to continue and the optimal policy is followed for the remaining $k - 1$ throws, then the expected payout of the game equals $\sum_{j=1}^{6} f_{k-1}(j)/6$. Applying the law of total expectation gives the following recursion relation:

$$f_k(i) = \max\left\{i, \frac{1}{6}\sum_{j=1}^{6} f_{k-1}(j)\right\}.$$

Starting with $f_0(i) = i$ for all i, successively calculate, for $k = 1,\ldots,5$, the value function $f_k(i)$ for $i = 1,\ldots,6$. The expected maximum payout for a game is

$$f_6(0) = \frac{1}{6}\sum_{j=1}^{6} f_5(j).$$

For every combination (k, i), it is necessary to remember the optimal decision, say $d_k(i)$, for which the maximum is reached in the recursive relation for $f_k(i)$. This "manual" of decisions gives the optimal policy.

Numerical Calculations

It is instructive to carry out the calculations for this particular problem. The algorithm is initiated with $f_0(i) = i$ for $i = 1,\ldots,6$, where $d_0(i) = $ "stop" for all i. The value function $f_k(i)$ is calculated successively for $k = 1,\ldots,5$:

$$f_1(i) = \begin{cases} 3.5 & \text{for } i = 1,2,3 \quad \text{with } d_1(i) = \text{"continue,"} \\ i & \text{for } i = 4,5,6 \quad \text{with } d_1(i) = \text{"stop,"'} \end{cases}$$

$$f_2(i) = \begin{cases} 4.25 & \text{for } i = 1,2,3,4 \quad \text{with } d_2(i) = \text{"continue,"} \\ i & \text{for } i = 5,6 \quad \text{with } d_2(i) = \text{"stop,"} \end{cases}$$

$$f_3(i) = \begin{cases} 4.667 & \text{for } i = 1,2,3,4 \quad \text{with } d_3(i) = \text{"continue,"} \\ i & \text{for } i = 5,6 \quad \text{with } d_3(i) = \text{"stop,"} \end{cases}$$

$$f_4(i) = \begin{cases} 4.944 & \text{for } i = 1,2,3,4 \quad \text{with } d_4(i) = \text{"continue,"} \\ i & \text{for } i = 5,6 \quad \text{with } d_4(i) = \text{"stop,"} \end{cases}$$

$$f_5(i) = \begin{cases} 5.130 & \text{for } i = 1,2,3,4,5 \text{ with } d_5(i) = \text{"continue,"} \\ i & \text{for } i = 6 \quad \text{with } d_5(i) = \text{"stop."} \end{cases}$$

Finally, the maximum expected payout is calculated to be $f_6(0) = 5.275$. This means that it is profitable to play this game if the amount that was bet is s with $0 < s < 5.275$. Then, according to the law of large numbers, the actual average gain per game will be arbitrarily close to $5.275 - s$ if the game is played often enough and the optimal policy is followed. The calculations of the dynamic programming algorithm show that the optimal policy has a simple structure and is characterized by the index numbers $s_1 = 4$, $s_2 = 5$, $s_3 = 5$, $s_4 = 5$, and $s_5 = 6$: if there are still k throws to go, then the policy requires one to stop if the last throw yielded s_k or more dots and to continue otherwise.

Verify that for the situation where two dice are rolled at most six times and every time the sum of the two dice is taken, the maximum expected payout per game is 9.474 and is reached by stopping when there are still k throws to go if the last throw yielded s_k or more dots, where $s_1 = s_2 = 8$, $s_3 = s_4 = 9$, and $s_5 = 10$.

5.8.3   Roulette: Betting Red or Black

Suppose that a gambler goes to the casino because he needs a certain amount of money by tomorrow morning. Gambling is his last resort to get the money. He decides to bet on red or black at the roulette table. Because of the time limit, he can only play a finite number of times, say n. Every time, he can bet any amount of money up to everything he has. The gambler doubles the money he bets with a given probability p and loses the money with probability $q = 1 - p$. He begins with a capital of A euros, and his objective is to end up with at least B euros, with $B > A$. What is the optimal betting policy to maximize the probability of obtaining the money the gambler needs in no more than n bets?

This problem can be interpreted as a stochastic sequential decision problem. The state is equal to i if the current capital is i euros. For state i, the possible decisions are $d = 0, 1, \ldots, i$, where decision d corresponds to betting d euros. To solve this problem using dynamic programming, for $k = 1, \ldots, n$, define the value function

$f_k(i) = $ maximum probability of obtaining a capital of at least
$\qquad\quad B$ euros if one can still bet k times and the current
$\qquad\quad$ capital is i euros.

The following reasoning is used to find a recursive relation for $f_k(i)$. Suppose that the gambler bets d euros when there are still k bets to go and then follows the optimal policy for the remaining $k - 1$ bets. Under the condition that the next state is equal to s, the probability of reaching the goal is $f_{k-1}(s)$. The next state is either $i + d$ or $i - d$, with respective probabilities p and q. So, in the situation where there are still k bets to go and the current capital is equal to i euros, the gambler can reach his goal with probability $pf_{k-1}(i + d) + qf_{k-1}(i - d)$ if he bets d euros and then follows the optimal policy for the remaining $k - 1$ bets. Maximizing this probability over all possible d gives the recursion

$$f_k(i) = \max_{d=0,1,\ldots,i} \{pf_{k-1}(i + d) + qf_{k-1}(i - d)\}.$$

Every decision d for which the maximum is reached in this recursive relation is an optimal decision for the situation where there are still k bets to place and the current capital is equal to i euros. Calculating the value function recursively leads to the optimal betting policy. The recursion is started with $f_0(i) = 1$ for $i \geq B$ and $f_0(i) = 0$ for $i < B$.

If the probability p of winning satisfies $p \leq \frac{1}{2}$, then the optimal policy is to simply bet the entire capital i if $i < B/2$ and to bet $B - i$ if the current capital is greater than or equal to $B/2$. This result's proof requires very profound mathematics. An intuitive explanation for the optimality of the all-or-nothing policy for the case of an unfavorable game is that this exposes the gambler's money to the casino's house advantage as little as possible and thus maximizes the probability of winning. For the numerical example $A = 1$, $B = 5$, $p = 0.5$, $n = 7$, this recursion equation gives the maximum probability $f_7(1) = 0.1953$ of obtaining 5 euros if one begins with 1 euro and no more than 7 bets are allowed.

5.8.4 *The Kelly Strategy for an Investment Problem*

As an introduction to the investment problem, first consider the following scenario. During the internet boom on the stock exchange at the end of the last century, countless dot-com companies went public. Especially in the first week, the share price of a new dot-com company could fluctuate considerably. Suppose that, on average, the share prices of half of the new dot-com companies on the stock exchange increase by 80% in the first week and those of the other half decrease by 60%. So every dollar invested in a new dot-com company has expected value after one week equal to $0.5 \times \$1.8 + 0.5 \times \$0.4 = \$1.1$, an expected profit of 10%. Consider an investor who has an initial capital of \$10 000 to invest in these types of companies. Suppose that she uses the following strategy for the next 52 weeks: at the beginning of every week, invest the entire capital in a new dot-com company, and the end of the week, sell the shares. What is the most likely value of her capital after 52 weeks? The reader is invited to stop reading and make an estimate. Many think that the most likely value is around 15 thousand dollars and consider the probability of having less than the initial capital of 10 thousand dollars after 52 weeks nearly equal to zero. However, the most likely value of the investor's capital after 52 weeks is \$1.95. This is immediately clear if one takes into account that the number of increases in 52 weeks is binomially distributed with parameters $n = 52$ and $p = 0.5$. So the most likely value of the capital after 52 weeks is $(1.8)^{26}(0.4)^{26} \times \$10000 = \$1.95$. If the investor reinvests her entire capital every time, then the probability that the final capital is no more than \$1.95 is equal to the binomial probability 0.555, and the probability that the capital is higher than the initial amount of 10 thousand dollars is equal to 0.058. The 10% average profit per week is misleading: the factor 1.8×0.4 is decisive, and this factor is less than 1. When investing over a longer period, however, it is optimal to invest the same fixed fraction of the capital every time. In the example above, it is optimal to invest the same fraction $\alpha^* = \frac{5}{24}$ of the capital every time. In that case, one can calculate that the probability of a final capital of no more than \$1.95 is zero, while the probability of a final capital higher than the initial amount of 10 thousand dollars is almost 70%. The value $\alpha^* = \frac{5}{24}$ follows from a general formula known as the Kelly criterion. This formula concerns the situation where there is a series of independent investment opportunities, where with each investment, the payout is f_1 times the invested amount with probability p and f_2 times the invested amount with probability $1 - p$, with values for p, f_1, and f_2 that satisfy

$$f_1 > 1, \quad 0 \le f_2 < 1, \quad \text{and} \quad pf_1 + (1-p)f_2 > 1.$$

In the example above, we have $p = 0.5$, $f_1 = 1.8$, and $f_2 = 0.4$. When we apply the Kelly criterion, the long-term growth of the capital is at its maximum and the capital ultimately exceeds any value. We can make this result plausible by considering the situation where we have a series of N investment opportunities and the goal is to maximize the expected value of the logarithm of the end capital. We do not explain the choice of a logarithmic utility function; it can be justified using economic considerations. Define the value function as follows:

$L_k(x)$ = maximum expected utility value for the end capital if there
are k more opportunities to invest at capital x

for $k = 0, \ldots, N$ and $x > 0$, where $L_0(x) = \ln(x)$. We assume that the capital is "infinitely divisible," that is, every time, any fraction of the current capital may be invested. We represent the decision in each state as the fraction of the current capital that is invested. We are now dealing with a dynamic programming problem for which the set of states is a "continuous" set, as is the set of possible decisions. However, this does not matter for the principle of dynamic programming. We now prove the following result. We reason as usual to find a recursive relation for $L_k(x)$. Suppose that one invests a fraction α of the current capital x and then trades optimally in the remaining $k-1$ investment opportunities. The capital when $k-1$ periods remain is equal to $x - \alpha x + f_1 \times \alpha x$ with probability p and equal to $x - \alpha x + f_2 \times \alpha x$ with probability $1 - p$. Applying the law of total expectation then gives that the expected value of the logarithm of the final capital is

$$pL_{k-1}\left([1 + (f_1 - 1)\alpha]x\right) + (1 - p)L_{k-1}\left([1 + (f_2 - 1)\alpha]x\right).$$

Maximizing this expression over α leads to the recursion

$$L_k(x) = \max_{0 \le \alpha \le 1}\{pL_{k-1}\left([1 + (f_1 - 1)\alpha]x\right) + (1 - p)L_{k-1}\left([1 + (f_2 - 1)\alpha]x\right)\}.$$

We now prove the following result. For all $k = 1, \ldots, N$ and $x > 0$, the decision α^* for which the maximum is reached in the optimality equation and the value function $L_k(x)$ for α^* are given by

$$\alpha^* = \min\left\{\frac{pf_1 + (1 - p)f_2 - 1}{(f_1 - 1)(1 - f_2)}, 1\right\} \quad \text{and} \quad L_k(x) = kc + \ln(x),$$

where $c = p\ln(1 + (f_1 - 1)\alpha^*) + (1 - p)\ln(1 + (f_2 - 1)\alpha^*) > 0$. In other words, the optimal strategy is to always invest the same fraction a^* of the current capital, regardless of the size of the capital and the number of investment opportunities that are still to come. This result is truly remarkable. Its proof uses induction, beginning at $n = 1$. Since $L_0(x) = \ln(x)$ and $\ln(ab) = \ln(a) + \ln(b)$, we can write the expression for $L_1(x)$ as

$$L_1(x) = \ln(x) + \max_{0 \le \alpha \le 1}\{p\ln(1 + (f_1 - 1)\alpha) + (1 - p)\ln(1 + (f_2 - 1)\alpha)\}.$$

Consider the function $h(\alpha) = p\ln(1 + (f_1 - 1)\alpha) + (1 - p)\ln(1 + (f_2 - 1)\alpha)$ for $0 \le \alpha \le 1$. The solution of $h'(\alpha) = 0$ is $\alpha_0 = \frac{pf_1 + (1-p)f_2 - 1}{(f_1 - 1)(1 - f_2)}$. The function $h(\alpha)$ is concave in α with $h(0) = 0$. This means that on the interval $(0, 1)$, the function $h(\alpha)$ takes on its maximum at $\alpha^* = \min(\alpha_0, 1)$. We also see that $L_1(x) = \ln(x) + c$. Suppose that the statement has been proved for $n = 1, \ldots, k - 1$. If we substitute $L_{k-1}(x) = (k - 1)c + \ln(x)$ into the recursion equation for $L_k(x)$, we obtain

$$L_k(x) = \ln(x) + (k - 1)c + \max_{0 \le \alpha \le 1}\{p\ln(1 + (f_1 - 1)\alpha) + (1 - p)\ln(1 + (f_2 - 1)\alpha)\}.$$

The equation for $L_k(x)$ is the same as that for $L_1(x)$ up to an additive constant $(k-1)c$. Consequently, the value α for which the recursive equation for $L_k(x)$ takes on its maximum is also given by $\alpha = \alpha^*$ and $L_k(x) = (k-1)c+\ln(x)+c = kc+\ln(x)$. This concludes the proof.

5.8.5 The Dice Game Pig

A popular dice game in the United States is Pig. Two persons take turns. The player whose turn it is rolls one die at a time but may roll the die several times during his turn. If he rolls a 1, his turn is over; otherwise, the player must choose whether to roll the die again or stop. If the player's turn is over because he rolls a 1, all points scored that turn are lost, and his total score remains what it was at the beginning of the turn. If the player decides to end his turn without rolling a 1, the total number of points scored during the turn is added to the player's total score. The winner is the first player to score 100 or more points. A coin toss determines which player begins.

We only consider the one-player version with goal to minimize the expected number of turns needed to reach the 100 points.[1] We say that the state is $(i, 0)$ if it is the beginning of a turn and i more points are needed to reach the goal of $G = 100$ points (so the current total score is $100 - i$ points). When a turn is in progress, the state of the system is defined by $s = (i, k)$ with i the number of points still needed at the beginning of the current turn and $k\ (\geq 2)$ the number of points scored in the current turn without rolling a 1. In every state (i, k), there are two decisions, "stop" and "roll," where, of course, the decision "roll" is the only possible one in state $(i, 0)$. The value function $V(s)$ is defined by

$V(s) =$ minimum expected value of the number of new turns to still play
to reach 100 points from state s.

The goal is to find $V(100, 0)$ with the corresponding optimal policy. The optimality equation for $V(s)$ is obtained through the usual argument in dynamic programming. For a state (i, k) with $k = 0$, applying the law of total expectation gives

$$V(i,0) = 1 + \frac{1}{6}V(i,0) + \sum_{r=2}^{6}\frac{1}{6}V(i,r).$$

For $k \geq 2$ and $i - k > 0$, we have

$$V(i,k) = \min\left\{V(i - k,0),\ \frac{1}{6}V(i,0) + \sum_{r=2}^{6}\frac{1}{6}V(i,k+r)\right\},$$

where $V(i,l) = 0$ for (i,l) with $i - l \leq 0$. The first term on the right-hand side of the optimality equation for $V(i,k)$ corresponds to the decision to "stop" in state (i,k) and then follow the optimal policy from state $(i - k,0)$ on, and the second term corresponds to the decision to "roll" in state (i, k) and then follow the optimal

---

[1]The two-player case may be analyzed by analogy.

222   *Operations Research: Introduction to Models and Methods*

policy from the resulting state. A closer look at the optimality equation for $V(s)$ shows that this equation is not recursive due to the absence of a time index that changes by 1 each turn. In dynamic programming, there is a standard method for solving optimality equations like the one described above, namely the successive substitution method. Begin with a randomly selected function $V_0(s)$, in general $V_0(s) = 0$ for all s. Then, define the functions $V_1(s), V_2(s), \ldots$ recursively by

$$V_n(i,0) = 1 + \frac{1}{6}V_{n-1}(i,0) + \sum_{r=2}^{6} \frac{1}{6}V_{n-1}(i,r),$$

and

$$V_n(i,k) = \min\left\{ V_{n-1}(i-k,0), \frac{1}{6}V(i,0) + \sum_{r=2}^{6} \frac{1}{6}V_{n-1}(i,k+r) \right\}.$$

A basic result from the theory of dynamic programming shows that

$$\lim_{n\to\infty} V_n(s) = V(s) \quad \text{for all } s.$$

In practice, one stops the calculations as soon as $\sum_s |V_m(s) - V_{m-1}(s)|$ is sufficiently small, say less than 0.001 (note that we are in fact dealing with a nested system of optimality equations that can be solved separately for each of $i = 1, 2, \ldots, 100$ in turn). For the minimum expected value of the number of turns needed to reach 100 points, we find $V(100,0) = 12.545$. The corresponding optimal policy is quite complex. Surprisingly, in this problem, a simple heuristic can be given that is almost as good as the optimal policy. The heuristic is to end a turn as soon as 20 or more points have been obtained, keeping in mind that one stops at i or more points if i more points are needed at the beginning of the turn. For the "stop at 20" heuristic, the expected number of turns is 12.637, which only differs by 0.7% from the minimal value.[2] The "stop at 20" rule can be deduced with a classic recipe for heuristics: pretend that there is only one step left to do. Suppose that in the current turn, we have obtained k points and roll the die again; then the direct expected increase in points is equal to $\frac{5}{6} \times 4$ and the expected number of points lost is equal to $\frac{1}{6} \times k$. The first k for which $\frac{1}{6} \times k \geq \frac{5}{6} \times 4$ is $k = 20$.

5.8.6   *Optimal Stopping and the One-Stage-Look-Ahead Rule*

In Section 5.8.2 we already considered an instance of an optimal stopping problem. In this section the optimal stopping problem in stochastic dynamic optimization is put in a more general context and the optimality of the appealing one-stage-look-ahead rule is discussed. Let us consider a random process that is observed at time

---

[2]For the "stop-at-20" heuristic, define $H(s)$ as the expected value of the number of turns required to reach 100 points starting from state s. Then the function $H(s)$ can be recursively calculated in a similar way as the value function $V(s)$. Note that in the equation for $H((i,k))$ there is only one term rather than the minimum of two terms, since only one decision is possible in each state (i,k). This leads to $H(100,0) = 12.637$.

points $t = 0, 1, 2, \ldots$ to be in one of the states of a finite or countably infinite set I. In each state there are no more than two possible actions: $a = 0$ (stop) and $a = 1$ (continue). When in state i and choosing the stopping action $a = 0$, a terminal reward $R(i)$ is received. Suppose that there is a termination state that is entered upon choosing the stopping action. Once this state is entered the system stays in that state and no further costs or rewards are made thereafter. When in state i and choosing action $a = 1$, a continuation cost $c(i)$ is incurred and the process moves on to the next state according to the transition probabilities p_{ij} for $i, j \in I$. The following assumption is made:

(A) $\sup_{i \in I} R(i) < \infty$ and $c(i) \geq 0$ for all $i \in I$.
(B) There is a nonempty set S_0 consisting of the states in which stopping is mandatory and having the property that the process will reach the set S_0 within a finite expected time when always the continuation action $a = 1$ is chosen in the states $i \notin S_0$, whatever the starting state is.

The goal is to find a stopping rule that maximizes the expected total reward earned up to stopping of the process, interpreting the continuation cost $c(i)$ is as a negative reward. This rule can be found by solving the optimality equation

$$V(i) = \max\{R(i), -c(i) + \sum_{j \in I} p_{ij} V(j)\} \quad \text{for } i \notin S_0,$$

where $V(i) = R(i)$ for $i \in S_0$. The reader is asked to take for granted that the maximal total expected reward $V(i)$ for starting state i is well-defined for all i and satisfies the optimality equation. Also, it is stated without proof that the stopping rule that chooses in each state i the action maximizing the right side of the optimality equation is optimal among all conceivable stopping rules. The optimality equation can be numerically solved by iterating

$$V_n(i) = \max\{R(i), -c(i) + \sum_{j \in I} p_{ij} V_{n-1}(j)\}$$

with $V_0(i) = R(i)$. However, under certain conditions this numerical work can be avoided when the so-called one-stage-look-ahead rule is optimal. This is an intuitively appealing rule that looks only one step ahead, as the name says. What does this rule look like? Define the set B as

$$B = \{i \notin S_0 : R(i) \geq -c(i) + \sum_{j \in I} p_{ij} R(j)\}.$$

That is, B consists of the states in which it is at least as good as to stop now as to continue one more step and then stop. Clearly, it cannot be optimal to stop in a state $i \notin B$, since in that case it would be strictly better to continue one more step and then stop. The *one-stage-look-ahead rule* is now defined as the rule that stops in state $i \notin S_0$ only if $i \in B$. The following condition is now introduced:

(C) The set B is closed so that once the process enters the set B it remains in B. That is, $p_{ij} = 0$ for $i \in B$ and $j \notin B$.

Under this condition it can be shown that the one-stage-look-ahead rule is overall optimal. An interesting application is the following problem.

Devil's penny problem

Someone puts $n + 1$ closed boxes in random order in front of you. One of these boxes contains a devil's penny and the other n boxes contain given dollar amounts $a_1, \ldots, a_n$. You may open as many boxes as you wish, but they must be opened one by one. You can keep the money from the boxes you have opened as long as you have not opened the box with the devil's penny. Once you open this box, the game is over and you lose all the money gathered so far. What is an optimal stopping rule when you want to maximize the expected value of your gain?

This problem can be put in the framework of the optimal stopping model. The state space is finite and is taken as

$$I = \{0\} \cup \{(k, b_1, \ldots, b_k) : 1 \le k \le n - 1\} \cup \{(1,0)\}.$$

State 0 means that you have opened the box with the devil's penny. The state $(k, b_1, \ldots, b_k)$ means that there are still $k+1$ unopened boxes including the box with the devil's penny, where dollar amounts $b_1, \ldots, b_k$ are contained in the k unopened boxes not having the devil's penny (each b_i is one of the $a_1, \ldots, a_n$). State $(1, 0)$ means that the only unopened box is the box with the devil's penny. In state $(k, b_1, \ldots, b_k)$ you have gathered so far $A - \sum_{i=1}^{k} b_i$ dollars, where $A = \sum_{i=1}^{n} a_i$ is the original total dollar amount in the n boxes. In state 0 the process stops and a terminal reward $R(0) = 0$ is received. In the other states there are two possible actions $a = 0$ and $a = 1$. The action $a = 1$ means that you continue and open one other box. There is no cost associated with this action. If you take the stopping action $a = 0$ in state $s = (k, b_1, \ldots, b_k)$, you receive a terminal reward of $R(s) = A - \sum_{i=1}^{k} b_i$ (and $R(s) = A$ for $s = (1,0)$). Then the process moves on to state 0 with probability $\frac{1}{k+1}$ and to state $(k - 1, b_1, \ldots, b_{l-1}, b_{l+1}, \ldots, b_k)$ with the same probability $\frac{1}{k+1}$ for $1 \le l \le k$. Assumption (B) is satisfied for the set $S_0 = \{0\}$.

To analyze the one-stage-look-ahead rule, let for state $s = (k, b_1, \ldots, b_k)$ the random variable $D(s)$ denote the amount by which your current gain $G = A - \sum_{j=1}^{k} b_j$ would change when you would decide to open one other box when you are in state s. Then,

$$\mathbb{E}[D(s)] = \frac{1}{k+1} \sum_{j=1}^{k} b_j - \frac{1}{k+1} G = \frac{1}{k+1}(A - G) - \frac{1}{k+1} G.$$

Thus, for state $s = (k, b_1, \ldots, b_k)$,

$$\mathbb{E}[D(s)] \le 0 \quad \text{if and only if} \quad G \ge \frac{1}{2} A,$$

where $G = A - \sum_{j=1}^{k} b_j$ is your current gain. Let

$$B = \{0\} \cup \{s : E[D(s)] \le 0\}.$$

It is immediate that continuation action $a = 1$ in state $s \in B$ can only lead to a worse state $t \in B$. Thus, we can conclude that the one-stage-look-ahead rule is optimal among the class of all conceivable control rules. The one-stage-look-ahead rule prescribes to stop in state $s = (k, b_1, \ldots, b_k)$ only if the dollar amount $G = A - \sum_{i=1}^{k} b_i$ gathered so far is at least half of the original total dollar amount A. It is quite remarkable that the optimal stopping rule depends only on the total dollar amount in the boxes and not on the distribution of the total dollar amount over the boxes. In the case that there are two boxes with a devil's penny, an examination of the analysis shows that it is optimal to stop as soon as the dollar amount you have gathered is at least one third of the original total dollar amount in the boxes.

A special case of the devil's penny problem is the following game. You play repeatedly a game with 11 cards: ace, two, three, $\ldots$, nine, ten, and joker. Each card counts for its face value, where the ace counts for 1. The randomly ordered cards are turned over one by one. Your score is the sum of the face values of the cards turned over as long as the joker has not shown up. You can stop any moment you wish. If the joker appears, the game is over and your end score is zero. What is a stopping rule that maximizes the average score per game? This is the devil's penny problem with $a_i = i$ for $i = 1, \ldots, 10$. The optimal stopping rule is to stop as soon your cumulative score is 28 or more. The average score per game is 15.453. This can be argued as follows. Under the optimal rule you use at least 4 cards and at most 7 cards, where your end score is no more than 37. Let $a_c(p)$ be the number of combinations of c non-joker cards so that the total number of points of the c cards is p but less than 28 if any of the c cards is removed. Then, noting that the joker is not among c randomly picked cards with probability $(11 - c)/11$, the expected value of your end score is $\sum_{c=4}^{7} \sum_{p=28}^{37} p \left[a_c(p)/\binom{10}{c} \right] \times [(11 - c)/11]$. The $a_c(p)$ can be found by enumeration. This gives the expected value 15.453 for the end score (the standard deviation is 15.547 and the probability of picking the joker is 0.5). You are asked to derive the optimal stopping rule and its expected end score when there are two jokers next to the cards with face values 1 to 10.

5.9 Exercises

5.1 Consider the shortest-path model as discussed in Section 5.2. Suppose that in addition to the costs for passing through the edges, a cost of $2\frac{1}{2}$ is incurred every time a change of direction occurs in the path. First define the value function with appropriate state variables, and then define a dynamic programming algorithm to calculate the minimum-cost path. Apply the algorithm to the data in Figure 5.1.

5.2 Suppose that in the Manhattan network of Figure 5.1, heavily-loaded trucks go from A to B. The numbers $R(x, y)$ and $U(x, y)$ along the road segments indicate the total weight in tons the road segment in question can tolerate. Explain that the value function $f(x, y)$ defined as the heaviest weight possible from point (x, y) to the endpoint B satisfies the recursive relation

$$f(x,y) = \max\left\{\min\left\{R(x,y),\ f(x+1,y)\right\},\ \min\left\{U(x,y),\ f(x,y+1)\right\}\right\}$$

and calculate the maximum-load path.

5.3 Solve the shortest-path problem from Exercise 3.2 in Chapter 3 with dynamic programming. Define the value function and the recursion relation.

5.4 In the Wild West, traveling by stagecoach was risky. A trip from Santa Fé to Fort Thomas took three days. The possible routes are given in Figure 5.6. The numbers along the possible day trips give the probability, as a percentage, of a robbery along the route in question. The beloved country singer Lilian Song from Santa Fé must perform in Fort Thomas. She asks her manager to determine the route the stagecoach should take to maximize the probability of arriving safely in Fort Thomas.

   (a) Define the value function and the recursion relation, and then carry out the calculations for the manager.
   (b) Once the safest path has been determined, Lilian, capricious as ever, changes her mind and decides to follow the path for which the greatest distance traveled in one day is as small as possible (now, the numbers along the arcs indicate the distances in units of 10 kilometers). Again, define the value function and the recursion relation, and calculate the optimal path.

Fig. 5.6   Stagecoach problem.

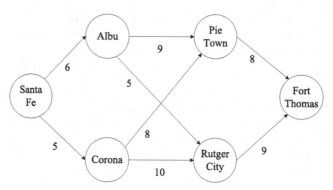

5.5 A group of four persons is standing on the southern bank of a river. The group must cross the bridge to get to the other side. However, it is pitch dark, and the group carries only one flashlight. The flashlight is indispensable to cross the rather rickety bridge. As no more than two persons can cross the bridge at the same time, several crossings are required. Not all of the four persons can cross the bridge at the same speed: person 1 needs 1 minute, person 2 needs 2 minutes, person 3 needs 5 minutes, and person 4 needs 10 minutes. Of course, the crossing time of a group consisting of two persons is equal to the time of the slowest of the two. Use dynamic programming to determine how the river crossing should take place to minimize the total time needed until everyone is on the northern bank of the river (use the obvious fact that after a crossing, the fastest person from the group on the northern bank returns to the other bank if there are still people there).

5.6 Consider the shortest-path problem for the Manhattan network in Figure 5.2. In Section 5.2, we formulated a so-called "backward" dynamic programming algorithm. For deterministic dynamic programming, it is also possible to formulate a "forward" algorithm. Define the value function $g(x, y)$ as the length of the shortest path from the starting point $A = (0, 0)$ to the "endpoint" (x, y). The aim is to find $g(n, m)$ and the corresponding shortest path. Explain why the following recursion holds:

$$g(x, y) = \min\{g(x - 1, y) + R(x - 1, y), \, g(x, y - 1) + U(x, y - 1)\}.$$

This recursion is initiated with $g(1, 0) = R(1, 0)$ and $g(0, 1) = U(0, 1)$. Apply this "forward" algorithm to the numerical example from Figure 5.1.

5.7 Every day, the manager of a power plant must decide which generators to use. The power plant has two different electricity generators. The required amount of electricity fluctuates throughout the day, which is divided into six periods. The number of megawatts requested is equal to 2000 megawatt (mW) in period 1, 4000 mW in period 2, 6000 mW in period 3, 9000 mW in period 4, 7000 mW in period 5, and 3000 mW in period 6. At the beginning of each period, generators can be switched on or off. A generator that is not used during a period will be switched off at the beginning of that period. Both generators are turned off for testing at the end of the day. The maximum capacities of generators 1 and 2 are 4000 mW and 6000 mW, respectively, for each period. The following costs are incurred. Every time generator j is started up after it has been turned off, a setup cost of C_j euros is made, where $C_1 = 500$ and $C_2 = 300$. In addition, there is an operational cost of a_j euros for each period that generator j is used, where $a_1 = 1000$ and $a_2 = 1100$. What schedule minimizes the total daily cost? (*Hint*: Take (k, i) to be the state with k the period of the day and i the status of the two generators.)

5.8 A popular game in which two persons play against each other is as follows. There are $m = 20$ matches on a table. The first player picks up 1, 2, or 3 matches. Next, the opponent picks up 1, 2, or 3 matches. Then it is the first player's turn, and so on. The player who picks up the last match loses. What policy should the first player follow to win? Use dynamic programming to calculate that the first player must force his opponent to pick up matches from a stack of 1, 5, 9, 13, or 17 matches. (*Hint*: Define the state as $(k, 1)$ when k matches are left and it is the first player's turn and as $(k, 2)$ when there are k matches and it is the opponent's turn. Define $f_k(i)$ as the "probability" that the first player wins from state (k, i) when both players follow the optimal policy, and note that $f_k(2) = 1 - f_k(1)$.)

5.9 An old Chinese game for two persons is as follows. On the table are two stacks of stones, say one stack of r stones and one stack of s stones for given values of r and s. The players take turns. The player whose turn it is takes one or more stones from both stacks or stones from only one of the two stacks. If she takes stones from both stacks, then she must take the same number from both. Other than that, the player is free to decide how many stones to remove. The player who removes the last stone wins. Develop a dynamic programming algorithm to calculate which positions lead to a loss in an optimal policy.

5.10 How can one pay the amount of C cents using the least number of coins when a sufficient number of each of the coin types $k = 1, \ldots, r$ is available and a coin of type k has value w_k cents with $w_1 = 1$? Formulate this as a shortest-path problem.

5.11 A sawmill has been commissioned to supply a_1 planks of length L_1, a_2 planks of length L_2, ..., and a_r planks of length L_r, where $L_1 < L_2 < \cdots < L_r$. To make planks of a given length, the sawmill must calibrate the machine, which costs C_i for planks of length L_i. The sawmill can also meet the demand for planks of a certain length by supplying longer planks, which must all be of the same length. The cost per plank of length L_i is c_i. How can the sawmill meet the demand at minimum cost? Formulate this as a shortest-path problem in an acyclic network.

5.12 A factory needs a special machine for each of the next seven years and plans to buy a new machine now. The machine is inspected at the end of every year and then either refurbished for use in the following year or replaced by a new one. The overhaul cost and residual value depend on the age of the machine and are given in the table below. A machine that has been in use for five years cannot be refurbished and must be replaced. A new machine costs €10 000. Determine an optimal maintenance-replacement schedule using a shortest-path problem in an acyclic network.

Age	1	2	3	4	5
Overhaul Cost	€800	€900	€1100	€1600	
Residual Value	€6000	€5000	€4000	€1500	€500

5.13 Consider the production-stock problem for Jones Chemical from Section 5.5.2. This time, suppose that no more than two production runs can be carried out. Calculate an optimal production schedule.

5.14 Suppose that boxes of different sizes are needed to package certain products. There are nine different sizes, numbered $i = 1, \ldots, 9$, but a box of size i may also be used instead of a box of size j if $j < i$. At most three different sizes of boxes can be ordered. In all, d_i boxes of size i are needed; the price per box of size i is a_i for $i = 1, \ldots, 9$. Determine the purchasing schedule using a shortest-path problem.

5.15 The skipper of the barge the Lorelei IV has a remaining capacity of 80 tons for cargo for his journey from Rotterdam to Basel on Monday of next week. Six companies have offered cargo to the skipper, with each company offering an unlimited number of units of cargo. The table below shows the weight in tons and the profit per unit of freight. How should the barge be loaded to maximize the profit on the journey? Formulate this as a knapsack problem and solve it with dynamic programming.

Company	1	2	3	4	5	6
Weight	18	28	30	36	38	46
Profit	25	40	48	60	65	80

5.16 Use dynamic programming to solve the following knapsack problem:

$$\begin{aligned} \text{Maximize} \quad & 11x_1 + 8x_2 + 15x_3 \\ \text{subject to} \quad & 4x_1 + 3x_2 + 5x_3 \le 8 \\ \text{and} \quad & x_1, x_2, x_3 \ge 0 \text{ and integer.} \end{aligned}$$

5.17 Give a dynamic programming algorithm for the following 0-1 knapsack problem, where the a_j, the b_j, and B are given positive integers:

$$\text{Maximize } \sum_{j=1}^{n} a_j x_j$$
$$\text{subject to } \sum_{j=1}^{n} b_j x_j \leq B$$
$$\text{and} \qquad x_j \in \{0,1\} \text{ for } 1 \leq j \leq n.$$

5.18 **(a)** A manufacturer must decide how many microchips to include in each of the circuits A, B, and C in an electronic system to maximize the probability that the system works over a given period of time. The system only works if in each circuit, at least one chip works. Both the circuits and the chips within a circuit work independently of one another. The probability that a chip fails within the scheduled time period is 0.2, 0.3, and 0.25 for the circuits A, B and C. In all, no more than six chips can be included, and no circuit can have more than two. How can the manufacturer design the most reliable system?

   (b) Use dynamic programming to solve the following problem:

$$\text{Maximize } x_1^{1/2} x_2 x_3^2$$
$$\text{subject to } 0.5x_1 + x_2^2 + 4x_3 \leq 15$$
$$\text{and} \qquad x_1, x_2, x_3 \geq 0 \text{ and integer.}$$

5.19 An airline must transport 575 passengers. The airline has three types of aircraft at its disposal. Each plane of type i has a capacity of c_i passengers, where $c_1 = 125$, $c_2 = 150$, and $c_3 = 175$. The cost $a_i(d)$ of deploying d planes of type i is shown in the table below. Begin by defining the value function and recursion relation in general terms. Then, calculate the least expensive transportation schedule.

	$d = 1$	$d = 2$	$d = 3$	$d = 4$	$d = 5$
$i = 1$	10	14	20	27	35
$i = 2$	12	18	25	34	-
$i = 3$	14	19	27	35	-

5.20 There are five districts in Weatherland, namely Floodstate, Stormstate, Hailstate, Snowstate, and Rainstate, with respective populations of 9061, 7179, 5259, 3319, and 1182. A distribution of the $R = 26$ available seats for a board of directors must be made over these five districts. Ideally, district j should receive Rf_j seats, where f_j is the fraction of the total population living in district j. However, the number of seats for each district must be an integer. Use dynamic programming to calculate a seat allocation for which the greatest difference between the ideal and actual numbers of delegates over the five districts is as small as possible. (*Hint*: See this as a five-step decision problem, where in the jth step, the seat allocation for district j is decided.) Also carry out the calculations for the situation where there are $R = 27$ and $R = 28$ seats to be distributed. Is the calculated allocation house monotonous? (By definition, the allocation is house monotonous if a district cannot be allocated fewer seats when the number of available seats on the board increases.[3])

---

[3]Seat allocation methods that must meet the criteria of house monotony and other criteria are discussed in the article M. L. Balinski and H. P. Young, The quota method of apportionment, *The American Mathematical Monthly*, 82 (1975), 701–736.

5.21 Consider Exercise 5.6 again. Now, assume that after picking up 1, 2, or 3 matches, each player flips a coin. If the outcome is heads, the player must pick up an additional match. Is there a simple optimal policy?

5.22 In the popular TV quiz *Think First, Act Later*, the candidate is asked to answer a number of questions. There are at most K questions. Experience shows that the probability of answering question k correctly is equal to p_k for $k = 1, \ldots, K$. The quiz applies the all-or-nothing principle. A wrong answer means that the candidate is eliminated and wins nothing. If the candidate answers question k correctly, he can either stop and win a_k thousand euros, or continue on to the next question. What policy maximizes the expected gain? Formulate a dynamic programming algorithm. Solve this problem for the numerical values $p_k = 0.95, 0.90, 0.85, 0.75, 0.50, 0.70$, and 0.75 and $a_k = 5, 15, 25, 35, 50, 75$, and 100 for $k = 1, \ldots, 7$.

5.23 Suppose that in Exercise 5.22, the candidate may answer one question incorrectly. If he stops after question k, the prize is either a_k or a_{k-1} thousand euros, depending on whether he answered all questions correctly or missed one. Solve the problem again.

5.24 A banker considers a number of investment opportunities. On each occasion, the banker may invest in at most one of the two projects A and B. An investment in the risky project A gives a profit of 100% with probability p_A and is lost with probability $1 - p_A$. The banker cannot lose money by investing in project B. An investment in project B yields a profit of 100% with probability p_B ($< p_A$) and recovers the investment without profit with probability $1 - p_B$. There are at most N investment opportunities. Each time, the banker can invest any integer amount in euros, up to the capital at that time; the profit from previous investments may be reinvested. What investment policy should the banker follow to maximize the probability of obtaining S euros if the current capital is equal to R euros? Formulate a dynamic programming algorithm. Solve the problem for $p_A = 0.5$, $p_B = 0.3$, $N = 5$, $R = 1$, and $S = 5$.

5.25 A company has decided to stop selling a particular spare part after this coming year. At the beginning of the year, 2000 units of the part are in stock. Inventory remaining at the end of the year will be destroyed. The inventory level can be increased at the beginning of each quarter by increasing the production. The demand in each quarter in units of thousands is equal to 1, 2, 3, and 4 with probabilities 0.2, 0.4, 0.3, and 0.1, where the demand quantities in the different quarters are independent of one another. Demand that cannot be delivered from stock is lost. The selling price of one part is 10 euros. Each production run involves a fixed production cost of 2500 euros, and the variable production cost is 3 euros per unit. There are no further costs. For what production schedule is the expected net profit the highest?

5.26 An investor has an option for buying one share of a certain fund at a fixed price c on one of the next N days. She is not obligated to use the option. The investor makes a profit of $s - c$ if she buys the share on a day that the price is equal to s. The share's current price is s_0, and the price changes every day by 1, 0, or -1 with respective probabilities p, q, and r with $p + q + r = 1$. What policy maximizes the expected profit? Formulate a dynamic programming algorithm. Solve the problem for the numerical values $c = 10$, $N = 7$, $s_0 = 10$, $p = 0.3$, $q = 0.2$, and $r = 0.5$.

5.27 (a) Consider a game where the player may toss a fair coin K times, where K is fixed beforehand. After each toss, the player can decide to stop. The payout is the fraction of the tosses that have come up tails at the moment the player stops. Calculate the maximum expected profit for $K = 10, 25, 50, 100$, and 1000.

(b) Consider a game where the player repeatedly draws a card from a pack of 26 red and 26 black cards without returning it. He can stop at any moment. The payout is the number of red cards drawn minus the number of black cards drawn. What is the maximum expected yield?

5.28 A gang of robbers is holding a meeting at its secret hideout. Outside, a police agent is waiting who discovered the hideout by accident and whose sole aim is to arrest the gang leader. The agent knows that the villains will come out one by one for security reasons and that as soon as she arrests one, the other villains will be warned and will escape. The agent wants to maximize the probability that she arrests the leader of the gang, which consists of N members. Fortunately, the agent knows that the leader is the tallest of the gang. What policy must the agent follow to maximize her probability of arresting the gang leader? Give a DP formulation and calculate the optimal policy for $N = 10$, 25, 50, and 100.

5.29 A family that is selling their house receives successive bids numbered $1, 2, \ldots, N$, where N is fixed. Once a bid has been made, the family must immediately decide whether or not to accept it (in the case of the very last bid, they have no choice and must accept). There is no way to backtrack to an earlier bid that has been rejected. The amounts of the bids are independent of one another, and every bid is a draw from a given probability distribution $\{a_j\}$. Give a DP formulation, first for maximizing the expected value of the accepted bid, and then for maximizing the probability of accepting the highest bid offered in the N bids.

5.30 A military aircraft carrying bombs must pass n strongholds before reaching the target on which the bombs are to be dropped. At each stronghold, there is a given probability of p that the bomber faces an enemy plane. The bomber has r missiles on board and has probability $1 - \alpha^i$ of surviving the attack of an enemy plane if it fires i missiles at the enemy plane. How should the bomber use the missiles to maximize the probability of reaching the final target?

5.31 Consider the one-player version of the game Pig from Section 5.8.5. Use dynamic programming to calculate the maximum probability of reaching 100 points in no more than N turns for $N = 7$, 10, 15, and 20.

5.32 Consider the following variant of the game Pig from Section 5.8.5. A player now rolls two dice at the same time, and the player may roll the dice several times during one turn. If the player rolls a 1, his turn is over, and he loses the points obtained in that round, and if the player rolls two 1's, then the total score is set back to zero. For the one-player case, use dynamic programming to answer the following questions:

(a) What is the minimum expected number of turns needed to obtain 100 points?
(b) What is an obvious heuristic approach? To answer this, calculate the expected number of turns needed to obtain 100 points.

5.33 In the finals of a TV show, two candidates play a game against each other. The game consists of a number of rounds. In each round, the candidates must simultaneously press a button without being able to see what button the other presses. They each push a button to indicate how many dice must be tossed for them, where they can choose from 1 to 20 dice. For each candidate, the specified number of dice is tossed (say, by a computer), and the total number of points added to the candidate's current total is the sum of the dots that have been thrown provided that there is no 1; otherwise, no points are added to the candidate's total. After every turn, each candidate is told both candidates' totals. The first candidate to obtain more than 100 points wins; if the candidates both obtain more than 100 points in the same round, the one with the most points wins. If the scores are the same, then fate is the deciding factor.

(a) Suppose that one candidate follows the five-dice heuristic. In this policy, five dice are thrown every time, except that $\lceil i/2 \rceil$ are thrown if only i more points are needed with $1 \leq i \leq 9$. What is the opponent's probability of winning if she plays optimally?

(b) For the one-player version of the problem, use dynamic programming to determine the minimum expected number of rounds needed to obtain 100 points. Also determine the expected number of rounds for the five-dice heuristic (explain this heuristic!). Calculate the maximum probability that the player obtains 100 points in at most N rounds for $N = 7, 10, 15,$ and 20.

5.34 In the game Flip and Flop, the player begins with 12 tokens numbered $1, 2, \ldots, 12$, with the numbers showing. Each turn, the player tosses two dice and chooses one combination of tokens with sum equal to the score of the throw and turns them over. If at some point, such tokens cannot be found, the player loses. The player wins when all tokens are turned over. Find the win probability and an optimal policy. Also do this if the player is allowed to skip turning over tokens m times, with $m = 1, 5,$ and 10.

5.35 The dice game Threes goes as follows. The player repeatedly tosses a number of dice at the same time, beginning with five dice. After tossing the dice, the player decides whether to freeze one or more of the dice. A die that is frozen counts for zero if the number of dots showing is three and else counts for the number of dots. The player continues with the remaining dice until all are frozen. What is the minimum expected value of the total score of the frozen dice?

5.36 In the TV show *The Weakest Link*, a team of players is given N questions, where each question comes from one of the sets $i = 1, 2, \ldots, 9$. After every correctly answered question, the team can decide to first secure the accumulated profit in the current chain of correctly answered questions before hearing the next question. If the team does this after a chain of length i, then an amount of c_i is added to the amount already secured in the safe and a new chain of questions starts. Every new chain begins with a question from set 1; if this question is answered correctly, the chain continues with a question from set 2, and so on. The house rule is that the amount in the safe may never exceed a given maximum of W; that is, if there is already a sum b in the safe and the team stops after i correct answers, the amount in the safe is equal to $\min(b + c_i, W)$. The c_i satisfy $0 < c_1 < c_2 < \cdots < c_9 = W$. If the chain is not stopped and the next question is answered wrong, then the money built up in the chain is lost, and a new chain starts provided that the number of questions asked has not reached N. Suppose that for every question k, the probability that the team answers correctly is p_k. Formulate a DP algorithm.

References

1. E. V. Denardo, *Dynamic Programming: Models and Applications*, Prentice-Hall, Englewood Cliffs, New Jersey, 1982.
2. S. E. Dreyfus and A. M. Law, *The Art and Theory of Dynamic Programming*, Academic Press, New York, 1977.

Chapter 6

Inventory Management

Good inventory management can save much money. Companies often have 40% or more of their capital invested in inventory. The two fundamental questions to be answered in inventory management are

- When should one order?
- How much should one order?

Quantitative methods play an essential role in answering these two questions. In this chapter, we discuss a number of basic methods that are frequently used in practice.

Inventory usually consists of more than one type of product. Often, not all products are of the same importance. Many lot sizing problems show the well-known *80/20 rule*: 20 percent of the products account for 80 percent of the turnover. The most common procedure is to classify the products into three groups A, B, and C based on their annual revenue. Group A typically contains 5 to 10 percent of the products and accounts for about 50 percent of annual sales in euros. Relatively much attention is paid to the products in group A. Group B represents 40 to 50 percent of the products and corresponds to approximately 40 percent of the annual sales in euros. The control over the inventory of the products in group B is generally modest. Group C contains the remaining products. Usually, relatively little attention is paid to the inventory management of the products in group C, as not much money is involved. The boundaries of this classification are not always clear-cut. The main point is that the products in group A are more important than the products in group B, which in turn are more important than the products in group C. Once the main classification has been determined, a suitable model for the inventory management of each group can be selected.

In the following sections, we discuss a number of scientific inventory models that have proved to be useful in real-world situations. The simplest inventory model is that where the demand can be predicted exactly and is constant in time and the stock can be replenished at any time. This model is known as the economic order quantity (EOQ) model. In Section 6.1, we derive the EOQ formula for the optimal order quantity for this model. This formula plays a vital role in many more complex lot sizing problems. In Section 6.2, we discuss an inventory model with a deterministic demand that fluctuates over time. For this model, we derive the Silver–Meal heuristic. The news vendor model is the subject of Section 6.3. This is the basic model for stochastic inventory situations with seasonal goods. In Section 6.4, we discuss stochastic inventory models with either continuous or periodic inventory review. The emphasis is on minimizing the total cost subject to a service requirement concerning the fraction of the demand that must be delivered directly from stock. The (s, Q), (R, S), and (R, s, S) review policies are discussed. These are simple policies that are often used in real-world situations.

6.1   The EOQ Model and the EPQ Model

This section first considers the EOQ inventory model in its simplest form. We derive the EOQ formula for the optimal order quantity. We also discuss the sensitivity of this formula to changes in the data. Then, we adjust the EOQ formula to the case of quantity discount. The concept of an exchange curve is also discussed. This concept is handy for making inventory decisions when no information is available on the exact holding and ordering costs. Finally, we discuss the economic production quantity (EPQ) model, in which the stock is gradually replenished through production.

6.1.1   *The Formula for the EOQ Inventory Model*

The EOQ model is the most basic inventory model. The assumptions of the EOQ model are as follows:

- The demand is deterministic and constant over time and causes the inventory level to decrease continuously.
- The demand must be met immediately.
- There are no restrictions on the size of the replenishment order.
- The time needed to replenish the stock is negligible (that is, the lead time, the time between placing a replenishment order and receiving it, is zero).
- The complete replenishment order is delivered at the same time.
- The only relevant costs are the ordering and holding costs.

This is an idealized inventory model, but the resulting EOQ formula for the optimal order quantity is nevertheless very useful in practice for more realistic inventory models. We analyze the EOQ model based on the average cost per unit of time, as is usual for most inventory models. The ordering and holding costs are specified as follows. Every time an order is placed, certain fixed costs are incurred that do not depend on the quantity that is ordered. Define

K = fixed ordering cost for placing a replenishment order.

The fixed ordering cost includes, for example, the cost of paperwork and the costs for the transportation, receipt, and inspection of the goods. The holding cost depends on the size of the inventory. It is generally defined as a percentage of the value in cash represented by the inventory. The capital that is invested in inventory is not available for investment for other purposes (unsold inventory is dead money). The loss of revenue from the capital invested in inventory usually represents the major part of the holding cost. Other relevant inventory costs are, for example, costs related to storage and insurance. Define

$v =$ unit purchase cost,

$r =$ annual interest factor, so that vr is the holding cost for keeping one unit in storage for one year.

Now that we have discussed the underlying cost structure, we can analyze the model. We introduce the following notation:

$D =$ annual demand for the product.

$Q =$ order quantity.

To determine the optimal value of Q, we need an expression for the total annual cost as a function of Q. We call this function $TC(Q)$. The annual number of orders is equal to D/Q. It follows that

$$\text{annual ordering and purchase costs} \;=\; \frac{D}{Q}K + Dv.$$

Since the lead time is equal to zero, an order will only be placed when the inventory level has dropped to zero. The inventory level therefore varies linearly between Q and 0; see Figure 6.1. This implies that the average inventory level is equal to $Q/2$, and consequently that

$$\text{annual holding cost} \;=\; \frac{Q}{2}vr.$$

Fig. 6.1 Inventory level pattern.

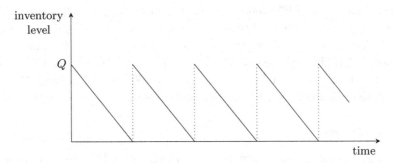

The total annual cost is therefore

$$TC(Q) = \frac{D}{Q}K + Dv + \frac{Q}{2}vr. \tag{6.1}$$

The value of Q that minimizes $TC(Q)$ is found by setting the derivative of $TC(Q)$ equal to zero. This leads to the equation $-KD/Q^2 + vr/2 = 0$. Solving this equation leads to the famous *EOQ formula*:

$$Q^* = \sqrt{\frac{2KD}{vr}}. \qquad (6.2)$$

The value Q^* indeed minimizes $TC(Q)$, as the second derivative of $TC(Q)$ is positive (verify). Formula (6.2) is also known as *Camp's formula*. At the optimal reorder quantity Q^*, the ordering frequency is equal to

$$\frac{D}{Q^*} = \sqrt{\frac{vrD}{2K}}$$

orders per year. The formulas show that if demand increases by a factor of f, the optimal order quantity and order frequency increase by a factor of $\sqrt{f}$.

Illustration

As an illustration, consider the following example. A distributor of Johnny Cola soft drinks supplies a large number of supermarkets. The demand for soft drinks is relatively constant and is approximately equal to 1500 crates per week. The purchase cost per crate is €7.50. The fixed ordering cost is estimated at €50. The distributor sets the annual holding cost at 20% of the average capital invested in inventory. What is the optimal order quantity?

Solution. Take a year as unit of time and apply formula (6.2) with $K = 50$, $D = 52 \times 1500 = 78\,000$, $v = 7.50$, and $r = 0.20$. This gives

$$Q^* = \sqrt{\frac{2 \times 50 \times 78\,000}{7.5 \times 0.2}} = 2280.4, \text{ say } 2280 \text{ crates.}$$

The optimal order quantity is therefore 2280 crates. It follows from formula (6.1) that the minimum annual ordering and holding costs equal €3420.53 (excluding the annual purchase cost of €585\,000 that the distributor cannot influence).

Now, suppose that the lead time L of an order is not negligible, but is exactly one week. What changes? The optimal order quantity remains the same. The only difference is that the replenishment order must now be placed when the inventory has dropped to $L \times D$ with $L = \frac{1}{52}$ and $D = 78\,000$. At the moment the order is delivered, the inventory level is exactly equal to zero.

Sensitivity Analysis

In practice, the estimates of the cost parameters r and K are not always very precise. The following question is therefore important. How sensitive is the true minimum cost to errors in the data? If we have incorrect data, we calculate an order quantity that differs from the actual optimal order quantity at minimum cost. We can obtain insight into the question above by assuming that the data are known exactly and then comparing the total annual cost for a nonoptimal Q with the total annual cost for the optimal order quantity Q^*. When comparing the total annual costs, it is not necessary to take the annual purchase cost of Dv into account. After all, this cost is the same for all values of Q. Therefore, define the function $\overline{TC}(Q)$ as $\overline{TC}(Q) = TC(Q) - Dv$. To compare the function value $\overline{TC}(Q)$ with $\overline{TC}(Q^*)$ for a value of Q near the optimal Q^*, it is useful to switch from the variable Q to the variable p defined by $Q = (1+p)Q^*$. Now, consider the relative percentage of the costs

$$RK(p) = \frac{\overline{TC}(Q) - \overline{TC}(Q^*)}{\overline{TC}(Q^*)} \times 100.$$

The quantity $RK(p)$ gives the relative deviation from the theoretically minimum ordering and holding costs if the order quantity $Q = (1+p)Q^*$ is used instead of the economic order quantity Q^*. After some calculation, the substitution of $Q^* = \sqrt{2KD/vr}$ in the expression for $\overline{TC}(Q)$ and $\overline{TC}(Q^*)$ gives (verify!):

$$\frac{\overline{TC}(Q) - \overline{TC}(Q^*)}{\overline{TC}(Q^*)} = \frac{\sqrt{\tfrac{1}{2}KDvr}/(1+p) + (1+p)\sqrt{\tfrac{1}{2}KDvr}}{\sqrt{\tfrac{1}{2}KDvr} + \sqrt{\tfrac{1}{2}KDvr}} - 1$$

$$= \tfrac{1}{2}\left\{1/(1+p) + (1+p)\right\} - 1.$$

This gives

$$RK(p) = \frac{50p^2}{1+p}.$$

This expression shows that the deviation from the theoretically minimum cost is relatively small even for values of p that differ significantly from zero. For example, an order quantity that is 30% ($p = 0.3$) more than the optimal order quantity Q^* has a total cost that is only 3.46% above the minimum.

A practical consequence of the flatness of the total cost curve around the minimum Q^* is that no excessive effort needs to be made to determine the *exact* values of the fixed ordering cost K and the interest factor r for the holding cost.

6.1.2 *Quantity Discount*

In real-world problems, it is not unusual to receive a quantity discount when the product is purchased in large quantities. The EOQ formula can be adapted to take this situation into account. Suppose that we have the following discount structure. For an order quantity Q, the price per unit v of the product is given by

$$v = \begin{cases} v_0 & \text{if } Q < Q_b, \\ v_0(1-d) & \text{if } Q \geq Q_b, \end{cases} \tag{6.3}$$

where Q_b is a given threshold and the discount factor d is between 0 and 1. The discount is expressed as a fraction d saved off the regular price v_0 if a quantity Q is ordered that is greater than or equal to the threshold Q_b. Note that the discount is given on the entire order. This discount structure is the most common one in practice.

Under the discount structure, the total annual cost $TC(Q)$ is given, as a function of Q, by

$$TC = \begin{cases} KD/Q + Dv_0 + v_0rQ/2 & \text{for } 0 < Q < Q_b, \\ KD/Q + Dv_0(1-d) + v_0(1-d)rQ/2 & \text{for } Q \geq Q_b. \end{cases}$$

If we draw the graphs of the functions $KD/Q + Dv_0 + v_0rQ/2$ and $KD/Q + Dv_0(1-d) + v_0(1-d)rQ/2$, then we can easily verify that the function $TC(Q)$ has its minimum at one of the three points $(2KD/v_0r)^{\frac{1}{2}}$, Q_b, and $(2KD/v_0(1-d)r)^{\frac{1}{2}}$. The following algorithm finds the optimal value for Q.

Step 1. Calculate the economic order quantity in case the discount holds,

$$Q^*_{disc} = \sqrt{\frac{2KD}{v_0(1-d)r}}.$$

If $Q^*_{disc} \geq Q_b$, then Q^*_{disc} is the optimal value of Q; otherwise, go to Step 2.
Step 2. Calculate the economic order quantity in case the discount does not hold,

$$Q^*_{reg} = \sqrt{\frac{2KD}{v_0r}}.$$

Compare the cost $TC(Q^*_{reg})$ with $TC(Q_b)$. If $TC(Q^*_{reg})$ is less than $TC(Q_b)$, then the optimal value for Q is equal to Q^*_{reg}; otherwise, the optimal value for Q is equal to Q_b.

This algorithm can easily be extended to the case of different thresholds and increasing discount percentages. The optimal order quantity is always equal to a threshold or to a feasible economic order quantity. We will not go into detail.

Numerical Illustration

To illustrate the algorithm, we consider the example from Section 6.1.1. Suppose that a discount of €0.75 per crate is given if the order quantity is 2500 crates or more. So $d = 1/10$ and $Q_b = 2500$. The calculations are as follows:

Step 1. $Q^*_{disc} = 2404 < Q_b$.
Step 2. $Q^*_{reg} = 2280$, $TC(Q^*_{reg}) = €588\,421$, and $TC(Q_b) = €529\,748$. Since $TC(Q_b)$ is less than $TC(Q^*_{reg})$, the optimal order quantity is Q_b, so 2500 crates.

6.1.3 The Exchange Curve

In many real-world situations, it can be difficult (or expensive) to obtain good approximations for the fixed ordering and holding costs. This is, in particular, the case when many different products are ordered. Suppose that we have n different products, numbered $i = 1, \ldots, n$. The EOQ model applies to each individual product, except that we do not have precise information on the ordering and holding costs. The only available information is the following:

$$D_i = \text{annual demand for product } i,$$
$$v_i = \text{purchase cost per unit of product } i.$$

How can we compare the different order decisions in a meaningful way when no information is available about the cost parameters? A meaningful comparison can be based on two aggregated performance measures, namely the average inventory investment (AII) and the total annual number of orders (ANO). Let Q_i be the order quantity for product i. Since the average inventory level of product i is equal to $Q_i/2$, it follows that

$$AII = \sum_{i=1}^{n} v_i \frac{Q_i}{2}.$$

The annual number of orders for product i is equal to D_i/Q_i; hence,

$$ANO = \sum_{i=1}^{n} \frac{D_i}{Q_i}.$$

For every choice of order quantities $Q_1, \ldots, Q_n$, we now have

$$AII \times ANO \geq \frac{1}{2}\left(\sum_{i=1}^{n} \sqrt{D_i v_i}\right)^2.$$

To see this, we use the Cauchy–Schwarz inequality. This inequality states that

$$\left(\sum_{i=1}^{n} a_i^2\right) \times \left(\sum_{i=1}^{n} b_i^2\right) \geq \left(\sum_{i=1}^{n} a_i b_i\right)^2.$$

If we apply this inequality with $a_i = \sqrt{v_i Q_i/2}$ and $b_i = \sqrt{D_i/Q_i}$, we find the inequality above.

The right-hand side of the inequality above is a constant γ that does not depend on the values of the Q_i. So for each order quantity, the aggregated performance measure can be displayed as a point in the plane to the right of the hyperbolic curve $AII \times ANO = \gamma$; see Figure 6.2. This curve is called the *exchange curve*. The exchange curve tells us how a given order strategy $(Q_1, \ldots, Q_n)$ can be improved if it leads to a point P that is not on the curve. It follows directly from Figure 6.2 that the order strategy corresponding to a point P^* on the thick, dashed part of the curve is an improvement on the order strategy $(Q_1, \ldots, Q_n)$ corresponding to point P. For every point on the hashed part of the curve, both the average inventory investment and the total annual number of orders are less than they are for point P. The question is now, of course, whether the best ordering strategy can be found at a given point on the exchange curve. The answer to this question is "yes."

Let P^* be a given point on the curve with coordinates AII^* and ANO^*. Define the number λ^* by

$$\lambda^* = AII^*/ANO^*$$

Fig. 6.2   The exchange curve.

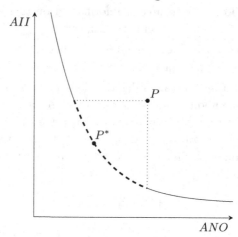

and the order quantities $Q_1^*, \dots, Q_n^*$ by

$$Q_i^* = \sqrt{\frac{2\lambda^* D_i}{v_i}} \quad \text{for } i = 1, \dots, n.$$

If we can prove that

$$\sum_{i=1}^{n} v_i Q_i^* / 2 = AII^* \text{ and } \sum_{i=1}^{n} D_i / Q_i^* = ANO^*,$$

then we have shown that the order quantities $Q_1^*, \dots, Q_n^*$ correspond to the point P^* on the exchange curve. Substituting the values of $Q_1^*, \dots, Q_n^*$ gives

$$\sum_{i=1}^{n} v_i Q_i^* / 2 = \left(\frac{1}{2}\lambda^*\right)^{\frac{1}{2}} \sum_{i=1}^{n} \sqrt{D_i v_i}, \qquad \sum_{i=1}^{n} D_i / Q_i^* = (2\lambda^*)^{-\frac{1}{2}} \sum_{i=1}^{n} \sqrt{D_i v_i}.$$

We have $AII^* \times ANO^* = \frac{1}{2}\left(\sum_{i=1}^{n} \sqrt{D_i v_i}\right)^2$ because $P^* = (AII^*, ANO^*)$ is on the exchange curve. This gives

$$\sum_{i=1}^{n} \sqrt{D_i v_i} = \sqrt{2 AII^* \times ANO^*}.$$

Substituting this and the equality $\lambda^* = AII^*/ANO^*$ into the relations for $\sum_{i=1}^{n} v_i Q_i^*/2$ and $\sum_{i=1}^{n} D_i Q_i^*$ given above leads to the desired result.

6.1.4   *The EPQ Production Model*

In the EOQ inventory model from Section 6.1.1, it is assumed that the entire replenishment order is received at the same time. This assumption is unrealistic in situations where the goods are not ordered externally but produced internally. The EPQ production model assumes that the good is produced at a rate of p units per unit of time. The demand for the product is deterministic and equal to D units per unit of time. By assumption, no shortage may occur. A production run can be started at any time. The cost of a production run of size Q is $K + vQ$, where $K > 0$ is the fixed setup cost and v is the variable production cost per unit of product. The holding cost per unit of product per unit of time is vr, where r is an interest factor. In the case of a fixed size Q for each production run, the evolution of the inventory level is shown in Figure 6.3.

Fig. 6.3   Inventory level evolution for the EPQ model.

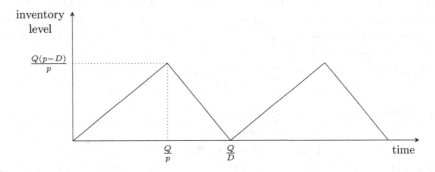

During a production run, the inventory level increases continuously by $p - D$ per unit of time. When the production run is complete, the inventory level is therefore $(p - D)\frac{Q}{p}$, after which it decreases to zero at a rate of D per unit of time. The total holding cost in a production run is therefore vr times the area of the triangle in Figure 6.3 (a production cycle is the time interval between the starting times of two consecutive production runs). This area is equal to $\frac{1}{2}(p - d)\frac{Q}{p} \times \frac{Q}{D}$. Combining this with the fact that the number of production cycles per unit of time is equal to $\frac{D}{Q}$ gives the following formula for the total cost per unit of time:

$$TC(Q) = \frac{D}{Q}\left[\frac{1}{2}vr(p - d)\frac{Q}{p} \times \frac{Q}{D} + K + vQ\right]$$
$$= \frac{1}{2}\frac{vrQ(p - d)}{p} + \frac{KD}{Q} + vD.$$

If we set the derivative of the function $TC(Q)$ equal to zero, we find that the optimal size Q^* of a production run is

$$Q^* = \sqrt{\frac{2KDp}{vr(p - D)}}.$$

The EOQ formula has therefore been corrected with the factor $\sqrt{p/(p - D)}$. This factor goes to 1 when the production rate p becomes very high. As in the EOQ model from Section 6.1.1, a small deviation from the optimal production size Q^* has little influence on the total cost per unit of time. Verify that the cost function

$\overline{TC}(Q) = TC(Q) - vD$ satisfies

$$\frac{\overline{TC}(Q)}{\overline{TC}(Q^*)} = \frac{1}{2}\left(\frac{Q}{Q^*} + \frac{Q^*}{Q}\right).$$

This means that for $Q = (1+q)Q^*$, the relative deviation from the minimum cost is again equal to $50q^2/(1+q)\%$.

Illustration

The Yolan Factory produces the popular smartphone model Veronica. The demand for this phone is constant in time and is $52\,000$ units per year. A total of 2000 units can be produced in one week. A production run of size Q costs $1060 + 20Q$ euros. The annual holding cost is 11% of the variable production cost of the phones in stock. How often should a production run start, and what size should it be?

If we take a week as unit of time, the data are

$$D = 1000, \ p = 2000, \ K = 1060, \ v = 20, \ \text{and} \ r = \frac{0.11}{52}.$$

This gives the optimal production quantity

$$Q^* = \sqrt{\frac{2 \times 1060 \times 1000 \times 2000}{20 \times (0.11/52) \times (2000 - 1000)}} \approx 10\,000.$$

The length of a production run is $Q^*/p = 5$ weeks. After a production run is completed, the next one starts $(1 - \frac{D}{p})Q^*/D = 5$ weeks later.

Material Requirements Planning

Both the EOQ inventory model and the EPQ production model assume that the demand is *independent*, that is, that it comes from external sources. Independent demand typically occurs in the case of end products. Making the end product often requires a large number of components. The demand for these components is determined not externally but internally. The inventory management of components with an internally determined and therefore dependent demand is usually done with a technique known as MRP *(material requirements planning)*. We will not discuss this technique in more detail; multiple books have been written about it. However, we do explain the need for MRP technology. Consider the example above concerning the production of phones. Suppose that the production of this end product requires sensors that are purchased from a supplier. The delivery time of the sensors is exactly one week. If an EOQ inventory model were used for the sensors, with (suppose) an optimal order quantity of $20\,000$ sensors, then the supply would be replenished every 20 weeks with $20\,000$ units. A production run lasts 5 weeks and requires $10\,000$ sensors. After that, it takes 5 weeks before the next run starts. It should be clear that the use of the EOQ inventory model for the sensors leads to unnecessarily high stocks of sensors. The reason is that the EOQ model ignores the dependency between the demand for sensors and the production process. In this situation, it is wiser to each time order $10\,000$ sensors one week before a production run starts.

6.2 The Silver–Meal Heuristic

In many real-world lot sizing problems, the demand is not constant but varies over time. A typical example is the situation of contractual delivery, where a contract requires specific quantities of a given product to be delivered to the customer at agreed times.

The assumptions of the dynamic inventory model are as follows:

- The stock of a product can only be replenished by production at the beginning of the given periods $j = 1, \ldots, N$.
- For every period j, the demand D_j is known. This demand must be met within that period. No ordering beforehand or back-ordering is allowed.
- A stock replenishment in period j is available for the demand in that period and for subsequent periods.
- There is no restriction on the size of a stock replenishment, and sufficient storage space is available.
- The variable production cost of v per unit does not depend on the quantity that is produced.
- The influenceable costs are the fixed production cost (setup cost) and the linear holding cost. A fixed cost of $K > 0$ is incurred for each stock replenishment. The holding cost in each period is $h > 0$ per unit of stock present at the end of the period, where h is usually given by $h = v \times r$ with r an interest factor for the capital invested in inventory.

Silver–Meal Heuristic

The dynamic inventory model can be solved exactly using integer programming or dynamic programming; see Sections 2.2.6 and 5.5.3. In practice, the Silver–Meal heuristic is often used to find a suboptimal solution, certainly when applying material requirements planning. This heuristic is easy to understand, requires little computation time, and usually provides a very good solution that differs little in cost from the optimal solution.

The Silver–Meal heuristic uses the following insights. Each production quantity must be *exactly* enough for the total delivery in a number of consecutive periods. Otherwise, savings could be made on the holding cost by postponing the production of the "fractional" part until the next time. As a result, production will only take place if nothing is left in stock and the production quantity will then be exactly sufficient to cover the demand for a specific number of periods. How do we determine the production quantity? The idea behind this heuristic is to choose the production quantity such that, over the periods in question, the average cost per period is as small as possible. For the costs, we take the fixed setup cost and the holding cost, but not the variable production costs. After all, the total variable production costs are the same for every production schedule and are always equal to v times the total demand. Suppose that at the beginning of period 1, a quantity of $D_1 + \cdots + D_T$ is produced to exactly cover the demand in periods $1, \ldots, T$. The fixed setup cost is K, and the total holding cost over the periods $1, \ldots, T$ is $hD_2 + 2hD_3 + \cdots + (T-1)hD_T$ (keep in mind that the quantity D_j produced for the demand in month j is kept in stock for $j - 1$ months). So the average cost per period over the first T periods given by

$$A(T) = \frac{1}{T} \left\{ K + h \sum_{j=1}^{T} (j-1)D_j \right\}.$$

The Silver–Meal heuristic takes the first value of T for which

$$A(T+1) > A(T).$$

In other words, we calculate $A(T)$ for successive values of T until the average cost per period goes up for the first time. Let T^* be the first value of T for which this occurs. The Silver–Meal heuristic tells us to produce $D_1 + \ldots + D_{T^*}$ units at the beginning of period 1. Period $T^* + 1$ is then renumbered as period 1, period $T^* + 2$ as period 2, and so on, and we apply the Silver–Meal heuristic again.

We illustrate the Silver–Meal heuristic with the production-stock problem of Jones Chemical we previously solved with exact methods in Sections 2.2.6 and 5.5.3. In this example, the planning period has length 6 and the data are $K = 500$, $h = 4$, $D_1 = 100$, $D_2 = 75$, $D_3 = 90$, $D_4 = 60$, $D_5 = 40$, and $D_6 = 85$. We take a month as unit of time and first calculate

$$A(1) = \frac{500}{1} = 500, \quad A(2) = \frac{500 + 4 \times 75}{2} = 400,$$

$$A(3) = \frac{500 + 4 \times 75 + 2 \times 4 \times 90}{3} = 506.67.$$

Since $T^* = 2$, the Silver–Meal heuristic tells us to produce $D_1 + D_2 = 175$ tons right before the beginning of period 1. We then calculate

$$A(1) = \frac{500}{1} = 500, \quad A(2) = \frac{500 + 4 \times 60}{2} = 370,$$

$$A(3) = \frac{500 + 4 \times 60 + 2 \times 4 \times 40}{3} = 353.33,$$

$$A(4) = \frac{500 + 4 \times 60 + 8 \times 40 + 3 \times 4 \times 85}{4} = 520.$$

In this case, we have $T^* = 3$, and therefore the Silver–Meal heuristic tells us to produce $D_3 + D_4 + D_5 = 190$ tons at the beginning of period 3. Since period 6 is the only remaining period, we produce $D_6 = 85$ tons at the beginning of period 6. The production schedule of this heuristic is therefore to produce 175 tons at the beginning of July, 190 tons at the beginning of September, and 85 tons at the beginning of December. In this numerical example, the production schedule found using the heuristic is optimal. This is less coincidental than might seem. Extensive numerical experiments have shown that in many applications of the dynamic inventory model, the Silver–Meal heuristic provides a solution with a cost no more than 1 to 2% above the theoretically minimum cost.

6.3 The News Vendor Problem

The news vendor problem is the prototype of an inventory problem with uncertain demand. The problem takes on many forms in real-world situations, such as a news vendor who must choose how many newspapers to order at the beginning of the day, a Christmas tree seller who must decide how many Christmas trees to stock at the beginning of December, or a fashion buyer who must decide what type of clothing to buy for the coming season. We first use a numerical example to provide insight into the effects of uncertainty in the demand for the product. Suppose that a news vendor sells newspapers daily at the entrance to the train station. The demand for newspapers varies from day to day, but the average daily demand is constant, equal to 200 newspapers. The daily demand is uniformly distributed between 150 and 250 newspapers. Early in the morning, the news vendor therefore orders $Q = 200$ newspapers from the publisher at the price of €1.00 per newspaper. The paper is sold for €2.50 a piece. For each copy left over at the end of the day, the vendor receives €0.25 from the publisher. If the vendor sells all his newspapers before the end of the day, he quits and goes home.

Question: What is the average long-term net profit per day for the news vendor?

Although everyone realizes that the actual net profit fluctuates from day to day, many argue as follows: on average, the number of papers sold per day equals the average demand. So in the long run, the average net profit per day equals $200 \times (2.50 - 1.00) = 300$ euros. This answer is incorrect! The fallacy of the averages has been committed. In a situation of uncertainty, it is generally not so that the average value of the output depends only on the average values of the input. The average value of the uncertain demand does not suffice; one needs to consider its entire probability distribution. The formulas we deduce below show that in the numerical example above, the average net profit per day is less than €300 (why?), namely equal to €274.75.

In general, the assumptions of the news vendor problem are as follows:

- The (perishable) product can only be kept in stock for one time period.
- The product can only be ordered at the beginning of the period and can therefore not be reordered if the stock runs out before the end of the period.
- The demand for the product is stochastic and has a given probability distribution.
- The cost and revenue structure is given by

$$v = \text{purchase price (in €/unit)},$$
$$p = \text{selling price (in €/unit)},$$
$$s = \text{residual value of the remaining stock (in €/unit)},$$
$$b = \text{understocking cost (in €/unit)},$$

where $v > s$ and $p > v$. The time sequence is crucial in the news vendor problem: an order must be placed before the demand for the product can be observed. In the following subsections, we discuss how to choose the order quantity in this situation of uncertainty, both for the case of discretely distributed demand and for the case of continuously distributed demand.

6.3.1 *Discretely Distributed Demand*

The demand X for the product in the coming period has a discrete probability distribution $f_j = \mathbb{P}(X = j)$ for $j \in \mathbb{Z}_{\geq 0}$. The product is ordered before the demand is observed. In the following theorem, we deduce the expected value of the net profit for an order quantity of Q. This deduction is instructive in itself and can be applied to all kinds of variants of the news vendor problem.

Theorem 6.1. *The expected value $P(Q)$ of the net profit for an order quantity of Q is given by*

$$P(Q) = -vQ + \sum_{j=0}^{Q} [pj + s(Q - j)]\, f_j + \sum_{j=Q+1}^{\infty} [pQ - b(j - Q)]\, f_j.$$

The function $P(Q)$ is at its maximum for the smallest Q with

$$\sum_{j=0}^{Q} f_j \geq \frac{p - v + b}{p - s + b}. \tag{6.4}$$

Proof. We start with a given order quantity Q. A crucial step is to pretend that we know the demand in the coming period and define the function $w(j)$ by

$$w(j) = \text{net profit if the realized demand in the coming time period is } j.$$

It is clear that for the resulting function $w(j)$, we have

$$w(j) = \begin{cases} -vQ + pj + s(Q - j) & \text{if } j \leq Q, \\ -vQ + pQ - b(j - Q) & \text{if } j > Q. \end{cases}$$

Now, let the random variable X denote the total demand in the coming time period. The *actual* net profit to be made in the coming period is a random variable given by $W = w(X)$. Since X takes on the values $i \in \mathbb{Z}_{\geq 0}$ with probabilities f_i, the function $W = w(X)$ takes on the values $w(i)$ for $i \in \mathbb{Z}_{\geq 0}$ with probabilities f_i. This gives $P(Q) = \mathbb{E}[W] = \sum_{j=0}^{\infty} w(j) f_j$. Splitting the summation over $j \leq Q$ and $j > Q$ and substituting the expression for $w(j)$ lead to the desired formula for $P(Q)$.[1]
To find the minimum of the function $P(Q)$, consider the difference

$$\Delta(Q) = P(Q + 1) - P(Q)$$

---

[1] In the same way that we derived $\mathbb{E}[W]$, we find that the probability distribution of W is given by $\mathbb{P}(W > w) = \sum_{j : w(j) > w} f_j$.

as a function of Q. After some simple calculations (verify!), we find

$$\Delta(Q) = -v + p\,(Q+1)f_{Q+1} + s\sum_{j=0}^{Q} f_j - p(Q+1)f_{Q+1}$$

$$+ (p+b)\sum_{j=Q+1}^{\infty} f_j = (p-v+b) - (p-s+b)\sum_{j=0}^{Q} f_j.$$

Since $p > v$ and $s < v$, we have $p - v + b > 0$ and $p - s + b > p - v + b$. This means that the function $\Delta(Q)$ is first positive and then negative. In other words, $P(Q)$ is first increasing in Q and then decreasing in Q. Let Q^* be the smallest value of Q with

$$(p - v + b) - (p - s + b)\sum_{j=0}^{Q} f_j \le 0;$$

then $P(Q)$ is at its maximum for $Q = Q^*$. This proves the theorem. $\qquad\square$

As an application of the relation (6.4), we consider the following numerical example for the news vendor problem:

$$v = 1,\ p = 2.5,\ s = 0.25,\ b = 0,\ \text{and } f_j = \frac{1}{101} \text{ for } j = 150, \dots, 250.$$

The value Q^* of Q that maximizes the average net profit per day is the smallest Q with

$$\sum_{j=150}^{Q} \frac{1}{101} \ge \frac{2.5 - 1}{2.5 - 0.25} = \frac{2}{3}.$$

This gives the order quantity $Q^* = 217$. For this order quantity, the expected value of the net profit in one day is equal to $P(Q^*) = 274.75$ euros; that is, in the long run, the average net profit per day is 274.75 euros. Now, suppose that at the beginning of every day, a "clairvoyant" can tell a news vendor the exact demand for that day before the vendor puts in his order. What is the *value of this information* for the vendor? This information is not worth more than an average of $1.5 \sum_{j=150}^{250} j f_j - 274.75 = 300 - 274.75 = 25.25$ euros per day to the news vendor. After all, if the vendor were to know that the demand on the coming day is exactly equal to j, he would buy exactly j newspapers and make a profit of $(2.5 - 1)j = 1.5j$ euros.

Marginal Analysis

The news vendor formula (6.4) has the following equivalent formulation:

$$\sum_{j=0}^{Q} f_j \ge \frac{c_u}{c_0 + c_u}, \tag{6.5}$$

where

c_0 = overstocking cost for every unsold unit of the product,
c_u = understocking cost for every unit of demand that cannot be met.

In the previous formulation, for every unsold unit, an amount of $v - s$ from the purchase cost v is not recovered, while for every unit of demand that cannot be met, a loss of profit of $p - v$ and a penalty cost of b are made, that is,

$$c_0 = v - s \quad \text{and} \quad c_u = p - v + b.$$

In many applications, it is simpler to work directly with the overstocking cost c_0 and the understocking cost c_t without using v, p, s, and b.

The news vendor formula (6.5) gives a plausible balance between the risk of a shortage and the risk of a surplus in terms of the overstocking cost and the understocking cost. The news vendor formula (6.5) can be derived directly using *marginal analysis*. Suppose that the order quantity Q is increased by a single unit. A simple probabilistic argument is used to calculate the effect of the additional unit on the total expected net profit. The understocking cost decreases by c_t if the demand is greater than Q; the probability of this is $\mathbb{P}(X > Q)$. On the other hand, the overstocking cost increases by c_0 if the demand is less than Q; the probability of this is $\mathbb{P}(X \leq Q)$. So the expected decrease in the understocking cost is $c_u \mathbb{P}(X > Q)$, and the expected increase in the overstocking cost is $c_0 \mathbb{P}(X \leq Q)$ if the order quantity is increased from Q to $Q + 1$. As Q increases, at a certain point, the expected increase in the overstocking cost is greater than the expected decrease in the understocking cost. After all, $\mathbb{P}(X \leq x)$ is increasing in x and $\mathbb{P}(X > x)$ is decreasing in x. Intuitively, one may expect that the optimal value of Q is the smallest value of Q for which

$$c_0 \mathbb{P}(X \leq Q) \geq c_u \mathbb{P}(X > Q),$$

that is, $c_0 \mathbb{P}(X \leq Q) \geq c_u [1 - \mathbb{P}(X \leq Q)]$. This gives the news vendor formula again. The direct approach with marginal analysis can also be applied to other inventory problems of the same type as the news vendor problem. We illustrate this with an overbooking problem for an airline.

Overbooking Problem

An airline can sell as many seats as it wants for a particular flight with room for 175 passengers. Every booking passenger pays the same amount of F euros. The airline decides to accept more reservations than there are seats to avoid, as much as possible, empty seats due to booked passengers who do not show up. Booked passengers show up independently of one another with a probability of 93%. A passenger who does not show up is refunded 80% of the amount of F euros she paid. Passengers who show up to be turned away because the flight is full are refunded the amount paid and given a compensation of $f \times F$ euros. What is the optimal value Q of the number of accepted bookings when the airline wants to maximize the total expected net revenue for the flight? This value is the smallest value of Q with

$$0.07 \times 0.2F + 0.93[F \times \mathbb{P}(X < 175) - (f \times F) \times \mathbb{P}(X \geq 175)] \leq 0,$$

where the random variable X represents the number of passengers from the first Q reservations who show up. So X is binomially distributed with parameters $n = Q$ and $p = 0.93$. This relation can be deduced directly using marginal analysis. To determine the change in the expected net profit if the number of bookings is increased from Q to $Q + 1$, we consider the probability given that the $(Q + 1)$st passenger does or does not show up. If this passenger does not show up, the net profit increases by $0.2F$, while if she does show up, the net profit increases by F if fewer than 175 passengers from the first Q reservations show up and decreases by $f \times F$ otherwise.

6.3.2 *Continuously Distributed Demand*

Although the derivation of the expected net profit given in the case of a discretely distributed demand holds as such in the case of a continuously distributed demand, we give an alternative derivation for the latter case. This derivation has the advantage of providing insight into the formulas for the special case of normally distributed demand.

Suppose that the stochastic demand X for the (perishable) product in the coming period has a probability distribution function $F(x) = \mathbb{P}(X \leq x)$ with probability density $f(x)$ and expected value $\mu = \mathbb{E}[X]$. An order of size Q must be placed *before* the demand can be observed.

Again, define

$$P(Q) = \mathbb{E}[\text{net profit in the coming period}].$$

For $P(Q)$, we have the basic formula

$$P(Q) = (p - v)Q + (s - p)\mathbb{E}[\text{surplus}] - b\mathbb{E}[\text{shortage}], \tag{6.6}$$

where $\mathbb{E}[\text{surplus}]$ and $\mathbb{E}[\text{shortage}]$ are given by

$$\mathbb{E}[\text{surplus}] = \int_0^Q (Q - x) f(x) \, dx, \tag{6.7}$$

$$\mathbb{E}[\text{shortage}] = \mathbb{E}[\text{surplus}] + \mu - Q. \tag{6.8}$$

We have the following elegant derivation of this formula. The basic formula for $P(Q)$ is

$$P(Q) = -vQ + p\mathbb{E}[\min(X, Q)] + s\mathbb{E}[\text{surplus}] - b\mathbb{E}[\text{shortage}].$$

Using the notation $a^+ = \max(a, 0)$, we have

$$\mathbb{E}[\text{surplus}] = \mathbb{E}[(Q - X)^+], \quad \mathbb{E}[\text{shortage}] = \mathbb{E}[(X - Q)^+].$$

Based on the relations $(a - b)^+ = a - \min(a, b)$ (verify!) and $\mathbb{E}[\text{surplus}] = \mathbb{E}[(Q - X)^+]$, we can rewrite the basic formula for $P(Q)$ as (6.6). It remains to verify (6.7) and (6.8). From the relation $\mathbb{E}[(X - Q)] = \mathbb{E}[(X - Q)^+] - \mathbb{E}[(Q - X)^+]$ (verify!) follows

$$\mu - Q = \mathbb{E}[\text{shortage}] - \mathbb{E}[\text{surplus}].$$

This gives (6.8). Applying the substitution rule $\mathbb{E}[g(X)] = \int_0^\infty g(x)f(x)\,dx$ with $g(x) = (Q - x)^+$ gives (6.7), thus concluding the derivation.

If we set the derivative of the (concave) function $P(Q)$ equal to zero, a simple calculation shows that $P(Q)$ is at its maximum for the smallest value of Q with

$$F(Q) = \frac{p - v + b}{p - s + b}. \tag{6.9}$$

This formula is consistent with formula (6.4) for the discrete case.

If the news vendor problem is formulated in terms of the overstocking cost c_0 and the understocking cost c_t, then for an order quantity of Q, we have

$$\mathbb{E}[\text{costs in the coming period}] = c_0\mathbb{E}[\text{surplus}] + c_t\mathbb{E}[\text{shortage}].$$

This cost function is at its lowest for the smallest value of Q with

$$F(Q) = \frac{c_t}{c_0 + c_t}. \tag{6.10}$$

In the news vendor problem, there is a clear relationship between the probability of understocking and the understocking cost c_t if we assume that the overstocking cost c_0 is given. For an order quantity of Q, the probability of a shortage is equal to $1 - F(Q)$. The relation (6.10) then shows that for the optimal order quantity Q, we have

$$\text{probability of a shortage} = \frac{c_0}{c_0 + c_t}.$$

Normally Distributed Demand

An important case is that where X is (approximately) *normally* distributed with expected value μ and standard deviation σ. If, instead of the variable Q, we take the variable k with

$$Q = \mu + k\sigma,$$

then the news vendor formulas (6.9) and (6.10) reduce to

$$\Phi(k) = \frac{p - v + b}{p - s + b} \quad \text{and} \quad \Phi(k) = \frac{c_t}{c_0 + c_t}, \tag{6.11}$$

where $\Phi(k)$ is the well-known standard normal distribution function. The value of k can be found using a table for the standard normal distribution or with statistical software. In inventory management software packages, the value of k is typically calculated using a so-called rational approximation for the inverse function $\Phi^{-1}(y)$. This useful approach is discussed in Appendix B. The formulas in (6.11) follow directly from (6.9) and (6.10). To see this, note that $(X - \mu)/\sigma$ has the standard normal distribution and rewrite $F(Q) = \mathbb{P}(X \leq Q)$ as

$$F(Q) = \mathbb{P}\left(\frac{X - \mu}{\sigma} \leq \frac{Q - \mu}{\sigma}\right) = \Phi\left(\frac{Q - \mu}{\sigma}\right).$$

In case the demand X has an $N(\mu, \sigma^2)$ distribution, insightful formulas can be given for the expected surplus and shortage for the order quantity Q. If we write $Q = \mu + k\sigma$, then we have

$$\mathbb{E}[\text{shortage}] = \sigma I(k) \quad \text{and} \quad \mathbb{E}[\text{surplus}] = \sigma k + \sigma I(k),$$

where $I(k)$ is the so-called *normal loss function* defined by

$$I(k) = \frac{1}{\sqrt{2\pi}} \int_k^\infty (z - k)e^{-\frac{1}{2}z^2}\, dz.$$

This normal loss function can easily be calculated numerically; see Appendix B. The formulas above clearly show that a better forecast of the demand (that is, a lower σ) leads to a decrease in both the expected shortage and the expected surplus. The derivation of the formulas for the expected shortage and surplus is not complicated. Let Z be a standard normally distributed random variable; then the order quantity Q satisfies

$$\mathbb{E}[\text{shortage}] = \mathbb{E}[(X - Q)^+] = \sigma \mathbb{E}\left[\left(\frac{X - \mu}{\sigma} - \frac{Q - \mu}{\sigma}\right)^+\right]$$

$$= \sigma \mathbb{E}\left[\left(Z - \frac{Q - \mu}{\sigma}\right)^+\right] = \sigma I\left(\frac{Q - \mu}{\sigma}\right) = \sigma I(k).$$

By (6.8), we have $\mathbb{E}[\text{surplus}] = \mathbb{E}[\text{shortage}] + Q - \mu$. For $Q = \mu + k\sigma$, this gives $\mathbb{E}[\text{surplus}] = \sigma I(k) + \sigma k$.

Repurchase Agreements

A seasonal producer can increase the quantity purchased by the sales channel by offering to take back unsold items at the end of the season at a certain percentage of the selling price. In the book trade, it is not unusual for a publisher to make an agreement with large bookshops to take back unsold copies of certain books. If the remaining copies' residual value increases for the sales channel, not only will the sales channel's profit increase, but the sales channel will also order a larger quantity from the producer. So, taking back unsold copies by the producer can mean an increase in profit for both the seller and the producer. Let us illustrate this with the following example. The national retailer Dirk van der Brink has signed a contract with the producer of a plush mascot for the next UEFA European Championship for the exclusive sale of the mascot. The retailer will sell the mascot in the two months preceding the soccer competition. The mascots are made in the Far East, and the retailer therefore has to indicate well in advance how many mascots it wants to buy. The manager estimates that the demand for the mascots will be normally distributed with an expected value of 50 thousand mascots and a standard deviation of 15 thousand mascots. The production cost for the producer is €0.50 per mascot. The retailer buys the mascots from the producer for €5.25 each and sells them for €12.50 each. After the European Championship, it will be impossible to sell the leftover mascots, which will therefore have no more market value. If the residual value of every unsold mascot is zero for the retailer, the latter will buy

$$Q^* = \mu + k\sigma = 53\,028$$

mascots, where $\mu = 50\,000$, $\sigma = 15\,000$, and the factor $k = 0.20189$ is determined by the equation $\Phi(k) = (12.5 - 5.25)/12.5 = 0.58$. For this order quantity, the retailer's expected profit is equal to 289 207 euros and the producer's is $Q^*(5.25 - 0.50) = 251\,885$ euros. Now, suppose that the producer signs a contract with the retailer to pay s euros for every unsold mascot. In this situation, the producer's expected profit is $(5.25 - 0.5)Q^* - s\mathbb{E}[\text{surplus}]$ for an order quantity of Q^* by the retailer. Table 6.1 gives the retailer's optimal order quantity Q^* for some values of s, as well as the corresponding values of the expected profit of both the retailer and the producer.

Table 6.1   The expected profits for varying residual values.

s	Q^*	Expected Surplus	$\mathbb{E}[\text{profit}]$ Retailer	$\mathbb{E}[\text{profit}]$ Producer	$\mathbb{E}[\text{profit}]$ Combined
0	53 028	7 620	289 207	251 885	541 092
0.5	53 962	8 173	293 152	252 234	545 386
1	54 995	8 810	297 394	252 416	549 810
2	57 458	10 438	306 972	252 050	559 022
3	60 747	12 831	318 521	250 056	568 577

The table shows that as the repurchase value increases, the retailer's expected profit increases. Another way to increase the expected profits for both the retailer and the producer is to draw up a contract with a flexible order quantity in which the producer allows the retailer to increase the order quantity after observing part of the demand; see Exercise 6.13.

6.3.3   *The News Vendor Problem with Multiple Products*

Consider the inventory model for the news vendor problem with n perishable products that must be purchased before the beginning of the coming period, before the demand for the products can be observed. However, a limited budget of B is available to purchase the products. The demand sizes of the products are independent of one another, and the demand for product i has a continuous probability distribution with probability distribution function $F_i(x)$ and probability density $f_i(x)$ for $i = 1, \ldots, n$. The purchase cost for each unit of product i is v_i, the selling cost for each unit is p_i, the residual value for each unsold unit is s_i, and the understocking cost is b_i for each unit of shortage, where $v_i, p_i, s_i, b_i > 0$, $v_i > s_i$, and $p_i > v_i$. How much of each of the products should be bought to maximize the total expected net profit subject to the constraint that the total purchase costs do not exceed the budget B? This leads to the following optimization problem:

$$\text{Maximize} \quad \sum_{i=1}^{n} P_i(Q_i)$$

$$\text{subject to} \quad \sum_{i=1}^{n} v_i Q_i \leq B$$

$$\text{and} \qquad Q_i \geq 0,$$

where Q_i is the order quantity for product i and, based on the relations (6.6)–(6.8), the function $P_i(Q_i)$ is given by

$$P_i(Q_i) = (p_i - v_i + b_i)Q_i - b_i \mu_i - (p_i - s_i + b_i) \int_0^{Q_i} (Q_i - x) f_i(x) \, dx.$$

This optimization problem can be solved effectively using the Lagrangian method. Before we give the algorithm for this specific application, let us discuss the Lagrangian problem in a more general form. Consider the optimization problem

$$\max_{\mathbf{x} \in X} \quad f(\mathbf{x}) \quad \text{subject to} \quad g(\mathbf{x}) \leq \mathbf{b},$$

where $\mathbf{x} = (x_1, \ldots, x_n)$ is a point in n-dimensional space and X is, for example, given by $X = \{\mathbf{x} \mid \mathbf{x} \geq 0\}$. Assume that the optimal solution $\mathbf{x}_0$ of the problem $\max_{\mathbf{x} \in X} f(\mathbf{x})$ without the constraint $g(\mathbf{x}) \leq \mathbf{b}$ satisfies $g(\mathbf{x}_0) > \mathbf{b}$ (otherwise, this solution is also optimal for the problem with the constraint). The *Lagrangian method* can then be applied. This method includes the constraint in the objective function and determines

$$\max_{\mathbf{x} \in X} L(\mathbf{x}, \lambda) \quad \text{with} \quad L(\mathbf{x}, \lambda) = f(\mathbf{x}) + \lambda[\mathbf{b} - g(\mathbf{x})]$$

for a fixed number $\lambda > 0$. Now, suppose that for some $\lambda = \lambda^*$, the point $\mathbf{x}(\lambda)$ that maximizes the Lagrangian function $L(\mathbf{x}, \lambda)$ satisfies $g(\mathbf{x}(\lambda^*)) = \mathbf{b}$. Then $\mathbf{x}(\lambda^*)$ is also the point that maximizes the function in the original optimization problem. After all, for every feasible point $\mathbf{x}$, the inequalities $\mathbf{b} - g(\mathbf{x}) \geq 0$ and $\lambda^* > 0$ imply

$$f(\mathbf{x}(\lambda^*)) = f(\mathbf{x}(\lambda^*)) + \lambda^*[\mathbf{b} - g(\mathbf{x}(\lambda^*)]$$
$$\geq f(\mathbf{x}) + \lambda^*[\mathbf{b} - g(\mathbf{x})] \geq f(\mathbf{x}).$$

Under certain regularity conditions on $f(\mathbf{x})$ and $g(\mathbf{x})$, the desired value λ^* of the *Lagrange multiplier* λ can be found through bisection. This is the case when $h(\lambda) = g(\mathbf{x}(\lambda))$ is a continuous function of λ with $h(0) > \mathbf{b}$ and $h(\lambda) < \mathbf{b}$ for λ sufficiently large. One can see λ as a penalty for violating the constraint $g(\mathbf{x}) \leq \mathbf{b}$. The optimal Lagrange multiplier λ^* also has an important interpretation. Under certain niceness conditions, λ^* gives the *maximum increase* of the maximum objective value for a small increase of the right-hand side coefficient $\mathbf{b}$.

How do we apply this to the news vendor problem? To this end, it is important to note that the Lagrange function $L(Q_1, \ldots, Q_n, \lambda)$ can be written as

$$L(Q_1, \ldots, Q_n, \lambda) = \sum_{i=1}^{n} P_i(Q_i, \lambda) + \lambda B$$

where $P_i(Q_i, \lambda)$ is the same as $P_i(Q_i)$ except that we have replaced v_i with $v_i(1+\lambda)$ (verify!). This means that for fixed $\lambda \geq 0$, the $Q_i(\lambda)$ that maximize the Lagrange function can be found by solving the equation

$$F_i(Q_i(\lambda)) = \frac{p_i - v_i(1+\lambda) + b_i}{p_i - s_i + b_i}$$

for $i = 1, \ldots, n$, where $F_i(x)$ is the probability distribution function for the demand for product i. Note that we take $Q_i(\lambda) = 0$ if the equation above does not have a positive zero.

Algorithm 6.1.

Step 1. Calculate the $Q_i(0)$ for $\lambda = 0$. If $\sum_{i=1}^{n} v_i Q_i(0) \leq B$, then the optimal solution has been found. Otherwise, choose a $\lambda > 0$ and go to Step 2.

Step 2. Calculate the $Q_i(\lambda)$ for the current value of λ, where $Q_i(\lambda)$ is zero if the equation for $F_i(Q_i(\lambda))$ has no positive solution.

Step 3. Compare $\sum_{i=1}^{n} v_i Q_i(\lambda)$ to B. If the two are equal, then the optimal solution has been found. If $\sum_{i=1}^{n} v_i Q_i(\lambda) < B$, then repeat Step 2 with a lower value of λ and otherwise repeat it with a higher value of λ.

A useful method to adjust the value of λ is the bisection method. The idea behind this method is simple. Choose two values λ_0 and λ_1 such that the budget constraint applies as a $>$ inequality for λ_0 and as a $<$ inequality for λ_1. As the next value for λ, try $\lambda_2 = (\lambda_0 + \lambda_1)/2$. Depending on whether the budget constraint applies as a $<$ inequality or a $>$ inequality for λ_2, the optimal value λ^* is between λ_0 and λ_2 or between λ_2 and λ_1. By repeating this step over and over, we can approximate the optimal value of λ as closely as we wish. The bisection method converges quickly since the interval in which λ^* is located is cut in half every time.

6.4   Stochastic Inventory Models

In this section, we consider stochastic inventory models in which the stock must be managed over a very long period of time. We assume that the model's parameters do not change significantly during this time period. We consider the situation with stochastic demand and positive lead times. In this situation, it is not possible to prevent the product from being sold out. A shortage occurs when the demand during the lead time exceeds the stock on the shelves at the reorder time. The two basic questions we want to answer are:

1. When should one order? (reorder point)
2. How much should one order? (order quantity)

The reorder point is expressed in units of product. If the inventory level has dropped to this reorder point, new stock is ordered. An important term is that of *safety stock*. The safety stock level determines the service delivered to the customers. In the situation where the inventory can be replenished at any time, the safety stock level is the difference between the reorder point and the expected demand during the lead time (in stochastic inventory models with periodic inventory review, the concept of safety stock is slightly more subtle). An increase in the safety stock level decreases the probability of the product being sold out but increases the average inventory level. In inventory management, one typically seeks a balance between the service to customers and the holding and ordering costs.

In the following subsections, we discuss, among other things, the (s, Q) inventory model with continuous inventory review and the (R, S) inventory model with periodic inventory review. Before we discuss these inventory models, which are widely used in practice, we introduce a number of basic concepts. In stochastic inventory models, one needs to indicate what happens to the demand that occurs while the system is out of stock. We distinguish two cases:

Back-ordering. The demand that occurs when the system is out of stock is delivered at a later time when sufficient stock is available again.

No back-ordering. The demand that occurs when the system is out of stock is lost.

We use the following inventory concepts:

On-hand inventory. This is the inventory that is physically present on the shelves. This inventory level is always nonnegative.

Net inventory = (on-hand inventory) − (back-ordered demand). This quantity is only negative if there are still outstanding orders to be filled (and in that case, the on-hand inventory is zero). In the model with lost sales, the net inventory is always equal to the on-hand inventory.

Economic inventory = (net inventory) + (ordered quantity). The economic inventory is equal to the net inventory plus the total ordered quantity in the pipeline.

The term *inventory position* is also often used instead of economic inventory. Inventory management must be based on the economic inventory and not on the net inventory. If the latter were used for inventory control, then, for example, a replenishment order could be placed today while a large replenishment order comes in tomorrow.

6.4.1   *The (s, Q) Continuous Review Inventory Model*

The widely used (s, Q) inventory model is applicable in the following situation:

- The economic inventory is continuously reviewed and inventory can be reordered at any time.
- The individual demand transactions are so small that the inventory level can be seen as a continuous variable.
- A replenishment order of quantity Q is placed whenever the economic inventory decreases to the reorder point s.
- The lead time of an order is a positive constant L.[2]
- The demanded quantities in disjoint time intervals can be treated as independent random variables.

These assumptions represent, in one way or another, approximations of reality. Nevertheless, the model and the heuristic solution have proved to be extremely useful in practice. In the heuristic analysis, it is not necessary to specify the stochastic demand process completely; instead, it is sufficient to know the probability distribution of the random variable

$$X_L = \text{total demand during the lead time.}$$

We introduce the following notation:

$$f_L(x) = \text{probability density of the demand during the lead time,}$$
$$\mu_L = \text{expected value of the demand during the lead time,}$$
$$\sigma_L = \text{standard deviation of the demand during the lead time.}$$

In fact, we will only need μ_L and σ_L for the heuristic solution. In practice, μ_L and σ_L are calculated based on collected data on the demand. Suppose that μ_1 and σ_1 are the expected value and standard deviation of the demand during one week. If the lead time is fixed and equal to L weeks, then we have

$$\mu_L = L\mu_1 \quad \text{and} \quad \sigma_L = \sqrt{L}\sigma_1.$$

We first consider the (s, Q) *back-order model*. An exact analysis of this model is rather complicated and does not lead to practical results. We therefore give a heuristic analysis. This heuristic analysis is based on common sense and leads to results that are sufficiently precise for practical purposes. The analysis also has the advantage that it can be applied to other stochastic inventory models.

For a given (s, Q) rule with $s > 0$, we deduce approximations for the following performance measures:

- the probability that the system is out of stock during the lead time
- the fraction of the demand that is delivered directly from stock.

The analysis is based on the following basic result:

For the (s, Q) back-order inventory model, we have

net inventory right before a replenishment order is received $= s - X_L$.

---

[2]The assumption can be weakened to stochastic lead times provided that the probability that replenishment orders overlap is negligible.

This result's proof is simple. Choose a replenishment order and mark it. Since the lead time is constant, the replenishment orders will arrive in the same order as they were placed. So all inventory that is still in the pipeline at the time the marked order is placed arrives before the marked order. The economic inventory was s at the time the marked order was placed. Since excess demand is delivered later, it now follows that the net inventory right before the marked order arrives is equal to s minus the total demand during the lead time of the marked order.

The Probability of Running out of Stock

By the basic result above, it is reasonable to approximate the probability that the system runs out of stock during the lead time by $\mathbb{P}(X_L > s)$. This leads to the approximation

$$\text{probability of running out of stock during the lead time} \approx \int_s^\infty f_L(x)\,dx.$$

Fraction of the Demand Delivered Directly

To calculate this fraction, we use the fact that the stochastic process that describes the course of the economic stock is a so-called regenerative stochastic process. Probabilistically, as it were, this stochastic process starts over every time the economic inventory drops to the level s and is then replenished to $s + Q$. The long-term behavior of the inventory process can be analyzed in terms of the behavior of the process between two consecutive regeneration points. For the regeneration periods, it is more convenient to choose the times when replenishment orders are received rather than the times when replenishment orders are placed. We define a *cycle* as the time interval between the moments when two consecutive replenishment orders come in. The further analysis is based on the following basic formula from the theory of regenerative stochastic processes: the long-term fraction of the demand that is not delivered directly from stock is equal to the following constant with probability 1:

$$\frac{\mathbb{E}[\text{demand per cycle that cannot be delivered directly}]}{\mathbb{E}[\text{total demand in a cycle}]}. \tag{6.12}$$

We do not prove this appealing formula, which is based on the strong law of large numbers from probability theory. The denominator in the formula follows by observing that in the back-order model, the total demand is ultimately met. In the long term, the average demand per cycle must therefore equal the average quantity of stock received per cycle. The latter is equal to Q. So

$$\mathbb{E}[\text{total demand during a cycle}] = Q.$$

To determine the numerator in the formula above, we note that

$$\mathbb{E}[\text{demand per cycle that cannot be delivered directly}] = S_1 - S_2$$

with

$$S_1 = \mathbb{E}[\text{shortage at the end of a cycle}],$$
$$S_2 = \mathbb{E}[\text{shortage at the beginning of a cycle}].$$

The end of a cycle is the instant right before a replenishment order arrives, while the beginning of a cycle is the instant right after a replenishment order arrives. By the basic result above, the net inventory right before a replenishment order arrives is $s - X_L$. This means that the expected shortages right before and right after a replenishment order arrives are given by

$$S_1 = \mathbb{E}[(X_L - s)^+] \quad \text{and} \quad S_2 = \mathbb{E}[(X_L - s - Q)^+], \tag{6.13}$$

where x^+ is the usual notation for $x^+ = \max(x, 0)$. If we apply the basic formula $\mathbb{E}[g(X_L)] = \int_0^\infty g(x)f_L(x)dx$ with $g(x) = 0$ for $0 \le x \le s$ and $g(x) = x - s$ for $x > s$, respectively $g(x) = 0$ for $0 \le x \le s + Q$ and $g(x) = x - s - Q$ for $x > s + Q$, then we obtain

$$S_1 = \int_s^\infty (x - s)f_L(x)\,dx \quad \text{and} \quad S_2 = \int_{s+Q}^\infty (x - s - Q)f_L(x)\,dx. \tag{6.14}$$

This gives the desired formula

fraction of the demand that cannot be delivered directly from stock

$$= \frac{1}{Q}\left\{ \int_s^\infty (x - s)f_L(x)\,dx - \int_{s+Q}^\infty (x - s - Q)f_L(x)\,dx \right\}.$$

Cost Minimization Subject to Service Requirement

Suppose that the following information on the cost structure is available:

$K =$ fixed cost associated with a replenishment order (in €),
$v =$ purchase cost per unit (in €),
$r =$ interest factor, so that vr is the holding cost (in €),
     per unit of inventory per unit of time.

In most real-world situations, it is difficult to specify the costs associated with shortages. In practice, therefore, a service-level requirement is often used instead of penalty costs. Typical service requirements are

P_1 : the probability of no shortage during the lead time of a replenishment order must be at least a given number α with $0 < \alpha < 1$;
P_2 : the fraction of the demand that is delivered directly from stock must be at least equal to a given number β with $0 < \beta < 1$.

To measure the service to the customer, the second service measure is usually more suitable than the first, as we will see later in an example. The service measure P_2 is called the *fill rate*.

In the practice of inventory management, the aim is often to minimize the average holding and ordering costs per unit of time subject to one of the service requirements P_1 or P_2. It is possible to deduce an approximation formula for the average cost per unit of time as a function of s and Q. Theoretically, s and Q should be calculated simultaneously by minimizing the cost function subject to the service requirement. This is rather laborious from a calculatory point of view. However, empirical studies have shown that in practical cases, a much simpler approach works almost as well. The simpler approach is a sequential approach, where first, the order quantity Q is determined with the EOQ formula and then, the reorder point s is determined based on the set service requirement.

Sequential Approach

Step 1. Calculate the order quantity Q from the EOQ formula

$$Q^* = \sqrt{\frac{2\mu_1 K}{vr}},$$

where μ_1 is the average demand per unit of time.

Step 2. Determine the reorder point s based on the set service requirement. For the requirement P_1, the reorder point s is calculated using the equation

$$\int_s^\infty f_L(x)\,dx = 1 - \alpha. \tag{6.15}$$

For the requirement P_2, the reorder point s is calculated using the equation

$$\int_s^\infty (x - s) f_L(x)\,dx - \int_{s+Q^*}^\infty (x - s - Q^*) f_L(x)\,dx = (1 - \beta)Q^*. \tag{6.16}$$

It turns out that the sequential approach works very well as long as the economic order quantity Q^* is greater than the standard deviation of the demand during the lead time. A warning regarding formula (6.16) is in order here. In the literature, the second integral in (6.16) is often ignored. This integral represents the term S_2 defined as the expected shortage right after an order has arrived. Intuitively, it seems reasonable to ignore the term S_2. However, numerical experiments indicate that for the calculation of the reorder point, it may be misleading to neglect this term if σ_L/μ_L is not small (say, $\sigma_L/\mu_L > 0.5$) or β is not close to 1 (say, $\beta < 0.9$).

In general, it is not an easy task to solve equation (6.16) numerically. Fortunately, for the important case of normally distributed demand, a straightforward solution procedure exists.[3]

Normally Distributed Demand

In real-world situations, it is often reasonable to model the demand during the lead time of an order as normally distributed. If the demand comes from a large number of independent customers, using the normal distribution is justified by the

---

[3] Another density for the demand for which the equation (6.16) is relatively easy to solve is the gamma distribution. This distribution is a popular one for modeling demand with the property $\sigma_L/\mu_L \geq 0.5$.

central limit theorem. Assume that the demand during the lead time is normally distributed with expected value μ_L and standard deviation σ_L. This assumption requires that σ_L/μ_L not be too large (say, $\sigma_L/\mu_L \leq 0.5$) because otherwise there would be a significant probability of a negative demand.

To simplify the equations (6.15) and (6.16), it is helpful to use the following representation:

$$s = \mu_L + k\sigma_L.$$

The factor k is called the *safety factor* and $k\sigma_L(= s - \mu_L)$ is called the *safety stock*. If we denote the distribution function of the standard normal distribution by $\Phi(k)$ and observe that $(X_L - \mu_L)/\sigma_L$ is standard normally distributed, then it follows that $\mathbb{P}(X_L > s) = 1 - \Phi(k)$. This means that (6.15) can be simplified to

$$1 - \Phi(k) = 1 - \alpha. \tag{6.17}$$

In software packages, k is often calculated using a rational approximation for the inverse function $\Phi^{-1}(y)$; see Appendix B.

We need the normal loss function $I(z) = (1/\sqrt{2\pi}) \int_z^\infty (x - z) e^{-\frac{1}{2}x^2}\, dx$ to solve (6.16) in the case of a normally distributed demand. To rewrite (6.16) in terms of $I(z)$, we first note that the function $I(z)$ has the probabilistic interpretation:

$$I(z) = \mathbb{E}[(U - z)^+],$$

where U is a standard normally distributed random variable and $x^+ = \max(x, 0)$. If the demand X_L is $N(\mu_L, \sigma_L^2)$ distributed, then for every constant a, we have

$$\mathbb{E}[(X_L - a)^+] = \sigma_L\, \mathbb{E}\left[\left(\frac{X_L - \mu_L}{\sigma_L} - \frac{a - \mu_L}{\sigma_L}\right)^+\right] = \sigma_L\, I\left(\frac{a - \mu_L}{\sigma_L}\right).$$

This means that we can simplify the expressions in (6.13) and (6.14) to

$$S_1 = \sigma_L\, I\left(\frac{s - \mu_L}{\sigma_L}\right) \quad \text{and} \quad S_2 = \sigma_L\, I\left(\frac{s + Q - \mu_L}{\sigma_L}\right).$$

Equation (6.16) can now be written as

$$\sigma_L\, I\left(\frac{s - \mu_L}{\sigma_L}\right) - \sigma_L\, I\left(\frac{s + Q^* - \mu_L}{\sigma_L}\right) = (1 - \beta)Q^*.$$

For practical purposes, this equation can be further simplified by keeping in mind that the term S_2 gives the expected shortage right after a replenishment order arrives. For a normally distributed demand, the second term in the equation above may be ignored provided that the required service level β is sufficiently high (say, $\beta \geq 0.9$). This leads to the well-known formula

$$\sigma_L\, I(k) = (1 - \beta)Q^* \tag{6.18}$$

from which k must be solved for the calculation of the reorder point $s = \mu_L + k\sigma_L$. This formula is one of the *most important* formulas in inventory management. In inventory management software packages, k is calculated from a rational approximation for the inverse function $I^{-1}(y)$; see Appendix B.

The relations (6.17) and (6.18) and the representation $s = \mu_L + k\sigma_L$ lead to a few interesting conclusions. Under the service requirement concerning the probability of no shortage during the lead time, the reorder point s remains the same if the order quantity changes, while under the service requirement concerning the fraction of the demand that can be delivered directly from stock, the reorder point s decreases if the order quantity increases. From the relation $\sigma_L = \sqrt{L}\sigma_1$, we can further conclude that the required safety stock decreases by only a factor of $\sqrt{f}$ if the supplier's lead time decreases by a factor of f.

Numerical Illustration

A wholesaler sells a particular type of racing bike. The annual demand for these bikes is approximately normally distributed with an average of 2600 bikes and a standard deviation of 200 bikes. The wholesaler orders the bicycles directly from the factory. The lead time for an order is almost constant and equals 3 weeks. Placing an order costs €225. The purchase cost per bike is €200. The annual cost of keeping a bike in stock is estimated at 15% of the purchase cost of the bike. Since the bikes are of an exclusive type, customers are prepared to wait for their bikes if they are not in stock. The wholesaler manager has decided to strive for a high level of service. At least 99% of the demand must be delivered directly from stock. The manager uses an (s, Q) inventory policy. How should he choose s and Q in order to achieve the required service level at minimum cost?

Solution. We use the sequential approach. We first calculate the order quantity. If we take a year as unit of time, we can apply the EOQ formula with $\mu_1 = 2600$, $K = 225$, $v = 200$, and $r = 0.15$. This gives

$$Q^* = \sqrt{\frac{2 \times 2600 \times 225}{200 \times 0.15}} = 197.5, \text{ say } 198 \text{ bikes.}$$

We then calculate the reorder point using formula (6.18). Since the lead time L is equal to $3/52$ year, we find

$$\mu_L = \frac{3}{52} \times 2600 = 150 \quad \text{and} \quad \sigma_L = \sqrt{\frac{3}{52}} \times 200 = 48.038.$$

Applying equation (6.18) then gives

$$48.038 \times I(k) = (1 - 0.99) \times 198,$$

that is, $I(k) = 0.0412$. The solution to this is $k = 1.347$. The reorder point s is therefore

$$s = 150 + 1.347 \times 48.038 = 214.7, \text{ say } 215 \text{ bikes.}$$

The wholesaler therefore places an order for 198 bicycles every time the economic inventory has dropped to 215 bikes.

Let us conclude by showing that the probability of not depleting the stock during the lead time is a less suitable measure for the service to the customer. To this end, we solve equation (6.17) with $\alpha = 0.99$. This gives $k = 2.326$ or the reorder point $s = 150 + 2.326 \times 48.038 = 261.7$, say 262 bikes. In other words, the safety stock $k\sigma_L$ increases from 65 to 112. In order to reach the same service level of 99%, the service measure with respect to not depleting the stock requires a safety stock that is approximately 72% higher than the previously calculated safety stock. This is a considerable increase. The fraction of the demand that can be delivered directly from stock is 0.9992 if the (s, Q) rule is used with $s = 262$ and $Q = 198$. Using the probability of not depleting the stock during the lead time as service measure can lead to an unintended high service level to the customer and therefore to an unnecessarily high inventory level.

Cost Minimization With Penalty Costs

Suppose that instead of a service requirement, penalty costs are considered for demand that cannot be delivered directly from stock. Assume that in addition to the fixed ordering cost K for a replenishment order and holding cost $h = vr$ per unit of inventory per time unit, a penalty cost of b is incurred for each unit of product that is delivered late. A heuristic solution for how to choose s and Q to minimize the total cost is as follows:

Step 1. Determine the order quantity Q from the EOQ formula $Q^* = \sqrt{\frac{2K\mu_1}{vr}}$, where μ_1 is the average demand per unit of time.

Step 2. Determine the reorder point s using

$$F_L(s) = 1 - \frac{vrQ^*}{b\mu_1}, \tag{6.19}$$

assuming $vrQ^*/b\mu_1 < 1$, where $F_L(x)$ is the probability distribution function of the total demand during the lead time. If the demand during the lead time is normally distributed, use $s = \mu_L + k\sigma_L$ to simplify this formula to

$$\Phi(k) = 1 - \frac{vrQ^*}{b\mu_1}. \tag{6.20}$$

In a way similar to that done for the news vendor problem, the formula for the reorder point s can be found using *marginal analysis*. Suppose that for a given value of the order quantity Q, the reorder point increases from s to $s + \Delta$ with Δ small. The average increase of the holding cost per unit of time is then approximately equal to $vr\Delta$. What is the average saving on the penalty cost per unit of time? In every cycle in which the total demand during the lead time is greater than s, approximately $b\Delta$ is saved in penalty costs. The fraction of the cycle for which the demand during the lead time is greater than s is equal to $1 - F_L(s)$. The average number of cycles per unit of time is μ_1/Q in the back-order model (because the average demand per cycle is Q). This means that an increase of the reorder point from s to $s + \Delta$ leads to an average decrease in the penalty cost of about

$$\frac{b\Delta\mu_1}{Q}[1 - F_L(s)]$$

per unit of time. As a function of s, this decrease itself decreases as s increases and therefore at some point becomes less than the average increase of $vr\Delta$ in the holding cost. This suggests that one should choose the reorder point s according to

$$\frac{b\Delta\mu_1}{Q}[1 - F_L(s)] = vr\Delta.$$

This leads to formula (6.19) for the reorder point s.

The deduction above shows that (6.19) and (6.20) can also be used to determine the reorder point s for a given penalty cost if the order quantity Q^* is different from that determined by the EOQ formula. Furthermore, in situations where the penalty cost is difficult to specify, the formulas (6.19) and (6.20) can be used to determine the implied value of the penalty cost s for a given (s, Q) inventory policy.

Numerical Illustration

Consider, again, the bike wholesaler from the previous numerical illustration. Suppose that a bike that cannot be delivered directly from stock is delivered later at a discount of €100 off the selling price. What (s, Q) strategy is then used in the situation without direct service requirement? The order quantity Q is also 198 bikes in the model with penalty costs. It follows from formula (6.20) that the safety factor k of the reorder point s is determined by

$$\Phi(k) = 1 - \frac{200 \times 0.15 \times 198}{100 \times 2600} = 0.97715.$$

This gives $k = 1.9982$, so the reorder point is $s = 150 + 1.9982 \times 48.038 = 246$ bikes. For the (s, Q) policy with $s = 246$ and $Q = 198$, it should be noted that the fraction of the demand that is delivered directly from stock is equal to $\frac{1}{Q}\sigma_L I(k) = 0.998$. An interesting question is for which discount on the selling price for a bike that is delivered late the reorder point is the same as the $s = 215$ in the situation of a service requirement of 99% of the demand being delivered directly. To $s = 215$ corresponds a safety factor of $k = 1.353$ with $\Phi(k) = 0.912$. Solving b from $1 - 5940/(b \times 2600) = 0.912$ gives $b = 25.96$. A service requirement of 99% of the demand being delivered directly therefore amounts to an approximate discount of 26 euros on a bike that is delivered later.

Stochastic Lead Time

In the above, we have always assumed that the supplier's lead time is deterministic. A closer look at the heuristic analysis makes it clear that the analysis also holds if the lead time is stochastic, provided that the probability that two or more outstanding replenishment orders overlap is negligible. The results found earlier then also apply to the case of a stochastic lead time. In particular, the formulas (6.17), (6.18), and (6.20) remain valid when μ_L and σ_L are taken equal to

$$\mu_L = \mathbb{E}[L]\mu_1 \quad \text{and} \quad \sigma_L = \sqrt{\mathbb{E}[L]\sigma_1^2 + \sigma^2(L)\mu_1^2},$$

where $\mathbb{E}[L]$ and $\sigma(L)$ are the expected value and the standard deviation of the stochastic lead time L. Variability in the lead time has a great influence on the safety stock. Smaller safety stocks are sufficient if the variability of the supplier's lead time is reduced. As an illustration, consider, once again, the bicycle wholesaler. Suppose that the factory's lead time is not exactly 3 weeks, but 2 or 4 weeks, each with probability $\frac{1}{2}$. What is the reorder point s subject to the service requirement of 99% of the demand being delivered directly from stock? For μ_L and σ_L, we find the values (the unit of time is one year)

$$\mu_L = \frac{3}{52} \times 2600 = 150, \ \sigma_L = \sqrt{\frac{3}{52} \times 40000 + \frac{1}{(52)^2} \times (2600)^2} = 52.257.$$

Next, the safety factor k is solved from $52.257 \times I(k) = 0.01 \times 198$. This gives $k = 2.288$, so $s = 150 + 2.288 \times 52.257 \approx 277$. The reorder point s therefore increases from 215 to 270 if the lead time is not 3 weeks but 2 or 4 weeks, each with probability $\frac{1}{2}$. The safety stock increases by 55 bikes, and therefore nearly doubles.

6.4.2 The (s, Q) Model with Lost Sales

An exact analysis of the inventory model with lost sales is even more difficult than that of the back-order model. However, the heuristic analysis in the previous subsection requires only minor adjustments for the model with lost sales. The basic result concerning the net inventory right before the replenishment arrives was crucial in the analysis of the back-order model. What is the corresponding result for the model with lost sales? It is tempting to say that the net inventory right before the replenishment arrives is exactly equal to $(s - X_L)^+$. However, this need not hold if other replenishment orders were outstanding when the relevant replenishment order was placed (verify!). Nevertheless, it is reasonable to take $(s - X_L)^+$ as an approximation for the net inventory right before a replenishment arrives, especially if we assume that s and Q are such that lost sales do not occur too often.

The probability of running out of stock during the lead time of an order is again approximated by $P(X_L > s)$, so

$$\text{probability of running out of stock during the lead time} \approx \int_s^{\infty} f_L(x)\,dx.$$

A subtler argument is required for the fraction of sales that are lost. The starting point is again formula (6.12). The numerator is approximated by

$$\mathbb{E}[\text{amount of lost sales per cycle}] \approx \mathbb{E}[(X_L - s)^+].$$

To obtain the numerator of (6.12), we note that

$$\mathbb{E}[\text{total demand per cycle}] = \mathbb{E}[\text{amount of lost sales per cycle}]$$
$$+ \mathbb{E}[\text{amount of delivered sales per cycle}].$$

The first term on the right-hand side is approximately equal to $\mathbb{E}[(X_L - s)^+]$, and the second term must be equal to the order quantity Q. We therefore find

$$\mathbb{E}[\text{total demand in a cycle}] \approx \mathbb{E}[(X_L - s)^+] + Q.$$

It now follows that

$$\text{fraction of sales that are lost} \approx \frac{\int_s^\infty (x - s) f_L(x)\, dx}{\int_s^\infty (x - s) f_L(x)\, dx + Q}. \tag{6.21}$$

Suppose that the order quantity Q is given and that the aim is to find the reorder point s such that at least a fraction β of the demand can be delivered directly from stock, where $0 < \beta < 1$ is a given service value. If we set the expression (6.21) equal to $1 - \beta$, then it immediately follows that the reorder point s is found by solving the following equation:

$$\int_s^\infty (x - s) f_L(x)\, dx = \frac{1 - \beta}{\beta} Q.$$

For a normally distributed demand, this equation simplifies to

$$\sigma_L\, I(k) = \frac{(1 - \beta)}{\beta} Q, \tag{6.22}$$

where s and k are related through $s = \mu_L + k\sigma_L$.

Equation (6.22) is the same as equation (6.18) for the back-order model, except that $1 - \beta$ has been replaced by $(1 - \beta)/\beta$. It should be clear that for β close to 1, the model with lost sales is practically the same as the back-order model. In the example in the previous subsection, the assumption of lost sales would have led to the same value for the reorder point.

Cost Minimization With Emergency Purchase Cost

Suppose that we do not assume a service requirement, but assume that for every unit of demand that cannot be delivered from stock, an emergency purchase is made at a penalty cost of b on top of the regular purchase cost of v. For the heuristic solution, the EOQ formula for the order quantity Q^* remains the same, but formula (6.19) is adjusted to

$$F_L(s) = 1 - \frac{vrQ^*}{vrQ^* + b\mu_1}.$$

6.4.3   *The (R, S) Periodic Review Inventory Model*

In the (s, Q) inventory model from Sections 6.4.1 and 6.4.2, a replenishment order could be placed at any time. In this subsection and the next, we consider two important inventory policies for stochastic inventory models in which the inventory can only be replenished periodically (for example, at the beginning of every week). In the (R, S) inventory model, the stock is replenished at every inventory review. The assumptions for this inventory model are as follows:

- The inventory position is reviewed every R period, where R is a given positive integer.
- At every inventory review, the economic inventory is replenished to the level S, where S is positive.
- The lead time of a replenishment order is L periods, where L is a given non-negative integer.
- The sizes of the demand for the product in time periods $t = 1, 2, \ldots$ are independent random variables that have the same probability density $f_1(x)$ with expected value μ_1 and standard deviation σ_1.

In the analysis of the (R, S) periodic review model, we restrict ourselves to determining the long-term fraction of the demand that is delivered directly from stock. The analysis runs parallel to the analysis for the (s, Q) continuous review model. First, we introduce some notation. We define the random variable T_k as

$$T_k = \text{total demand in } k \text{ consecutive periods.}$$

We denote the probability density of T_k by $f_k(x)$ and the expected value and standard deviation of T_k by μ_k and σ_k. As a result of the assumption that the sizes of the demand are independent, we have

$$\mu_k = k\mu_1 \text{ and } \sigma_k = \sqrt{k}\sigma_1 \text{ for } k \geq 1.$$

For a given (R, S) rule, let

$$\beta(R, S) = \text{long-term fraction of the demand that is delivered}$$
$$\text{directly from stock.}$$

We first analyze the back-order model. The heuristic analysis is again based on the course of the inventory process during one cycle, where a cycle is the time period between the arrival of two consecutive replenishment orders. The key to the analysis is the result that right before a replenishment order arrives, the net inventory has the same distribution as $S - T_{R+L}$. To see this, fix one of the replenishment orders. The economic inventory right before placing the replenishment order has the same distribution as $S - V$, where V is the total demand since the previous inventory review. So V has the same distribution as T_R. The constant lead time assumption implies that all stock that is already on order right before the replenishment order is placed arrives before the replenishment order itself arrives. So, for the back-order model, the net inventory right before the replenishment order arrives has the same distribution as $S - V - W$, where W is the total demand during the lead time of the replenishment order. The random variable W has the same distribution as T_L and is independent of V. So $V + W$ has the same distribution as T_{R+L}. This shows that the net inventory right before the replenishment order arrives (end of the cycle) has the same distribution as $S - T_{R+L}$. The size of the replenishment order has the same distribution as the demand T_R in the R periods that precede the period in which the replenishment order is placed. It is now immediately apparent that the net inventory right after the replenishment order arrives (beginning of the cycle) has the same distribution as $S - T_L$. This gives

$$\mathbb{E}[\text{shortage at the beginning of a cycle}] = \mathbb{E}[(T_L - S)^+]$$

and

$$\mathbb{E}[\text{shortage at the end of a cycle}] = \mathbb{E}[(T_{R+L} - S)^+].$$

Since $\beta(R, S)$ is equal to 1 minus the quotient of $\mathbb{E}[\text{shortage that is added during a cycle}]$ and $\mathbb{E}[\text{total demand during a cycle}]$, it follows that

$$\beta(R, S) = 1 - \frac{\mathbb{E}[(T_{R+L} - S)^+] - \mathbb{E}[(T_L - S)^+]}{R\mu_1},$$

that is,

$$\beta(R, S) = 1 - \frac{1}{R\mu_1} \left[\int_S^\infty (x - S) f_{R+L}(x)\, dx - \int_S^\infty (x - S) f_L(x)\, dx \right].$$

If we want to choose the replenishment level S such that the fraction of the demand that is met directly is at least β for a given $0 < \beta < 1$, then S must be solved from the equation $\beta(R, S) = \beta$, that is

$$\int_S^\infty (x - S) f_{R+L}(x)\, dx - \int_S^\infty (x - S) f_L(x) = (1 - \beta) R\mu_1.$$

Empirical studies show that the second term on the left-hand side of the equation may be ignored if $\sigma_{R+L}/\mu_{R+L} \le 0.5$ and $\beta \ge 0.9$. If we also assume that the demand is normally distributed and write $S = \mu_{R+L} + k\sigma_{R+L}$, then the equation above can be simplified to

$$\sigma_{R+L} I(k) = (1 - \beta) R\mu_1 \tag{6.23}$$

in analogy with the equation (6.18), where $\sigma_{R+L} = \sqrt{R + L}\, \sigma_1$ and $I(k)$ is the normal loss function from Appendix B. The difference with (6.18) is that in the periodic review model, the safety stock $k\sigma_{R+L}$ must provide protection against fluctuations in the demand until the next inventory review plus the demand during the subsequent lead time of the replenishment order, whereas in the continuous review model, the safety stock only needs to provide protection against fluctuations in the demand during the lead time.

The approximation formula (6.23) was deduced for the back-order model. For the inventory model with lost sales, formula (6.23) needs a slight adjustment. By analogy with (6.22), the approximation formula then becomes

$$\sigma_{R+L} I(k) = \frac{(1 - \beta)}{\beta} R\mu_1. \tag{6.24}$$

The approximation formulas (6.23) and (6.24) can also be used when the demand is not normally distributed, as long as the coefficient of variation σ_{R+L}/μ_{R+L} of the total demand during the review time plus the lead time is less than 0.5. For $\sigma_{R+L}/\mu_{R+L} \le 0.5$, it turns out that the value of the replenishment level S for which the service requirement is met is insensitive to the form of the demand's probability distribution as long as the first two moments of the demand are the same. Further, we note that the approximation formulas (6.23) and (6.24) can also be used if the lead time L is stochastic, provided that the probability that order deliveries overlap is small for the relevant values of S. If we denote the demand in the stochastic lead time L by ξ, then μ_{R+L} must be replaced by $R\mu_1 + \mathbb{E}[L]\mu_1$ and σ_{R+L} by $\sqrt{R\sigma_1^2 + \sigma^2(\xi)}$ with $\sigma^2(\xi) = \mathbb{E}[L]\sigma_1^2 + \sigma^2(L)\mu_1^2$.

6.4.4   The (R, s, S) Inventory Model

The difference between the (R, s, S) model and the (R, S) model is that a replenishment order is not necessarily placed at every inventory review. In the (R, s, S) inventory model, a replenishment order to bring the economic inventory up to S is only placed if the economic inventory is less than s at inventory review; otherwise, nothing is ordered. We assume $0 \le s < S$. In real-world applications, the value of $S - s$ is often based on cost considerations (for example, $S - s$ is chosen according to the EOQ formula). Now, suppose that $S - s$ is given and that the aim is to choose the reorder point s such that the fraction of demand delivered directly from stock is at least β for a given value of β. In case $S - s \ge 1.5\mu_R$ and $\beta \ge 0.9$, a simple approximation formula for the reorder point s can be given if the demand per period is normally distributed. In this situation, one can show that for the back-order model, the reorder point s can be approximated by

$$s = \mu_{R+L} + k\sigma_{R+L},$$

where k is the solution to the equation

$$\sigma_{R+L}^2 J(k) = (1 - \beta)2\mu_R \left\{ S - s + \frac{\sigma_R^2 + \mu_R^2}{2\mu_R} \right\}. \tag{6.25}$$

Here, $J(k) = (1/\sqrt{2\pi}) \int_k^\infty (x - k)^2 e^{-\frac{1}{2}x^2}\, dx$. This function is called the normal quadratic loss function.[4] For the inventory model with lost sales, equation (6.25) is adjusted by replacing $1 - \beta$ with $(1 - \beta)/\beta$. In the derivation of the heuristic solution, an important role is played by the probability distribution of the quantity by which the economic inventory is lower than s when a replenishment order is necessary. Assuming that $S - s$ is sufficiently large in relation to μ_R, we have that a good approximation to this probability distribution does not depend on $S - s$ and is given by the famous balance distribution $\frac{1}{\mu_R} \int_0^x (1 - F_R(y))\, dy$ from renewal theory, where $F_R(x)$ is the probability distribution function of the demand in the time between two inventory reviews.

In the case that the demand per period is not normally distributed but we do have $\sigma_{R+L}/\mu_{R+L} \le 0.5$, the requested reorder point s can also be approximated well by using the equation above with the normal quadratic loss function $J(k)$. The

---

[4]The function $J(k)$ is calculated using $J(k) = (1 + k)^2[1 - \Phi(k)] - k\phi(k)$ with $\phi(k)$ the standard normal density function.

reorder point s turns out to be highly insensitive to more than the first two moments of the demand when $\sigma_{R+L}/\mu_{R+L} \leq 0.5$. The adjustment to the formulas for the case of a stochastic lead time is analogous to that for the (R, S) inventory model.

Numerical Illustration

Let us illustrate the quality of the approximation formula (6.25) for the reorder point s in the (R, s, S) inventory model with back-ordering using some numerical examples in which we take $R = 1$ and $L = 2$. For the probability distribution of the demand per period, we take a Poisson distribution ($\sigma_1^2/\mu_1 = 1$) and a negative binomial distribution with $\sigma_1^2/\mu_1 = 3$, where μ_1 is successively taken to be 8, 16, and 32. The required service level β is taken to be 0.90, 0.95, and 0.99. In each numerical example, the value of $S - s$ is taken to be the integer closest to $\sqrt{2K\mu_1/h}$, where $K = 36$ represents the fixed ordering cost of a replenishment order and $h = 1$ the holding cost per unit of inventory for every period the unit is in stock. Table 6.2 gives the values of s calculated using the approximation formula (6.25), rounded to the nearest integer. The table also gives the exact values $\beta(s, S)$ for the fraction of the demand that is delivered directly from stock under the (s, S) policy in question. This exact value is calculated using a Markov-chain method. The analysis

Table 6.2   Approximation of the reorder point s for $S - s = \sqrt{2K\mu_1/h}$.

		$\mu_1 = 8$		$\mu_1 = 16$		$\mu_1 = 32$	
		(s, S)	$\beta(s, S)$	(s, S)	$\beta(s, S)$	(s, S)	$\beta(s, S)$
$\beta = 0.90$	Poisson	(19,43)	0.910	(39,73)	0.911	(78,126)	0.911
	neg bin	(23,47)	0.916	(43,77)	0.909	(83,131)	0.902
$\beta = 0.95$	Poisson	(22,46)	0.957	(43,77)	0.953	(85,133)	0.960
	neg bin	(26,50)	0.947	(48,82)	0.948	(92,140)	0.950
$\beta = 0.99$	Poisson	(27,51)	0.992	(50,84)	0.989	(97,145)	0.994
	neg bin	(33,57)	0.985	(59,93)	0.988	(108,156)	0.989

above focused on the determination of the reorder point s such that the set service requirement is met. The choice of $S - s$ was based on the EOQ formula for a given fixed ordering cost $K(= 36)$ and linear holding cost $h(= 1)$. We already mentioned in Section 6.1.1 that this simple square root formula for the order quantity in the deterministic inventory model can also be used surprisingly well in more complex stochastic inventory models. To illustrate this, we have also determined the (s, S) policies for the numerical examples above that minimize the average ordering and holding costs per period subject to the service requirement on the fraction of the demand that must be delivered directly from stock. These optimal (s, S) policies are calculated using a combination of dynamic programming and Markov chains. Table 6.3 gives these optimal (s, S) policies with the corresponding values of $\beta(s, S)$. A comparison of the results in Tables 6.2 and 6.3 confirms how well the combination of the EOQ formula and the approximation formula (6.25) works. In all cases, the average cost per period of the heuristically determined (s, S) policy does not differ by more than 4% from the average cost per period of the optimal (s, S) policy.

Table 6.3   The optimal (s, S) policies.

		$\mu_1 = 8$		$\mu_1 = 16$		$\mu_1 = 32$	
		(s, S)	$\beta(s, S)$	(s, S)	$\beta(s, S)$	(s, S)	$\beta(s, S)$
$\beta = 0.90$	Poisson	(19,43)	0.910	(39,71)	0.911	(77,125)	0.902
	neg bin	(22,48)	0.909	(42,77)	0.902	(83,131)	0.902
$\beta = 0.95$	Poisson	(22,45)	0.956	(43,71)	0.953	(85,131)	0.951
	neg bin	(26,52)	0.951	(49,82)	0.953	(93,138)	0.950
$\beta = 0.99$	Poisson	(27,49)	0.991	(51,76)	0.991	(98,141)	0.990
	neg bin	(35,60)	0.990	(61,92)	0.990	(111,154)	0.991

6.5   Exercises

6.1 A large bookshop sells 200 copies of the classic cookbook Grandma's Secrets every month, with demand spread evenly over the month. The purchase cost is €15 per book. The bookshop's total annual holding cost is 24% of the capital invested in inventory. The fixed ordering cost is estimated at €35.

(a) What is the optimal order quantity? Also calculate the minimum annual ordering and holding costs.
(b) What is the deviation from the theoretical minimum cost if the book can only be ordered by multiples of 50?
(c) When should an order be placed if the lead time is half a month?

6.2 A specialist store sells 3500 Golden Tip ballpoint pens a year, with demand spread evenly over the year. The store estimates the annual holding cost at 18% of the capital invested in inventory. The purchase cost of the ballpoint pen is €7.00 if at least 350 pens are ordered; otherwise, the cost is €7.50. Every time the shop places an order with the supplier, it costs €25. What is the optimal order quantity?

6.3 Jim Finch is a truffle trader who regularly travels between Amsterdam and Bangkok to buy truffles. The trip takes one week and has a fixed cost of €2450. The purchase cost of truffles is €100 per 100 grams. In Amsterdam, Jim Finch has a constant and guaranteed sales volume of 50 units of 100 grams per week, and Jim has committed himself to always meet the demand strictly on time. The profit Jim makes is €500 per unit. The weekly holding cost is 0.5% of the purchase cost of the truffles in stock.

(a) When should Jim replenish his stock and by how much?
(b) What is the optimal replenishment quantity if Jim is given a 20% discount off the purchase price if he buys at least 750 units of 100 grams of truffles?

6.4 The famous Shihimutu bicycle factory has agreed to deliver a specified number of racing bikes to the Bonesto cycling team at the end of the months of February through July for the coming cycling season. The Bonesto cycling team has demanded that the bikes be delivered no later but also no sooner than the end of the month in question.

The table below gives the numbers of racing bikes to be delivered.

Month	February	March	April	May	June	July
Bikes	2	4	10	6	4	2

The Shihimutu bicycle factory has no stock of racing bikes at the beginning of February. The factory can carry out a production run every month for any number of racing bikes. If a production run is carried out in a given month, the racing bikes are ready and available for delivery at the end of the month in question. The fixed cost of a production run in a given month is ¥200 000 regardless of the number of bikes produced. We can ignore the variable costs for making a racing bike because all requested bikes are delivered. A holding cost of ¥20 000 per bike is incurred for every month the bike is kept in stock.

(a) Use the Silver–Meal heuristic to calculate a (suboptimal) production schedule.

(b) Now, suppose that the demand for bikes over the next six months is *constant* in time and that at the end of each of the six months exactly 5 bikes must be delivered to the Bonesto cycling team. What is the *optimal* production schedule?

6.5 The truffle trader Jim Finch from Exercise 6.3 has decided to stop his trade. However, he has promised a loyal customer, Jo Bryant, that at the beginning of each of the next five weeks, he would deliver the respective quantities of $D_1 = 120$, $D_2 = 180$, $D_3 = 360$, $D_4 = 130$, and $D_5 = 270$ grams of truffles. He does not need to fly to Bangkok to purchase the truffles; he can go to an address in the eastern part of the Netherlands. This supplier charges a fixed ordering cost of €100 and purchase cost of €20 per gram of truffles. Jim's weekly holding cost is 0.5% of the purchase cost of the truffles he has in stock that week. Calculate how Jim should buy truffles for the coming five weeks if he is currently out of stock. Apply the Silver–Meal heuristic.

6.6 At the beginning of every month, a pharmacist orders a number of strips of a particular expensive medication. The purchase cost per strip is 100 euros, and the selling cost per strip is 400 euros. The medicine has a limited shelf life. Medicine that is not sold during the month is no longer good at the end of the month and must be thrown away. If demand exceeds the stock, the difference is purchased elsewhere at a cost of 350 euros per strip. The monthly demand for the medicine is uncertain and has a given probability distribution with probabilities 0.3, 0.1, 0.2, 0.2, 0.05, 0.05, 0.05, and 0.05 of a demand for, respectively, 3, 4, 5, 6, 7, 8, 9, or 10 strips. What order quantity maximizes the average net profit per month?

6.7 For a temporary residential complex, solar panels are installed on the roof to meet the energy requirements. It is not possible to specify in advance how many of these expensive solar panels are necessary. Based on previous experience, it is estimated that 5, 6, 7, 8, 9, or 10 panels are necessary to meet the energy demand with respective probabilities 0.40, 0.20, 0.10, 0.25, 0.03, and 0.02. The installation cost of a solar panel is €2500 per panel during the construction of the residential complex and €7500 per panel if it turns out later that not enough solar panels have been installed. What number of solar panels minimizes the expected total cost?

6.8 At the beginning of every day, Jane Snip buys copies of a local newspaper from the local distributor. This costs her 25 cents per newspaper, and she sells the newspapers for 50 cents each. If at some point, Jane runs out of newspapers, she cannot order more. If she has not sold all newspapers at the end of the day, the distributor refunds her 10 cents per newspaper. Jane estimates that the number of newspapers she sells per day is normally distributed with an average of 400 and a standard deviation of 50. How many newspapers should Jane buy every day? What is the probability that Jane sells all her newspapers on a given day?

6.9 The buyer for a department store intends to buy a certain kind of ski jacket for the coming winter season. The ski jackets cost €30 each, and the department store sells them for €75 each. Every jacket that is not sold by the end of the winter season is discounted to €20 and sold during a special spring sale. Experience has shown that all discounted jackets are sold. The buyer estimates that the total demand for the ski jacket is approximately normally distributed with an expected value of $\mu = 750$ and standard deviation $\sigma = 150$. Calculate the optimal order quantity and the expected net profit, the expected shortage, and the expected surplus.

6.10 A medical association's yearly congress is concluded with a buffet offered by a pharmaceutical company. The hotel where the congress is held requires that the number of participants be specified one week before the start of the buffet. The buffet costs €100 per person. If fewer people show up than registered in advance, the pharmaceutical company will still pay for the full number of registrations. If more people show up, they can all participate, but each additional participant costs €150. If we assume that the total number of participants is uniformly distributed between 350 and 500, how many participants should be registered in advance?

6.11 Every year, the University of Irvine renews the maintenance contract with the supplier of the printers used by the university. The supplier has made the following offer for the coming year. The university may enter into a maintenance contract for a specified number of repairs in return for payment of the agreed number times €50. Every time this agreed number is exceeded, the university pays €100. No refund is made if the agreed number is not reached. For the coming year, the university estimates that the number of repairs for which a mechanic must be called is Poisson distributed with an expected value of 400. What agreement should the university make with the supplier to minimize the expected cost of repairs? What is the probability that a mechanic will be called in more often than planned? Carry out the calculations by approximating the Poisson distribution with a normal distribution.

6.12 The production of the G–50 aircraft is stopped. The factory must decide how many spare parts of a certain type should be produced in the last production run. The remaining operational life span of the aircraft is uncertain. It is estimated to be 5, 6, or 7 years, each with probability 1/3. Past experience has shown that the demand for the spare parts can be approximated by a Poisson distribution with an average of 25 per year. The parts cost €200 each. If the parts are not used during the life span of the aircraft, they become worthless. If there is demand for the spare part while there is no more stock, the part must be purchased elsewhere for €1000. How many spare parts should the factory produce? (*Hint*: Calculate the expected value and standard deviation of the total demand for the spare parts in the operational life span and approximate the demand by a normal distribution.)

6.13 In Section 6.3.2, we showed, in the case of seasonal goods, how the combined seller's and producer's expected profit can increase by contractually agreeing that the producer will take back any remaining items at the end of the season. Another possibility is a contract with flexible order quantity. In this contract, the sales channel commits to buy at least Q units before the season begins, and the producer promises that, during the season, the sales channel may buy additional units up to a maximum of γQ units for a given value of $\gamma \geq 0$ (the producer therefore makes $(1 + \gamma)Q$ units).

(a) Suppose that the sales channel estimates that the demand during the season is normally distributed with expected value μ and standard deviation σ. For given values of Q and γ, determine the expected quantity bought by the sales channel, the expected quantity sold by the sales channel, the sales channel's expected surplus, the sales channel's expected shortage, and the producer's expected surplus.

(b) Consider the mascot example discussed at the end of Section 6.3.2. For each of the values $\gamma = 0$, $\gamma = 0.5$, $\gamma = 1$, $\gamma = 1.5$, $\gamma = 2$, and $\gamma = 2.5$, calculate the value of Q that maximizes the retailer's expected profit, and for this value of Q, calculate both the retailer's and the producer's expected profit (assume that the residual value of each mascot is zero for both the retailer and the producer).

6.14 The Madame de Pompadour tea salon has built up a name for itself with its delicious and always fresh pastries. The tea salon's owner has an ordering policy whereby fresh pastries are purchased twice a day. Pastries are first bought early in the morning and then purchased again after the salon has been open for three hours. The salon is open for nine hours every day. In the first three hours, the demand for pastries is Poisson distributed with an expected value of 50 pastries. Experience shows that the demand for pastries in the remaining six hours the salon is open has a Poisson distribution with expected value $2d$ if the realized demand in the first three hours is d. The owner uses this data in her purchase policy, which is determined by two parameters Q_1 and Q_2 with $Q_2 \geq Q_1$: early in the morning, Q_1 pastries are bought, and three hours after the salon is opened, the stock of pastries is refilled to Q_2. The pastries are sourced from a nearby bakery with a large stock, so that the lead time is negligible. The purchase price of a pastry is €2.25 at the beginning of the morning and €2.50 during the day. The selling price of a pastry is €7.50. Any pastries left at the end of the day have no residual value and are thrown away.

(a) For given values of Q_1 and Q_2, determine formulas for a day's expected surplus, expected shortage, and expected net profit.

(b) Numerically determine the values of Q_1 and Q_2 that maximize the expected net profit per day. Compare the maximum profit with the maximum expected net profit if pastries can only be purchased at the beginning of the morning, for a demand that is Poisson distributed over the day, with an expected value of 150.

6.15 Nobel Airlines flies daily from Amsterdam to Palermo. The price of a ticket for this popular regular service is €100. The airplane has a capacity of 150 passengers. Experience shows that the number of passengers that want to make a reservation for a given flight has a Poisson distribution with expected value 170. The airline accepts more reservations than there are seats to cover itself against passengers who do not show up. The probability that a passenger with reservation fails to show up is 10%. Any passenger who does not show up is refunded only half of the ticket price. If more passengers show up than there are seats, every passenger who cannot be seated is refunded the full amount of the ticket and receives an additional compensation of €425. Suppose that for every flight, the airline accepts the first Q bookings and refuses bookings above this maximum.

(a) For a given value Q, determine a formula for the average net profit per flight, and use a computer to calculate the value of Q that maximizes the average net profit per flight.

(b) Let N_Q be binomially distributed with parameters $n = Q$ and $p = 0.9$. Use marginal analysis to show that the optimal value of Q does not depend on the distribution of the demand for seats and is the smallest value of Q for which

$$0.1 \times 50 + 0.9[100\mathbb{P}(N_Q < 150) - 425\mathbb{P}(N_Q \geq 150)] \leq 0.$$

(*Hint*: Subject to the condition that the demand is greater than Q, consider the two cases with the additional condition that the $(Q + 1)$st reservation does or does not show up.)

6.16 Consider again the example at the end of Section 6.3 regarding the purchase of mascots for the next UEFA European Championship. Now, suppose that at the time the order must be placed, it is unknown whether the Dutch team has qualified for the championship or not (a decisive game still has to be played). At the time of ordering, experts estimate the probability of the Dutch team qualifying to be $\frac{2}{3}$. The demand for mascots during the championship will depend on whether the Dutch team qualifies or not. If it does, the demand for mascots is estimated to be normally distributed with an expected value of 50 thousand and a standard deviation of 15 thousand, while if the Dutch team does not qualify, the demand is estimated to be normally distributed with an expected value of 20 thousand mascots and a standard deviation of 5000 mascots. Recalculate the optimal order quantity and the expected values of the profits for the sales channel and the producer for different repurchase values from the producer for each unsold mascot.

6.17 A consultant is asked to compare different overbooking strategies for a flight of an airplane with 50 business-class seats and 200 economy-class seats. Business-class tickets cost 500 euros and economy-class tickets 250 euros. For both classes, so many people are interested that any desired number of tickets can be sold. A business-class passenger has probability 10% of not showing up for the flight, while this probability is only 5% for an economy-class passenger. Passengers decide independently of one another whether or not to show up for a flight. A business-class passenger who fails to show up receives a full refund of the 500 euros paid, while an economy-class passenger who fails to show up does not get a refund. The consultant settles on an overbooking of O_1 passengers in business class and O_2 passengers in economy class. Every business-class passenger who shows up but for whom there is no seat in business class gets his ticket refunded and receives a compensation of 300 euros (a business-class passenger never settles for a seat in economy class). An economy-class passenger who is told at the airport that the economy class is full is given a business-class seat if one is available; otherwise, the passenger is refunded the amount paid plus a compensation of 150 euros.

(a) Determine a mathematical formula for the expected value of the net profit for the flight as a function of O_1 and O_2.

(b) Calculate the optimal values of O_1 and O_2.

6.18 Judith Martin is an enterprising lady who lives in Venlo, North Limburg. She realizes that many people use the airport that is 50 km across the border in Germany or need transportation from the airport to Venlo. She has therefore bought a luxurious 10-person van to transport passengers to and from the airport. There are four rides a day: a morning and an evening ride to the airport and a morning and an evening ride from the airport to Venlo. After careful consideration, Judith formulates a number of basic principles:

- Reservations for a ride can be made up to 12 hours in advance for the regular fare of 35 euros, with a direct payment of a deposit of 10 euros. This deposit is

not refunded if the passenger does not show up. Judith accepts reservations for
up to Q places, where Q is greater than or equal to the capacity of the bus.

- For each ride, the number of passengers that want to make a reservation up to
  12 hours in advance is uniformly distributed between 7 and 14; that is, each of
  the eight possible values $7, \ldots, 14$ is evenly likely. Of course, Judith only accepts
  reservations up to the reservation limit Q.
- The probability that a person who has made a reservation actually shows up for
  the ride is 0.8. Anyone who does not show up loses the 10 euro deposit. The
  passengers behave independently of one another in whether or not they show up
  for a ride.
- If more passengers show up for a ride than the number 10 of available seats,
  Judith must arrange alternative transportation for the passengers she cannot
  take in her van. For every passenger transferred, this entails a cost of 75 euros
  (and therefore a loss of $75 - 35 = 40$ euros).
- In addition to the passengers with a reservation, last-minute passengers can show
  up for the ride without a reservation: 0 such passengers with probability 0.30, 1
  with probability 0.45, and 2 with probability 0.25. Last-minute passengers can
  only travel if there is room and pay 50 euros for the ride.
- Judith always rides to (and from) the airport, even if no passengers show up for
  the ride. The cost per ride for Judith is 100 euros.

(a) Determine the value of the reservation limit Q that maximizes the expected net
profit per ride. (*Hint*: For the strategy where Judith accepts up to Q reservations, first deduce a formula for the quantity $w_r(Q)$ defined as the expected value
of Judith's net profit for a realization r of the number of bookings accepted in
advance.) For the optimal value of Q, also calculate the probability of refusing
a passenger with a reservation and the probability of the van not being full.

(b) Use marginal analysis to verify that the optimal value of the reservation limit Q
is the same as that found in part (a) if the number of persons who want to make
a reservation in advance has a Poisson distribution.

6.19 Snack bar "The Bridge" is widely known for its Thai spring rolls. The total demand
for these spring rolls per week is approximately normally distributed with an expected
value of 700 and a standard deviation of 75 spring rolls. For the inventory management, an (s, Q) policy is followed, where every replenishment consists of $Q = 500$
spring rolls. The lead time of a replenishment order is two days. How should the
reorder point s be chosen to satisfy at least 99% of the demand?

6.20 The discount store "Select" intends to use an (s, Q) policy to manage the stock of
a particular model TV. The monthly demand for the TV is approximately normally
distributed with an expected value of 150 and a standard deviation of 40 TVs. Buyers
who do not find the TV in stock are willing to wait until the stock is replenished.
The discount store sources the TVs directly from the factory. The lead time of a
replenishment order is one half month. The fixed ordering cost is €80. The purchase
cost per TV is €250. The annual holding cost is 24% of the capital invested in
inventory. How should the store choose s and Q to minimize the average cost subject
to the service requirement that an average of 99% of the demand must be met directly?

6.21 The Northern-European distributor of Piralley tires is planning to switch to an (s, Q)
inventory policy for the tire inventory management. The weekly demand for tires is
normally distributed with an expected value of 2500 tires and a standard deviation of
500 tires. If the distributor reorders from the factory, he must count on a lead time

that is either 2 weeks, with probability $\frac{2}{3}$, or 3 weeks, with probability $\frac{1}{3}$. The fixed ordering cost is €2000, and the purchase cost per tire is €50. For the tires in stock, the distributor counts on an annual holding cost of 25% of the purchase value of the tires. What (s, Q) strategy should the distributor use to minimize the costs subject to the service requirement that the probability of running out of stock during the factory's lead time is less than 5%? What value does this service requirement imply for the penalty cost for a tire delivered late?

6.22 Consider the (s, Q) inventory model with back-ordering. Suppose that s and Q are such that shortages do not occur frequently. Under this assumption, explain that the average on-hand inventory level can be approximated well by $s - \mu_L + \frac{1}{2}Q + \frac{1}{2}\sigma_L I\left(\frac{s-\mu_L}{\sigma_L}\right)$ if the demand during the lead time is normally distributed. Furthermore, verify that for the (R, S) inventory model with back-ordering, the corresponding approximation formula is given by $S - \mu_{R+L} + \frac{1}{2}\mu_R + \frac{1}{2}\sigma_{R+L} I\left(\frac{S-\mu_{R+L}}{\sigma_{R+L}}\right)$. (*Hint:* Take the average of the inventory levels at the start and at the end of a cycle).

References

1. S. Axsäter, *Inventory Control*, 2nd ed., Springer, New York, 2006.
2. S. Chopra and P. Meindl, *Supply Chain Management: Strategy, Planning, and Operation*, Prentice-Hall, Inc., Upper Saddle River, New Jersey, 2001.
3. E. A. Silver, D. Pyke, and R. Peterson, *Inventory Management and Production Planning and Scheduling*, 3rd ed., John Wiley & Sons, Inc., Hoboken, New Jersey, 1998.
4. P. H. Zipkin, *Foundations of Inventory Management*, Mc Graw Hill, New York, 2000.

Chapter 7

Discrete-Time Markov Chains

7.1 Introduction

Markov chains are used to analyze dynamic stochastic systems whose state changes over time. For many such systems, it is reasonable to assume that the probability of going from one state to another depends only on the current state of the system and is not influenced by additional information on how the system came to be in its current state. This memoryless property is the characteristic property of a Markov chain. In applications, one can satisfy this property by selecting the appropriate state variable(s). Viewing Markov chains from the point of view of states and state transitions is enlightening. Although the Markov model is a simple probabilistic model, it is one of the most useful models from probability theory for analyzing practical problems in a wide variety of areas such as inventory management, maintenance management, property insurance, and telecommunication. The essence of the discrete-time Markov model is well illustrated by the following example.

Example 7.1 (A Random Walk). A drunk starts walking in the middle of a town square; see Figure 7.1. He makes a random walk over the square. With each step, he moves one unit of length in one of the four directions east, west, north, and south. As long as the drunk has not reached the edge of the square, he randomly chooses one of the four directions. When he reaches an edge and is not in one of the four corners, he randomly chooses one of the three directions that keep him on the square. In a corner, he randomly chooses one of the two directions that keep him on the square. What is the expected number of steps the drunk needs to return to his starting point?

For $n = 0, 1, \ldots$, define the random variable X_n by

$$X_n = \text{ position of the drunk after the } n\text{th step}$$

with $X_0 = (0,0)$; then $\{X_n, n = 0, 1, \ldots\}$ is a stochastic process with a discrete time parameter and state space $I = \{(x,y) \mid -L \leq x, y \leq L\}$, where L and $-L$ represent the edges of the square. In this stochastic process, the drunk's successive positions (states) are not independent of one another but correlated. However, the dependence is of a special nature that is characteristic of many applications

Fig. 7.1   A random walk.

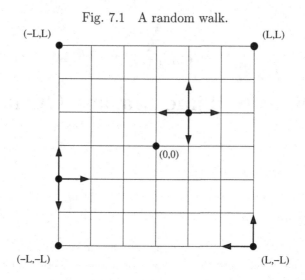

of stochastic processes. Only the drunk's current position, and not the previous ones, determine the drunk's position after the next step. The current position gives sufficient information to determine the next position; additional information about the path the drunk followed to get to the current position is not relevant for determining the probability distribution of the next position. If the drunk's current position is a point (x, y) that is not on the edge, then each of the four points $(x + 1, y)$, $(x - 1, y)$, $(x, y + 1)$, and $(x, y - 1)$ has probability $\frac{1}{4}$ of being the drunk's position after the next step. Something similar applies to the points on the edges. For example, from position (y, L) with $y \notin \{L, -L\}$, the drunk goes to one of the three positions $(y + 1, L)$, $(y - 1, L)$, and $(y, L - 1)$ with equal probability $\frac{1}{3}$. From, for example, the corner point (L, L), the drunk goes to one of the two positions $(L - 1, L)$ and $(L, L - 1)$ with equal probability $\frac{1}{2}$. In the next section, we see that for the random walk, the process $\{X_n\}$ is a discrete-time Markov chain.

7.2   The Discrete-Time Markov Chain

Suppose that $\{X_n, \, n = 0, 1, \ldots\}$ is a stochastic process with a discrete state space I. For example, X_n can be the state at time $t = n$ of a stochastic system whose state changes dynamically over time. The state space I may be finite or countably infinite.

Definition 7.1 (Discrete-time Markov chain). *The stochastic process* $\{X_n,$ $n = 0, 1, \ldots\}$ *is called a* discrete-time Markov chain *if we have*

$$\mathbb{P}(X_{n+1} = i_{n+1} \mid X_0 = i_0, X_1 = i_1, \ldots, X_{n-1} = i_{n-1}, X_n = i_n)$$
$$= \mathbb{P}(X_{n+1} = i_{n+1} \mid X_n = i_n) \tag{7.1}$$

for every $n = 0, 1, \ldots$, *for all possible states* $i_0, i_1, \ldots, i_{n+1} \in I$.

The probability $\mathbb{P}(X_{n+1} = i_{n+1}|X_0 = i_0, X_1 = i_1, \ldots, X_n = i_n)$ should be interpreted as follows: it is the probability that at the next time $t = n+1$, the process is in state i_{n+1} given that at the *current* time $t = n$, the process is in state i_n and the current state was reached by successively passing through the states $i_0, i_1, \ldots, i_{n-1}$. In words, Definition 7.1 states that a stochastic process is a Markov chain if only the current state determines the further course of the process and additional information on how the current state has been reached is irrelevant for the further course. This memoryless property (7.1) is called the *Markov property*. In the stochastic process $\{X_n\}$ in Example 7.1, the Markov property was trivially fulfilled by taking the drunk's position as the state. Now, in Example 7.1, suppose that the drunk never chooses the same direction for his next step as that of the previous step but randomly chooses one of the other possible directions that keep him on the square. For the Markov property to remain fulfilled, a third state variable r must be added to the state variables x and y indicating the drunk's position, where r is the direction of the previous step and can only take on the values E, W, N, and S. If we define the random variable X_n in terms of the state variables x, y, and r as the state after the nth step, then $\{X_n, n = 0, 1, \ldots\}$ is a Markov chain (the initial state is $X_0 = (0,0)$). For example, if (x, y, S) is the current state of the process, with (x, y) an interior point of the square, then each of the three states $(x + 1, y, E)$, $(x - 1, y, W)$, and $(x, y + 1, N)$ has the same probability of being the next state.

From now on, we only consider Markov chains with time-homogeneous transition probabilities; that is, we assume that $\mathbb{P}(X_{n+1} = j \mid X_n = i)$ does not depend on the time parameter n. The transition probabilities $\mathbb{P}(X_{n+1} = j \mid X_n = i)$ are denoted by

$$p_{ij} = \mathbb{P}(X_{n+1} = j \mid X_n = i) \quad \text{for } i, j \in I.$$

The probabilities p_{ij} are called the *one-step transition probabilities* and satisfy

$$p_{ij} \geq 0 \quad \text{for } i, j \in I \quad \text{and} \quad \sum_{j \in I} p_{ij} = 1 \quad \text{for } i \in I.$$

A Markov chain $\{X_n, n = 0, 1, \ldots\}$ is entirely determined by the probability distribution of the initial state X_0 and the one-step transition probabilities p_{ij}. In applications of the Markov model, the trick is to

a. choose the state variable(s) such that the Markov property holds and
b. determine the one-step transition probabilities.

Once this modeling step is done, it is a matter of applying the existing theory to calculate all sorts of things for the specific application. The modeling step is often the most difficult, namely choosing the state variable(s) such that they contain sufficient information. To choose the appropriate state variable(s), one might consider which variables one would use to write a simulation program for the probability problem in question. To show how to model a problem as a Markov chain in practical situations, we give four examples. The first example concerns the Ehrenfest model of gas diffusion, a Markov model that has played an important role in physics.

Example 7.2 (The Ehrenfest model). Together, two compartments A and B contain r particles. At each time point, exactly one of the particles is chosen randomly and transferred from the compartment it is in to the other one. What stochastic process describes the number of particles in each of the compartments?

As state, choose the number of particles in compartment A. Note that if one of the compartments contains i particles, the other one contains $r - i$ particles. Define the random variable X_n as

$$X_n = \text{number of particles in compartment } A \text{ after the } n\text{th step.}$$

Due to the physical construction of the model with independent choices of the particles, the process $\{X_n\}$ has the Markov property and is therefore a Markov chain. The state space is $I = \{0, 1, \ldots, r\}$. The probability of going from state i to state j in one step is zero unless $|i - j| = 1$. The one-step transition probability $p_{i,i+1}$ is translated into the probability that a randomly selected particle belongs to compartment B and $p_{i,i-1}$ is translated into the probability that a randomly selected particle belongs to compartment A. So, for $1 \le i \le r - 1$, we have

$$p_{i,i+1} = \frac{r - i}{r} \quad \text{and} \quad p_{i,i-1} = \frac{i}{r}.$$

Furthermore, $p_{01} = p_{r,r-1} = 1$. The other p_{ij} are zero.

Example 7.3 (An absent-minded professor). Every day, an absent-minded professor takes the car to go to the university in the morning and return home at the end of the day. At any given time, his driver's license is either at home or at the university. If his driver's license is at his departure point, the professor takes it along with probability 0.5. What stochastic process describes whether the professor has his driver's license with him when he drives from his home to the university or vice versa?

The first thought may be to define two states 1 and 0, where state 1 describes the situation where the professor has his driver's license with him during the car ride, and state 0 describes the situation where he does not. However, these two states do not suffice for a Markov model: state 0 does not contain enough information to predict whether the professor will have his driver's license with him for his next ride. To give the probability distribution of the next state, one needs to know where the professor's driver's license currently is. Including this necessary information in the description of the states gives a Markov model. We say that the process is in state 1 if the professor has his driver's license with him during the ride, in state 2 if he does not have his driver's license with him and it is at his departure point, and in state 3 if he does not have his driver's license with him and it is at his destination. Now, define the random variable X_n as

$$X_n = \text{state during the } n\text{th ride home or to the university.}$$

The process $\{X_n\}$ has the property that every current state contains sufficient information to give the probability distribution of future states. So the process $\{X_n\}$ is a Markov chain with state space $I = \{1, 2, 3\}$. Next, we determine the one-step transition probabilities p_{ij}. For example, the probability p_{32} translates into the probability that the professor will not have his driver's license with him on his next ride given that the driver's license is at the departure point of that ride. This gives $p_{32} = 0.5$. Likewise, $p_{31} = 0.5$. Similarly, we find $p_{23} = 1$ and $p_{11} = p_{12} = 0.5$. The other p_{ij} are zero.

The following two problems concern an inventory problem and an insurance problem. The modeling of these two problems is somewhat more difficult than that of the previous examples.

Example 7.4 (An (s, S) inventory model with lost sales). The hardware store Hergarden carries a pipe wrench in its assortment. The demand for this pipe wrench is stable over time. The total demand per month is Poisson distributed with a given expected value of λ. The demand quantities in consecutive weeks are independent of one another. A customer who asks for a pipe wrench when it is not in stock goes to another store. The hardware store's owner can only replenish the stock at the beginning of the month and uses an (s, S) inventory policy for the replenishment. Under this policy with given parameters $0 < s \leq S$, the stock is replenished to S at the beginning of the month if the on-hand inventory level is less than s; otherwise, no replenishment takes place. The time required for a replenishment is negligible. What is the average inventory level right before a replenishment? What is the frequency with which the stock is replenished, and what is the average number of lost sales every month?

To answer these questions, we need a Markov chain. As the state, choose the inventory level right before a review moment, and define X_n as the state at the beginning of the nth month; then the inventory process $\{X_n, n = 0, 1, \ldots\}$ is a discrete-time Markov chain with state space I. After all, the Markov property holds: only the inventory level at the beginning of the current month and the demand during the current month determine the state at the beginning of the next month; how the inventory level evolved in the past is irrelevant. The one-step transition probabilities p_{ij} of the Markov chain $\{X_n\}$ are determined as follows. Distinguish between the cases $i < s$ and $i \geq s$.

Case (a): $i \geq s$. In this case, the stock level right before the inventory review is equal to i, and we have

$$p_{ij} = \mathbb{P}(\text{the total demand for the coming month is } i - j)$$

$$= e^{-\lambda} \frac{\lambda^{i-j}}{(i-j)!} \quad \text{for } 1 \leq j \leq i.$$

This formula does not hold for $j = 0$. Indeed,

$$p_{i0} = \mathbb{P}(\text{the demand for the coming month is greater than or equal to } i)$$

$$= \sum_{k=i}^{\infty} e^{-\lambda}\frac{\lambda^k}{k!} = 1 - \sum_{k=0}^{i-1} e^{-\lambda}\frac{\lambda^k}{k!},$$

which can also be seen using $p_{i0} = 1 - \sum_{j \neq 0} p_{ij}$. Of course, $p_{ij} = 0$ for $j > i$.
Case (b): $i < s$. In this case, the stock level right after the inventory review is equal to S. The same translation as in Case (a) then gives (verify!):

$$p_{ij} = e^{-\lambda}\frac{\lambda^{S-j}}{(S-j)!} \quad \text{for } 1 \leq j \leq S \quad \text{and} \quad p_{i0} = 1 - \sum_{k=0}^{S-1} e^{-\lambda}\frac{\lambda^k}{k!}.$$

Example 7.5 (An insurance problem). A transport company has taken out the following special insurance contract for its fleet. The transport company may save up the damage claims incurred during the year until the end of the year. Only at the end of the year does the company have to decide whether or not to claim the total damage from the past year. The company must pay the insurance premium at the beginning of every year. There are four possible insurance rate classes $i = 1, \ldots, 4$. The insurance premium in class i is P_i for $i = 1, \ldots, 4$, where $P_1 > P_2 > P_3 > P_4$. If, in the current year, the transport company is in rate class i and does not claim any damage at the end of the year, then at the beginning of the following year, the company is in rate class $i + 1$, where $P_{i+1} = P_4$ if $i = 4$. If the company does claim damage at the end of the year, it returns to rate class 1 at the beginning of the following year. In rate class i, the company has a deductible of r_i, so is liable to pay the first r_i euros of the total claim. The total claims over consecutive years are assumed to be independent of one another and satisfy a common probability distribution function $F(s)$ with probability density $f(s)$. What is a good claim strategy for the transport company that minimizes the average costs per year?

Obviously, a claim strategy is determined by four parameters $\alpha_1, \ldots, \alpha_4$: if the current rate class is i, claim the total damage at the end of the year if it is higher than α_i; otherwise, do not claim anything. Of course, α_i is taken greater than r_i for $i = 1, \ldots, 4$. We can find the optimal values for $\alpha_1, \ldots, \alpha_4$ in the standard manner using a minimization procedure once we have a formula to calculate the average cost per year for a *given* $(\alpha_1, \ldots, \alpha_4)$-strategy. Such a formula can be found using Markov chain theory. For a given $(\alpha_1, \ldots, \alpha_4)$-strategy, we must first define an appropriate Markov chain. After some thought, it should be clear that the state must be the rate class at the beginning of the year. If we define

$$X_n = \text{rate class at the beginning of the } n\text{th year,}$$

then $\{X_n, n = 0, 1, \ldots\}$ is a Markov chain with state space $I = \{1, 2, 3, 4\}$ (verify!). To find the one-step transition probabilities, translate $\mathbb{P}(X_{n+1} = j \mid X_n = i)$ to the transport company's situation. We have

$$p_{i,i+1} = \mathbb{P}(\text{the total damage claim in the coming year is at most } \alpha_i)$$
$$= F(\alpha_i) \quad \text{for } i = 1, 2, 3$$

and

$$p_{i1} = \mathbb{P}(\text{the total damage claim in the coming year is greater than } \alpha_i)$$
$$= 1 - F(\alpha_i) \quad \text{for } i = 1, 2, 3, 4.$$

Once the company is in rate class 4, it remains in that class as long as no damage is claimed, so that

$$p_{44} = F(\alpha_4).$$

Of course, the other p_{ij} are zero.

7.3   Time-Dependent Behavior

As already mentioned, a Markov chain $\{X_n\}$ is completely fixed by the initial state and the one-step transition probabilities. These probabilities also determine the probability that the process will go from state i to state j in the next n steps (*transient probability*). The n-step transition probabilities are defined by

$$p_{ij}^{(n)} = \mathbb{P}(X_n = j \mid X_0 = i) \quad \text{for } i, j \in I$$

for $n = 1, 2, \ldots$. Note that $p_{ij}^{(1)} = p_{ij}$.

Theorem 7.1 (Chapman-Kolmogorov equations). *For all* $n, m = 1, 2, \ldots$, *we have*

$$p_{ij}^{(n+m)} = \sum_{k \in I} p_{ik}^{(n)} p_{kj}^{(m)} \quad \text{for } i, j \in I. \tag{7.2}$$

Proof. If we condition on the state of the process at time $t = n$, the law of total probability gives

$$\mathbb{P}(X_{n+m} = j \mid X_0 = i)$$
$$= \sum_{k \in I} \mathbb{P}(X_{n+m} = j \mid X_0 = i, X_n = k) \mathbb{P}(X_n = k \mid X_0 = i)$$
$$= \sum_{k \in I} \mathbb{P}(X_{n+m} = j \mid X_n = k) \mathbb{P}(X_n = k \mid X_0 = i)$$
$$= \sum_{k \in I} \mathbb{P}(X_m = j \mid X_0 = k) \mathbb{P}(X_n = k \mid X_0 = i),$$

which gives the desired equation. Note that the Markov property is used in the second equality and that the last equality uses the assumption that the one-step transition probabilities are time independent. $\qquad \square$

In words, the theorem states that the probability of going from state i to state j in $n + m$ steps is obtained by multiplying the probability of going from state i to some other state k in n steps with the probability of going from state k to state j in m

steps and then taking the sum over all possible states k. This formulation helps to remember the equations. In particular, for every $n = 1, 2, \ldots$, we have

$$p_{ij}^{(n+1)} = \sum_{k \in I} p_{ik}^{(n)} p_{kj}, \quad \text{for } i, j \in I.$$

This means that the n-step transition probabilities $p_{ij}^{(n)}$ can be calculated recursively from the one-step transition probabilities. It follows from the recursion formula that the $p_{ij}^{(n)}$ for $i, j \in I$ are the elements of the n-fold matrix product $\mathbf{P}^n$, where $\mathbf{P}$ is the matrix with elements the one-step transition probabilities p_{ij}. Indeed, if we denote by $\mathbf{P}^{(n)}$ the matrix with elements $p_{ij}^{(n)}$, then in matrix form, the recursion formula above is $\mathbf{P}^{(n+1)} = \mathbf{P}^{(n)} \times \mathbf{P}$ for all $n \geq 1$. Since $\mathbf{P}^{(1)} = \mathbf{P}$, we have $\mathbf{P}^{(2)} = \mathbf{P}^{(1)} \times \mathbf{P} = \mathbf{P} \times \mathbf{P} = \mathbf{P}^2$. By repeated substitution, we then find $\mathbf{P}^{(m)} = \mathbf{P}^m$ for all $m = 1, 2 \ldots$. This is a simple but very useful result.

Example 7.6 (The weather as a Markov chain). On the Wadden Islands to the north of the Dutch mainland, the weather is classified every day as sunny, cloudy, or rainy. The next day's weather depends only on the current weather and not on that of the previous days. If it is sunny today, then it will be sunny, cloudy, or rainy tomorrow with respective probabilities 0.70, 0.10, and 0.20. For a cloudy day, the respective probabilities for the following day are 0.50, 0.25, and 0.25, and for a rainy day, they are 0.40, 0.30, and 0.30. An interesting question is what fraction of the days are sunny, cloudy, or rainy over a longer period of time. We begin with an easier question, namely the probability that it will be sunny three days from now if it is rainy today. For this, we define a Markov chain $\{X_n\}$ with three states 1, 2, and 3. On day n, the process is in state 1 if the day is sunny, in state 2 if the day is cloudy, and in state 3 if the day is rainy. The matrix $\mathbf{P}$ of one-step transition probabilities is

$$
\begin{array}{cccc}
\text{from}\backslash\text{to} & 1 & 2 & 3 \\
1 & \begin{pmatrix} 0.70 & 0.10 & 0.20 \\ 2 & 0.50 & 0.25 & 0.25 \\ 3 & 0.40 & 0.30 & 0.30 \end{pmatrix}
\end{array}
$$

To determine the probability that it will be sunny three days from now, we need the matrix product $\mathbf{P}^3$:

$$\mathbf{P}^3 = \begin{pmatrix} 0.6015000 & 0.1682500 & 0.2302500 \\ 0.5912500 & 0.1756250 & 0.2331250 \\ 0.5855000 & 0.1797500 & 0.2347500 \end{pmatrix}.$$

This shows that the probability of it being sunny three days from now if it is rainy today is $p_{31}^{(3)} = 0.5855$. We can also ask what the weather's probability distribution is many days from now. Intuitively, one expects that this probability distribution does not depend on the current weather. The following calculations support this:

$$\mathbf{P}^5 = \begin{pmatrix} 0.5963113 & 0.1719806 & 0.2317081 \\ 0.5957781 & 0.1723641 & 0.2318578 \\ 0.5954788 & 0.1725794 & 0.2319418 \end{pmatrix},$$

$$\mathbf{P}^{12} = \begin{pmatrix} 0.5960265 & 0.1721854 & 0.2317881 \\ 0.5960265 & 0.1721854 & 0.2317881 \\ 0.5960265 & 0.1721854 & 0.2317881 \end{pmatrix} = \mathbf{P}^{13} = \mathbf{P}^{14} = \cdots .$$

In this example, as $n \to \infty$, the n-step transition probabilities $p_{ij}^{(n)}$ converge to a limit that does not depend on the initial state. We see that many days from now, the weather will be sunny, cloudy, or rainy with probabilities 0.5960, 0.1722, and 0.2318, respectively. It should be clear that these probabilities also give the fractions of the days that are sunny, cloudy, or rainy over a longer period of time.

7.3.1 *Transient and Recurrent States*

A *path* from state i to state j is a sequence of transitions that starts in i and ends in j such that each transition in the sequence has positive probability. State j is called *reachable* from state i for the Markov chain $\{X_n\}$ if there is a path from i to j, that is, for some $n \geq 1$ there exist states $i_1, i_2, \ldots, i_{n+1}$ such that $i = i_1$, $j = i_{n+1}$, and $p_{i_k,i_{k+1}} > 0$, $k = 1, \ldots, n$. Two states i and j *communicate* if j is reachable from i and i is reachable from j. State i is an absorbing state if $p_{ii} = 1$. A set of states $I' \subset I$ is a *closed set* if no state outside I' is reachable from any state in I'.

A state i is called *recurrent* if the probability of ultimately returning from state i to itself is equal to 1. As a consequence, the Markov chain returns to i infinitely often. If this probability is less than 1, state i is called a *transient state*. For a transient state i there exists a state j that is reachable from i, but state i is not reachable from j. As a consequence, the Markov chain will eventually not return to state i. Observe that recurrence and transience are *class properties*: if i and j communicate then both i and j are either recurrent or transient. For example, for the three-state Markov chain with one-step transition probabilities $p_{11} = \frac{1}{3}$, $p_{12} = \frac{2}{3}$, $p_{22} = p_{23} = \frac{1}{2}$, $p_{32} = p_{33} = \frac{1}{2}$, and $p_{ij} = 0$ for the other probabilities, state 1 is a transient state, and states 2 and 3 are recurrent. Intuitively, it should be clear that in a Markov chain with a finite number of states, after a while, the system will only pass through recurrent states. For a Markov chain with a finite state space, the set of recurrent states is always nonempty. However, this need not be the case for a Markov chain with a countably infinite state space. An example is the Markov chain with the countably infinite state space $I = \{0, 1, \ldots\}$ and one-step transition probabilities $p_{i,i+1} = 1$ for all i. This Markov chain only has transient states. The theory of Markov chains with a countably infinite state space is much more complex than that of Markov chains with a finite state space, as we will illustrate for the random walk in Example 7.7.

To make recurrence and transience more precise, consider the return probability

$$f_i = \mathbb{P}\left(X_n = i \text{ for some } n \geq 1 \mid X_0 = i\right).$$

Then state i is recurrent if $f_i = 1$ and transient if $f_i < 1$. If state i is transient the expected number of visits of the Markov chain to state i is finite, and

$$\mathbb{P}(\text{exactly } k \text{ returns to state } i \text{ over } n = 1, 2, \ldots \mid X_0 = i) = (1 - f_i)f_i^k, \quad k \geq 0.$$

We can also evaluate the expected number of visits to state j starting from state i.

Theorem 7.2. *For every pair of states $i, j \in I$, we have, for $n = 1, 2, \ldots$*

$$\mathbb{E}[\text{number of visits to state } j \text{ during the times } t = 1, \ldots, n \mid X_0 = i] = \sum_{t=1}^{n} p_{ij}^{(t)}.$$

Proof. The proof of this result is instructive. Fix $i, j \in I$. For every $t \geq 1$, let

$$I_t = \begin{cases} 1 & \text{if } X_t = j, \\ 0 & \text{otherwise.} \end{cases}$$

The number of visits to state j during the times $t = 1, \ldots, n$ is given by the random variable $A_j(n) = \sum_{t=1}^{n} I_t$. By observing that

$$\mathbb{E}[I_t \mid X_0 = i] = 1 \times \mathbb{P}(I_t = 1 \mid X_0 = i) + 0 \times \mathbb{P}(I_t = 0 \mid X_0 = i)$$
$$= \mathbb{P}(X_t = j \mid X_0 = i) = p_{ij}^{(t)},$$

we see that $\mathbb{E}[\sum_{t=1}^{n} I_t \mid X_0 = i] = \sum_{t=1}^{n} \mathbb{E}[I_t \mid X_0 = i] = \sum_{t=1}^{n} p_{ij}^{(t)}$, proving the stated result. $\qquad\square$

As an illustration, consider Example 7.6 once more. What is the expected number of sunny days in the next seven days if today is cloudy? The answer is that the expected number is equal to $\sum_{t=1}^{7} p_{21}^{(t)} = 4.05$ days.

From Theorem 7.2 we readily obtain the following characterization of recurrence and transience.

Corollary 7.1. *State i is*

$$\text{recurrent if } \sum_{t=1}^{\infty} p_{ii}^{(t)} = \infty, \quad \text{transient if } \sum_{t=1}^{\infty} p_{ii}^{(t)} < \infty.$$

Proof. Assume $X_0 = i$. Then $\sum_{n=1}^{\infty} p_{ii}^{(t)}$ is the expected number of visits to state i over an infinite time period. $\qquad\square$

Let m_{ii} be the expected return time to state i, starting in i.[1] We have the following three cases:

- $f_i < 1$, hence $m_{ii} = \infty$: state i is transient;
- $f_i = 1$, but $m_{ii} = \infty$: state i is null recurrent;
- $f_i = 1$, and $m_{ii} < \infty$: state i is positive recurrent.

Example 7.7 (A random walk on an infinite state space). Consider the random walk on state space $I = \mathbb{Z}$ with transition probabilities $p_{i,i+1} = p = 1 - p_{i,i-1}$, $i \in \mathbb{Z}$. All states communicate so I is either a recurrent or a transient class.

---

[1] A set of equations to determine the expected return time is presented in Section 7.3.2.

The n-step transition probabilities are readily obtained (verify!). In particular, for $n = 1, 2, \ldots$, we have $p_{00}^{(2n-1)} = 0$, and

$$p_{00}^{(2n)} = \binom{2n}{n} p^n (1 - p)^n = \frac{(2n)!}{n!n!} (p(1 - p))^n.$$

From Stirling's formula, $n! \sim n^n e^{-n} \sqrt{2\pi n}$, we obtain

$$p_{00}^{(2n)} \sim \frac{(4p(1 - p))^n}{\sqrt{\pi n}}.$$

As $4p(1 - p) < 1$ if $p \neq 1/2$ and $4p(1 - p) = 1$ if $p = 1/2$ we obtain that

$$\sum_{n=1}^{\infty} p_{00}^{(2n)} = \infty \quad \text{if and only if} \quad p = 1/2.$$

Hence, the *symmetric* random walk that has $p = \frac{1}{2}$ is null recurrent and the *asymmetric* random walk that has $p \neq \frac{1}{2}$ is transient.

For the symmetric random walk in higher dimensions, the famous Pólya's theorem says in flowery language: *A drunk man will find his way home, but a drunk bird may get lost forever.* More formally, a symmetric random walk on the d-dimensional lattice $\mathbb{Z}^d$ is null recurrent for $d = 1, 2$, and transient for $d > 2$.

7.3.2 Mean First Passage Times

The *mean first passage time* from state i to state j is the expected number of steps before the Markov chain first reaches state j, given that the Markov chain is currently in state i. It is denoted m_{ij}, $i, j \in I$. We can determine m_{ij} using first step analysis. To this end, observe that with probability p_{ij} the Markov chain makes one step from i to j and with probability p_{ik}, $k \neq j$, the Markov chain makes a step from i to k and then has to make on average m_{kj} steps from k to j and thus $1 + m_{kj}$ steps from i to j. Hence

$$m_{ij} = p_{ij} \times 1 + \sum_{k \neq j} p_{ik} \times (1 + m_{kj}), \quad i, j \in I.$$

Now observe that $\sum_k p_{ik} = 1$. Then

$$m_{ij} = 1 + \sum_{k \neq j} p_{ik} m_{kj}, \quad i, j \in I.$$

This system of equations with number of unknowns equal to the number of equations may be solved using standard methods.

Example 7.8 (Gambler). Suppose that a gambler goes to the casino because he needs a certain amount of money. Gambling is his last resort to get the money. He decides to bet on red or black at the roulette table. Every time, he bets one euro. The gambler doubles the money he bets with a given probability p and loses the money with probability $q = 1 - p$. He begins with a capital of A euros, and his objective is to end up with at least B euros, with $B > A$. Clearly, the gambler stops betting as soon as either his capital reaches 0 or B euros.[2]

The gambler's fortune is a Markov chain with state space $I = \{0, \ldots, B\}$ with transition probabilities

$$p_{i,i+1} = p, \quad p_{i,i-1} = 1 - p, \qquad i = 1, \ldots, N - 1,$$
$$p_{00} = 1, \quad p_{NN} = 1.$$

Observe that states 0 and B are absorbing states.

Let $B = 5$ and the current fortune is $i < 5$. What is the expected number of rounds the gambler can play before either being broke or reaching his objective? To answer this question, we create a new absorbing state $*$ that combines the absorbing states 0 and 5. We can now find m_{i*} from

$$m_{1,*} = 1 + p\, m_{2,*},$$
$$m_{2,*} = 1 + (1 - p)\, m_{1,*} + p\, m_{3,*},$$
$$m_{3,*} = 1 + (1 - p)\, m_{2,*} + p\, m_{4,*},$$
$$m_{4,*} = 1 + (1 - p)\, m_{3,*}.$$

Suppose $p = 1/2$, then the solution is:

$$m_{1,*} = 4, \quad m_{2,*} = 6, \quad m_{3,*} = 6, \quad m_{4,*} = 4.$$

The probability that the gambler who starts with A euros goes broke at or before time t is $p_{A0}^{(t)}$ and the probability that the gambler reaches his objective at or before time t is $p_{AB}^{(t)}$. Thus, the probability that the gambler reaches his objective is $\sum_{t=1}^{\infty} p_{AB}^{(t)}$. Such probabilities are the topic of the next section.

7.3.3 *Absorbing Markov Chains*

The transient probabilities $p_{ij}^{(n)}$ also come in handy when there are absorbing states. Recall that state i is absorbing if $p_{ii} = 1$. The Markov chain model with absorbing states is an extremely useful model with interesting applications. Many probability problems can be formulated in terms of an absorbing Markov chain.

Example 7.9 (Success runs). If we repeatedly toss a fair die, what is the probability distribution of the number of throws needed to toss three sixes in a row?

The trick for this type of problem is to define a Markov chain with an absorbing state. Choose a Markov chain with the four states 0, 1, 2, and 3, where state 3 is absorbing and corresponds to the situation where the last three throws are sixes. State 0 corresponds to the first toss of the die or to the situation where the last throw did not give a six. For $i = 1, 2$, we say that the process is in state i if the last i throws, but not the one before that, have been sixes. Moreover, after a toss that puts the process in state 3, only fictitious throws are carried out, leaving the

---

[2]The gambler was also considered in Section 5.8.3, where an optimal betting strategy is derived. Here we focus on the Markov chain for the case the gambler can bet one euro in each round.

process in state 3. If we define X_n as the state after the nth throw, then $\{X_n\}$ is a Markov chain with state space $I = \{0, 1, 2, 3\}$. This Markov chain's one-step transition probabilities are given by

$$p_{i0} = \frac{5}{6}, \quad p_{i,i+1} = \frac{1}{6} \quad \text{for } i = 0, 1, 2, \quad p_{33} = 1, \quad \text{otherwise } p_{ij} = 0.$$

The number of throws needed to toss three sixes in a row is higher than r only if state 3 does not occur in the first r steps of the Markov chain. For every r, we have

$$\mathbb{P}(X_k \neq 3 \text{ for } k = 1, \ldots, r \mid X_0 = 0) = \mathbb{P}(X_r \neq 3 \mid X_0 = 0)$$
$$= 1 - \mathbb{P}(X_r = 3 \mid X_0 = 0),$$

where the first equality uses that state 3 is absorbing, so that $X_r \neq 3$ means that in every step k with $k < r$, state 3 also did not occur. This leads to the desired result

$\mathbb{P}(\text{more than } r \text{ throws are needed to toss three sixes in a row for the first time})$
$= 1 - p_{03}^{(r)}$ for $r = 1, 2, \ldots$.

For example, this probability has the values 0.8283, 0.3788, and 0.1425 for $r = 50$, 250, and 500, respectively.

Longest run in Monte Carlo casino

A reasoning similar to that used in Example 7.9 can be used to analyze the following situation. On August 18, 1913, a memorable event took place at the Monte Carlo casino: 26 times in a row, the roulette ball fell on black. Suppose that, up to that date, the roulette wheel had been spun about 8 million times since the casino's opening in 1861. What is the probability for the European-style roulette that in n spins with $n \approx 8 \times 10^6$, the ball falls on the same color red or black 26 or more times in a row? (A European roulette wheel contains the numbers 0 through 36, where the 0 is green and half of the numbers 1 through 36 are red and half are black.) To calculate this probability, we can define a Markov chain with 27 states $0, 1, \ldots, 26$. For $i = 1, \ldots, 26$; the state i corresponds to the situation where the ball has fallen on the same color in the last i spins of the roulette wheel but not the time before that. State 0 corresponds to the situation where the ball has fallen on the zero (or the game begins). State 26 is taken to be absorbing (imagine that the roulette game is only played fictitiously once state 26 is reached). Now, define X_n as the state after the nth spin of the roulette wheel; then $\{X_n\}$ is a Markov chain with state space $I = \{0, 1, \ldots, 26\}$. The one-step transition probabilities are given by

$$p_{00} = \frac{1}{37}, \quad p_{01} = \frac{36}{37},$$
$$p_{i,i+1} = p_{i1} = \frac{18}{37}, \quad p_{i0} = \frac{1}{37} \quad \text{for } i = 1, \ldots, 25,$$
$$p_{26,26} = 1, \quad p_{ij} = 0 \text{ otherwise.}$$

The same reasoning as in Example 7.9 gives (verify!)

$$\mathbb{P}(\text{more than } n \text{ spins of the roulette wheel are needed to}$$
$$\text{get red or black 26 times in a row}) = 1 - p_{0,26}^{(n)}.$$

To find the order of magnitude of the desired probability $p_{0,26}^{(n)}$ for $n = 8$ million, it is not very efficient to multiply the 27×27 matrix $\mathbf{P} = (p_{ij})$ with itself a total of 8 million times. This may be calculated much more efficiently observing that

$$\mathbf{P}^2 = \mathbf{P} \times \mathbf{P}, \ \mathbf{P}^4 = \mathbf{P}^2 \times \mathbf{P}^2, \ \mathbf{P}^8 = \mathbf{P}^4 \times \mathbf{P}^4, \text{ and so on.}$$

For $s = 23$, the power $n = 2^s$ is approximately equal to 8 million. So to find $p_{0,26}^{(n)}$ for $n = 2^{23}$, it suffices to carry out 23 multiplications of a 27×27 matrix with itself. This gives the numerical value 0.061 for the probability $p_{0,26}^{(n)}$.

Example 7.10 (Chess). The Dutch child prodigy Rico Forest plays a match against the Russian world chess champion. The match lasts until either Forest, or the world champion has won two consecutive games. For both the first game and any game that ends in a draw, the probability that Forest wins the next game is 0.4, the probability that the world champion wins it is 0.3, and the probability that it ends in a draw is 0.3. After a win by Forest, the probabilities of these outcomes for the next game are 0.5, 0.25, and 0.25, and after a loss by Forest, they are 0.3, 0.5, and 0.2. What is the probability that the match lasts longer than 10 games? What is the expected duration of the match, and what is the probability that Forest wins the match?

To answer these questions, we use a Markov chain with two absorbing states. The Markov chain has five states 0, $(1, K)$, $(2, K)$, $(1, D)$, and $(2, D)$. State 0 corresponds to the match beginning or the previous game ending in a draw. In state $(1, K)$, Forest has won the last game but not the one before that, and in state $(2, K)$, Forest has won the last two games. The meanings of the states $(1, D)$ and $(2, D)$ are analogous. The states $(2, K)$ and $(2, D)$ are, of course, absorbing. The matrix $\mathbf{P}$ of the one-step transition probabilities is given by

from/to	0	$(1, K)$	$(1, D)$	$(2, K)$	$(2, D)$
0	0.3	0.4	0.3	0	0
$(1, K)$	0.25	0	0.25	0.5	0
$(1, D)$	0.2	0.3	0	0	0.5
$(2, K)$	0	0	0	1	0
$(2, D)$	0	0	0	0	1

The match only lasts more than 10 games if during the first 10 steps, the Markov chain never enters the state $(2, K)$ or the state $(2, D)$. The same argument as in Example 7.9 tells us that the probability of this is equal to

$$1 - p_{0,(2,K)}^{(10)} - p_{0,(2,D)}^{(10)} = 1 - 0.5332 - 0.4349 = 0.0319.$$

We give the n-step transition probabilities for $n = 10$ and $n = 30$, where we note that for $n = 30$, all $p_{ij}^{(n)}$ have already converged up to four decimal places:

$$\mathbf{P}^{10} = \begin{pmatrix} 0.0118 & 0.0109 & 0.0092 & 0.5332 & 0.4349 \\ 0.0066 & 0.0061 & 0.0051 & 0.7094 & 0.2727 \\ 0.0063 & 0.0059 & 0.0049 & 0.3165 & 0.6663 \\ 0.0000 & 0.0000 & 0.0000 & 1.0000 & 0.0000 \\ 0.0000 & 0.0000 & 0.0000 & 0.0000 & 1.0000 \end{pmatrix},$$

$$\mathbf{P}^{30} = \begin{pmatrix} 0.0000 & 0.0000 & 0.0000 & 0.5506 & 0.4494 \\ 0.0000 & 0.0000 & 0.0000 & 0.7191 & 0.2809 \\ 0.0000 & 0.0000 & 0.0000 & 0.3258 & 0.6742 \\ 0.0000 & 0.0000 & 0.0000 & 1.0000 & 0.0000 \\ 0.0000 & 0.0000 & 0.0000 & 0.0000 & 1.0000 \end{pmatrix} = \mathbf{P}^{31} = \ldots .$$

The probability distribution of the number of games required can, of course, be used to calculate the expected value of the duration of the match. However, a simpler approach also exists to calculate this expected value. The idea is, as in dynamic programming, to parametrize with respect to the starting state and to define the quantity μ_s as the expected number of games still to be played when the current state is s. The goal is to calculate μ_0. Conditioning on the outcome of the next game gives the system of linear equations

$$\mu_0 = 1 + 0.3\mu_0 + 0.4\mu_{(1,K)} + 0.3\mu_{(1,D)},$$
$$\mu_{(1,K)} = 1 + 0.25\mu_0 + 0.25\mu_{(1,D)} + 0.5\mu_{(2,K)},$$
$$\mu_{(1,D)} = 1 + 0.2\mu_0 + 0.3\mu_{(1,K)} + 0.5\mu_{(2,D)},$$

where $\mu_{(2,K)} = \mu_{(2,D)} = 0$. The solution to this system of three linear equations in three unknowns leads to $\mu_0 = 3.079$.

The probability that Forest wins the match can be calculated as the limit $\lim_{n\to\infty} p_{0,(2,K)}^{(n)}$ (why?). A simpler and more practical method to calculate this probability is through a system of linear equations. Define f_s as the probability that Forest wins the match given the current state s. The desired probability f_0 results from the solution to the following system of linear equations (condition on the outcome of the next game):

$$f_0 = 0.3f_0 + 0.4f_{(1,K)} + 0.3f_{(1,D)},$$
$$f_{(1,K)} = 0.25f_0 + 0.25f_{(1,D)} + 0.5f_{(2,K)},$$
$$f_{(1,D)} = 0.2f_0 + 0.3f_{(1,K)} + 0.5f_{(2,D)},$$

where $f_{(2,K)} = 1$ and $f_{(2,D)} = 0$. The solution to this system gives $f_0 = 0.5506$.

Tabu Probabilities

The trick of introducing an absorbing state can also be used when calculating so-called tabu probabilities (probabilities that certain states are avoided). As an illustration, consider Example 7.6 about the weather. Suppose that today is sunny and that we want to know the probability of it not raining in the next five days. A simple way to answer this question is to make state 3 (rain) absorbing. We adjust the matrix $\mathbf{P} = (p_{ij})$ of one-step transition probabilities given in Example 7.6 by replacing the third row by $(0, 0, 1)$:

$$\mathbf{Q} = \begin{pmatrix} 0.70 & 0.10 & 0.20 \\ 0.50 & 0.25 & 0.25 \\ 0 & 0 & 1 \end{pmatrix}.$$

If we calculate

$$\mathbf{Q}^5 = \begin{pmatrix} 0.2667 & 0.0492 & 0.6841 \\ 0.2458 & 0.0454 & 0.7087 \\ 0 & 0 & 1 \end{pmatrix},$$

we see that the desired probability of no rain in the next five days is given by $1 - q_{13}^{(5)} = 1 - 0.6841 = 0.3159$.

Expected time before absorption

One may also ask the question: how many periods do we expect to spend in each of the transient states before absorption takes place? For transient states i and j, let s_{ij} be the expected number of time steps that the Markov chain spends in state j, given that it starts in state i. Order the states such that $T = \{1, 2, \ldots, \ell\}$ is the set of transient states of this Markov chain, and $\mathbf{P}_T$ the corresponding transition probability matrix. That is, $\mathbf{P}_T$ specifies only the transition probabilities from transient states into transient states, so some of its row sums are less than 1 (otherwise, T would be a closed set). To calculate s_{ij}, let $\delta_{ij} = 1$ when $i = j$ and 0 otherwise, and condition on the first transition:

$$s_{ij} = \delta_{ij} + \sum_{k \in I} p_{ik} s_{kj}$$

$$= \delta_{ij} + \sum_{k=1}^{\ell} p_{ik} s_{kj} \tag{7.3}$$

where the last equality follows from the fact that a transition from a recurrent to a transient state is impossible, so that $s_{kj} = 0$ for recurrent states k. Now, let $\mathbf{S}$ be the matrix with elements s_{ij} for $i, j \in T$. In matrix notation, (7.3) can be written as $\mathbf{S} = \mathbf{I} + \mathbf{P}_T \mathbf{S}$, where $\mathbf{I}$ is the $\ell \times \ell$ identity matrix. This is equivalent to $(\mathbf{I} - \mathbf{P}_T)\mathbf{S} = \mathbf{I}$, and by multiplying both sides by $(\mathbf{I} - \mathbf{P}_T)^{-1}$ we obtain $\mathbf{S} = (\mathbf{I} - \mathbf{P}_T)^{-1}$. Hence, the quantities s_{ij} can be obtained by inverting the matrix $\mathbf{I} - \mathbf{P}_T$.[3] The matrix $\mathbf{S}$ is often referred to as the Markov chain's *fundamental matix*.

Example 7.8 (Continued). Suppose the gambler begins with a capital of $A = 2$ euros, has the objective to end up with at least $B = 5$ euros, and the probability p

---

[3]The transition probability matrix $\mathbf{P}_T$ is substochastic, i.e., $\mathbf{P}_T$ has at least one row with row sum strictly less than 1, which implies that $\mathbf{I} - \mathbf{P}_T$ is invertible.

of doubling the money he bets equals 0.3. What is the expected number of times that the gambler owns 4 euros before he stops betting?

The transition probability matrix on the transient states $T = \{1, 2, 3, 4\}$ is

$$\mathbf{P}_T = \begin{pmatrix} 0 & 0.3 & 0 & 0 \\ 0.7 & 0 & 0.3 & 0 \\ 0 & 0.7 & 0 & 0.3 \\ 0 & 0 & 0.7 & 0 \end{pmatrix}.$$

Inverting $\mathbf{I} - \mathbf{P}_T$ gives

$$\mathbf{S} = (\mathbf{I} - \mathbf{P}_T)^{-1} = \begin{pmatrix} 1.4006 & 0.5723 & 0.2173 & 0.0652 \\ 1.3354 & 1.9078 & 0.7245 & 0.2173 \\ 1.1833 & 1.6904 & 1.9078 & 0.5723 \\ 0.8283 & 1.1833 & 1.3354 & 1.4006 \end{pmatrix}.$$

The expected number of times that the gambler owns 4 euros before he stops betting is $s_{2,4} = 0.2173$.

7.4 Equilibrium and Stationary Probabilities

We observed in Example 7.6 that $\lim_{n \to \infty} p_{ij}^{(n)}$ converged to a limit that does not depend on the initial state i. Now suppose that this limit exists for all $i, j \in I$ and is independent of the initial state i, and denote the limit by π_j, i.e.,

$$\pi_j = \lim_{n \to \infty} p_{ij}^{(n)} \quad \text{for all } i, j \in I$$

and moreover assume that the state space I is finite, then

$$\pi_j = \sum_{k \in I} \pi_k p_{kj} \quad \text{for all } j \in I. \tag{7.4}$$

This immediately follows by taking $n \to \infty$ in $p_{ij}^{(n+1)} = \sum_{k \in I} p_{ik}^{(n)} p_{kj}$, the Chapman-Kolmogorov equations (7.2) for $m = 1$, and reversing the order of the limit and sum. This brings us to calling the $\{\pi_j\}$ the *equilibrium distribution*. We may also consider a stationary distribution.

Definition 7.2. *A probability distribution $\{\eta_j, j \in I\}$ is called a* stationary distribution *of the Markov chain if*

$$\eta_j = \sum_{k \in I} \eta_k p_{kj} \quad \text{for all } j \in I.$$

The reason behind this name is as follows. Suppose that the initial state X_0 is determined by drawing lots with probabilities

$$\mathbb{P}(X_0 = j) = \eta_j \quad \text{for all } j \in I.$$

From the point of view of an outsider who has this as only information and does not know which state has emerged from the draw, the state of the process at any future time n will be j with probability

$$\mathbb{P}(X_n = j) = \eta_j \quad \text{for all } j \in I.$$

It is easy to show this using induction. Suppose that for $k = 0, \ldots, m$, we know that $\mathbb{P}(X_k = j) = \eta_j$ for all j; then

$$\mathbb{P}(X_{m+1} = j) = \sum_{k \in I} \mathbb{P}(X_{m+1} = j \mid X_m = k) \mathbb{P}(X_m = k)$$

$$= \sum_{k \in I} \mathbb{P}(X_m = k) p_{kj} = \sum_{k \in I} \eta_k p_{kj} = \eta_j.$$

The next sections consider the limiting behavior and investigate in more detail when the equilibrium distribution $\{\pi_j\}$ from (7.4) is a stationary distribution.

7.4.1 *Limiting Behavior*

Natural questions are whether the n-step transition probabilities $p_{ij}^{(n)}$ always have a limit as $n \to \infty$ and, if so, whether the limit is also independent of the initial state i. The answer to these questions is not always "yes." The easiest way to see this is through counterexamples. Suppose that a Markov chain has state space $I = \{1, 2\}$ with one-step transition probabilities $p_{12} = p_{21} = 1$ and $p_{11} = p_{22} = 0$. Then $p_{ij}^{(n)}$ alternates between 0 and 1 for $n = 1, 2, \ldots$ and therefore has no limit as $n \to \infty$. The reason this limit does not exist lies in the periodicity of the state transitions.

Definition 7.3 (Periodic Markov chain). *A Markov chain is called* periodic *if there exist at least two disjoint subsets $R_1, \ldots, R_d$ of the state space such that a state in R_k always transitions to a state in R_{k+1} for $k = 1, \ldots, d$ (with $R_{d+1} = R_1$). If this is not the case, then the Markov chain is called* aperiodic.

The random walk of Example 7.7 may return to state 0 after an even number of steps, only. Here R_1 contains the even numbers and R_2 the odd numbers.

Even if the limit of the $p_{ij}^{(n)}$ exists, it is not necessarily independent of the initial state. Consider the Markov chain with state space $I = \{1, 2\}$ and one-step transition probabilities $p_{11} = p_{22} = 1$ and $p_{12} = p_{21} = 0$. Then $p_{11}^{(n)} = 1$ and $p_{21}^{(n)} = 0$ for all $n \geq 1$, so that $\lim_{n \to \infty} p_{i1}^{(n)}$ depends on the initial state i. This dependence occurs when the Markov chain has two or more disjoint closed sets. Recall that a set $I' \subset I$ is closed if $p_{ij} = 0$ for $i \in I'$ and $j \notin I'$. The set I' is an *irreducible set* if it is closed and all its states communicate. Two irreducible sets must be disjoint (verify!). The state space may therefore be decomposed into disjoint irreducible sets $I_1, I_2, \ldots$, and a non-irreducible set I^*.

Definition 7.4 (Irreducible Markov chain). *A Markov chain is called* irreducible *if every state k is reachable from every other state j; that is, for all $j, k \in I$, there is an $n \geq 1$ such that $p_{jk}^{(n)} > 0$.*

A Markov chain is irreducible if its state space is an irreducible set.

The examples above show that conditions must be imposed on the Markov chain when the limiting behavior of the Markov chain is studied. For a Markov chain with a *finite* state space I, one can show that $\lim_{n\to\infty} p_{ij}^{(n)}$ exists for all i, j if the Markov chain is aperiodic, where the limit is moreover independent of the initial state i if the Markov chain does not contain two or more disjoint closed sets. In the case of Example 7.6, these conditions hold and therefore, as $n \to \infty$, the matrix $\mathbf{P}^n$ converges to a limiting matrix in which all rows are the same. In applications, however, the aperiodicity does not always hold. For instance, in Example 7.2, the Markov chain is periodic with period 2 (from an even state, the process always goes to an odd state, and vice versa). In applications of Markov chains in operations research, whether or not a Markov chain is aperiodic is rarely relevant. In these applications, the important concept is that of the so-called Cesàro limit[4]

$$\lim_{n\to\infty} \frac{1}{n} \sum_{t=1}^{n} p_{ij}^{(t)}.$$

With or without periodicity, this limit always exists. If $\lim_{n\to\infty} p_{ij}^{(n)}$ exists, then the Cesàro limit is equal to the ordinary limit. We can make this result plausible as follows. Suppose that a profit structure is placed on the process, where a profit of 1 is made in one of the states and no profit in the other states. More precisely, fix a state $j = r$ and suppose that every time the process goes to state r, a profit of 1 is made, whereas a transition to another state gives no profit. By Theorem 7.2, we know that $\sum_{t=1}^{n} p_{ir}^{(t)}$ is the total expected profit up to and including time $t = n$ when the initial state is i. Intuitively, the long-run average expected profit per unit of time is well defined. In other words, $\lim_{n\to\infty} \frac{1}{n} \sum_{t=1}^{n} p_{ir}^{(t)}$ exists. It immediately follows from the discussion above that this limit can also be seen as the long-run expected fraction of the time that the process is in state r for an initial state i.

For a Markov chain with a finite state space, assuming that there are no two disjoint closed sets of states, not only is $\lim_{n\to\infty} \frac{1}{n} \sum_{t=1}^{n} p_{ij}^{(t)}$ independent of the initial state i, we also have that the equilibrium distribution is a true probability distribution. For a Markov chain with an infinite state space, the assumption is not strong enough. This follows from the counterexample given earlier: $I = \{0, 1, 2, \ldots\}$ and $p_{i,i+1} = 1$ for all $i \in I$. This Markov chain has only transient states with $\lim_{n\to\infty} p_{ij}^{(n)} = 0$ for all $i, j \in I$. In the next subsection, we give a practical assumption that excludes such cases for Markov chains with an infinite state space.

7.4.2 *Balance Equations*

In this subsection, we make the following assumption for the Markov chain.

---

[4]For a sequence of numbers $(a_1, a_2, \ldots)$ for which $\lim_{n\to\infty} a_n$ does not exist, the limit $\lim_{n\to\infty} \frac{1}{n} \sum_{k=1}^{n} a_k$ often does exist. The latter is called the Cesàro limit. If the ordinary limit exists, the Cesàro limit also exists and is equal to the ordinary limit.

Assumption 7.1. There is a state r that can ultimately be reached from every initial state i with probability 1, and the expected value of the number of steps needed to return from state r to itself is finite.

This assumption is satisfied in almost every practical application concerning the equilibrium of the Markov chain. For a Markov chain with a finite state space I, the assumption is automatically satisfied when the Markov chain has no two disjoint closed sets. Under Assumption 7.1, one can prove that for all $i, j \in I$,

$$\pi_j = \lim_{n \to \infty} \frac{1}{n} \sum_{t=1}^{n} p_{ij}^{(t)} \tag{7.5}$$

is independent of the initial state i, and moreover $\sum_{j \in I} \pi_j = 1$. Thus, π_j is the expected fraction of time $\lim_{n \to \infty} \frac{1}{n} \sum_{t=1}^{n} p_{ij}^{(t)}$ that the process is in state j. The fraction π_j is related to the mean first passage time to reach state j from state j, m_{jj}, as

$$\pi_j = \frac{1}{m_{jj}}.$$

The interpretation of $\lim_{n \to \infty} \frac{1}{n} \sum_{t=1}^{n} p_{ij}^{(t)}$ as the *expected* fraction of the time that the process is in state j can be extended to the stronger interpretation:

the *actual* long-run fraction of the time the process is in state j
$= \pi_j$ with probability 1,

independently of the initial state $X_0 = i$. We do not prove this extremely important interpretation, nor the result that $\lim_{n \to \infty} p_{ij}^{(n)}$ exists and π_j is also given by $\lim_{n \to \infty} p_{ij}^{(n)}$ if the Markov chain satisfies Assumption 7.1 and is *aperiodic*.

Definition 7.5 (Ergodic Markov chain). *An irreducible, aperiodic, positive recurrent Markov chain with stationary distribution is called* ergodic.

A Markov chain is ergodic if Assumption 7.1 is satisfied and the state space I is a single irreducible set. We prove the weaker interpretation of π_j as the expected fraction of time that the process is in state j.

Theorem 7.3. *Let the Markov chain be ergodic, then*

$\pi_j = $ *the* expected *long-run fraction of the time the process is in state* j.

Proof. Recall the definition of I_t as the indicator that $X_t = j$ and $A_j(n)$ as the number of visits to state j by time n in the proof of Theorem 7.2. Then

$$\lim_{n \to \infty} \frac{\mathbb{E}[A_j(n)|X_0 = i]}{n} = \lim_{n \to \infty} \frac{\mathbb{E}[\sum_{t=1}^{n} I_t|X_0 = i]}{n}$$

$$= \lim_{n \to \infty} \frac{\sum_{t=1}^{n} \mathbb{P}(X_t = j|X_0 = i)}{n}$$

$$= \lim_{n \to \infty} p_{ij}^{(n)} = \pi_j.$$

$\square$

The next question is how to calculate the numbers π_j for $j \in I$. The answer is by solving a system of linear equations, as we already observed for the Markov chain with finite space in (7.4). The following result holds.

Theorem 7.4. *If the Markov chain is ergodic, the π_j for $j \in I$ form the unique equilibrium distribution of the Markov chain. The π_j are the unique solution to the system of linear equations*

$$\pi_j = \sum_{k \in I} \pi_k p_{kj}, \quad \text{for all } j \in I, \tag{7.6}$$

$$\sum_{j \in I} \pi_j = 1. \tag{7.7}$$

Proof. Ergodicity implies that $\pi_j = \lim_{n \to \infty} p_{ij}^{(n)}$ and that the equilibrium distribution is unique. Just like the motivation for (7.4), we may now use the Chapman-Kolmogorov equations to obtain (7.6) $\qquad\qquad\qquad\square$

The equations (7.6) are called the *balance equations* or *global balance equations*, while (7.7) is called the *normalizing equation*. The balance equations determine the numbers π_j for $j \in I$ up to a *multiplicative constant*; adding the normalizing equation determines them uniquely. The numbers π_j for $j \in I$ are called the *equilibrium probabilities* of the Markov chain. The simplest way to remember the balance equations is by keeping in mind that they arise by multiplying the *row vector* $\boldsymbol{\pi} = (\pi_i, i \in I)$ with the *column vectors* of the matrix $\mathbf{P}$ of one-step transition probabilities

$$\boldsymbol{\pi} = \boldsymbol{\pi}\mathbf{P}.$$

The balance equations (7.6) may be interpreted as *flow out* = *flow in* equations. To this end, using that $\sum_{k \in I} p_{jk} = 1$ we may rewrite (7.6) as

$$\sum_{k \in I} \pi_j p_{jk} = \sum_{k \in I} \pi_k p_{kj}, \quad \text{for all } j \in I.$$

The left-hand side is the flow out of state j to any of the states $k \in I$ and the right-hand side the flow into state j from any of the states $k \in I$. In equilibrium, the flow out of state j must balance the flow into state j.

Example 7.6 (Continued). The equilibrium probabilities π_1, π_2, and π_3 of the Markov model in this example give the fractions of days in a year that the weather is sunny, cloudy, or rainy, respectively. These probabilities are found by solving the system of linear equations

$$\pi_1 = 0.70\pi_1 + 0.50\pi_2 + 0.40\pi_3,$$
$$\pi_2 = 0.10\pi_1 + 0.25\pi_2 + 0.30\pi_3,$$
$$\pi_3 = 0.20\pi_1 + 0.25\pi_2 + 0.30\pi_3,$$
$$\pi_1 + \pi_2 + \pi_3 = 1,$$

where one of the first three equations may be left out to have as many equations as unknowns. The solution is

$$\pi_1 = 0.5960, \ \pi_2 = 0.1722, \ \pi_3 = 0.2318,$$

which corresponds to that found with the matrix products calculated earlier. It is sunny 59.60% of the time, cloudy 17.22% of the time, and rainy 23.18% of the time.

Example 7.3 (Continued). For the absent-minded professor, the balance equations are

$$\pi_1 = 0.5\pi_1 + 0.5\pi_3,$$
$$\pi_2 = 0.5\pi_1 + 0.5\pi_3,$$
$$\pi_3 = \pi_2.$$

The solution to these equations together with $\pi_1 + \pi_2 + \pi_3 = 1$ is $\pi_1 = \pi_2 = \pi_3 = \frac{1}{3}$. Based on the equation $\pi_2 + \pi_3 = \frac{2}{3}$, we can state that in the long run, the professor will not have his driver's license with him 66.7% of the time. In other words, the probability that on any given day, a driver's license check leads to the professor being fined for not having it with him is $\frac{2}{3}$.

7.4.3   *Markov Chains with a Cost Structure*

Many practical problems concern Markov chains with a cost or revenue structure. Suppose that every time the Markov chain makes a transition to state j, a cost of $c(j)$ is incurred. What is the long-run average cost per unit of time? Intuitively, it should be clear that this cost is equal to $\sum_{j \in I} c(j)\pi_j$ because π_j can be interpreted as the average number of transitions to state j per unit of time. To give a mathematical proof that this cost formula is correct, we need a technical assumption in addition to the earlier assumption that there is a state r that is ultimately reached from the initial state i with probability 1. The additional assumption is that $\sum_{j \in I} |c(j)| \pi_j < \infty$ and that for every initial state $i \in I$, the total cost made until the first return to state r has a finite expected value. This assumption is automatically satisfied if the Markov chain has a finite state space and no two disjoint closed sets. The following main theorem holds.

Theorem 7.5. *The actual long-run average cost per unit of time is given by*

$$\lim_{n \to \infty} \frac{1}{n} \sum_{t=1}^{n} c(X_t) = \sum_{j \in I} c(j)\pi_j \quad \text{with probability 1,}$$

independently of the initial state $X_0 = i$.

We do not give the proof of this so-called *ergodic theorem*. It relies on the famous renewal-reward theorem from the theory of regenerative stochastic processes. The proof moreover shows that the theorem also applies to the case of continuous costs between the state transitions rather than *direct* costs $c(j)$. If, in the continuous case, we define the cost function $c(j)$ by

$$c(j) = \mathbb{E}[\text{cost between the times } t = n - 1 \text{ and } t = n \mid X_{n-1} = j],$$

then it is also true that the *actual* long-run average cost per unit of time is equal to $\sum_{j \in I} c(j)\pi_j$ with probability 1. This is an extremely useful observation for real-world applications. In applications, it can also be helpful to interpret certain performance measures as average costs per unit of time.

Example 7.5 (Continued). In this example, the equilibrium probability π_j gives the long-run fraction of years that the transport company is in the insurance rate class j when a fixed $(\alpha_1, \ldots, \alpha_4)$-strategy is used. The equilibrium probabilities $\pi_1, \ldots, \pi_4$ are the unique solution to the balance equations

$$\pi_1 = \sum_{k=1}^{4} [1 - F(\alpha_k)]\pi_k,$$

$$\pi_2 = F(\alpha_1)\pi_1,$$

$$\pi_3 = F(\alpha_2)\pi_2,$$

$$\pi_4 = F(\alpha_3)\pi_3 + F(\alpha_4)\pi_4$$

together with the normalizing equation $\pi_1 + \cdots + \pi_4 = 1$. For a fixed claim strategy, the actual long-run average cost per year is equal to $\sum_{j=1}^{4} c(j)\pi_j$ with probability 1, where c_j is defined as

$$c(j) = \text{premium for class } j + \mathbb{E}[\text{unclaimed damages in}$$
$$\text{the coming year when the company is in rate class } j].$$

How does one find the expected value of the unclaimed damages at the end of the coming year? Apply the law of total expectation:

$$\mathbb{E}[X] = \int_0^\infty \mathbb{E}[X \mid S = s]f(s)ds.$$

If the realized value of the total annual damage S is equal to s, then the unclaimed damages are given by s if $s \le \alpha_j$ and by r_j if $s > \alpha_j$. Taking the average of these two costs over the probability density $f(s)$ of the random variable S gives

$$c(j) = P_j + \int_0^{\alpha_j} sf(s)ds + \int_{\alpha_j}^\infty r_j f(s)ds \quad \text{for } j = 1, \ldots, 4.$$

The average cost per year under the $(\alpha_1, \ldots, \alpha_4)$-strategy is therefore

$$g(\alpha_1, \ldots, \alpha_4) = \sum_{j=1}^{4} \left[P_j + \int_0^{\alpha_j} sf(s)ds + r_j(1 - F(\alpha_j)) \right] \pi_j.$$

The next step to determine the optimal values of $\alpha_1, \ldots, \alpha_4$ is routine. A numerical procedure to minimize a function of several variables must be used. In every function evaluation in this procedure, in general, numerical integration is used for the integrals in $g(\alpha_1, \ldots, \alpha_4)$. As an illustration, consider the following example: $P_1 = 10000$, $P_2 = 7500$, $P_3 = 6000$, $P_4 = 5000$, $r_1 = 1500$, $r_2 = 1000$, $r_3 = 750$, and $r_4 = 500$. The expected value of the damage D in a given year is set to 5000, and the coefficient of variation c_D of the damages D is successively taken to be $c_D = 1$, 2, and 5. To study the effect of the form of the probability density on the damage D, we consider both a gamma distribution and a lognormal distribution for the damage D with the same first two moments. Table 7.1 gives the optimal claim limits $\alpha_1^*, \ldots, \alpha_4^*$ with the corresponding average cost per year.

Table 7.1   The optimal claim limits and the minimal costs.

	Gamma			Lognormal		
	$c_D^2 = 1$	$c_D^2 = 4$	$c_D^2 = 25$	$c_D^2 = 1$	$c_D^2 = 4$	$c_D^2 = 25$
α_1^*	5908	6008	6280	6015	6065	6174
α_2^*	7800	7908	8236	7931	7983	8112
α_3^*	8595	8702	9007	8717	8769	8890
α_4^*	8345	8452	8757	8467	8519	8640
g^*	9058	7698	6030	9174	8318	7357

7.5   Markov Chain Monte Carlo Method[5]

Markov chain Monte Carlo (MCMC) methods are powerful simulation techniques to sample from a (multi-dimensional) probability density π_j that is known up to a multiplicative constant, where it is not feasible to compute the constant directly. In Bayesian statistics one often encounters the situation that obtaining the posterior density requires the computation of a multiplicative constant in the form of a (multi-dimensional) integral. Unlike Monte Carlo sampling methods that are able to draw independent samples from the distribution, Markov Chain Monte Carlo methods draw samples where each sample is dependent on the previous sample. These samples come from a Markov chain that has π_j as equilibrium distribution. MCMC has revolutionized applied mathematics and is used in many areas of science including statistics, computer science, physics and biology.

7.5.1   *Reversible Markov Chains*

Let $\{X_n\}$ be an ergodic Markov chain with state space I. The Markov chain $\{X_n\}$ has unique equilibrium distribution $\{\pi_j\}$. The quantity $\pi_j p_{jk}$ can be interpreted as the long-run average number of transitions per unit time from state j to state k. Many Markov chains have the much stronger property that the rate of transitions from j to k is equal to the rate of transitions from k to j for any $j, k \in I$. An example of such a Markov chain is the random walk of Example 7.7 on the integers with transitions $+1$ or -1 at each step. In this Markov chain the rate of transitions from $i-1$ to i is equal to the rate of transitions from i to $i-1$ for any i because between any two transitions from $i-1$ to i there must be a transition from i to $i-1$ and conversely. As a consequence, $\pi_{i-1}p_{i-1,i} = \pi_i p_{i,i-1}$.

Definition 7.6. *An ergodic Markov chain $\{X_n\}$ with state space I is said to be* reversible *if the equilibrium probabilities π_j satisfy the so-called* detailed balance equations

$$\pi_j p_{jk} = \pi_k p_{kj} \quad \text{for all } j, k \in I. \tag{7.8}$$

A probability distribution satisfying the detailed balance equations must be an equilibrium distribution of the Markov chain. This is easy to prove.

---

[5]This section contains more specialized material.

Theorem 7.6. *Let $\{X_n\}$ be an ergodic Markov chain with state space I. If $\{a_j, j \in I\}$ is a probability distribution satisfying*

$$a_j p_{jk} = a_k p_{kj} \quad \text{for all } j, k \in I,$$

then $\{a_j, j \in I\}$ is the unique equilibrium distribution of the Markov chain.

Proof. To prove this result, sum both sides of the equation $a_j p_{jk} = a_k p_{kj}$ over $k \in I$. Together with $\sum_{k \in I} p_{jk} = 1$, this gives $a_j = \sum_{k \in I} a_k p_{kj}$ for all $j \in I$. By Theorem 7.4, $\{a_j, j \in I\}$ is a the unique equilibrium distribution of the Markov chain. $\qquad\qquad\square$

An interesting question is the following. Let $\{a_j, j \in I\}$ be a probability mass function on a finite set of states I with $a_j > 0$ for all j. Can we construct a Markov chain on I with $\{a_j, j \in I\}$ as unique equilibrium distribution? The answer is yes. The physical construction of the Markov chain proceeds as follows: if the current state of the process is j, then choose at random one of the other states as candidate state for the next state. Suppose the chosen candidate state is k. If $a_k > a_j$, then the next state of the Markov chain is state k; otherwise, the next state is either k with probability a_k/a_j or the current state j with probability $1 - a_k/a_j$. Thus, letting $N = |I|$ denote the number of states, we have constructed a Markov chain $\{X_n\}$ on I with one-step transition probabilities

$$p_{jk} = \begin{cases} \frac{1}{N-1}\min\{\frac{a_k}{a_j}, 1\} & \text{for } k \neq j, \\[2mm] 1 - \sum_{l \neq j} p_{jl} & \text{for } k = j. \end{cases}$$

It is immediate that the Markov chain $\{X_n\}$ is irreducible (and aperiodic). Also, the Markov chain satisfies the reversibility condition: it holds for any $j, k \in I$ with $j \neq k$ that

$$a_j p_{jk} = \frac{1}{N-1}\min\{a_k, a_j\} = a_k \frac{1}{N-1}\min\left\{\frac{a_j}{a_k}, 1\right\} = a_k p_{kj}.$$

It now follows from Theorem 7.6 that the Markov chain has $\{a_j\}$ as unique equilibrium distribution.

Simulated Annealing Algorithm

Reversible Markov chains form the basis of the simulated annealing algorithm. This is a search method for determining the absolute minimum of an often complicated function on a finite but very large domain. The main idea of the algorithm is to move from one point to another following a probability distribution so that the search procedure can also escape from a local minimum. We only give a rough sketch of the algorithm. Suppose that $c(i)$ is a given function on a finite set I. For every point $i \in I$, choose a local set of neighbors $N(i)$ with $i \notin N(i)$ such that $j \in N(k)$ if $k \in N(j)$. For the sake of convenience, assume $|N(j)| = M$ for all j. Now, define the following Markov chain on I. If the current state is j, randomly choose a candidate state k from $N(j)$. The Markov chain's next state is k if $c(k) < c(j)$; otherwise, it is k with probability $e^{-c(k)/T}/e^{-c(j)/T}$ and the current state j with probability $1 - e^{-c(k)/T}/e^{-c(j)/T}$. Here $T > 0$ is a control parameter. In other words, the Markov chain is defined by the one-step transition probabilities

$$
p_{jk} = \begin{cases} \frac{1}{M} \min\left\{e^{-c(k)/T}/e^{-c(j)/T}, 1\right\} & \text{for } k \in N(j), \\ 1 - \sum_{\ell \neq j} p_{j\ell} & \text{for } k = j, \end{cases}
$$

and $p_{jk} = 0$ otherwise.[6] This Markov chain is reversible:

$$
e^{-c(j)/T} p_{jk} = e^{-c(k)/T} p_{kj} \quad \text{for all } j, k \in I.
$$

The proof is simple. Take $k \neq j$. For $k \notin N(j)$, we have $p_{jk} = p_{kj} = 0$, and for $k \in N(j)$, we have

$$
e^{-c(j)/T} \frac{1}{M} \min\left\{1, \frac{e^{-c(k)/T}}{e^{-c(j)/T}}\right\} = \frac{1}{M} \min\{e^{-c(j)/T}, e^{-c(k)/T}\}
$$

$$
= e^{-c(k)/T} \frac{1}{M} \min\left\{1, \frac{e^{-c(j)/T}}{e^{-c(k)/T}}\right\}.
$$

Suppose that the sets $N(i)$ are sufficiently correlated for the Markov chain to be irreducible. It then follows from Theorem 7.6 that the Markov chain's unique equilibrium distribution is given by

$$
\pi_i = \frac{1}{A} e^{-c(i)/T} \quad \text{for } i \in I \quad \text{with } A = \sum_{k \in I} e^{-c(k)/T}.
$$

If the function $c(i)$ takes on its absolute minimum at a unique point m, then it follows from $\pi_m = 1/(1 + \sum_{k \neq m} e^{-(c(k)-c(m))/T})$ that $\pi_m \to 1$ as $T \to 0$ (verify that $\sum_{i \in M} \pi_i \to 1$ as $T \to 0$ with M the set of points at which the function $c(i)$ takes on the absolute minimum). This result forms the basis for the simulated annealing algorithm. In this algorithm, one begins with a high value of the so-called cooling temperature T and then lowers the parameter T in every iteration n of the algorithm, for example by setting $T = C/\ln(n+1)$ for a constant $C > 0$. The algorithm is initiated with a state $X_0 = i_0$. The states $X_1 = i_1, X_2 = i_2, \ldots, X_N = i_N$, with N large, are then generated one after the other, and the absolute minimum of the function $c(i)$ is approximated by $\min_{k=0,1,\ldots,N} c(i_k)$.

An interesting application of the simulated annealing algorithm is to the traveling salesman problem. Suppose that starting from city 0, a salesman must visit the cities $1, \ldots, r$, where each of these cities may be visited only once, and the salesman must then return to the starting point 0. Suppose that a nonnegative cost $c(i, j)$ is made when going directly from city i to city j. A permutation $\mathbf{x} = (x_1, \ldots, x_r)$ of the integers $1, \ldots, r$ then gives a route, meaning that the route goes from city 0 to city x_1, from city x_1 to city x_2, and so on, to ultimately return from city x_r to city

---

[6] If $|N(j)|$ is not the same for all j, take $M = \max_j |N(j)|$.

0. The total cost of such a route is $c(\mathbf{x}) = \sum_{i=1}^{r+1} c(x_{i-1}, x_i)$, with $x_0 = x_{r+1} = 0$. In the simulated annealing algorithm, for each route $\mathbf{x}$, one could choose the set of neighbors $N(\mathbf{x})$ to be the set of routes created by exchanging two elements in the permutation $(x_1, \ldots, x_r)$. Every set of neighbors $N(\mathbf{x})$ then consists of the same number of elements $\binom{r}{2}$.

7.5.2 The Metropolis-Hastings Algorithm

The Metropolis-Hastings algorithm is an example of a Markov chain Monte Carlo method. The algorithm will be explained for the case of a discrete probability distribution, but the basic idea of the algorithm can directly be generalized to the case of a continuous probability distribution.

Let us explain how the method can be used to attack the following basic problem arising amongst others in Bayesian inference. How to calculate $\sum_{s \in S} h(s)\pi(s)$, where $h(s)$ is a given function and the probability mass function $\pi(s)$ on the very large but finite set $S \subset I$ is known up to a multiplicative constant that cannot be computed directly? The idea is to construct a Markov chain that has $\pi(s)$ as its unique equilibrium distribution and to simulate a sequence $s_1, s_2, \ldots, s_m$ of successive states of this Markov chain for large m. Then $\sum_{s \in S} h(s)\pi(s)$ can be estimated by $\frac{1}{m} \sum_{k=1}^{m} h(s_k)$, by the ergodic Theorem 7.5.

The algorithm uses candidate-transition functions $\bar{p}(t \mid s)$ for $s \in S$.[7] This function is to be interpreted as saying that when the current state is s the candidate for the next state is t with probability $\bar{p}(t \mid s)$. Thus you first choose, for each $s \in S$, a probability mass function $\{\bar{p}(t \mid s), t \in S\}$. These functions must be chosen in such a way that the Markov matrix with the $\bar{p}(t \mid s)$ as one-step transition probabilities is irreducible. The idea is to next adjust these transition probabilities in such a way that the corresponding Markov chain has $\{\pi(s), s \in S\}$ as its unique equilibrium distribution. The detailed balance equations (8.11) are the key to this idea. If the candidate-transition functions $\bar{p}(t \mid s)$ already satisfy the detailed balance equations

$$\pi(s)\bar{p}(t \mid s) = \pi(t)\bar{p}(s \mid t) \quad \text{for all } s, t \in S,$$

you are done: the Markov chain with the $\bar{p}(t \mid s)$ as one-step transition probabilities has $\{\pi(s), s \in S\}$ as its unique equilibrium distribution. What should you do when the detailed balance equations are not fully satisfied? The answer is to modify the transition probabilities by rejecting certain transitions. To work out this idea, fix two states s and t for which the detailed balance equation is not satisfied. It is no restriction to assume that $\pi(s)\bar{p}(t \mid s) > \pi(t)\bar{p}(s \mid t)$. Otherwise, reverse the roles of s and t. If $\pi(s)\bar{p}(t \mid s) > \pi(t)\bar{p}(s \mid t)$, then, loosely speaking, the process moves from s to t too often. How could you restore this? A simple trick to reduce the number of transitions from s to t is to use an acceptance probability $\alpha(t \mid s)$: the

[7]For clarity of presentation, Section 7.5.2 and Section 7.5.3 use the notation $\bar{p}(t \mid s)$ rather than $\bar{p}_{st}$ and $\pi(s)$ rather than π_s as the states in these sections often are multidimensional or contain indices.

process is allowed to make the transition from s to t with probability $\alpha(t \mid s)$ and otherwise the process stays in the current state s. The question remains how to choose $\alpha(t \mid s)$. The choice of $\alpha(t \mid s)$ is determined by the requirement

$$\pi(s)[\bar{p}(t \mid s)\alpha(t \mid s)] = \pi(t)[\bar{p}(s \mid t)\alpha(s \mid t)].$$

Taking $\alpha(s \mid t) = 1$ for transitions from t to s, you get

$$\alpha(t \mid s) = \frac{\pi(t)\bar{p}(s \mid t)}{\pi(s)\bar{p}(t \mid s)}.$$

Therefore, for any $s, t \in S$, the acceptance probability is defined by

$$\alpha(t \mid s) = \min\left\{\frac{\pi(t)\bar{p}(s \mid t)}{\pi(s)\bar{p}(t \mid s)}, 1\right\}.$$

The one-step transition probabilities to be used in the algorithm are now defined as

$$\bar{p}^{MH}(t \mid s) = \begin{cases} \bar{p}(t \mid s)\alpha(t \mid s) & \text{for } t \neq s, \\ 1 - \sum_{t \neq s} \bar{p}(t \mid s)\alpha(t \mid s) & \text{for } t = s. \end{cases}$$

The Markov chain with these one-step transition probabilities satisfies the detailed balance equations $\pi(s)\bar{p}^{MH}(t \mid s) = \pi(t)\bar{p}^{MH}(s \mid t)$ for all s, t. Therefore this Markov chain has $\{\pi(s), s \in S\}$ as its unique equilibrium distribution. It is important to note that for the construction of the Markov chain it suffices to know the $\pi(s)$ up to a multiplicative constant because the acceptance probabilities involve only the ratio's $\pi(s)/\pi(t)$.

Summarizing, the Markov chain operates as follows. If the current state is s, a candidate state t is generated from the probability mass function $\{\bar{p}(t \mid s), t \in S\}$. If $t \neq s$, then state t is accepted with probability $\alpha(t \mid s)$ as the next state of the Markov chain; otherwise, the Markov chain stays in state s.

Algorithm 7.1 (Metropolis-Hastings).

Step 0. Choose probability mass functions $\{\bar{p}(t \mid s), t \in S\}$ for $s \in S$ such that the Markov matrix with the $\bar{p}(t \mid s)$ as elements is irreducible. Choose a starting state s_0. Let $n := 1$.

Step 1. Generate a candidate state t_n from the probability mass function $\{\bar{p}(t \mid s_{n-1}), t \in S\}$. Calculate the acceptance probability

$$\alpha = \min\left\{\frac{\pi(t_n)\bar{p}(s_{n-1} \mid t_n)}{\pi(s_{n-1})\bar{p}(t_n \mid s_{n-1})}, 1\right\}.$$

Step 2. Generate a random number u from $(0, 1)$. If $u \leq \alpha$, accept t_n and let $s_n := t_n$; otherwise, $s_n := s_{n-1}$.

Step 3. $n := n + 1$. Repeat step 1 with s_{n-1} replaced by s_n.

Note that when the chosen probability densities $\bar{p}(t \mid s)$ are symmetric, that is, $\bar{p}(t \mid s) = \bar{p}(s \mid t)$ for all $s, t \in S$, then the acceptance probability α in Step 1 reduces to

$$\alpha = \min\left\{\frac{\pi(t_n)}{\pi(s_{n-1})}, 1\right\}.$$

In applications of the algorithm, one typically wants to estimate $\mathbb{E}[h(X)]$ for a given function $h(x)$, where X is a random variable with $\pi(s)$ as its probability density. If the states $s_1, s_2, \ldots, s_m$ are generated by the Metropolis-Hastings algorithm for a sufficiently large m, then $\mathbb{E}[h(X)] = \sum_{s \in S} h(s)\pi(s)$ is estimated by

$$\frac{1}{m}\sum_{k=1}^{m} h(s_k).$$

This estimate is based on a law of large numbers result for Markov chains, saying that $\lim_{m \to \infty} \frac{1}{m}\sum_{k=1}^{m} h(s_k) = \mathbb{E}[h(X)]$. A heuristic explanation is as follows. The probability $\pi(s)$ can be interpreted as the long-run fraction of transitions into state s and so, for m large, $\pi(s) \approx \frac{m(s)}{m}$, where $m(s)$ is the number of times that state s occurs among the sequence $s_1, s_2, \ldots, s_m$. This gives $\sum_{s \in S} h(s)\pi(s) \approx \frac{1}{m}\sum_{s \in S} h(s)m(s) = \frac{1}{m}\sum_{k=1}^{m} h(s_k)$.

A practical approach to construct an approximate confidence interval for $\mathbb{E}[h(X)]$ is the batch-means method that will be discussed in Section 11.5. How the starting state s_0 should be chosen and how many iterations should be done are empirical questions. A simple approach is to use multiple dispersed initial values to start several different chains. One suggestion for an initial value is to start the chain close to the mode of the target density if this density is unimodal. Diagnosing convergence to the target density is an art in itself. To diagnose the convergence of an average $\frac{1}{t}\sum_{k=1}^{t} h(s_k)$, one can look at a plot of this average as function of t.

Remark 7.1 (Probability densities). It is important to point out that the Metropolis-Hastings algorithm directly extends to the case of a probability density $\pi(s)$ on a (multi-dimensional) continuous set S, where you want to calculate $\int_{s \in S} h(s)\pi(s)\, ds$ for the case that the density $\pi(s)$ is only known up to a multiplicative constant.

Implementation aspects

What are the best options for the proposal densities $\bar{p}(t \mid s)$? There are two general approaches: independent chain sampling and random walk chain sampling. These approaches will be briefly discussed.[8]

(a) In independent chain sampling the candidate state t is drawn independently of the current state s of the Markov chain, that is, $\bar{p}(t \mid s) = g(t)$ for some proposal density $g(x)$.

---

[8]For more details the reader is referred to S. Chib and E. Greenberg, "Understanding the Metropolis-Hastings algorithm," *The American Statistician*, **49** (1995): 327–335.

(b) In random walk chain sampling the candidate state t is the current state s plus a draw from a random variable Z that does not depend on the current state. In this case, $\bar{p}(t \mid s) = g(t - s)$ with $g(z)$ the density of the random variable Z. If $g(z) = g(-z)$ for all z, the proposal density is symmetric and, as noted before, the acceptance probability reduces to $\alpha(t \mid s) = \min\{\pi(t)/\pi(s), 1\}$.

It is very important to have a well-mixing Markov chain, that is, a Markov chain that explores the state space S adequately and sufficiently fast. In other words, by the choice of the proposal densities one tries to avoid that the Markov chain stays in small regions of the state space for long periods of time. The variance of the proposal density can be thought of as a tuning parameter to get better mixing. It affects both the acceptance probability and the magnitude of the moves of the state. It is a matter of experimentation to find a tradeoff between these two features. In random walk chain sampling a rule of thumb is to choose the proposal density in such a way that on average about 50% of the candidate states are accepted. In independent chain sampling it is important that the tail of the proposal density $g(s)$ dominates the tail of the target density $\pi(s)$.

Let us illustrate the implementation aspects with the following example.

Example 7.11. Let the density $\pi(x)$ of the positive random variable X be proportional to $x^{-2.5}e^{-2/x}$ for $x > 0$. Suppose that this target density is simulated by the Metropolis-Hastings algorithm with independent chain sampling, where both the uniform density on $(0, 1,000)$ and the Pareto density $1.5x^{-2.5}$ for $x > 1$ are taken as proposal density. How goes the mixing of the state of the Markov chain? What are the estimates for $\mathbb{E}[X]$?

Solution. It is instructive to make a plot of the first 500 (say) states simulated by the Metropolis-Hastings algorithm. Such a plot is given in Figure 7.2, where the left figure corresponds to the uniform proposal density and the right figure to the Pareto proposal density. In each of the simulation studies the starting state was chosen as $s_0 = 1$. In the simulation with the Pareto density, we simulated the shifted density $\pi_{shift}(x) = \pi(x-1)$ for $x > 1$. It is directly seen from the figure that the mixing of the state is very bad under the uniform density, though this proposal density covers nearly all the mass of the target density. A bad mixing means that the candidate state is often rejected. This happens for the uniform density because the variance of this proposal density is very large, which makes that the candidate state is often far away from the current state and this in turn implies a very small acceptance probability. The Pareto density exhibits the same tail behavior as the target density $\pi(x)$ and gives an excellent mixing of the state of the Markov chain. The shortcoming of the uniform density as proposal density also appears from the fact that our simulation study with one million runs gave the estimate of 3.75 for $\mathbb{E}[X]$, where the exact value of $\mathbb{E}[X]$ is 4. The simulation study with the Pareto density as proposal density gave the estimate of 3.98 for $\mathbb{E}[X]$.

Fig. 7.2 Mixing of the state of the Markov chain.

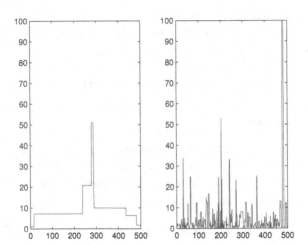

7.5.3 *The Gibbs Sampler*

The Gibbs sampler is a special case of Metropolis-Hastings sampling and is used to simulate from a multivariate density whose univariate conditional densities are fully known. This sampler is frequently used in Bayesian statistics. To introduce the method, consider the random vector $(X_1, \ldots, X_d)$ with the multivariate density function

$$\pi(x_1, \ldots, x_d) = \mathbb{P}(X_1 = x_1, \ldots, X_d = x_d).$$

The univariate conditional densities of $(X_1, \ldots, X_d)$ are denoted by

$$\pi_k(x \mid x_1, \ldots, x_{k-1}, x_{k+1}, \ldots, x_d)$$
$$= \mathbb{P}(X_k = x \mid X_1 = x_1, \ldots, X_{k-1} = x_{k-1}, X_{k+1} = x_{k+1}, \ldots, X_d = x_d).$$

The key to the Gibbs sampler is the assumption that the univariate conditional densities are fully known. The Gibbs sampler generates random draws from the univariate densities and defines a Markov chain with $\pi(x_1, \ldots, x_d)$ as its unique equilibrium distribution. In each iteration, one component of the state vector is randomly chosen and adjusted. The algorithm is a special case of the Metropolis-Hastings method with

$$\bar{p}(\mathbf{y} \mid \mathbf{x}) = \frac{1}{d} \mathbb{P}\left(X_k = y \mid X_j = x_j \text{ for } j = 1, \ldots, d \text{ with } j \neq k\right),$$

for two states $\mathbf{x}$ and $\mathbf{y}$ such that $\mathbf{x} = (x_1, \ldots, x_{k-1}, x_k, x_{k+1}, \ldots, x_d)$ and $\mathbf{y} = (x_1, \ldots, x_{k-1}, y, x_{k+1}, \ldots, x_d)$. Since $\bar{p}(\mathbf{x} \mid \mathbf{y}) = c\pi(\mathbf{x})$ and $\bar{p}(\mathbf{y} \mid \mathbf{x}) = c\pi(\mathbf{y})$ with $c^{-1} = d \, \mathbb{P}(X_k = y \mid X_j = x_j \text{ for } j \neq k)$, the reversibility property $\bar{p}(\mathbf{x} \mid \mathbf{y})\pi(\mathbf{y}) = \bar{p}(\mathbf{y} \mid \mathbf{x})\pi(\mathbf{x})$ is satisfied, and so the acceptance probability is always 1. The Gibbs sampler is as follows:

Algorithm 7.2 (Gibbs).

Step 0. Choose a starting state $\mathbf{x} = (x_1, \ldots, x_d)$.

Step 1. Generate a random integer k from $\{1, \ldots, d\}$. Simulate a random draw y from the conditional density $\pi_k(x \mid x_1, \ldots, x_{k-1}, x_{k+1}, \ldots, x_d)$. Define state $\mathbf{y}$ by $\mathbf{y} = (x_1, \ldots, x_{k-1}, y, x_{k+1}, \ldots, x_d)$.

Step 2. The new state $\mathbf{x} := \mathbf{y}$. Return to step 1 with $\mathbf{x}$.

Standard Gibbs sampler

In practice one usually uses the standard Gibbs sampler, where in each iteration all components of the state vector are adjusted rather than a single component. Letting $\mathbf{x}^{(n)} = (x_1^{(n)}, \ldots, x_d^{(n)})$ denote the state vector obtained in iteration n, the next iteration $n+1$ proceeds as follows:

$$x_1^{(n+1)} \text{ is a random draw from } \pi_1(x \mid x_2^{(n)}, x_3^{(n)}, \ldots, x_d^{(n)}),$$
$$x_2^{(n+1)} \text{ is a random draw from } \pi_2(x \mid x_1^{(n+1)}, x_3^{(n)}, \ldots, x_d^{(n)}),$$
$$\vdots$$
$$x_d^{(n+1)} \text{ is a random draw from } \pi_d(x \mid x_1^{(n+1)}, x_2^{(n+1)}, \ldots, x_{d-1}^{(n+1)}).$$

This gives the new state vector $\mathbf{x}^{(n+1)} = (x_1^{(n+1)}, \ldots, x_d^{(n+1)})$. The standard Gibbs sampler also generates a sequence of states $\{\mathbf{x}^{(k)}, k = 1, \ldots, m\}$ from a Markov chain having $\pi(x_1, \ldots, x_d)$ as equilibrium distribution. The technical proof is omitted.

The expectation of any function h of the random vector $(X_1, \ldots, X_d)$ can be estimated by $(1/m) \sum_{k=1}^{m} h(\mathbf{x}^{(k)})$ for a Gibbs sequence $\{\mathbf{x}^{(k)}, k = 1, \ldots, m\}$ of sufficient length m. In particular one might use the Gibbs sequence to estimate the marginal density of any component of the random vector, say the marginal density $\pi_1(x)$ of the random variable X_1. A naive estimate for $\pi_1(x)$ is based on the values $x_1^{(1)}, \ldots, x_1^{(m)}$ from the Gibbs sequence, but a better approach is to use the other values from the Gibbs sequence together with the explicit expression for the univariate conditional density of X_1. Formally, by the law of conditional expectation, $\pi_1(x) = \mathbb{E}[\pi_1(x \mid X_2, \ldots, X_d)]$. Thus, a better estimate for $\pi_1(x)$ is given by

$$\hat{\pi}_1(x) = \frac{1}{m} \sum_{k=1}^{m} \pi_1(x \mid x_2^{(k)}, \ldots, x_d^{(k)}).$$

This estimate uses more information than the estimate based only on the individual values $x_1^{(1)}, \ldots, x_1^{(m)}$ and will typically be more accurate.

Example 7.12. The bivariate density $\pi(x, y)$ of the random vector (X, Y) is proportional to

$$\binom{r}{x} y^{x+\alpha-1}(1-y)^{r-x+\beta-1} \quad \text{for } x = 0, 1, \ldots, r, \quad 0 \le y \le 1,$$

where r, α and β are given positive integers. Assuming the data $r = 16$, $\alpha = 2$, and $\beta = 4$, we are interested in estimating the expected value, the variance and the marginal density of X. Since the univariate conditional densities of X and Y can be explicitly determined, we can use the Gibbs sampler. This method also works for the case of a random vector (X, Y) with a discrete component X and a continuous component Y. How do we find the univariate conditional densities? Therefore note that $\pi_1(x \mid y)$ is the ratio of the joint density of (X, Y) and the marginal density of Y. The marginal density of Y is given by $\sum_{u=0}^{r} \pi(u, y)$. Thus, for any fixed y, it follows from $\pi_1(x \mid y) = \pi(x, y)/\sum_{u=0}^{r} \pi(u, y)$ that

$$\pi_1(x \mid y) = \frac{\binom{r}{x} y^x (1-y)^{r-x}}{\sum_{u=0}^{r} \binom{r}{u} u(1-u)^{r-x}} = \binom{r}{x} y^x (1-y)^{r-x}$$

for $x = 0, 1, \ldots, r$. Hence $\pi_1(x \mid y)$ is the binomial(r, y) density for any fixed y. In the same way, we obtain from $\pi_2(y \mid x) = \pi(x, y)/\int_0^1 \pi(x, u)\, du$ that for any fixed x the conditional density $\pi_2(y \mid x)$ is given by

$$\pi_2(y \mid x) = \frac{y^{x+\alpha-1}(1-y)^{r-x+\beta-1}}{\int_0^1 u^{x+\alpha-1}(1-u)^{r-x+\beta-1}\, du} = y^{x+\alpha-1}(1-y)^{r-x+\beta-1}$$

for $0 \le y \le 1$. Hence $\pi_2(y \mid x)$ is the beta$(x + \alpha, r - x + \beta)$ density.

A Gibbs sequence $(x_0, y_0), (x_1, y_1), \ldots, (x_m, y_m)$ of length m is generated as follows by using the standard Gibbs sampler. Choose an integer x_0 between 0 and r. The other elements of the sequence are iteratively obtained by generating alternately a random draw y_j from the beta density $\pi_2(y \mid x_j)$ and a random draw x_{j+1} from the binomial density $\pi_1(x \mid y_j)$. Codes to simulate from the binomial density and the beta density are widely available. In our simulation we have generated $m = 250{,}000$ observations for the state vector (x, y). The first histogram in Figure 7.3 gives

Fig. 7.3   Simulated and exact histogram for $\pi_1(x)$.

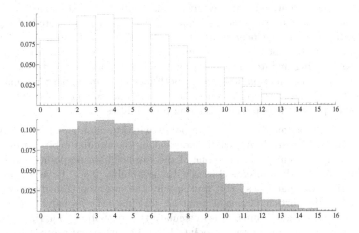

the simulated histogram for the marginal density $\pi_1(x)$ of the random variable X. By comparison, the second histogram gives the exact values of $\pi_1(x)$ (a direct computation of the proportionality constant for $\pi_1(x)$ is possible). The simulated histogram is based on the estimate

$$\hat{\pi}_1(x) = \frac{1}{m} \sum_{k=1}^{m} \binom{r}{x} y_k^x (1 - y_k)^{r-x}$$

for $\pi_1(x)$, using the explicit expression for $\pi_1(x \mid y)$. On basis of $\hat{\pi}_1(x)$, the estimates 5.35 and 11.20 are found for $\mathbb{E}[X]$ and $\text{var}(X)$, where the exact values of $\mathbb{E}[X]$ and $\text{var}(X)$ are 5.33 and 11.17.

7.6   Exercises

7.1 Two compartments A and B each contain r particles. Of these $2r$ particles, r are of type 1 and r are of type 2. Every unit of time, exactly at the same time, one particle from each compartment is randomly taken and switched to the other compartment. Formulate a Markov chain that describes the number of particles of type 1 in compartment A and specify the one-step transition probabilities.

7.2 Consider the following modification of Example 7.2. If the professor's driver's license is at his departure point, he takes it along with probability 0.75 if he is leaving his home and with probability 0.5 if he is leaving the university. Formulate a Markov chain that describes whether or not the professor has his driver's license with him during his ride and specify the one-step transition probabilities.

7.3 Every day on Rainbow Island, it is either sunny or rainy the entire day. The next day's weather depends on both the current weather and the previous day's weather. If the last two days were sunny, then the next day is sunny with probability 0.9. This probability is 0.45 if the last two days were rainy. If today is sunny and yesterday was rainy, then tomorrow will be sunny with probability 0.7. This probability is 0.5 if today is rainy and yesterday was sunny. Formulate a Markov chain that describes the weather on Rainbow Island and specify the one-step transition probabilities.

7.4 To increase a production system's reliability, two identical production machines have been connected in parallel. At any given time, only one of the machines is needed. At the end of the day, the machine that has been used is inspected. Regardless of how long the machine has been in use, the probability that the inspection leads to the conclusion that the machine requires maintenance is $1/10$. A revision takes exactly three days. During maintenance, the production is carried out by the other machine if it is available. Production stops when both machines are undergoing maintenance. Suppose that there are two mechanics. Formulate a Markov chain for this production system and specify the one-step transition probabilities.

7.5 In the same way as on a clock, twelve points with the numbers $1, \ldots, 12$ are evenly spaced on a circular orbit. A particle moves along the circular orbit. At every step, it jumps from its current position to one of the two neighboring points, clockwise with probability 0.45 and counterclockwise with probability 0.55. Give a Markov chain that describes the particle's orbit and specify the one-step transition probabilities. Suppose that the particle begins at point 12. Calculate the probability that it is back at point 12 after eight jumps. Also calculate this probability for the case that the particle starts at any of the twelve points $1, \ldots, 12$ with equal probability $\frac{1}{12}$.

7.6 **(a)** Consider Exercise 7.2 again. It is Sunday evening, and the professor's driver's license is at home. An unannounced traffic check has been planned for the coming Friday night at the university's exit. Calculate the probability that the professor will be fined on Friday.

(b) Consider Exercise 7.3 again. Suppose that today and yesterday have been rainy. What is the probability that it will be sunny five days from now, and what is the probability that it will be sunny five days from now and the day after that? What is the expected number of sunny days in the next 14 days? What is the long-run fraction of days that it is sunny?

7.7 In a closed system, the number of individuals present in a particular organism is constant and equal to 1000. Each individual is in one of the three stages 1, 2, and 3. At each time point, an individual's stage can change, or the individual can die. The place of an individual who dies is directly taken in by a new individual in stage 1. The individuals behave independently of one another. If an individual is now in stage 1, then at the next time, it will be in stage 1 with probability 0.1, in stage 2 with probability 0.7, in stage 3 with probability 0, and dead with probability 0.2. For an individual in stage 2, these probabilities are, respectively, 0, 0.5, 0.4, and 0.1, and for an individual in stage 3, they are 0, 0, 0.9, and 0.1. Suppose that there are currently 500 individuals in stage 1, 300 individuals in stage 2, and 200 individuals in stage 3. What is the expected number of individuals in stage i five time units from now for $i = 1, 2, 3$? What is the equilibrium distribution?

7.8 A training program consists of three consecutive parts, each taking one month. Of the beginning participants, on average 50% pass the first part right after the first month and continue on to the second part, 30% drop out during the first month, and 20% must repeat the first part. Of the last group, 70% pass the first part after the second attempt, and the remaining 30% drop out. Of the participants in the second part, 80% pass after the first attempt, 10% drop out, and the remaining 10% pass the second part after a second attempt. Every participant in the third part passes it. Calculate the probability that a participant successfully completes the program.

7.9 Calculate the probabilities that in 16, 32, 64, 128, or 256 throws of a die, the same number of dots is thrown three times in a row. What are the probabilities for throwing three sixes in a row?

7.10 A fair coin is tossed repeatedly, and two players bet on which combination of outcomes of three consecutive throws will appear first. Player A begins by choosing a combination and takes one of the four combinations HHH, HHT, HTH, and HTT. Player B parries HHH with THH, HHT with THH, HTH with HHT, and HTT with HHT (the first outcome in player B's combination is the opposite of the second outcome in player A's combination, and player B's last two outcomes are player A's first two). For each of the four cases, what is the probability that player B wins?

7.11 In Lotto 6/49, each ticket contains six different numbers from 1 to 49. A draw also consists of six different numbers from 1 to 49. A player receives a payout if several numbers on her ticket match drawn numbers, where the height of the payout increases with the number of matches. What is the probability that in the next 1240 draws of Lotto 6/49, there is a number that is not drawn in a continuous period of 72 draws? (This was a "record gap" in the UK National Lottery in November 2000.) Calculate an approximation for this probability by first using an absorbing Markov chain to calculate the exact value of the probability that in the next 1240 draws, a given number (for example, the number 7) is not drawn in 72 consecutive draws.

7.12 Joe Dalton desperately wants to raise his bankroll of $600 to $1,000 in order to pay his debts before midnight. He enters a casino to play European roulette. He decides to bet on red each time using bold play, that is, Joe bets either his entire bankroll or the amount needed to reach the target bankroll, whichever is smaller (a bet on red is won with probability $\frac{18}{37}$). Thus the stake is $200 if his bankroll is $200 or $800 and the stake is $400 if his bankroll is $400 or $600. What is the probability that Joe will reach his goal? What is the expected number of bets of Joe?

7.13 A famous TV show is The Tonight Show with Jimmy Fallon. In this show Jimmy plays the Egg Russian Roulette game with a guest of the show. The guest is always a celebrity from sports or film. The guest and Jimmy take turns picking an egg from a carton and smashing it on their heads. The carton contains a dozen eggs, four of which are raw and the rest are boiled. Neither Jimmy nor the guest knows which eggs are raw and which are boiled. The first person who has cracked two raw eggs on their head loses the game. The guest is the first to choose an egg. Use an absorbing Markov chain in order to calculate the probability that the guest will lose the game and the expected length of the game. Do the same for the case that the carton contains three raw eggs and nine boiled eggs.

7.14 In Exercise 7.2, calculate the long-run percentage of the number of rides that the professor has his driver's license with him. In Exercise 7.4, calculate the long-run fraction of the time that production is stopped.

7.15 A gambling machine is set up in such a way that a player who wins a game wins the next one with probability 0.25; if he loses the first game, the probability of winning the next one is 0.50. The player bets €1.00 per game and receives €2.50 if he wins. Is this a "fair" game?

7.16 Verify that a doubly stochastic Markov chain with a finite state space and no two disjoint closed sets has the discrete uniform distribution as equilibrium distribution. A Markov chain is called doubly stochastic if for every column j of the matrix (p_{ij}), the sum of all column elements is 1. Use this result to show the following: if the random variable S_n is the sum of the outcomes of n throws of a die, then for every integer $r \geq 2$, the probability that S_n is divisible by r converges to $1/r$ as $n \to \infty$. (*Hint:* Define the random variable X_n as $X_n = S_n$ (modulo r).)

7.17 When repeatedly tossing a coin, what are the long-run frequencies that three consecutive tosses show HTH and HTT, respectively?

7.18 Consider the (s, S) inventory model from Example 7.4. Suppose $s = 5$ and $S = 10$ and that the total demand in a week is Poisson distributed with expected value 4. For an infinitely long planning period, calculate the average on-hand inventory at the end of the week, the average order frequency, and the average demand lost per week.

7.19 Consider Exercise 7.3 again. The local entrepreneur Jerry Woodhouse has a restaurant on the island. On a sunny day, his turnover in euros is $N(\mu_1, \sigma_1^2)$ distributed with $\mu_1 = 1000$ and $\sigma_1 = 200$, while on a rainy day, his turnover is $N(\mu_2, \sigma_2^2)$ distributed with $\mu_2 = 500$ and $\sigma_2 = 75$. What is his long-run average turnover per day?

7.20 A control panel contains two components connected in parallel. Both components are switched on. The control panel functions properly as long as one of the components works. The components are identical and work independently of each other. Each component can break down. The failure rate of a component increases with the component's lifetime. The failure rate r_i gives the probability that a component breaks down next week if the component has worked the past i weeks. Past observations give the failure rates $r_0 = 0$, $r_1 = 0.05$, $r_2 = 0.07$, $r_3 = 0.12$, $r_4 = 0.25$, $r_5 = 0.50$, and $r_6 = 1$. Every failing component is replaced at the beginning of the following week. Furthermore, any six-week-old component is replaced. Suppose that the control panel breaking down costs 750 euros and that the replacement of a component costs 100 euros. Calculate the long-run fraction of the time in weeks that the control panel works and the long-run average cost per week.

7.21 Consider Markov chain $\{X_n\}$ with state space I and transition probabilities p_{ij}, $i, j \in I$. Assume that $\{X_n\}$ has equilibrium distribution π_i, $i \in I$. Let $A \subset I$. Show that

$$\sum_{i \in A} \sum_{j \in A^c} \pi_i p_{ij} = \sum_{i \in A} \sum_{j \in A^c} \pi_j p_{ji}.$$

7.22 Let $\{X_n\}$ be a finite-state Markov chain with one-step transition probabilities p_{ij}. Suppose that the Markov chain $\{X_n\}$ has no two or more disjoint closed sets. For fixed τ with $0 < \tau < 1$, define the perturbed Markov chain $\{\overline{X}_n\}$ by the one-step transition probabilities $\overline{p}_{ii} = 1 - \tau + \tau p_{ii}$ and $\overline{p}_{ij} = \tau p_{ij}$ for $j \neq i$. Explain why the Markov chain $\{\overline{X}_n\}$ is aperiodic. Verify that the Markov chain $\{\overline{X}_n\}$ has the same unique equilibrium distribution as the original Markov chain $\{X_n\}$.

7.23 In a small student city, there are three restaurants: an Italian restaurant, a Mexican restaurant, and a Thai restaurant. A student eating in the Italian restaurant on a given evening, will eat the next evening in the Italian restaurant with probability 0.10, in the Mexican restaurant with probability 0.35, in the Thai restaurant with probability 0.25, or at home with probability 0.30. The probabilities of switching are 0.4, 0.15, 0.25, and 0.2 when eating in the Mexican restaurant, 0.5, 0.15, 0.05, and 0.3 when eating in the Thai restaurant, and 0.40, 0.35, 0.25, and 0 when eating at home. What proportion of time a student will eat at home?

7.24 In a certain town, there are four entertainment venues. Both Linda and Bob are visiting every weekend one of these venues, independently of each other. Each of them visits the venue of the week before with probability 0.4 and chooses otherwise at random one of the other three venues. What is the long-run fraction of weekends that Linda and Bob visit a same venue and what is the limiting probability that they visit a same venue two weekends in a row?

7.25 In a long DNA sequence of a particular genome, it has been recorded how often each of the four bases A, C, G and T is followed by another base. The DNA sequence is modeled by a Markov chain with state space $I = \{A, C, G, T\}$ and one-step transition probabilities

from\to	A	C	G	T
A	0.340	0.214	0.296	0.150
C	0.193	0.230	0.345	0.232
G	0.200	0.248	0.271	0.281
T	0.240	0.243	0.215	0.302

What is the long-run frequency of the base A appearing? What is the long-run frequency of observing base A followed by another A?

7.26 Suppose that a conveyor belt is running at uniform speed and is transporting items on individual carriers equally spaced along the conveyor. Two work stations 1 and 2 are placed along the conveyor. Each work station can process only one item at a time and has no storage capacity. The processing time of an item at work station i has an exponential distribution with an expected value of $1/\mu_i$ time units, where $1/\mu_1 = 0.75$ and $1/\mu_2 = 1.25$. An item for processing arrives at fixed times $t = 0, 1, \ldots$. The item is lost when both work stations are occupied and is handled otherwise by an idle work station, where station 1 is the first choice when both stations are idle. What is the long-run fraction of items that are lost?

7.27 Simulate the probability distribution $\pi_1 = 0.2$ and $\pi_2 = 0.8$ with the proposal probabilities $\bar{p}(t \mid s) = 0.5$ for $s, t = 1, 2$ and see how quickly the simulated Markov chain converges.

7.28 The joint density of the continuous random variables X_1 and X_2 is proportional to

$$e^{-\frac{1}{2}(x_1^2 x_2^2 + x_1^2 + x_2^2 - 7x_1 - 7x_2)} \quad \text{for } -\infty < x_1, x_2 < \infty.$$

What are the expected value, the standard deviation and the marginal density of the random variable X_1? Use the Metropolis-Hastings algorithm with random walk chain sampling, where the increments of the random walk are generated from (Z_1, Z_2) with Z_1 and Z_2 independent $N(0, a^2)$ random variables. Experiment with several values of a (say, $a = 0.02, 0.2, 1,$ and 5) to see how the mixing in the Markov chain goes and what the average value of the acceptance probability is.

7.29 In an actuarial model the random vector (X, Y, N) has a trivariate density $\pi(x, y, n)$ that is proportional to

$$\binom{n}{x} y^{x+\alpha-1}(1-y)^{n-x+\beta-1} e^{-\lambda} \frac{\lambda^n}{n!}$$

for $x = 0, 1, \ldots, n$, $0 < y < 1$ and $n = 0, 1, \ldots$. The random variable N represents the number of policies in a portfolio, the random variable Y represents the claim probability for any policy and the random variable X represents the number of policies resulting in a claim. First verify that the univariate conditional density functions of X, Y and N are given by the binomial(n, y) density, the beta$(x + \alpha, n - x + \beta)$ density and the Poisson$(\lambda(1 - y))$ density shifted to the point x. Assuming the data $\alpha = 2$, $\beta = 8$ and $\lambda = 50$, use the Gibbs sampler to estimate the expected value, the standard deviation and the marginal density of the random variable X.

References

1. S. M. Ross, *Introduction to Probability Models*, 10th ed., Academic Press, New York, 2010.

2. H. C. Tijms, *Understanding Probability*, 3rd ed., Cambridge University Press, Cambridge, UK, 2012.

Chapter 8

Continuous-Time Markov Chains

8.1 Introduction

In discrete-time Markov chains the intervals between transitions are fixed. In continuous-time Markov chains, the state can change at any time, and the intervals between the state transitions have an exponential distribution. Where the one-step transition probability is a building block for the discrete-time Markov model, the so-called infinitesimal transition rate will be the building block for the continuous-time Markov model. The continuous-time Markov chain model is one of the most important models from probability theory and has numerous applications, among others in queueing, inventory and reliability problems. Many real-world problems can be modeled as a continuous-time Markov chain.

Example 8.1 (Filling station). At a filling station, a combustible product is stored in a separate tank. The arrivals of customers asking for the product follow a Poisson process with rate λ, and every client asks for one unit of the product.[1] Any demand that occurs when the station is out of stock is lost. Opportunities to replenish the stock occur at intervals generated by a Poisson process with rate μ. By assumption, this process is independent of the demand process and the evolution of the inventory level. For security reasons, the stock is only replenished when the tank is empty. In that case, the stock is replenished with Q units of the product. What is the average inventory level in the tank? What fraction of the time is the system out of stock?

In this example, the stochastic process that describes the evolution of the inventory level has the property that the intervals between changes in the inventory level are exponentially distributed. After all, in a Poisson process, the intervals between two consecutive events are exponentially distributed. The inventory level can change at any time. The stochastic process that describes the evolution of the inventory level therefore has a continuous time parameter.

---

[1]In a Poisson arrival process with rate λ, the customers come in one by one with independent interarrival times that are exponentially distributed with expected value $1/\lambda$. Appendix C provides details on the Poisson process.

8.2   The Continuous-Time Markov Chain

We give a constructive definition of a continuous-time Markov chain via the *Markov jump chain.*

Definition 8.1 (Markov jump chain). *Consider a stochastic dynamic system with a discrete state space I. The system jumps from state to state according to the following rules:*

(1) The time spent in state i is exponentially distributed with expected value $1/\nu_i$.
(2) When the process leaves state i, it jumps to another *state j with a given probability p_{ij} that is independent of the duration of the stay in state i.*

The convention is that $p_{ii} = 0$ for all i; in other words, when state i is left, it is left for a different state. This is a natural convention for applications and is not restrictive (a geometric sum of exponentially distributed times is again exponentially distributed). The following therefore applies to the state transitions:

$$\sum_{j \neq i} p_{ij} = 1 \qquad \text{for all } i \in I.$$

Moreover, it is understood that the number of state transitions during any finite time interval is finite with probability 1. This assumption is automatically satisfied when I is finite. Now, for every $t \geq 0$, define the random variable $X(t)$ as

$$X(t) = \text{state of the system at time } t,$$

where at the transition time, the state right after the jump is taken. The stochastic process $\{X(t), t \geq 0\}$ is then the prototype of a continuous-time Markov chain.

Definition 8.2 (Continuous-time Markov chain). *The stochastic process $\{X(t),\ t \geq 0\}$ is called a* continuous-time Markov chain *if it has a finite or countable state space I and we have*

$$P(X(t+s) = j | X(s) = i, X(u) = x(u), 0 \leq u < s)) = P(X(t+s) = j | X(s) = i)$$

for all $s, t \geq 0$, for every $x(u), 0 \leq u < s$, $i, j \in I$.

In words, this definition comes down to the fact that only the current state of the process determines the future evolution of the state and that how the current state was reached is not relevant. This is the Markov property. It suffices to note that under weak regularity conditions, it can be proved that every continuous-time Markov chain can be represented as Markov jump chain.

Example 8.1 (Continued). If we return to Example 8.1 and, for every $t \geq 0$, define

$$X(t) = \text{inventory level at time } t,$$

then the stochastic process $\{X(t), t \geq 0\}$ is a continuous-time Markov chain with state space $I = \{0, 1, \ldots, Q\}$. Indeed, the time spent in state i is exponentially distributed, with

$$\nu_i = \begin{cases} \lambda \text{ for } i = 1, 2, \ldots, Q, \\ \mu \text{ for } i = 0. \end{cases}$$

The transition probabilities p_{ij} also have a particularly simple form in this example:

$$p_{i,i-1} = 1 \text{ for } i = 1, 2, \ldots, Q,$$
$$p_{0Q} = 1, \text{ otherwise } p_{ij} = 0.$$

In applications of continuous-time Markov chains, in general, one does not work with the Markov jump chain representation but rather with the infinitesimal transition rates that concern the system's behavior over a very small time interval. To define this concept, we need the basic result that the exponential probability density has a *constant* failure rate. By this we mean the following: suppose that the life span of a machine has an exponential probability density with expected value $\frac{1}{\mu}$; then (see Appendix C)

$\mathbb{P}(\text{machine of age } t_0 \text{ fails in the upcoming period of time of length } \Delta t)$
$= \mu \Delta t + o(\Delta t) \qquad \text{for } \Delta t \text{ very small,}$

regardless of the value t_0 of the machine's current age. Here, $o(\Delta t)$ indicates a rest term that is negligibly small compared to Δt if Δt itself is very small (for example, $(\Delta t)^2$ is $o(\Delta t)$). A machine with an exponentially distributed life span therefore has a *constant* failure rate μ ("used is as good as new"). The exponential probability distribution has no memory. This memoryless property also holds for the Poisson process that is discussed extensively in Appendix C. For the purpose of this chapter, it suffices to note that the customers' arrival process is called a Poisson process with arrival rate λ if the customers arrive one by one and the interarrival times are independent random variables that have a common exponential probability density with expected value $1/\lambda$. This process has the following property: during a very short time interval of length Δt, either exactly one customer arrives (with probability $\lambda \Delta t + o(\Delta t)$), or no customer arrives (with probability $1 - \lambda \Delta + o(\Delta t)$); the probability of two or more arrivals is negligible with respect to Δt. The Poisson process also has no memory: the probability of a customer's arrival during an upcoming time interval $(t, t + \Delta t)$ does not depend on how long before the time t the last customer arrived.

Memoryless property

A random variable A with distribution $\mathbb{P}(A \leq a)$ has the memoryless property if for all t, h,

$$\mathbb{P}(A > t + h \mid A \geq t) = \mathbb{P}(A > h). \tag{8.1}$$

The exponentially distribution is the only continuous distribution that has the memoryless property. We will only show that the exponential distribution indeed satisfies (8.1). If A is exponentially distributed with mean $1/\lambda$ then

$$\mathbb{P}(A > t + h \mid A \geq t) = \frac{\mathbb{P}(A > t + h, A \geq t)}{\mathbb{P}(A > t)} = \frac{\mathbb{P}(A > t + h)}{\mathbb{P}(A \geq t)} = \frac{e^{-\lambda(t+h)}}{e^{-\lambda t}} = e^{-\lambda h}$$
$$= \mathbb{P}(A > h).$$

Infinitesimal Transition Rates

Now that we know about the constant failure rate associated with the exponential distribution, we return to the continuous-time Markov chain. Suppose that at time t, the process is in a given state i. The residual time in state i has an exponential probability distribution with expected value $1/\nu_i$ because of the memoryless property of the exponential distribution. The probability that the process leaves state i during the upcoming time interval $(t, t + \Delta t)$ is $\nu_i \Delta t + o(\Delta t)$ for Δt small by the failure rate representation of the exponential distribution. If the process leaves state i, then it jumps to state j with $j \neq i$ with probability p_{ij}. So the probability that during the upcoming time interval $(t, t + \Delta t)$, the process jumps to a new state j is $(\nu_i \Delta t) \times p_{ij} + o(\Delta t)$ for Δt small. The probability that the process remains in state i during the upcoming time interval $(t, t + \Delta t)$ is $1 - \nu_i \Delta t + o(\Delta t)$ for Δt small. Summarized as formulas, this is

$$\mathbb{P}(X(t + \Delta t) = j \mid X(t) = i) = (\nu_i p_{ij})\Delta t + o(\Delta t) \qquad \text{for } j \neq i$$

and

$$\mathbb{P}(X(t + \Delta t) = i \mid X(t) = i) = 1 - \nu_i \Delta t + o(\Delta t)$$

provided that Δt is small. Note that the probability of two or more jumps during a time interval of length Δt is of order $(\Delta t)^2$ and therefore $o(\Delta t)$ if Δt is small. From here on, this will be used implicitly. This representation of the continuous-time Markov chain $\{X(t), t \geq 0\}$ leads to the definition of the so-called *infinitesimal transition rates* q_{ij}:

$$q_{ij} = \nu_i p_{ij} \qquad \text{for } i, j \in I \text{ with } i \neq j.$$

The q_{ij} therefore have the following interpretation:

$\mathbb{P}$(in the upcoming short period of length Δt, the process jumps to
a new state j if the process is now in state i)
$= q_{ij}\Delta t + o(\Delta t) \qquad \text{for } j \neq i.$

The continuous-time Markov chain is uniquely determined by the infinitesimal transition rates. Note that it follows from the definition of the q_{ij} that

$$\nu_i = \sum_{j \neq i} q_{ij} \qquad \text{for all } i \in I.$$

In continuous-time Markov chains, the infinitesimal transition rates play the role of the one-step transition probabilities in discrete-time Markov chains. In applications, the q_{ij} are usually not calculated by first determining the ν_i and p_{ij}; instead, they are calculated directly from the interaction of two or more "Poisson-like" processes.[2]

Remark 8.1 (Equivalence of representations). The characterization of the continuous-time Markov chain via the infinitesimal transition rates and the Markov jump chain are equivalent. Starting from the Markov jump chain with exponentially distributed time with mean $1/\nu_i$ and probabilities p_{ij} to jump to state i we obtain the infinitesimal transition rates as $q_{ij} = \nu_i p_{ij}$. Conversely, starting from the infinitesimal transition rates q_{ij} we obtain the mean time in state i as $\nu_i = \sum_{j \neq i} q_{ij}$ and the transition probabilities as $p_{ij} = q_{ij}/\nu_i$.

The infinitesimal transition rates q_{ij} can also be interpreted as the rates of the exponentially distributed time before the Markov chain jumps from state i to state j. You can think of exponential random variables T_{ij} that are in competition for the sojourn time in state i. The next state is then determined by the minimum of these random variables. This minimum has an exponential distribution with mean $[\sum_{j \neq i} q_{ij}]^{-1} = 1/\nu_i$ and the probability that the transition to state j occurs is the probability $\mathbb{P}(T_{ij} = \min_{k \neq i} T_{ik}) = q_{ij}/\sum_{k \neq i} q_{ik} = p_{ij}$.

Example 8.1 (Continued). In this example, the inventory process $\{X(t)\}$ arises from the interaction between a Poisson demand process and a Poisson replenishment process. During a small time interval, the only possible state transitions are $i \to i-1$ for $i \neq 0$ and $i \to Q$ for $i = 0$. It follows from the Δt-representation of the Poisson process that

$$\mathbb{P}(X(t + \Delta t) = i - 1 \mid X(t) = i)$$
$$= \mathbb{P}(\text{a customer arrives during the interval } (t, t + \Delta t))$$
$$= \lambda \Delta t + o(\Delta t) \quad \text{for } i = 1, 2, \ldots, Q$$

$$\mathbb{P}(X(t + \Delta t) = Q \mid X(t) = 0)$$
$$= \mathbb{P}(\text{a replenishment takes place during the interval } (t, t + \Delta t))$$
$$= \mu \Delta t + o(\Delta t).$$

This gives

$$q_{i,i-1} = \lambda \text{ for } i = 1, 2, \ldots, Q$$
$$q_{0Q} = \mu, \quad \text{otherwise } q_{ij} = 0.$$

A handy tool to indicate the state transitions and the associated transition rates is the so-called *transition rate diagram*. Figure 8.1 gives the transition rate diagram for the continuous-time Markov chain $\{X(t)\}$ from Example 8.1. This transition rate diagram speaks for itself. In Section 8.7, we answer the questions asked in Example 8.1. To do this, we first develop the necessary theory in Sections 8.3 and 8.5.

---

[2]An important fact here is that the minimum of two independent, exponentially distributed random variables is again exponentially distributed.

Fig. 8.1   Transition rate diagram for the inventory process.

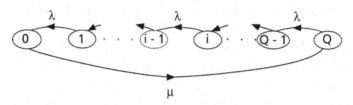

In applications of continuous-time Markov chains, the trick is to choose the right state variable(s) and set up the transition rate diagram. The calculations on the model are then "simply" a matter of applying the formulas that we deduce in the next sections. We conclude this section with another modeling problem.

Example 8.2. The arrivals of ships for unloading at a container terminal follow a Poisson process with rate λ. For practical purposes, we may assume that there is no limitation on the number of docks. There is a single container crane in use for unloading, which can handle only one ship at a time. The unloading time of each ship is exponentially distributed with expected value $1/\mu$. The container crane can break down while working on a ship, in which case the unloading stops until the crane has been fixed. It is assumed that the container crane can only malfunction while it is in operation. The length of time the crane works without breaking down is exponentially distributed with expected value $1/\delta$, while the time needed to resolve a malfunction is exponentially distributed with expected value $1/\beta$. What is the average number of ships in the port, and what is the average time a ship spends in the port?

To model this problem as a continuous-time Markov chain, we need two state variables:

$$X_1(t) = \text{number of ships in the port at time } t$$

and

$$X_2(t) = \begin{cases} 1 \text{ if the container crane is available at time } t, \\ 0 \text{ if the container crane is in repair at time } t. \end{cases}$$

The two-dimensional process $\{(X_1(t), X_2(t), t \geq 0\}$ is a continuous-time Markov chain (verify!). Its state space is given by

$$I = \{(i, 1) \mid i = 0, 1, \ldots\} \cup \{(i, 0) \mid i = 1, 2, \ldots\}.$$

Note that there is no state $(0, 0)$ (why?). The state space with the possible state transitions is shown schematically in Figure 8.2. The transition rate diagram shows the states and the transition rates. Let us explain some of the latter. For example, it follows from

$$\mathbb{P}(X(t + \Delta t) = (i, 1) \mid X(t) = (i, 0))$$
$$= \mathbb{P}(\text{the repair taking place at time } t \text{ ends during the interval } (t, t + \Delta t)$$
$$\text{and no ship arrives during that interval})$$
$$= \beta \Delta t (1 - \lambda \Delta t) + o(\Delta t) = \beta \Delta t + o(\Delta t)$$

Fig. 8.2 The transition rate diagram of the unloading problem.

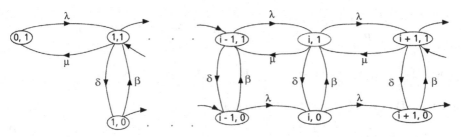

that $q_{(i,0)(i,1)} = \beta$ for $i = 1, 2, \ldots$. It follows from

$$\mathbb{P}(X(t + \Delta t) = (i + 1, 0) \mid X(t) = (i, 0))$$
$$= \mathbb{P}(\text{a new ship arrives during the interval } (t, t + \Delta t), \text{ and the repair taking}$$
$$\quad \text{place at time } t \text{ is not finished during that interval})$$
$$= \lambda \Delta t (1 - \beta \Delta t) + o(\Delta t) = \lambda \Delta t + o(\Delta t)$$

that $q_{(i,0)(i+1,0)} = \lambda$ for $i = 1, 2 \ldots$. The other transition rates shown in Figure 8.2 are found likewise. Above, instead of $\mathbb{P}$(the repair taking place at time t ends during the interval $(t, t + \Delta t)$ and no ship arrives during that interval), we could also have simply written $\mathbb{P}$(the repair taking place at time t ends during the interval $(t, t + \Delta t)$); up to a negligible term of order $(\Delta t)^2$, the two probabilities are the same (the probability that during a small time interval of length Δt both a repair ends and a ship arrives is of order $(\Delta t)^2$ and therefore negligible). We answer the questions from Example 8.2 in Section 8.7 using the equilibrium probabilities of the Markov chain.

8.3 Transient Probabilities

Consider a continuous-time Markov chain with state space I and infinitesimal transition rates q_{ij} for $i, j \in I$ with $j \neq i$. As observed in the previous section,

$$\nu_i = \sum_{j \neq i} q_{ij}$$

gives the parameter of the exponentially distributed time in state i. In this section, we assume that the state space I is *finite*. The *time-dependent* or *transient* probabilities $p_{ij}(t)$ are defined by

$$p_{ij}(t) = \mathbb{P}(X(t) = j \mid X(0) = i) \quad \text{for } i, j \in I \text{ and } t \geq 0.$$

In words, $p_{ij}(t)$ is the probability that t units of time from now, the process will be in state j given that it is now in state i. For a fixed i, the probabilities $p_{ij}(t)$ for $j \in I$ are calculated by solving a system of linear differential equations (there exists standard software for this). The equations in the following theorem are known as Kolmogorov's forward differential equations.

Theorem 8.1. *For every $i \in I$, we have*

$$\frac{dp_{ij}(t)}{dt} = \sum_{k \neq j} p_{ik}(t)q_{kj} - p_{ij}(t)\nu_j \quad \text{for } j \in I \text{ and } t > 0. \tag{8.2}$$

Proof. The derivation is instructive. For small Δt, we have

$$p_{ij}(t + \Delta t) = \mathbb{P}(X(t + \Delta t) = j \mid X(0) = i)$$
$$= \sum_{k \in I} \mathbb{P}(X(t + \Delta t) = j \mid X(t) = k, \ X(0) = i)\mathbb{P}(X(t) = k \mid X(0) = i)$$
$$= \sum_{k \in I} \mathbb{P}(X(t + \Delta t) = j \mid X(t) = k)p_{ik}(t),$$

where the second equality uses the law of total probability and the last equality uses that only the last observed state is relevant for the process's further course. Next, we split the summation over the k with $k \neq j$ and $k = j$ and apply the definition of the infinitesimal transition rates. This gives

$$p_{ij}(t + \Delta t) = \sum_{k \neq j} q_{kj} \Delta t\, p_{ik}(t) + (1 - \nu_j \Delta t)p_{ij}(t) + o(\Delta t),$$

that is,

$$\frac{p_{ij}(t + \Delta t) - p_{ij}(t)}{\Delta t} = \sum_{k \neq j} q_{kj}p_{ik}(t) - \nu_j p_{ij}(t) + \frac{o(\Delta t)}{\Delta t}.$$

If we let Δt go to zero, the desired result follows. $\square$

Just like the matrix notation for the one-step transition probabilities $\mathbf{P}$ of the discrete-time Markov chain with finite state space I, we may introduce the generator matrix $\mathbf{Q}$ with entries q_{ij} for the continuous-time Markov chain. Observe that $\mathbf{Q}$ has nonnegative entries q_{ij} for $i \neq j$, negative diagonal elements $-\nu_i = -\sum_{j \neq i} q_{ij}$ and row sums zero. With $\mathbf{P}(t)$ the matrix with entries $p_{ij}(t)$ the Kolmogorov forward equations (8.2) may then be written as

$$\frac{d\mathbf{P}(t)}{dt} = \mathbf{P}(t)\mathbf{Q} \quad t > 0. \tag{8.3}$$

Example 8.3 (On-off process). Suppose that a machine alternates between on (production) and off (repair). The on-times are independent random variables with an exponential distribution with expected value $1/\lambda$, and the off-times are independent random variables with an exponential distribution with expected value $1/\mu$. Moreover, the on-times are independent of the off-times. At time 0, the machine is on. What is the transient probability distribution of the machine's state?

The machine's state at time t is defined by

$$X(t) = \begin{cases} 1 \text{ if the machine is on at time } t, \\ 0 \text{ otherwise.} \end{cases}$$

The stochastic process $\{X(t), t \geq 0\}$ is a continuous-time Markov chain with state space $I = \{0, 1\}$. Figure 8.3 shows the transition rate diagram, with $q_{01} = \mu$ and $q_{10} = \lambda$.

Fig. 8.3   The transition rate diagram for the on-off process.

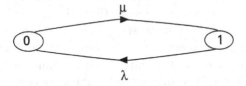

For the initial state $i = 1$, the differential equations from Theorem 8.1 are:

$$p'_{10}(t) = \lambda p_{11}(t) - \mu p_{10}(t),$$
$$p'_{11}(t) = \mu p_{10}(t) - \lambda p_{11}(t).$$

This is a simple system of differential equations that can be solved explicitly. Verify that the solution is given by

$$p_{10}(t) = \frac{\lambda}{\mu + \lambda} - \frac{\lambda}{\mu + \lambda} e^{-(\mu+\lambda)t} \qquad \text{for } t \geq 0,$$

$$p_{11}(t) = \frac{\mu}{\mu + \lambda} + \frac{\lambda}{\mu + \lambda} e^{-(\mu+\lambda)t} \qquad \text{for } t \geq 0.$$

Example 8.4 (Poisson process). Suppose customers arrive to a system with exponentially distributed interarrival times with mean $1/\lambda$. Let $X(t)$ denote the number of arrivals at time t, starting with $X(0) = 0$. The stochastic process $X(t)$ is a continuous-time Markov chain with state space $I = \{0, 1, \ldots\}$ and infinitesimal transition rates $q_{i,i+1} = \lambda$, $q_{ij} = 0$ for $j \neq i$ otherwise. It is a Poisson process, see Appendix C. The distribution $\mathbb{P}(X(t) = j)$ may be obtained from Kolmogorov's forward differential equations (8.2)

$$p'_{0,j}(t) = p_{0,j-1}(t)\lambda - p_{0,j}(t)\lambda,$$

with boundary condition $\mathbb{P}(X(0) = 0) = 1$. The solution is (verify!)

$$\mathbb{P}(X(t) = j) = p_{0j}(t) = \frac{(\lambda t)^j}{j!} e^{-\lambda t}, \quad t \geq 0.$$

8.4   The Uniformization Method

The Markov jump chain with state space I introduced in Definition 8.1 leaves state i after an exponentially distributed time with mean $1/\nu_i$ and then jumps to another state j $(j \neq i)$ with probability p_{ij}. Letting X_n denote the state of the process just after the nth state transition, the discrete-time stochastic process $\{X_n\}$ is called the *embedded Markov chain* with one-step transition probabilities p_{ij}. If $\nu_i = \nu$ for all i, the transition epochs are generated by a Poisson process with rate ν, and an expression for $p_{ij}(t)$ is directly obtained by conditioning on the number of Poisson events up to time t and using the n-step transition probabilities of the embedded Markov chain $\{X_n\}$. However, the leaving rates ν_i are in general not identical. Fortunately, there is a simple trick for reducing the case of non-identical leaving rates to the case of identical leaving rates. The uniformization method transforms the original continuous-time Markov chain with non-identical leaving rates into an equivalent stochastic process in which the transition epochs are generated by a Poisson process at a *uniform* rate. However, to achieve this, the discrete-time Markov chain describing the state transitions in the transformed process has to allow for self-transitions leaving the state of the process unchanged. The uniformization method is a powerful computational tool to solve Kolmogorov's equations for the time-dependent state probabilities and to calculate first passage time probabilities.

To formulate the uniformization method, choose a finite number ν with

$$\nu \geq \nu_i, \quad i \in I.$$

Define now $\{\bar{X}_n\}$ as the discrete-time Markov chain whose one-step transition probabilities $\bar{p}_{ij}$ are given by

$$\bar{p}_{ij} = \begin{cases} \frac{\nu_i}{\nu} p_{ij}, & j \neq i, \\ 1 - \frac{\nu_i}{\nu}, & j = i, \end{cases}$$

for any $i \in I$. Let $\{N(t), t \geq 0\}$ be a Poisson process with rate ν such that the process is independent of the discrete-time Markov chain $\{\bar{X}_n\}$. Define now the continuous-time stochastic process $\{\bar{X}(t), t \geq 0\}$ by

$$\bar{X}(t) = \bar{X}_{N(t)}, \quad t \geq 0. \tag{8.4}$$

In other words, the process $\{\bar{X}(t)\}$ makes state transitions at epochs generated by a Poisson process with rate ν and the state transitions are governed by the discrete-time Markov chain $\{\bar{X}_n\}$ with one-step transition probabilities $\bar{p}_{ij}$. Each time the Markov chain $\{\bar{X}_n\}$ is in state i, the next transition is the same as in the Markov chain $\{X_n\}$ with probability ν_i/ν and is a self-transition with probability $1-\nu_i/\nu$. The transitions out of state i are in fact delayed by a time factor of ν/ν_i, while the time itself until a state transition from state i is condensed by a factor of ν_i/ν. This heuristically explains why the continuous-time process $\{\bar{X}(t)\}$ is probabilistically identical to the original continuous-time Markov chain $\{X(t)\}$.

Another heuristic way to see that the two processes are identical is as follows. For any $i, j \in I$ with $j \neq i$

$$\mathbb{P}(\bar{X}(t + \Delta t) = j \mid \bar{X}(t) = i) = \nu \Delta t \times \bar{p}_{ij} + o(\Delta t)$$
$$= \nu_i \Delta t \times p_{ij} + o(\Delta t) = q_{ij} \Delta t + o(\Delta t)$$
$$= \mathbb{P}(X(t + \Delta t) = j \mid X(t) = i) \quad \text{for } \Delta t \to 0.$$

In the next theorem we give a formal proof that the two processes $\{X(t)\}$ and $\{\bar{X}(t)\}$ are probabilistically equivalent.

Theorem 8.2. *Suppose that the continuous-time Markov chain $\{X(t)\}$ is ergodic and ν exists such that $\nu \geq \nu_i$ for all $i \in I$. Then*

$$p_{ij}(t) = \mathbb{P}(\bar{X}(t) = j \mid \bar{X}(0) = i), \quad i, j \in I \text{ and } t > 0.$$

Proof. With $\mathbf{P}(t)$ the matrix with entries $p_{ij}(t)$, and $\mathbf{Q}$ the generator matrix, the Kolmogorov forward differential equations can be written as

$$\frac{\mathbf{P}(t)}{dt} = \mathbf{P}(t)\mathbf{Q}, \quad t > 0,$$

see (8.3). It is left to the reader to verify that the solution of this system of differential equations is given by

$$\mathbf{P}(t) = e^{t\mathbf{Q}} = \sum_{n=0}^{\infty} \frac{t^n}{n!} \mathbf{Q}^n, \quad t \geq 0. \tag{8.5}$$

The matrix $\bar{\mathbf{P}} = (\bar{p}_{ij})$, $i, j \in I$, can be written as $\bar{\mathbf{P}} = \mathbf{Q}/\nu + \mathbf{I}$, where $\mathbf{I}$ is the identity matrix. Thus

$$\mathbf{P}(t) = e^{t\mathbf{Q}} = e^{\nu t(\bar{\mathbf{P}} - \mathbf{I})} = e^{\nu t \bar{\mathbf{P}}} e^{-\nu t \mathbf{I}} = e^{-\nu t} e^{\nu t \bar{\mathbf{P}}} = \sum_{n=0}^{\infty} e^{-\nu t} \frac{(\nu t)^n}{n!} \bar{\mathbf{P}}^n.$$

On the other hand, by conditioning on the number of Poisson events up to time t in the $\{\bar{X}(t)\}$ process, we have

$$\mathbb{P}(\bar{X}(t) = j \mid \bar{X}(0) = i) = \sum_{n=0}^{\infty} e^{-\nu t} \frac{(\nu t)^n}{n!} \bar{p}_{ij}^{(n)},$$

where $\bar{p}_{ij}^{(n)}$ is the n-step transition probability of the discrete-time Markov chain $\{\bar{X}_n\}$. Together the latter two equations yield the desired result. $\quad\square$

Corollary 8.1. *The probabilities $p_{ij}(t)$ are given by*

$$p_{ij}(t) = \sum_{n=0}^{\infty} e^{-\nu t} \frac{(\nu t)^n}{n!} \bar{p}_{ij}^{(n)}, \quad i, j \in I \text{ and } t > 0, \tag{8.6}$$

where the probabilities $\bar{p}_{ij}^{(n)}$ can be recursively computed from

$$\bar{p}_{ij}^{(n)} = \sum_{k \in I} \bar{p}_{ik}^{(n-1)} \bar{p}_{kj}, \quad n = 1, 2, \ldots \tag{8.7}$$

starting with $\bar{p}_{ii}^{(0)} = 1$ and $\bar{p}_{ij}^{(0)} = 0$ for $j \neq i$.

This probabilistic result is extremely useful for computational purposes. The series in (8.6) converges much faster than the series expansion (8.5). The computations required by (8.6) are simple and transparent. For fixed $t > 0$ the infinite series can be truncated beforehand, since

$$\sum_{n=M}^{\infty} e^{-\nu t} \frac{(\nu t)^n}{n!} \bar{p}_{ij}^{(n)} \le \sum_{n=M}^{\infty} e^{-\nu t} \frac{(\nu t)^n}{n!}.$$

Example 8.5 (The Hubble space telescope). The Hubble space telescope is an astronomical observatory in space. It carries a variety of instruments including six gyroscopes to ensure stability of the telescope. The six gyroscopes are arranged in such a way that any three gyroscopes can keep the telescope operating with full accuracy. The operating times of the gyroscopes are independent of each other and have an exponential distribution with failure rate λ. Upon a fourth gyroscope failure, the telescope turns itself in a sleep mode. In the sleep mode further observations by the telescope are suspended. It requires an exponential time with mean $1/\mu$ to turn the telescope into the sleep mode. Once the telescope is into the sleep mode, the base station on earth receives a sleep signal. A shuttle mission to the telescope is next prepared. It takes an exponential time with mean $1/\eta$ before the repair crew arrives at the telescope and has repaired the stabilising unit with the gyroscopes. In the mean time the other two gyroscopes may fail. If the last gyroscope fails, a crash destroying the telescope will be inevitable. What is the probability that the telescope will crash in the next T years?

This problem can be analysed by a continuous-time Markov chain with an absorbing state. The transition diagram is given in Figure 8.4. The state labeled as the crash state is the absorbing state. The crash state is an absorbing state. Taking the value zero for the rate out of this state, we can apply the uniformization method to compute the first-passage time probability

$$Q(T) = \mathbb{P}(\text{ no crash will occur in the next } T \text{ years}$$
$$\text{when currently all six gyroscopes are working }).$$

Fig. 8.4   The transition rate diagram for the telescope.

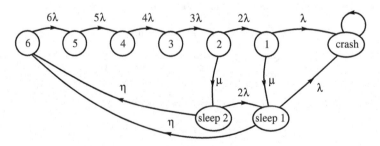

Taking one year as time unit, consider the numerical example with the data

$$\lambda = 0.1, \ \mu = 100 \text{ and } \eta = 5.$$

The uniformization method is applied with the choice $\nu = 100$ for the uniformized leaving rate ν (the value 0 is taken for the leaving rate of the crash state). The calculations yield the value 0.000504 for the probability $1 - Q(T)$ that a crash will occur in the next $T = 10$ years. Similarly, one can calculate that with probability 0.3901 the sleep mode will not be reached in the next 10 years. In other words, the probability of no shuttle mission in the next 10 years equals 0.3901. However, if one wishes to calculate the probability distribution of the number of required shuttle missions in the next 10 years, one must use the Markov reward model with lump rewards (assume a lump reward of 1 each time the process jumps from either state 2 or state 1 to the sleep mode). The interested reader can find a computational approach for the transient Markov reward model in Section 4.6 of reference 2 to this chapter.

8.5 Equilibrium Probabilities

While studying discrete-time Markov chains, we already saw that an accessibility condition is needed to guarantee that the Markov chain's equilibrium distribution does not depend on the initial state. For the continuous-time Markov chain $\{X(t), t \geq 0\}$, we make a similar assumption.

Assumption 8.1. There is a state r such that for any initial state i, the probability is 1 that there will ultimately be a transition to state r, and the expected time needed to return from state r to itself is finite.

This assumption is satisfied in almost all practical applications in operations research. Under this assumption, for every $j \in I$, the limiting probability

$$\pi_j = \lim_{t \to \infty} p_{ij}(t)$$

exists and is independent of the initial state i. The reason the ordinary limit always exists is that periodicity problems do not occur in continuous-time Markov chains because the times between state transitions have a continuous distribution. An irreducible continuous-time Markov chain satisfying Assumption 8.1 is called ergodic. The limiting probabilities for an ergodic Markov chain can be calculated by solving a system of linear equations.

Theorem 8.3. *Let the Markov chain be ergodic, then the limiting probabilities π_j for $j \in I$ are the unique solution to the linear equations*

$$\nu_j \pi_j = \sum_{k \neq j} \pi_k q_{kj} \quad for \ j \in I, \tag{8.8}$$

$$\sum_{j \in I} \pi_j = 1. \tag{8.9}$$

The equations (8.8) are called the *balance equations* or *global balance equations*, while (8.9) is called the *normalizing equation*.[3] These equations characterize the stationary distribution that for an ergodic Markov chain coincides with the limiting distribution. Later, we will give a physical interpretation of the balance equations that makes them simple to remember. We will then also see why the limiting probabilities π_j for $j \in I$ are also often called the *equilibrium probabilities*. We do not prove Theorem 8.3 in its full generality. For the case that I is finite, we only show that it is plausible that the π_j satisfy the equations. For this, we take $t \to \infty$ in the differential equations for the transient probabilities $p_{ij}(t)$ in Theorem 8.1. If we use that $p_{ij}(t) \to \pi_j$ and $p'_{ij}(t) \to 0$ as $t \to \infty$, it follows that $0 = \sum_{k \neq j} \pi_k q_{kj} - \pi_j \nu_j$ for all $j \in I$. The normalizing equation follows by taking $t \to \infty$ in $\sum_{j \in I} p_{ij}(t) = 1$.

Interpretation of the Balance Equations

We first give an intuitive interpretation of the equilibrium probability π_j. In words, it is the probability that an outsider who enters the system finds it in state j when the process has been running for a long time and the outsider has no knowledge of the past evolution of the system's state. In short, if the system has reached steady state, then the system is in state j with probability π_j. For applications in operations research, the following interpretation of the equilibrium probability π_j is often more useful:

actual long-run fraction of the time that the process is in state j

$$= \pi_j \quad \text{with probability 1.}$$

More precisely, defining the indicator variable $I_j(t)$ for $t \geq 0$ by

$$I_j(t) = \begin{cases} 1 & \text{if } X(t) = j, \\ 0 & \text{otherwise,} \end{cases}$$

then

$$\lim_{t \to \infty} \frac{1}{t} \int_0^t I_j(u) du = \pi_j \quad \text{with probability 1.}$$

We do not prove this strong law of large numbers. We can, however, easily show the weaker result that π_j is equal to the expected long-run fraction of the time that the process is in state j. For any initial state $X(0) = i$, we have

$$\mathbb{E}[I_j(t)] = 1 \times \mathbb{P}(I_j(t) = 1 \mid X(0) = i) + 0 \times \mathbb{P}(I_j(t) = 0 \mid X(0) = i)$$
$$= 1 \times \mathbb{P}(X(t) = j \mid X(0) = i) = p_{ij}(t),$$

so that

$$\lim_{t \to \infty} \frac{1}{t} \mathbb{E}\left[\int_0^t I_j(u) du \right] = \lim_{t \to \infty} \frac{1}{t} \int_0^t \mathbb{E}[I_j(u)] du$$

---

[3]When solving the equations numerically, any one of the balance equations can be left out to obtain a square system of linear equations.

$$= \lim_{t \to \infty} \frac{1}{t} \int_0^t p_{ij}(u)du = \pi_j$$

because the ordinary limit of $p_{ij}(t)$ as $t \to \infty$ is equal to π_j. Next, we can give a physical interpretation of the balance equations. We first observe the following. If the process is in state j, then the process is pulled out of state j with rate $\nu_j (= \sum_{\ell \neq j} q_{j\ell})$. The long-run fraction of the time that the process is in state j is denoted by π_j. This makes the following interpretation plausible:

the long-run average number of transitions per unit of time
from state j $= \pi_j \nu_j$.

On the other hand, if the process is in state k with $k \neq j$, then the process is pulled to state j with rate q_{kj}. The long-run fraction of the time that the process is in state k is equal to π_k. This makes the following interpretation plausible:

the long-run average number of transitions per unit of time
from state k *to* state $j = \pi_k q_{kj}$

for $j \neq k$. This leads to

the average number of transitions per unit of time
to state $j = \sum_{k \neq j} \pi_k q_{kj}$.

In the long run, the average number of transitions per unit of time *from* state j must equal the average number of transitions per unit of time *to* state j:

$$\sum_{k \neq j} \pi_j q_{jk} = \sum_{k \neq j} \pi_k q_{kj}. \qquad (8.10)$$

This equality gives the (global) balance equations for state j. Using the generator matrix $\mathbf{Q}$, and the row vector $\boldsymbol{\pi} = (\pi_i, \, i \in I)$, we may write the balance equations as

$$\boldsymbol{\pi} \mathbf{Q} = 0.$$

For a more physical interpretation of the balance equations (8.10), one can think of particles with a total mass of 1 that are distributed over "nodes" that correspond to the states $j \in I$ following the probability distribution $\{\pi_j, j \in I\}$. Particles leave node j with rate ν_j, and particles enter node j from another node k with rate q_{kj}. In steady state, the number of particles that enter node j must equal the number that leave it. In short, we have

rate out of state j = rate into state j.

This way, one can visualize the balance equations and write them down directly using the transition rate diagram. We illustrate this using Examples 8.1 and 8.2.

Example 8.1 (Continued). In this example, the equilibrium probability π_j represents the long-run fraction of the time that the on-hand inventory level is equal to j. The transition rate diagram in Figure 8.1 and the principle that the rate out of state j equals the rate into state j lead to the balance equations

$$\mu\pi_0 = \lambda\pi_1,$$
$$\lambda\pi_j = \lambda\pi_{j+1} \quad \text{for } j = 1, 2, \ldots, Q-1,$$
$$\lambda\pi_Q = \mu\pi_0.$$

This system of linear equations is solved together with the normalizing equation $\pi_0 + \pi_1 + \cdots + \pi_Q = 1$. We have

$$\text{long-run average on-hand inventory} = \sum_{j=0}^{Q} j\pi_j,$$

$$\text{long-run fraction of the time that the system is out of stock} = \pi_0.$$

What is the fraction of customers whose demand is lost, that is, what is the fraction of customers who find the system out of stock (state 0) when they arrive? Since the customers' arrival process is Poisson distributed, the fraction of customers who find the system in state 0 upon arrival is equal to the fraction of the time that the process is in state 0 and is therefore equal to π_0. This is a consequence of the general result that *Poisson arrivals see time averages* (PASTA) that is discussed in Section 9.3. If the state's evolution is described by a continuous-time Markov chain, we can also derive the result directly. The average number of customers per unit of time that find the system in state 0 upon arrival and whose demand is lost is equal to $\lambda\pi_0$. If we divide this average by the average number of customers who arrive per unit of time, we find that $\lambda\pi_0/\lambda = \pi_0$ is the fraction of customers whose demand is lost.

Example 8.2 (Continued). In this example, the equilibrium probabilities $\pi(i, 0)$ and $\pi(i, 1)$ represent the long-run fraction of the time that i ships are in the port and the crane is in repair or is working, respectively. The transition rate diagram in Figure 8.2 and the "rate out = rate in" principle lead to the balance equations

$$\lambda\pi(0, 1) = \mu\pi(1, 1),$$
$$(\lambda + \mu + \delta)\pi(i, 1) = \lambda\pi(i-1, 1) + \mu\pi(i+1, 1) + \beta\pi(i, 0) \quad \text{for } i \geq 1,$$
$$(\lambda + \beta)\pi(1, 0) = \delta\pi(1, 1),$$
$$(\lambda + \beta)\pi(i, 0) = \lambda\pi(i-1, 0) + \delta\pi(i, 1) \quad \text{for } i \geq 2.$$

This system of linear equations must be solved together with the normalizing equation $\sum_{i=0}^{\infty} \pi(i, 1) + \sum_{i=1}^{\infty} \pi(i, 0) = 1$. A solution exists only if the condition $\frac{1}{\mu} + \frac{\delta}{\mu} \times \frac{1}{\beta} < \frac{1}{\lambda}$ is satisfied, so that the length of the queue does not ultimately become infinite (note that $\frac{1}{\mu} + \frac{\delta}{\mu} \times \frac{1}{\beta}$ is the expected amount of time between the moment the unloading of a ship begins and the moment it ends). The system of linear equations for the equilibrium probabilities is not finite. To find the equilibrium probabilities numerically, this system must be truncated to a finite system in the unknowns $\pi(i, 0)$ and $\pi(i, 1)$ for $i \leq N$ with N sufficiently large such that

$\sum_{i=N}^{\infty} [\pi(i,0) + \pi(i,1)]$ is negligible. The choice of N is a matter of trial and error. The idea is to try different values of N until, say, the first four digits of the resulting probabilities no longer change. After the numerical calculations of the equilibrium probabilities, we find

$$\text{average number of ships in the port} = \sum_{i=1}^{\infty} i[\pi(i,0) + \pi(i,1)].$$

We denote this average by L. We denote the average time a ship spends in the port by W. It is not easy to calculate W directly from the equilibrium probabilities; it can be calculated indirectly using Little's law. This formula is a type of law of nature that holds for almost every queueing system:

> average number of customers in the system =
> (average arrival rate of customers per unit of time)
> $\times$ (average time a customer spends in the system).

A further explanation of this formula is given in Section 9.3. In the example in question, Little's law gives $L = \lambda W$.

Long-Run Average Cost per Unit of Time

In many applications, there is a continuous-time Markov chain with a cost structure. Suppose that $\{X(t), t \geq 0\}$ is a continuous-time Markov chain with state space I and transition rates q_{ij}. The Markov chain incurs a cost of $c(j)$ every unit of time the process is in state j and incurs a fixed cost of F_{jk} every time the process jumps from state j to state k. In addition to Assumption 8.1, we make the following assumption.

Assumption 8.2. For every initial state i, the total cost made before the first transition to state r has a finite value and, moreover, $\sum_{j \in I} |c(j)| \pi_j < \infty$ and $\sum_{j \in I} \pi_j \sum_{k \neq j} q_{jk} |F_{jk}| < \infty$.

For every $t > 0$, define the random variable $C(t)$ by

$$C(t) = \text{total cost up to time } t.$$

Under Assumptions 8.1 and 8.2, the following main theorem holds.

Theorem 8.4. *Under Assumptions 8.1 and 8.2, in the long run, the actual average cost per unit of time is equal to*

$$\lim_{t \to \infty} \frac{1}{t} \int_0^t C(u) \, du = \sum_{j \in I} c(j) \pi_j + \sum_{j \in I} \sum_{k \neq j} q_{jk} \pi_j F_{jk} \quad \text{with probability 1,}$$

independently of the initial state $X(0) = i$.

The first term on the right-hand side of the formula above can be seen by considering that π_j is the long-run fraction of the time that the process is in state j. The second term can be seen by observing that $q_{jk}\pi_j$ is the long-run average number of transitions from state j to state k per unit of time. We illustrate Theorem 8.4 using Example 8.1.

Example 8.1 (Continued). Suppose that the following cost structure has been added to the inventory model. For every unit of product, a holding cost of $h > 0$ is incurred for every unit of time the product is in stock, and a fixed cost of $K > 0$ is incurred for every replenishment. If we apply Theorem 8.4, we find that

$$\text{long-run average cost per unit of time} = h \sum_{j=0}^{Q} j\pi_j + K\mu\pi_0.$$

The second term follows by keeping in mind that a replenishment takes place only if there is a transition from state 0 to state Q. The average number of replenishments per unit of time is therefore $\mu\pi_0$.

8.6   Birth and Death Process

An important special case of a continuous-time Markov chain is the *birth and death process*. Suppose we have a system whose state at any time is the number of people present at that time. When there are i people present, births (new arrivals) occur at an exponential rate λ_i, while deaths (departures) occur at an exponential rate μ_i. That is, in state i, the time B_i until the next birth is exponentially distributed with mean $1/\lambda_i$, and the time D_i until the next death is exponentially distributed with mean $1/\mu_i$, and these times are independent of each other. Such a process is called a birth and death process, in which λ_i, $i = 0, 1, \ldots$, are called the birth rates and μ_i, $i = 1, 2, \ldots$, the death rates.

     A birth and death process is a continuous-time Markov chain with state space $I = \{0, 1, \ldots\}$ in which the only transitions possible from state i are to states $i - 1$ and $i + 1$ for $i > 0$ and to state 1 for $i = 0$. The Markov jump chain description of the birth and death process is:

$$\nu_0 = \lambda_0,$$
$$\nu_i = \lambda_i + \mu_i \quad \text{for } i > 0,$$
$$p_{0,1} = 1,$$
$$p_{i,i+1} = \mathbb{P}(B_i < D_i) = \frac{\lambda_i}{\lambda_i + \mu_i} \quad \text{for } i > 0,$$
$$p_{i,i-1} = \mathbb{P}(B_i > D_i) = \frac{\mu_i}{\lambda_i + \mu_i} \quad \text{for } i > 0.$$

The time until either a birth or a death occurs, which is the time until state i is left, is $\min(B_i, D_i)$ and is exponentially distributed with rate $\nu_i = \lambda_i + \mu_i$ for $i > 0$. Naturally, nobody can depart if there are no people present, so $\nu_0 = \lambda_0$ and correspondingly $p_{0,1} = 1$. Further, the transition from state i is to state $i + 1$ if a birth occurs before a death, and the probability that an exponential random variable with rate λ_i occurs earlier than an (independent) exponential random variable with rate μ_i is $\lambda_i/(\lambda_i + \mu_i)$. The last transition probability is explained analogously.

A birth and death process for which $\mu_i = 0$ for all i is called a *pure birth process*. A special case is the Poisson process: the pure birth process with constant birth rates $\lambda_i = \lambda$, see Appendix C.

8.6.1 *Equilibrium Distribution*

The infinitesimal transition rates of the birth and death process are

$$q_{ij} = \begin{cases} \lambda_i & \text{if } j = i + 1, \ i \geq 0 \text{ (birth rates)} \\ \mu_i & \text{if } j = i + 1, i \geq 1 \text{ (death rates)} \\ -\nu_i & \text{if } j = i, \ i \geq 0. \end{cases}$$

The balance equations are

$$\lambda_0 \pi_0 = \mu_1 \pi_1$$
$$\lambda_i \pi_i + \mu_i \pi_i = \lambda_{i-1} \pi_{i-1} + \mu_{i+1} \pi_{i+1} \quad \text{for } i = 1, 2, \ldots.$$

Combining the equation $\lambda_0 \pi_0 = \mu_1 \pi_1$ for $i = 0$ with the equation for $i = 1$, we observe that it must be that $\lambda_1 \pi_1 = \mu_2 \pi_2$. We can use induction to find that

$$\lambda_i \pi_i = \mu_{i+1} \pi_{i+1} \quad \text{for } i = 0, 1, \ldots. \tag{8.11}$$

These equations are called *detailed balance equations* as they relate the average number of transitions between neighboring states.

We may now solve the detailed balance equations. To this end, note that (8.11) implies that

$$\pi_{i+1} = \frac{\lambda_i}{\mu_{i+1}} \pi_i \quad \text{for } i = 0, 1, \ldots.$$

Iterating this equation gives that

$$\pi_i = \pi_0 \prod_{j=0}^{i-1} \frac{\lambda_j}{\mu_{j+1}} \quad \text{for } i = 1, 2, \ldots. \tag{8.12}$$

The birth and death process is ergodic if

$$G := 1 + \sum_{i=1}^{\infty} \prod_{j=0}^{i-1} \frac{\lambda_j}{\mu_{j+1}} < \infty.$$

In that case, $\pi_0 = G^{-1}$ and π_i in (8.12) is the equilibrium distribution.

8.6.2   *First Entrance Times*

For a birth and death process, a natural question is how long it takes, in expectation, to move from state i to state $j > i$. In other words, what is the expected *first entrance time* of state j if we start in state i? Let $T_{i,j}$ be the time to move from state i to state j, and let us first consider the expected time required to move from state i to $i + 1$. First, the remainder of the time spent in state i before jumping to any other state is exponentially distributed with mean $1/\nu_i$. If we subsequently jump to state $i + 1$, we are done. However, if we jump to state $i - 1$ instead, we will additionally incur the expected time to move from state $i - 1$ to state $i + 1$. Thus,

$$\mathbb{E}[T_{i,i+1}] = \frac{1}{\nu_i} + p_{i,i+1} \cdot 0 + p_{i,i-1}\mathbb{E}[T_{i-1,i+1}]$$

$$= \frac{1}{\lambda_i + \mu_i} + \frac{\mu_i}{\lambda_i + \mu_i}(\mathbb{E}[T_{i-1,i}] + \mathbb{E}[T_{i,i+1}]).$$

Taking $\mathbb{E}[T_{i,i+1}]$ to the left-hand side gives

$$\mathbb{E}[T_{i,i+1}] = \frac{1}{\lambda_i} + \frac{\mu_i}{\lambda_i}\mathbb{E}[T_{i-1,i}].$$

This recursive relation can be solved starting from $\mathbb{E}[T_{0,1}] = \frac{1}{\lambda_0}$. Finally, if $j > i$ we have $T_{i,j} = T_{i,i+1} + \cdots + T_{j-1,j}$.

8.7   Applications to Queueing Systems

In this section, we apply the theory of continuous-time Markov chains to two basic models in queueing theory. Section 8.7.1 deals with the $M/M/1$ queue, and Section 8.7.2 with the Erlang loss model. Both models are examples of birth and death processes.

8.7.1   *The Single-Server Queue*

Consider a single-server model where customer arrivals follow a Poisson process with arrival rate λ. The waiting room for the service station has an infinite capacity. An arriving customer who finds the server busy takes place in the waiting room until it is his turn to be served. The server can only help one customer at a time. The customers' service times are independent of one another and have a common exponential distribution with expected value $1/\mu$. The service process is independent of the arrival process. Assume $1/\mu < 1/\lambda$; that is, a customer's average service time is less than the average interarrival time between two consecutive customers. This model is called the $M/M/1$ queue; see Chapter 9. The following are interesting questions for this model: What is the average queue length? What is the average waiting time per customer? What are the probability distributions of the queue length and the customer's waiting time?

To answer these questions, we let

$$X(t) = \text{number of customers present at time } t.$$

Fig. 8.5   The transition rate diagram for the $M/M/1$ queue.

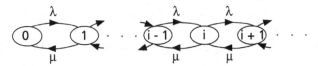

The stochastic process $\{X(t), t \geq 0\}$ is a continuous-time Markov chain with state space $I = \{0, 1, \ldots\}$. The transition rate diagram is given in Figure 8.5 (verify!). The equilibrium probability π_i gives the long-run fraction of the time that i customers are present in the system. Calculating the π_i follows the "rate out of state i = rate into state i" principle:

$$\lambda\pi_0 = \mu\pi_1 \quad \text{and} \quad (\lambda + \mu)\pi_i = \lambda\pi_{i-1} + \mu\pi_{i+1} \quad \text{for } i = 1, 2, \ldots \ .$$

This system of linear equations can be rewritten as a recursive relation. Substituting the first equality $\lambda\pi_0 = \mu\pi_1$ in the second $(\lambda + \mu)\pi_1 = \lambda\pi_0 + \mu\pi_2$ gives $\lambda\pi_1 = \mu\pi_2$. One can use induction to verify that

$$\mu\pi_i = \lambda\pi_{i-1} \quad \text{for } i = 1, 2, \ldots \ .$$

This recursive relation can even be solved explicitly. Repeatedly applying $\pi_i = (\lambda/\mu)\pi_{i-1}$ gives $\pi_i = (\lambda/\mu)^i\pi_0$ for $i = 1, 2, \ldots$. This last equation also holds for $i = 0$. Together with $\sum_{i=0}^{\infty} \pi_i = 1$, this leads to $1 = \pi_0 \sum_{i=0}^{\infty}(\lambda/\mu)^i = \pi_0/(1 - (\lambda/\mu))$, where the assumption $\lambda/\mu < 1$ is used. So $\pi_0 = 1 - \lambda/\mu$. This gives

$$\pi_i = \left(1 - \frac{\lambda}{\mu}\right)\left(\frac{\lambda}{\mu}\right)^i \quad \text{for } i = 0, 1, \ldots \ .$$

So the number of customers present has a *geometric* distribution. The symbol ρ is generally used for the quotient λ/μ:

$$\rho = \frac{\lambda}{\mu}.$$

The quantity ρ is called the *server utilization*. The explanation for this term lies in the fact that $\pi_0 = 1 - \rho$. In other words, the long-run fraction of the time that the server is busy is equal to $1 - \pi_0 = \rho$. The quantity ρ is also the average amount of work arriving per time unit: customers arrive at rate λ and require mean service duration $1/\mu$. Clearly, the average amount of work arriving to the system coincides with the fraction of time the server is busy.

Next, we define

L_q = average number of customers in the queue
(excluding the customer being served),

W_q = average waiting time per customer (excluding the service time).

These quantities can easily be calculated from the equilibrium probabilities. For the average queue length L_q, we have

$$L_q = \sum_{i=1}^{\infty}(i - 1)\pi_i.$$

The relation $\sum_{i=1}^{\infty} ix^i = \frac{x}{1-x}$ for $0 < x < 1$ and some simple calculations give

$$L_q = \frac{\rho^2}{1-\rho}.$$

The easiest way to calculate W_q is to apply Little's law $L_q = \lambda W_q$. Thus

$$W_q = \frac{\rho}{\mu(1-\rho)}.$$

Little's law is discussed in detail in Section 9.3. The existence of the explicit formulas for L_q and W_q is less important than the quantitative insight the formulas give. These formulas show that the average queue length and average waiting time per customer *does not* grow *linearly* as a function of the server utilization ρ and even increases *explosively* when ρ is *close to* 1. For example, if for an average service time of $1/\mu = 1$, the arrival rate λ increases by 5% from 0.9 to 0.945, then the average waiting time per customer increases by 91% from 9 to 17.2.

Probability Distribution at the Arrival Times

The equilibrium probability π_j gives the long-run fraction of the time that there are j customers in the system; it can also be viewed as the probability that an outsider encounters j customers if this outsider enters the system at a *random* time when the process has been in operation for a very long time and the outsider has no information on the system's evolution up to that point. For $j = 0, 1, \ldots$, define

a_j = long-run fraction of customers who find j other customers upon arrival.

In other words, a_j is the probability that an arriving customer finds j other customers if the process has been in operation for a very long time and the customer has no information on the system's evolution up to that point. To find a_j, we use the general result that in a continuous-time Markov chain, the quantity $\pi_j q_{jk}$ can be viewed as the average number of transitions per unit of time from state j to state k (see Section 8.5). In the $M/M/1$ queue, the average number of transitions per unit of time from state j to state $j+1$ is therefore equal to $\lambda \pi_j$. In other words, the average number of customers per unit of time who arrive and encounter j other customers is equal to $\lambda \pi_j$. Dividing this number by the total average number of customers arriving per unit of time (that is, λ) gives the fraction of customers who encounter j other customers when they arrive. This gives

$$a_j = \pi_j \quad \text{for } j = 0, 1, \ldots .$$

The customer-average probabilities a_j are therefore equal to the time-average probabilities π_j: in steady state, the probability that an arriving customer encounters j other clients is equal to the probability that there are j customers at any random time. This is no coincidence. The property that Poisson Arrivals See Time Averages (PASTA) found earlier holds in general for queueing systems with a Poisson arrival process for the customers; see also Section 9.3. If one can calculate the probability distribution of the number of customers present at a *random* time for such a queueing system using continuous-time Markov chains, one also has the probability distribution of the number of customers present *right before* an arrival time. This PASTA property is handy in all kinds of applications, especially in queueing problems.

Probability Distribution of the Waiting Time

The customer-average probabilities a_j allow us to calculate the probability distribution of a customer's waiting time. Define

$W_q(x)$ = long-run fraction of customers with waiting time at most x for $x \geq 0$.

In other words, $W_q(x)$ is the probability that in steady state, a customer waits no longer than a time x. Assuming that the customers are served in their order of arrival, we will show that

$$W_q(x) = 1 - \rho e^{-\mu(1-\rho)x} \qquad \text{for } x \geq 0.$$

To find the unconditional waiting time probability $W_q(x)$, we first calculate the conditional waiting time distribution for a customer who encounters j other clients upon arrival, with $j \geq 1$. This customer's waiting time is the sum of the residual service time of the customer being served and the $j - 1$ exponentially distributed service times of the other $j - 1$ customers. Because the exponential distribution has no memory, the residual service time has the same exponential probability distribution as a new service time. This means that the customer's conditional service time has the same distribution as the sum of j independent random variables with a common exponential distribution. It is well known in probability theory that this sum has an Erlang(j, μ) distribution. In other words,

$$\mathbb{P}(\text{waiting time of a customer} > x \mid \text{the customer encounters } j \text{ other}$$
$$\text{customers upon arrival}) = \sum_{k=0}^{j-1} e^{-\mu x} \frac{(\mu x)^k}{k!} \qquad \text{for } j \geq 1.$$

Averaging this expression over the probabilities $a_j = \pi_j$ and reversing the order of summation gives the unconditional waiting time probability $1 - W_q(x)$:

$$1 - W_q(x) = \sum_{j=1}^{\infty} \pi_j \sum_{k=0}^{j-1} e^{-\mu x} \frac{(\mu x)^k}{k!} = \sum_{k=0}^{\infty} e^{-\mu x} \frac{(\mu x)^k}{k!} \sum_{j=k+1}^{\infty} \pi_j$$

$$= \sum_{k=0}^{\infty} e^{-\mu x} \frac{(\mu x)^k}{k!} \sum_{j=k+1}^{\infty} (1 - \rho)\rho^j.$$

Keeping in mind that $\sum_{j=k+1}^{\infty} \rho^j = \frac{\rho^{k+1}}{1-\rho}$ and $\sum_{n=0}^{\infty} \frac{u^n}{n!} = e^u$, we see that this implies

$$1 - W_q(x) = \sum_{k=0}^{\infty} e^{-\mu x} \frac{(\mu x)^k}{k!} \rho^{k+1} = \rho e^{-\mu x} e^{\mu x \rho} = \rho e^{-\mu(1-\rho)x}.$$

8.7.2   The Erlang Loss Model

This model is one of the most important models from queueing theory. It can be described as follows. Consider a multiserver model where customer arrivals follow a Poisson process with arrival rate λ. The service station has *no waiting room*. There are c identical servers, where $c \geq 1$. An arriving customer who finds all servers busy leaves and has no influence on the system. A customer who arrives when not all servers are busy is immediately served by one of the free servers. The customers' service times are independent of one another and have a common exponential distribution with expected value $1/\mu$. An important performance measure for the model is the fraction of customers whose demand is lost. This queueing model is called the *Erlang loss model* (or the *M/G/c/c* queue) and was first considered by the Danish engineer A. K. Erlang in the early years of telephony (early 20th century). The Erlang distribution is also named after him. In Erlang's situation, the service station was a telephone exchange with c telephone lines (the servers), where a new incoming call (the customer) was lost if all lines were busy. Although the Erlang loss model is relatively old, it is still current and even now still has applications to various problems (see also the exercises).

To analyze the Erlang loss model, we define

$$X(t) = \text{number of busy servers at time } t.$$

The stochastic process $\{X(t), t \geq 0\}$ is a continuous-time Markov chain with state space $I = \{0, 1, \ldots, c\}$. The transition rate diagram is given in Figure 8.6.

Fig. 8.6   Transition rate diagram for the Erlang loss model.

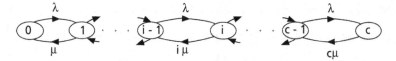

The transition rate $q_{i,i-1} = i\mu$ requires some explanation. Suppose that at a given time t, the process is in state i. The probability that the process transitions to state $i - 1$ during the coming interval $(t, t + \Delta t)$ is equal to the probability that one of the i busy servers finishes during the interval $(t, t + \Delta t)$. For $\ell = 1, \ldots, i$, define A_ℓ as the event that the ℓth server finishes during the coming interval $(t, t + \Delta t)$. Based on the failure rate representation of the exponential distribution, we have $\mathbb{P}(A_\ell) = \mu\Delta t + o(\Delta t)$ for $\ell = 1, \ldots, i$. The probability that two or more servers finish during the interval $(t, t + \Delta t)$ is of order $(\Delta t)^2$ and therefore $o(\Delta t)$. This gives

$$\mathbb{P}(A_1 \cup A_2 \cup \ldots \cup A_i) = \sum_{\ell=1}^{i} \mathbb{P}(A_\ell) + o(\Delta t) = i\mu\Delta t + o(\Delta t),$$

that is, $q_{i,i-1} = i\mu$. Another way to see this result is to consider that the minimum of i exponentially distributed service times has an exponential distribution with expected value $1/(i\mu)$. The transition rate diagram and the "rate out = rate in" principle lead to the balance equations

$$\lambda\pi_0 = \mu\pi_1,$$
$$(\lambda + i\mu)\pi_i = \lambda\pi_{i-1} + (i+1)\mu\pi_{i+1} \quad \text{for } i = 1, \ldots, c-1,$$
$$c\mu\pi_c = \lambda\pi_{c-1}.$$

Like the balance equations for the queueing model in Section 8.7.1, this system of balance equations can be rewritten as a recursive system. One can use induction to verify that

$$i\mu\pi_i = \lambda\pi_{i-1} \quad \text{for } i = 1, 2, \ldots, c.$$

This recursive system can also be solved explicitly. It follows from the equality $\pi_i = (\lambda/i\mu)\pi_{i-1}$ that $\pi_i = \left(\lambda/\mu\right)^i/i!\,)\pi_0$ for $i = 0, 1, \ldots, c$. With $\sum_{i=0}^{c}\pi_i = 1$, this leads to

$$\pi_i = \frac{(\lambda/\mu)^i/i!}{\sum_{k=0}^{c}(\lambda/\mu)^k/k!} \quad \text{for } i = 0, 1, \ldots, c.$$

This probability distribution is in fact a truncated Poisson distribution, as one can see by adding $e^{-\lambda/\mu}$ to both the numerator and the denominator. The equilibrium probability π_i gives the long-run fraction of the time that i servers are busy. The requested performance measure is the fraction of customers whose demand is lost. To calculate it, we again use the property that in steady state, the probability distribution of the number of customers right before an arrival time is equal to the probability distribution of the number of customers at a random time when the arrival process is Poisson distributed (Poisson arrivals see time averages). This last probability distribution is given by the π_i. The fraction of customers whose demand is lost is the fraction of customers who find the system in state c upon arrival, and this fraction is equal to the fraction of the time that the system is in state c. This gives

long-run fraction of customers whose demand is lost =

$$\pi_c = \frac{(\lambda/\mu)^c/c!}{\sum_{k=0}^{c}(\lambda/\mu)^k/k!}.$$

This is the famous Erlang loss formula.

Insensitivity Property

A significant result that we give without proof is that the formulas for the equilibrium probabilities π_i and in particular for the loss probability π_c given above also apply if a customer's service time has a *general* probability distribution. One only needs to replace $1/\mu$ in the formulas with the expected value of the service time. This means that in the Erlang loss model, the equilibrium distribution is *insensitive* for the form of the service time's probability distribution; only the service time's *expected value* matters. This insensitivity greatly extends the scope of application of the Erlang loss model.

If we let the number of servers c go to infinity, the Erlang loss model is transformed into the so-called $M/G/\infty$ queue with infinitely many servers, in which a customer is immediately assigned a server. Letting the value of c go to infinity in the formula given above for the π_i and keeping in mind that $\sum_{k=0}^{\infty} x^k/k! = e^x$, we obtain the Poisson distribution

$$\pi_i = e^{-\lambda/\mu} \frac{(\lambda/\mu)^i}{i!} \quad \text{for } i = 0, 1, \ldots.$$

The important insensitivity property also holds for the $M/G/\infty$ queue: the Poisson distribution for the equilibrium probabilities π_i holds regardless of the form of the service time's probability distribution and only requires the expected value $\mathbb{E}(S)$ of the service time S. In Section 9.4.7, we outline a probabilistic proof of the insensitivity of the $M/G/\infty$ queue.

8.7.3 *Recursive Calculation of the Equilibrium Probabilities*

In both the single-server queue and the Erlang loss model, it is possible to calculate the equilibrium probabilities recursively. This is no coincidence. In general, for a continuous-time Markov chain, the equilibrium distribution can be calculated recursively if the state space I is given by $I = \{0, 1, \ldots, N\}$ for some $N \leq \infty$ and if for every $i \in I$, the transition rates q_{ij} satisfy

$$q_{ij} = 0 \quad \text{for } j \leq i - 2.$$

In the continuous-time Markov chain formulation of many queueing systems, these conditions hold. Here is a probabilistic derivation of the recursive relation for the equilibrium probabilities. For every subset A of states with $A \neq I$, the long-run average number of transitions per unit of time from inside A to outside A is equal to the average number of transitions per unit of time from outside A to inside A. In other words,

$$\text{rate out of } A = \text{rate into } A.$$

Now, choose A well, namely equal to $A = \{i, i+1, \ldots, N\}$ for a fixed i with $i \neq 0$ so that $A \neq I$. The process can only leave the set of states A through state i, which happens at a rate of $q_{i,i-1}$. From a state $k \notin A$, the process is pulled into the set of states A at a rate of $\sum_{\ell \geq i} q_{k\ell}$. This therefore leads to the recursive relation

$$\pi_i q_{i,i-1} = \sum_{k=0}^{i-1} \pi_k \left(\sum_{\ell \geq i} q_{k\ell} \right) \quad \text{for } i = 1, 2, \ldots, N. \tag{8.13}$$

Because the balance equations uniquely determine the equilibrium probabilities up to a multiplicative constant, we can formulate the following simple algorithm:

Algorithm 8.1.

Step 1. Let $\bar{\pi}_0 := 1$.
Step 2. Successively calculate $\bar{\pi}_1, \bar{\pi}_2, \dots, \bar{\pi}_N$ from the recursive relation (8.13).
Step 3. Normalize by setting $\pi_i := \bar{\pi}_i / (\sum_{k=0}^{N} \bar{\pi}_k)$ for $i = 0, 1, \dots, N$.

This algorithm is generally much faster than solving a system of linear equations.

We illustrate the recursive approach using the $M/M/1$ queue from Section 8.7.1. In this model, in each transition, the value of a state can decrease by at most one. For the set $A = \{i, i+1, \dots\}$ with $i \neq 0$, the "rate out of A" is then equal to $\mu\pi_i$. The set A can only be entered from state $i - 1$, so that the "rate into A" is equal to $\lambda\pi_{i-1}$. This gives the recursion $\mu\pi_i = \lambda\pi_{i-1}$ for $i = 1, 2, \dots$.

The approach for the $M/M/1$ queue can immediately be extended to the multiserver $M/M/c$ queue. In the $M/M/c$ queue, there are c identical servers with exponentially distributed service times and a common queue for all servers. Verify that the recursive relation of the equilibrium probabilities is now given by

$$i\mu\pi_i = \lambda\pi_{i-1} \text{ for } 1 \leq i < c \text{ and } c\mu\pi_i = \lambda\pi_{i-1} \text{ for } i \geq c,$$

where we assume $\lambda < c\mu$. This recursion gives $\pi_i = \rho^{i-c+1}\pi_{c-1}$ for $i \geq c$ with $\rho = \lambda/(c\mu)$, so that

$$\sum_{i=c}^{\infty} \pi_i = \frac{\rho}{1-\rho}\pi_{c-1}.$$

This probability gives the long-run fraction of the time that all servers are busy and because of the property that Poisson arrivals see time averages also gives the fraction of customers who must wait. In Section 9.4.9, we return to the $M/M/c$ queue, which, among other things, is an important model for call centers.

8.8 Exercises

8.1 A familiar sight on the streets of Israel are the so-called sheruts. A sherut is a shared taxi that drives from a fixed spot in a city to another city. Consider a sherut stand that has room for only one sherut. The sherut has seven seats and leaves as soon as these seats are filled by passengers. After a sherut leaves, it takes an exponential time with expected value $1/\mu$ for a new sherut to arrive. Potential passenger arrivals at the stand follow a Poisson process with rate λ. A potential passenger only waits for a sherut if no more than seven other customers are already waiting; otherwise, the passenger goes to a competitor. Formulate a continuous-time Markov model to analyze the situation above. Specify the state variable(s) and the transition rate diagram.

8.2 A particular county has a central dispatcher for ambulance services. It is divided into two districts A and B, and each district is allocated one ambulance. Calls from districts A and B coming in at the central dispatcher follow independent Poisson processes with respective rates λ_A and λ_B. Ambulance 1 is the first-due for district A, and ambulance 2 is the first-due for district B. If a call comes in when the ambulance from the district in question is occupied while the other ambulance is idle, the latter is dispatched. A call that occurs when both ambulances are occupied is handled by an ambulance from another area. The service times (including travel time) of the different calls are independent of one another and are exponentially distributed with expected values $1/\mu_{iA}$ and $1/\mu_{iB}$, respectively, for calls from districts A and B handled by ambulance i. Formulate a continuous-time Markov model to analyze the situation above. Specify the state variable(s) and the transition rate diagram.

8.3 An assembly line for some product has two stations placed in series. A particular material must be processed successively at stations 1 and 2. The arrivals of units of the material at station 1 follow a Poisson process with rate λ. The processing times of a unit at stations 1 and 2 are independent of each other and have exponential distributions with respective expected values $1/\mu_1$ and $1/\mu_2$. Neither station 1 nor station 2 has a buffer to temporarily store an incoming unit when the station is occupied. This means that if station 1 is occupied, an incoming unit is rejected and must be handled elsewhere. When the assembly at station 1 is completed, the unit goes through directly to station 2 provided that station 2 is free. Otherwise, the unit remains at station 1 and blocks it until station 2 frees up. Formulate a continuous-time Markov model to analyze the situation above. Specify the state variable(s) and the transition rate diagram.

8.4 A production hall contains a fast machine and a slow machine to process orders. The arrival of orders follows a Poisson process with rate λ. An arriving order that finds both machines occupied is lost. If the fast machine is free, the order is assigned to it; if only the slow machine is free, the order goes to that machine. It is not possible to transfer an order from the slow machine to the fast machine if the latter frees up. An order's processing time is exponentially distributed with expected value $1/\mu_1$ on the fast machine and $1/\mu_2$ on the slow machine.

   (a) Formulate a continuous-time Markov chain. Specify the transition rate diagram and give the balance equations.
   (b) What is the fraction of the time that the fast, respectively slow, machine is in use? What is the fraction of incoming orders that are lost?

8.5 Car arrivals at a gas station follow a Poisson process with an average of ten customers per hour. A car only enters the gas station if no more than three cars are already present. The gas station has only one pump. A car's refueling time is exponentially distributed with an expected value of three minutes.

   (a) Calculate the average number of cars at the gas station.
   (b) What is the fraction of the time that all three spots at the gas station are occupied? What is the fraction of cars that decide not to stop at the gas station?

8.6 A small city has a one-man taxi company. This company has a stand at the local train station. Potential customers arrivals follow a Poisson process with an average of four customers per hour. When a customer arrives and finds the taxi waiting, the taxi immediately drives away with the customer. When the taxi returns to the stand and finds waiting customers, the driver picks up the customers up to a maximum of three and leaves. The length of a ride with i customers has an exponential distribution with expected value m_i minutes, where $m_1 = 10$, $m_2 = 12.5$, and $m_3 = 15$. A potential customer who arrives to find three waiting customers looks for transport elsewhere.

   (a) Define a suitable continuous-time Markov chain. Specify the transition rate diagram.
   (b) Calculate the fraction of the time that the taxi waits for customers and the fraction of potential customers who go elsewhere for transport.

8.7 Consider Exercise 8.1 again. What is the fraction of potential passengers who go elsewhere? What is the average waiting time for a customer who takes the sherut? Also answer these questions for the modified model in which a potential customer only waits if a sherut is present when he arrives.

8.8 Consider an unloading area for coal brought in by train from different coal mines. There is only one container crane on site for unloading, which can handle only one train at a time. The time needed to unload a train is exponentially distributed with expected value $1/\mu$ when the crane operates continuously. However, the crane can break down while working on a train. The length of time the crane works without malfunctioning is exponentially distributed with expected value $1/\alpha$, while the time needed to resolve a malfunction is exponentially distributed with expected value $1/\beta$. It is assumed that a malfunction can occur at any time and that the unloading of a train that is interrupted by a breakdown resumes at the point where it was interrupted. The coal transport involves N trains. As soon as a train is unloaded, it returns directly to the coal mine for a new load. The time it takes for the train to return to the unloading area with another load of coal is exponentially distributed with expected value $1/\lambda$.

   (a) Define a suitable continuous-time Markov chain to analyze the situation at the unloading area. Specify the transition rate diagram and give the balance equations to calculate the probability distribution of the number of trains in the unloading area.

   (b) What is the average number of trains that arrive at the unloading area per unit of time, and what is the average number of trains unloaded per unit of time? What is the fraction of trains that are unloaded immediately upon arrival? What is the fraction of the time that the crane is out of service due to a malfunction? What is the average amount of time a train spends at the unloading area?

8.9 Consider a communication system where the arrival of messages for transmission follows a Poisson process with rate λ. The system's transmission channel can send only one message at a time. The transmission time is exponentially distributed with expected value $1/\mu$. Messages arriving for transmission are temporarily stored in a buffer as long as no more than R messages are already present, with R a given control parameter. As soon as the number of messages in the buffer reaches R, new messages are rejected and routed elsewhere until the number of messages in the buffer has dropped to the value r, where $0 \le r < R$, after which messages are again admitted until the number of messages in the buffer reaches R again, and so on.

   (a) Formulate a suitable continuous-time Markov chain. Specify the transition rate diagram and draw up the balance equations.

   (b) What is the average number of messages in the buffer? What is the average number of messages arriving per unit of time and routed elsewhere?

8.10 At a gas station with one pump and room for five cars, arrivals follow a Poisson process with rate λ. An arriving car that finds i other cars present at the station drives on with probability $i/5$ and enters the gas station with probability $1 - i/5$. Only one car can refuel at a time, and the refueling time is exponentially distributed with expected value $1/\mu$.

   (a) Specify the transition rate diagram and balance equations of a continuous-time Markov chain drawn up to calculate the probability distribution of the number of cars at the gas station. What is the fraction of cars that decide to drive on, and what is the average waiting time per car that stays to refuel?

(b) Give an explicit expression for the equilibrium distribution of the number of cars at the gas station for the case that the station has room for an unlimited number of cars.

(c) Suppose that every driver who enters the gas station is prepared to wait only a certain amount of time for the pump to free up. This "patience time" is a sample from an exponential distribution with expected value $1/\theta$ and is independent for the different cars. If a driver's patience time has expired and the pump is not yet free, the driver leaves, regardless of the number of cars in front of him. Draw up the transition rate diagram for this new situation. What is the average number of cars per unit of time that leave the gas station without refueling?

8.11 The inventory level of a perishable product decreases due to demand and deterioration. Suppose that the demand is described by a Poisson process with a rate λ. The life spans of the units of the product are independent of one another and have a common exponential probability distribution with expected value $1/\mu$. The inventory strategy is as follows. Every time the product's inventory level drops to zero, the stock is replenished with Q units. The time needed for a replenishment is negligible. Calculate the average inventory level and the average number of replenishments per unit of time.

8.12 The arrivals of messages for transmission at a communication channel follow a Poisson process with a rate of λ messages per second. The channel can transmit only one message at a time and disposes of two transmission rates of σ_1 and σ_2 bites per second, with $0 < \sigma_1 < \sigma_2$. The message lengths are independent of one another and are exponentially distributed with expected value $1/\mu$ bits. Assume $1/(\sigma_2\mu) < 1/\lambda$. The transmission line may switch from one rate to the other at any time. The following control rule is used. The lower rate σ_1 is used whenever fewer than R messages are present; otherwise, the faster rate σ_2 is used. The following costs are involved. For every message, there is a holding cost of $h > 0$ per unit of time the message has not been fully sent. An operating cost of r_i is incurred per unit of time that the channel is in use and sends at rate σ_i.

(a) Determine the average cost per unit of time for a given value of R. Show that the equilibrium probabilities can be calculated recursively.

(b) Write a computer program to calculate the optimal value of R for the numerical data $\lambda = 0.8$, $\mu = 1$, $\sigma_1 = 1$, $\sigma_2 = 1.5$, $h = 1$, $r_1 = 5$, and $r_2 = 25$.

8.13 Consider the continuous review (s, Q) inventory model in which the customer arrivals follow a Poisson process with parameter λ, and every customer requests exactly one unit of the product. If the demand cannot be met because there is no on-hand inventory, the demand is back-ordered and sent as soon as stock becomes available. The (s, Q) policy is based on the economic inventory (see Chapter 6). As soon as the economic inventory drops to s, a replenishment order of size Q is placed. The lead time of a replenishment order is a constant $L > 0$.

(a) Use a continuous-time Markov-chain analysis to show that the equilibrium distribution of the economic inventory is uniformly distributed on $s + 1, \ldots, s + Q$.

(b) Use the result from part (a) to find the equilibrium distribution of the net inventory (*Hint*: The net inventory at time t is equal to the economic inventory at time $t - L$ minus the total demand during the interval $(t - L, t)$.)

(c) Calculate the average on-hand inventory, the average delay in delivery, and the fraction of the demand that is delivered late.

8.14 A parking lot has space for 10 cars. The arrivals of cars for long-term parking and short-term parking follow independent Poisson processes with respective rates of 4 and 6 cars per hour. Any car that arrives and finds all parking spaces occupied goes elsewhere. The parking time of cars in long-term parking is uniformly distributed between 1 and 2 hours, and that of cars in short-term parking is uniformly distributed between 20 and 60 minutes. Calculate the fraction of cars that find all parking spaces occupied upon arrival.

8.15 Containers are delivered to a stockyard following a Poisson process with an average of one container per hour. A container that arrives when the location is full is brought elsewhere. The time that a container is stored in the yard is uniformly distributed between 5 and 15 hours. How large should the stockyard's capacity be so that no more than 1% of the arriving containers find the yard full?

8.16 Consider the so-called base-stock $(S-1, S)$ inventory model with base inventory level S. Customers arrive following a Poisson process with rate λ. Every customer asks for exactly one unit of the product. Every time a demand is met from stock, a replenishment order of exactly one unit is placed so that the on-hand inventory plus the total quantity on order always equals S. Demand that occurs when the system is out of stock is lost. The lead times of the replenishment orders are independent and identically distributed random variables with expected value τ. Use the Erlang loss model to show that the long-run fraction of the time that the on-hand inventory level is j is equal to

$$\frac{(\lambda\tau)^{S-j}/(S-j)!}{\sum_{k=0}^{S}(\lambda\tau)^k/k!} \quad \text{for } j = 0, 1, \ldots, S.$$

What is the average on-hand inventory level, and what is the long-run fraction of the demand that is lost?

8.17 In an electronic system, a total of c units of a crucial component are connected in parallel to increase the system's reliability. The system runs as long as at least one of the c units works; otherwise, the system is down. Every unit is turned on and works unless it breaks down. The running times of the units are independent of one another and have a common exponential probability distribution with expected value $1/\alpha$. A broken unit is replaced by a new one. Only one unit can be replaced at a time, and the successive replacement times of units are independent, exponentially distributed random variables with expected value $1/\beta$.

   (a) Define a continuous-time Markov chain to calculate the fraction of the time that the system is out of order.

   (b) Does the fraction of the time that the system is out of order change if the replacement time has a general probability distribution with expected value $1/\beta$ instead of an exponential distribution? (*Hint:* Compare the transition rate diagram from part (a) with the transition rate diagram for the Erlang loss model.)

8.18 Consider Exercise 8.17 again, but now assume that there are sufficiently many mechanics to directly replace any broken units. Answer question (a) again using a continuous-time Markov chain. What is the effect on the fraction of the time that the system is down if the system's running time has a general probability distribution?

8.19 Suppose that a production facility has M operating machines and a buffer of B machines on standby. The machines in operation are subject to breakdowns. The running times of the operating machines are independent of one another and have a common exponential distribution with expected value $1/\lambda$. A malfunctioning machine is immediately replaced by a standby machine if one is available. The broken machine is immediately repaired. Suppose that there are sufficiently many mechanics. The repair time of a broken machine has an exponential distribution with expected value $1/\mu$.

(a) Define a continuous-time Markov chain to calculate the fraction of the time that fewer than M machines are in operation.

(b) Does the fraction calculated in part (a) change if the repair time has a general probability distribution with expected value $1/\mu$?

8.20 In a production-stock system, products for sale are made ready one by one at intervals that are independent of one another and exponentially distributed with expected value $1/\alpha$. As soon as a unit is ready, it is put on the shelf for sale. The arrivals of customers for the product follow a Poisson process with rate β. Assume $\alpha < \beta$. Every customer requests one unit of the product. If the product is in stock, the customer takes the unit of the product; otherwise, the customer goes elsewhere for the product (lost demand).

(a) Specify the transition rate diagram and balance equations of a continuous-time Markov chain drawn up to calculate the probability distribution of the on-hand inventory level. Determine the equilibrium probabilities explicitly.

(b) What is the fraction of the demand that is lost, and what is the average time that a unit is on the shelf?

(c) Now, assume that the on-hand inventory is perishable. The shelf life of every unit of inventory has an exponential distribution with expected value $1/\eta$, and the shelf lives of the different units are independent of one another. As soon as the shelf life of a unit has expired, it is removed from the shelves. Answer parts (a) and (b) again in this situation.

References

1. S. M. Ross, *Introduction to Probability Models*, 10th ed., Academic Press, New York, 2010.

2. H. C. Tijms, *A First Course in Stochastic Models*, John Wiley & Sons Ltd., Chichester, England, 2003.

Chapter 9

Queueing Theory

9.1 Introduction

Every one of us has waited in line at a bank, a supermarket, or a bus stop. Waiting is generally annoying, and much productivity is lost because of it. Queueing theory cannot help us eliminate waiting but does give us insight into how waiting times depend on the queueing system's design. This insight can be used to reduce the waiting times.

Queueing problems occur in many diverse real-life situations. In telephony and telecommunication problems, calls and messages wait for a free line; in seaports and airports, ships wait for loading and unloading facilities and planes for a runway to become available; in a production hall, machines wait for repair; and so on. The design of all kinds of systems raises questions such as "How many telephone lines are needed to guarantee a certain degree of service?", "Which appointment system should a hospital use to keep a patient's average waiting time below a given value?", "What is the effect on the downtime of machines in a production hall if the number of repairmen is increased?", and so on. In many cases, queueing theory can be useful in answering these types of questions.

In situations with uncertainty in the arrival pattern of customers and the lengths of the service times, it is inevitable that periods in which customers wait for service and periods in which the servers wait for customers alternate. One cannot suppress both forms of waiting at the same time. If the aim is to provide reasonable service to customers, then the system's load must not be too close to the maximum processing capacity. The price of good service is that the service station will, from time to time, be without work. A rule of thumb in queueing theory is that in the case of a single service station, on average, this station may not be filled to its full capacity more than 80% of the time if it is to provide reasonable service. Queueing theory will teach us that in a heavily loaded system, a *small* increase in the load can lead to an *enormous* increase in a customer's average waiting time. Nonlinear effects occur. A customer arrival rate that is twice as high does not generally lead to an average waiting time that is twice as long; the increase in the queue length is much higher. The cause of this phenomenon lies in the variability of the arrival times and the service times. The essence of waiting is beautifully expressed in the so-called waiting time paradox.

The Waiting Time Paradox[1]

Suppose that every 20 minutes, a bus is meant to arrive at a given bus stop. Someone who does not know the timetable goes to the bus stop hoping for the best. How long will he wait on average? Is it 10 minutes, half of 20 minutes? This is only correct if the buses arrive *exactly* every 20 minutes, without any variability. Otherwise, it is *not* correct, and the average waiting time is always longer than 10 minutes. This phenomenon can be clarified using a numerical example. Suppose that, on average, half of the time, the buses arrive after 10 minutes and half of the time, they arrive after 30 minutes. Then the average waiting time is not $\frac{1}{2} \times 5 + \frac{1}{2} \times 15 = 10$ minutes, but $\frac{3}{4} \times 15 + \frac{1}{4} \times 5 = 12.5$ minutes. After all, on average, a passenger will arrive during a long 30-minute interval 3 times out of 4 and during a short 10-minute interval 1 time out of 4. The explanation of the waiting time paradox therefore lies in the fact that the probability of arriving during a *long* interval is higher than that of arriving during a *short* interval. The same holds for checkout counters at the supermarket: the probability of arriving while the checker is helping a customer with a relatively long handling time is relatively high.

    A simple mathematical formula can be given for the waiting time paradox. For this, we need the notion of the *coefficient of variation* of a random variable. If the random variable T denotes the time between the departure of two consecutive buses, then the coefficient of variation c_T of T is defined as the quotient of the standard deviation of T and the expected value of T. So

$$c_T = \frac{\sigma(T)}{\mathbb{E}[T]}.$$

The coefficient of variation is dimensionless. It measures how variable the interdeparture time T is. If the interdeparture time T is constant, then $\sigma(T) = 0$ and therefore $c_T = 0$. The interdeparture time T is said to be strongly variable if $\sigma(T) > \mathbb{E}[T]$, that is, $c_T > 1$. If we assume that the buses' interdeparture times are independent of one another and have the same distribution as the random variable T, then one can prove that the average waiting time at the bus stop is given by

$$\frac{1}{2}\left(1 + c_T^2\right)\mathbb{E}[T]$$

for someone who arrives at the bus stop at a random point in time. Indeed, if the buses ride exactly on schedule ($c_T = 0$), then the average waiting time is equal to $\frac{1}{2}\mathbb{E}[T]$, as expected. Otherwise, the average waiting time is always longer than that. If the interdeparture times vary strongly ($c_T > 1$), then the average waiting time is even longer than the average interdeparture time.

---

[1]The waiting time paradox is related to the inspection paradox. For Poisson arrivals this paradox is detailed in Appendix C.

The Mathematics of Waiting

In queueing theory, which has its origins in the pioneering work the Danish engineer A.K. Erlang did in the field of telephony in the early twentieth century, a great number of queueing models have been analyzed mathematically. However, practical solutions have only been found for relatively simple models. In a queueing model, one is usually forced to make simplifying assumptions about the customer arrival process and the probability distribution of the service times if the model is to be solvable exactly. Nevertheless, the importance of a solution for a *simplified* model should not be underestimated. Such a solution can often serve as the basis for an approximate solution for a more complex model that is more closely in line with reality. If necessary, such an approximate solution can be improved by means of computer simulation. The latter is a widely used method for analyzing queueing problems. Although this method makes it possible to stick more strictly to reality, the value of simulation should not be overestimated. Whereas a model provides insight, a simulation mostly produces a great many numbers. It is not always easy to draw useful conclusions from all those numbers. Wherever possible, a combination of queueing theory and simulation seems more useful than merely carrying out a simulation.

In Section 9.2, we first discuss the basic elements of a queueing model, such as the customer arrival pattern and the service mechanism. In particular, we comment on the widely used Poisson arrival process. Not only is this process easy to handle mathematically, it also gives a good model description in many real-world situations. A sort of law of nature in queueing theory is Little's law, which gives a relationship between a customer's average waiting time and the average queue length. This formula and the fundamental property that "Poisson arrivals see time averages" are discussed in Section 9.3. In Section 9.4, we give several queueing models that have proved their worth in practice. These models are illustrated with a number of examples. Finally, Section 9.5 deals with networks of queues.

The Psychology of Waiting

Mathematical models give us insight into the causes of waiting and help us design systems that reduce waiting times. However, a psychological approach that enhances the customer's *perception* of the wait can be as effective as a real reduction of the waiting time. A basic psychological law for customer satisfaction is

$$satisfaction = perception - expectation.$$

This means the following:

> *If someone expects a certain level of service and experiences the service that is received as better, that person will be a satisfied customer. The customer's satisfaction regarding the wait can be influenced in two directions: by changing the customer's expectation and by changing the customer's perception of the wait.*

The following five basic principles can be used to influence the customer's satisfaction regarding the waiting times:

- *Waiting where the customer has some distraction seems shorter than simply waiting.* In New York, some large banks have light entertainment in the hall with, for example, live piano music. In Mexico City, street artists entertain drivers waiting at traffic lights with all kinds of acrobatic acts (and earn money from the drivers' waiting). At the Walt Disney amusement parks, the queues are also often entertained and, just as importantly, constantly kept in motion.
- *Waiting without information seems longer than waiting with information.* This is particularly true in situations where customers have no idea of the expected waiting time, for example in the case of a delayed plane or train.
- *Waiting where customers are not served in the order of arrival seems longer than waiting where customers experience the service order as fair* (assuming that the customers can see one another). This can be the psychological reason for having a single queue at a fast-food restaurant or at an airline's check-in desks.
- *Waiting seems longer when customers see that the servers are not doing their actual job.* This means, for example, that it is wise to keep servers who are engaged in other jobs out of sight of waiting customers.
- *A server's friendliness can do wonders to alleviate the suffering of waiting.*

The importance of the "perception of waiting" is shown by the following true story. At an airport in Texas, passengers complained strongly about long waiting times at baggage handling. Initially, engineers searched for expensive and complex adjustments to the baggage handling system, until someone came up with the brilliant idea of placing the baggage belt a little further away from the arrival gates. This simple solution was implemented, and almost all complaints disappeared even though the baggage handling time had not changed. In a similar way, the perception of waiting for elevators can be positively influenced by the use of mirrors near the elevators. Looking at oneself and others in the mirrors makes the wait seem shorter.

9.2   Basic Elements

For the description of a queueing model, a number of characteristics, such as the customer arrival pattern and the service mechanism, must be specified. In Section 9.2.1, we briefly discuss each of these characteristics. If, in the future, we speak of "customers" and "servers," these need not be persons. A customer may also be a machine, a ship, or a plane. Likewise, the term "server" must be interpreted broadly. In Section 9.2.2, we briefly discuss data analysis for the arrival process and service time distribution.

9.2.1 *Queue Characteristics*

Before a queueing model can be analyzed, a number of the model's characteristics must be specified.

The Arrival Process

The source or population of customers can be finite or infinite. If the population is finite but very large, the mathematical analysis is often simplified by assuming that the population is infinite.

An arrival process that assumes an infinite population and is useful to model many real-world situations is the *Poisson* process discussed in Appendix C. The Poisson process can be described as follows: the customer arrival times are unpredictable in the sense that the probability of a customer arrival during an upcoming time interval does not depend on how long ago the previous customer arrived. The Poisson arrival process is therefore *memoryless* and is the complete opposite of an arrival process in which customers arrive at fixed intervals (think of a system with appointments). In contrast to the Poisson process, in a deterministic arrival process, the time elapsed since the last arrival is crucial. Simple queueing formulas can generally only be given in case of a Poisson arrival process.

The Service Mechanism

In the first place, we must specify whether there are one or more servers. If there are several servers, we must indicate whether there is a single common queue or every server has a separate queue. The queue discipline that indicates in what order the customers are served is also important. The most common queue discipline is that where customers are served in order of arrival. In some situations, a priority rule is used where certain customers take precedence over others.

The probability distribution of a customer's service time is an important quantity. A common distribution is the *exponential distribution*. In this case, we have

$$\mathbb{P}(\text{service time is at most } t \text{ units of time}) = 1 - e^{-\mu t} \quad \text{for } t > 0.$$

The exponential distribution with parameter μ has expected value $1/\mu$. The exponential distribution is very popular in the literature because it is so manageable mathematically. An exponential service time has the characteristic property that the residual duration of an ongoing service has the same exponential distribution as the original service time regardless of how long the service has been in progress. So the exponential distribution is also "*memoryless*", see (8.1). In many situations, the exponential distribution provides a good description of the service time (for example, the duration of telephone calls), but in a great number of cases, the exponential distribution is not suitable for modeling the service time's distribution. In practice, the lognormal distribution is often used to model the latter. In general, the queueing formulas become rather complex when the service time is not exponentially distributed.

The Queue Capacity

In some systems, the queue capacity is assumed to be unbounded; that is, every arriving customer is admitted and waits to be served. Other systems have no queue capacity; that is, a prospective customer who finds all servers busy is not admitted into the system. These are called *loss systems*.

Between queueing systems with an infinite waiting room and loss systems, there are systems with a finite waiting room. In these finite-capacity systems, only those prospective customers are rejected who find all servers busy and all queue positions occupied.

Kendall's Notation

A convenient notation has been developed to describe queueing models. This notation is of the form $a/b/c$ or $a/b/c/d$ and is called Kendall's notation. The first symbol describes the arrival process, the second the service time distribution, and the third the number of available servers. The fourth symbol is used if the system does not have an infinite waiting room and gives the maximum number of customers who can be admitted into the system. Frequently used symbols for a are

M: the arrival process is a Poisson process,
D: the interarrival times are deterministic,

while commonly used symbols for b are

M: the service time is exponentially distributed,
D: the service time is deterministic,
G: the service time has a general probability distribution.

For example, the $M/M/1$ queue describes the situation where the customer arrivals follow a Poisson process, the customer service times are exponentially distributed, and there is one server.

9.2.2   *Data Analysis*

Suppose that the Poisson process appears to be a plausible model for the customer arrivals. How can one test whether the arrival data justify the use of the Poisson process? Roughly speaking, there are two types of tests: graphical tests and statistical ones. Graphical tests are informal and only give a first impression. Statistical tests are based on formal criteria. In this section, we briefly discuss the graphical method of the Q-Q plot and the statistical test of Kolmogorov–Smirnov.

Graphical Method: The Q-Q Plot

In a Poisson process, the customer interarrival times are independent random variables that have a common exponential distribution. A Q-Q plot is a first step to verifying whether it is reasonable to expect the data to come from an exponential distribution. More generally, a Q-Q plot is used to obtain an idea of the family of probability distributions in which to search for a distribution that fits the data. A Q-Q plot is a better visual aid than a histogram of the data. This holds, in particular, if the data come from a continuous distribution: the choice of the length of the histogram's subintervals can greatly influence the histogram's shape. The great advantage of the Q-Q plot is that for the main families of probability distributions, it is not necessary to know the relevant distribution's parameters beforehand. A Q-Q plot shows whether the percentiles of the empirical distribution set out against the percentiles of the proposed theoretical distribution approximately lie on a *straight line*. If $F(x)$ is a *continuous* probability distribution function, then the pth percentile of the theoretical distribution function $F(x)$ is defined as the smallest number x_p with

$$F(x_p) = p \quad \text{for } 0 < p < 1.$$

Assuming that $F(x)$ is strictly increasing, x_p is given by

$$x_p = F^{-1}(p).$$

For the empirical distribution of the data, the percentiles are determined as follows. Assuming that the data $X_1, \ldots, X_n$ come from a continuous probability distribution, order the data as

$$X_{(1)} < X_{(2)} < \cdots < X_{(n)}.$$

The proportion of the data that are less than or equal to $X_{(i)}$ is $\frac{i}{n}$. In other words, $X_{(i)}$ is the empirical distribution's $\frac{i}{n}$th percentile. For technical reasons, it is better to view $X_{(i)}$ as the empirical distribution's $\frac{i-0.5}{n}$th percentile. To check whether a theoretical probability distribution function $F(x)$ is a good fit for the data, plot the points

$$\left(X_{(i)}, F^{-1}\left(\frac{i - 0.5}{n} \right) \right) \quad \text{for } i = 1, \ldots, n$$

and check whether these approximately lie on a straight line. If this is not the case, the proposed theoretical distribution function can be excluded. Otherwise, there is a *provisional* indication that the theoretical distribution function can be used for the data. A statistical test then provides a more definitive answer. The Q-Q plot only gives a first impression. As mentioned before, for some important families of probability distributions, it is not necessary to know the distribution's parameters to make a Q-Q plot. Take, for example, the normal distribution with parameters μ and σ. The linear transformation $z = (x - \mu)/\sigma$ converts the distribution to the standard normal distribution; that is, the $N(\mu, \sigma^2)$ distribution's percentiles are a linear function of the $N(0,1)$ distribution's percentiles. To use a Q-Q plot to check whether a normal distribution fits the data, it therefore suffices to use the percentiles of the standard normal distribution. Likewise, for the family of the exponential distribution function $F(x) = 1 - e^{-\lambda x}$, there is no objection to choosing $\lambda = 1$ when making the Q-Q plot.

The Kolmogorov–Smirnov Test

This test is based on the maximum deviation between the theoretical distribution function and the empirical distribution function. The test can only be applied if the theoretical distribution function's parameters are specified. The theoretical distribution function is only accepted if the maximum deviation is sufficiently small. Critical values of the Kolmogorov–Smirnov test have been tabulated in the statistical literature. For continuous probability distributions, these critical values are independent of the distribution if the distribution's parameters have not been estimated from the data; otherwise, adjustments must be made depending on the form of the theoretical distribution function. For details, the reader is referred to the statistical literature.

The Kolmogorov–Smirnov test can be used to test whether the data on the arrival times during a *given* time interval $(0, T)$ come from a Poisson process. To do this, a relationship between the Poisson process and the uniform distribution is used rather than that between the Poisson process and the exponential distribution. In this case, it is not necessary to estimate a parameter. Suppose given that n arrivals have taken place during a given time interval $(0, T)$; then if the arrival process is a Poisson process, one can prove that the simultaneous probability distribution of the n arrival times is the same as the simultaneous probability distribution of the *ordered* values of n independent draws from the uniform distribution on $(0, T)$. The uniform distribution on $(0, T)$ has probability distribution function $F(t) = t/T$ for $0 \leq t \leq T$. The Kolmogorov–Smirnov test statistic for the Poisson process is then given by

$$\max_{t = t_1, \ldots, t_n} \left| \frac{A(t)}{A(T)} - \frac{t}{T} \right|,$$

where the data $t_1, \ldots, t_n$ are the n arrival times in $(0, T)$ and $A(t)$ is the number of arrivals in $(0, t]$. The Kolmogorov–Smirnov test is carried out conditionally on the number n of arrivals. Finally, we note that the Kolmogorov–Smirnov test is also very useful for testing whether, for example, a lognormal distribution can be fit to the available data on the customer service times.

9.3   Fundamental Queueing Results

In this section, we discuss some fundamental building blocks for queueing theory, such as Little's law, the property that "Poisson arrivals see time averages," and the regenerative stochastic process.

9.3.1   *Little's Law*

To introduce Little's law, we first consider some illustrative examples.

(1) At a hospital, on average, 25 new patients are admitted every day. A patient stays an average of four days. What is the average number of busy beds? Denote

by $\lambda = 25$ the average number of new patients admitted per day, by $W = 4$ the average number of days a patient stays at the hospital, and by L the average number of occupied beds. Of course, we have $L = \lambda W = 25 \times 4 = 100$ beds.

(2) A specialty shop sells an average of 100 bottles of a unique Mexican beer per week. On average, the shop has 250 bottles in stock. What is the average number of weeks a bottle is kept in stock? Denote by $\lambda = 100$ the average demand per week, by $L = 250$ the average number of bottles in stock, and by W the average number of weeks a bottle is kept in stock. The answer is then $W = L/\lambda = 250/100 = 2.5$ weeks.

These examples illustrate Little's law $L = \lambda W$. Little's law is a sort of law of nature in operations research and applies to almost all queueing systems. This important formula relates the average number of customers in the system (respectively, queue) to a customer's average *sojourn time*, that is, the time the customer spends in the system (respectively, waiting time). Let us consider an arbitrary queueing system without specifying the customer's arrival process, the evolution of the customer service time, or the number of available servers. However, we do assume that the system has an *infinite* waiting room and that every arriving customer enters and waits to be served. Define the following random variables:

$$
\begin{aligned}
L(t) \quad &= \text{number of customers in the system at time } t \\
&\quad \text{(including the customers in service),} \\
L_q(t) &= \text{number of customers in the queue at time } t \\
&\quad \text{(excluding the customers in service),} \\
V_n \quad &= \text{time the } n\text{th customers spends in the system} \\
&\quad \text{(including service time),} \\
W_n \quad &= \text{time the } n\text{th customer spends waiting in line} \\
&\quad \text{(excluding service time).}
\end{aligned}
$$

Next, we define the numbers

$$
\begin{aligned}
L &= \lim_{t \to \infty} \frac{1}{t} \int_0^t \mathbb{E}[L(u)]\, du && \text{(average number of customers in the system)} \\
L_q &= \lim_{t \to \infty} \frac{1}{t} \int_0^t \mathbb{E}[L_q(u)]\, du && \text{(average queue length)} \\
W &= \lim_{n \to \infty} \frac{1}{n} \sum_{k=1}^n \mathbb{E}[V_k] && \text{(average sojourn time per customer)} \\
W_q &= \lim_{n \to \infty} \frac{1}{n} \sum_{k=1}^n \mathbb{E}[W_k] && \text{(average waiting time per customer).}
\end{aligned}
$$

These limits exist under very general conditions. Roughly speaking, it suffices to require that the system becomes empty from time to time and then starts again, as it were (regenerates), where the expected time between the regeneration moments is finite. This condition is fulfilled in most real-world applications. In fact, under this assumption, we can give an even more appealing interpretation of the long-term averages defined above. The average number of customers in the system during the upcoming t units of time is a random variable, but it becomes less and less random as t increases, and it converges to the constant L with probability 1 when t becomes infinitely large. In other words, although the number of customers in the system fluctuates over time, the average long-run number of customers in the system is equal to the constant L with probability 1. This result is, in fact, based on the average cost formula for regenerative stochastic processes that we discuss in the next subsection. Analogous statements hold for L_q, W, and W_q. The statement that a certain quantity takes on a specific value with probability 1 is, of course, much more powerful than the statement that the quantity has that value as expected value.

We define the random variable $N(t)$ as

$$N(t) = \text{total number of arrivals up to time } t$$

and the arrival rate λ as

$$\lambda = \lim_{t \to \infty} \frac{\mathbb{E}[N(t)]}{t}.$$

The famous *Little's law* now states that

$$L = \lambda W \quad \text{and} \quad L_q = \lambda W_q.$$

This means that if one of the quantities L, L_q, W, or W_q has been calculated, the others directly follow using Little's law. Of course, we use the fact that there exists a simple relation between W and W_q. If $\mathbb{E}[B]$ is the average time a customer is in service, then we have the obvious relation

$$W = W_q + \mathbb{E}[B].$$

We can give an intuitive explanation for Little's law. Suppose that every incoming customer pays money to the queueing system's manager according to a fixed rule that is the same for every customer. Intuitively, the following principle is immediately clear:

> **Money principle:** *In the long run, the system's average yield per unit of time = (the average number of paying arrivals per unit of time) × (the average amount received per paying customer).*

Choose the payment rule as follows. Every customer pays €1 for every unit of time spent in the system. Then the amount a customer pays is equal to the customer's time in the system. In other words, in the long run, the average amount paid per customer is equal to W. On the other hand, the system receives j euros for every unit of time that j customers are present. In other words, the system's long-run average yield per unit of time is equal to the average number of customers in the system $(= L)$. The average number of arrivals per unit of time is defined to be λ. The arrival rate λ also gives the average number of paid arrivals per unit of time because every arriving customer is admitted into the system. This gives $L = \lambda W$. In the same way, one finds the formula $L_q = \lambda W_q$ by assuming that every customer pays €1 per unit of time spent in the queue. One can also choose a suitable payment rule to heuristically argue that

$$\text{the average number of busy servers} = \lambda\mathbb{E}[B],$$

assuming that every customer needs only one server and every server helps only one customer at a time.

The formulas given above require only a minor adjustment if the queueing system has a *bounded* waiting room and any customer who finds the waiting room full upon arrival is rejected. The quantities W_q and W are now defined as, respectively, the average waiting time and average time in the system per *admitted* customer. If we define

$$P_{\text{rej}} = \text{fraction of the customers who are rejected},$$

then the versions of Little's law given earlier remain valid if we replace λ by $\lambda(1 - P_{\text{rej}})$. After all, the average number of paying customers that now arrive per unit of time is equal to $\lambda(1 - P_{\text{rej}})$.

9.3.2 *Poisson Arrivals See Time Averages (PASTA)*

For a Poisson arrival process, for every time interval of length Δt with Δt small, the probability of an arrival during that interval is the same regardless of the course of the arrival process before this time interval. Roughly speaking, Poisson arrival times are completely random on the time axis. This is the intuitive explanation for the following characteristic property of a Poisson arrival process:

> *the long-run fraction of the customers who find the system in a particular state = the long-run fraction of the time the system is in that state.*

In other words,

> *if the arrival process is a Poisson process, the probability distribution of the system's state right before an arrival time is the same as the probability distribution of the system's state at an arbitrary time if the system has reached steady state.*

This property is known as *"Poisson Arrivals See Time Averages"* (the PASTA property) and means that if we can calculate one of the two probability distributions mentioned above, we automatically have the other one provided that the arrival process is a Poisson process.[2] The general proof of the PASTA property is quite deep; we do not give it here. For a queueing system with Poisson arrivals, a simple proof of the PASTA property can be given for the particular case that the evolution of the number of customers in the system is described by a continuous-time Markov

---

[2]The assumption of a Poisson arrival process is essential. The following counterexample with a deterministic arrival process shows this: Customers arrive exactly every minute at a service station, and a customer's service time is uniformly distributed between 1/4 and 3/4 of a minute. Then the fraction of the customers who find the server busy upon arrival is not equal to the fraction of the time the server is busy.

chain; see Section 8.7.1. However, by the PASTA property, the result that the equilibrium distribution of the number of customers present *right before* an arrival time is equal to the equilibrium distribution of the number of customers present at an *arbitrary* time also holds for queueing models with generally distributed service times if the customer arrivals follow a Poisson process.

An explicit proof of the PASTA property for a birth and death process is as follows.

Theorem 9.1 (PASTA for the birth and death process). *Consider a birth and death process $\{X(t)\}$ with constant birth rates $\lambda_j = \lambda$ for all $j \in I$. Then for all $i \in I$*

$$p_i = \text{long-run fraction of customers who find } i \text{ other customers upon arrival}$$
$$= \text{long-run fraction of the time that } i \text{ customers are present in the system}$$
$$= \pi_i.$$

Proof. Let $i \in I$. The probability that a birth in $(t, t + \Delta)$ finds the system in state i is equal to

$$\mathbb{P}(X(t) = i \mid \text{arrival in } (t, t + \Delta t)) = \frac{\mathbb{P}(X(t) = i, \text{ arrival in } (t, t + \Delta t))}{\mathbb{P}(\text{arrival in } (t, t + \Delta t))}$$
$$= \frac{\mathbb{P}(X(t) = i)\mathbb{P}(\text{arrival in } (t, t + \Delta t) \mid X(t) = i)}{\mathbb{P}(\text{Arrival in } (t, t + \Delta t))}$$
$$= \frac{\mathbb{P}(X(t) = i)\mathbb{P}(\text{arrival in } (t, t + \Delta t) \mid X(t) = i)}{\sum_{j \in I} \mathbb{P}(X(t) = j)\mathbb{P}(\text{arrival in } (t, t + \Delta t) \mid X(t) = j)}$$
$$= \frac{\mathbb{P}(X(t) = i)(\lambda_i \Delta t + o(\Delta t))}{\sum_{j \in I} \mathbb{P}(X(t) = j)(\lambda_j \Delta t + o(\Delta t))}, \tag{9.1}$$

where we have used Bayes' rule in the first and second equality, the law of total probability in the third equality and $\mathbb{P}(\text{arrival in } (t, t + \Delta t) \mid X(t) = i) = \lambda_i \Delta t + o(\Delta t)$ for the birth and death process in the fourth equality. For constant birth rates $\lambda_j = \lambda$ we obtain

$$\mathbb{P}(X(t) = i \mid \text{arrival in } (t, t + \Delta t)) = \mathbb{P}(X(t) = i),$$

which in equilibrium (taking the limit $t \to \infty$) gives $p_i = \pi_i$. $\qquad \square$

Observe that dividing (9.1) by Δt and taking the limit $t \to \infty$ and $\Delta t \to 0$ in (9.1) gives

$$p_i = \frac{\pi_i \lambda_i}{\sum_{j \in I} \pi_j \lambda_j},$$

which illustrates that the long-run fraction of customers who find i other customers upon arrival is obtained as the ratio of the arrival flow in state i and the total arrival flow.

The PASTA property and Little's law are illustrated by the following situation. A simulation is carried out for a rather complex inventory model. In this model, customers arrive following a Poisson process with rate λ, and every customer requests exactly one unit of a particular product. The customer's demand is lost if the system is out of stock. The simulation shows that, in the long run, the fraction of the time that the system is out of stock is equal to 0.05, while the average on-hand inventory level is 25 units. Based on this information, answer the following questions: What fraction of the demand is lost? What is the average time a unit of product is on-hand? The answer to the first question is 0.05 by the PASTA property. If we denote by W the average amount of time a unit is kept in stock, then W follows from Little's law $L = \lambda(1 - P_{\text{rej}})W$ with $L = 25$ and $P_{\text{rej}} = 0.05$.

9.3.3 Regenerative Stochastic Processes

A stochastic process $\{X(t)\}$ is a collection of random variables indexed by either a discrete or a continuous time parameter. Discrete-time Markov chains and continuous-time Markov chains are examples of stochastic processes. A stochastic process $\{X(t)\}$ is called *regenerative* if there exists a time T_1 with $\mathbb{E}[T_1] > 0$ such that, probabilistically, the continuation of the stochastic process from time T_1 on is a repetition of the process starting at time $T_0 = 0$. The existence of a regeneration time T_1 directly implies the existence of similar regeneration times $T_2, T_3, \ldots$ Each of the time intervals (T_0, T_1), (T_1, T_2), (T_2, T_3), $\ldots$ is called a *cycle* of the regenerative process. In most queueing systems, the queue length process $\{L_q(t), t \geq 0\}$ is regenerative with regeneration times the times at which a customer arrives and finds no other customers present; the waiting time process $\{W_n, n = 1, 2, \ldots\}$ is also regenerative with regeneration indices the indices of the customers who find no other customers present upon arrival. Now, suppose that the regenerative process $\{X(t)\}$ has a well-defined cost structure such that the costs in consecutive cycles are independent and identically distributed. If we assume that both the length of a cycle and the total cost within a cycle have finite expected values, then the following simple, but extremely useful, formula holds:

Average cost formula: *If $C(t)$ is the total cost during $[0, t)$, then the long-run average cost per unit of time is given by*

$$\lim_{t \to \infty} \frac{C(t)}{t} = \frac{\mathbb{E}[\text{total cost during a cycle}]}{\mathbb{E}[\text{length of a cycle}]} \quad \text{with probability 1.}$$

Let us outline a proof. Let $L_i = T_i - T_{i-1}$ be the length of the ith cycle and C_i the total cost during the ith cycle. Define the random variable $N(t)$ as the number of completed cycles in $[0, t)$; that is, $\sum_{i=1}^{N(t)} L_i \leq t < \sum_{i=1}^{N(t)+1} L_i$. If, for convenience, we assume that the total costs are nonnegative, then we also have $\sum_{i=1}^{N(t)} C_i \leq C(t) \leq \sum_{i=1}^{N(t)+1} C_i$. This gives

$$\frac{\sum_{i=1}^{N(t)} C_i}{\sum_{i=1}^{N(t)+1} L_i} \leq \frac{C(t)}{t} \leq \frac{\sum_{i=1}^{N(t)+1} C_i}{\sum_{i=1}^{N(t)} L_i} \quad \text{for every } t > 0.$$

The random variables $L_1, L_2, \ldots$ are independent and identically distributed, as are the random variables $C_1, C_2, \ldots$. By the strong law of large numbers,

$\lim_{n\to\infty}(1/n)\sum_{i=1}^{n} L_i = \mathbb{E}[L_1]$ with probability 1 and $\lim_{n\to\infty}(1/n)\sum_{i=1}^{n} C_i = \mathbb{E}[C_1]$ with probability 1. Moreover, we have $\lim_{t\to\infty} N(t) = \infty$. Dividing the numerator and denominator in both the left-hand side and the right-hand side of the displayed inequality by $N(t)$ and taking $t \to \infty$ gives the average cost formula.

As an illustration, consider the following queueing problem. In a queueing system with one server and no waiting room, customers arrive following a Poisson process with rate λ. A customer who finds the server busy at arrival is rejected and lost; otherwise, the customer is served immediately. A customer's service time is a random variable S with a finite expected value. What long-run fraction of the time is the server busy, and what fraction of the customers is lost? These questions can be answered using the average cost formula. For convenience, assume that at time 0, a customer arrives who finds the server idle. For every $t \geq 0$, define the random variable $I(t)$ to be 1 if the server is busy at time t, and set $I(t) = 0$ otherwise. For every $n = 1, 2, \ldots$, define the random variable I_n to be 1 if the nth arriving customer finds the server busy, and set $I_n = 0$ otherwise. Both the continuous-time process $\{I(t)\}$ and the discrete-time process $\{I_n\}$ are regenerative. A cycle begins every time an arriving customer finds the server idle. The fraction of the time that the server is busy can be seen as an average cost per unit of time by assuming a cost of 1 for every unit of time the server is busy. By the average cost formula, in the long run, the fraction of the time that the server is busy is equal to the expected time in a cycle that the server is busy divided by the expected length of a cycle. The expected time the server is busy during a cycle is, of course, $\mathbb{E}[S]$. The memoryless property of the Poisson process now implies that the expected time the server is idle during a cycle is equal to the expected value $1/\lambda$ of the interarrival time. It follows that, in the long run,

$$\text{the fraction of the time that the server is busy} = \frac{\mathbb{E}[S]}{\mathbb{E}[S] + 1/\lambda} \quad \text{with probability 1.}$$

Likewise, it follows from the average cost formula that, in the long run, the fraction of the customers who are lost is equal to the expected number of customers who are lost during a cycle divided by the expected number of customer arrivals during a cycle. The customer arrivals follow a Poisson process, so that during every time interval of length s, the expected number of arrivals is equal to λs. The law of conditional expectations then says that the expected number of rejected customers in the stochastic service time S is equal to $\lambda \mathbb{E}[S]$. In each cycle, only one customer is accepted, so that the expected number of customers arriving during a cycle is equal to $1 + \lambda \mathbb{E}[S]$. It follows that, in the long run,

$$\text{the fraction of the customers who are lost} = \frac{\lambda \mathbb{E}[S]}{1 + \lambda \mathbb{E}[S]} \quad \text{with probability 1.}$$

This expression for the fraction of the customers who are lost is equal to the expression $\frac{\mathbb{E}[S]}{\mathbb{E}[S]+1/\lambda}$ for the fraction of the time that the server is busy. In other words, in the long run, the fraction of the customers who find the server busy upon arrival is equal to the fraction of the time that the server is busy. This means that "Poisson arrivals see time averages." Another example in which the PASTA property is confirmed by the average cost formula is the inventory problem from Example 8.1 in Chapter 8. Use the average cost formula to verify that, in the long run, the fraction of the time that the system is out of stock is equal to $\frac{1/\mu}{1/\mu+Q/\lambda}$ and the fraction of the customers who find the system out of stock upon arrival is equal to $\frac{\lambda/\mu}{\lambda/\mu+Q}$.

9.4 Queueing Formulas

In this section, we discuss a number of simple queueing models that have proved their worth in practice. In each of these models, we assume that the customer service times are independent and identically distributed random variables that are also independent of the customers' arrival process.

Let us first introduce some notation. In the previous section, we have already defined the quantities L and L_q as the long-run average numbers of customers in the system and in the queue, respectively, and the quantities W and W_q as the long-run average amounts of time a customer is in the system and waits in line, respectively. Next, we define the probability distribution of the number of customers in the system and that of a customer's waiting time. For this, we define the random variable $T_j(t)$ as the total amount of time j customers are in the system during the time interval $(0, t)$. If we let t become very large, then by a deep theorem from probability theory, the fraction $T_j(t)/t$ becomes less and less variable and approaches a constant with probability 1. We denote this constant by π_j. So

$\pi_j =$ the long-run fraction of the time that j customers are present.

Likewise, the fraction of the customers with a waiting time less than or equal to a given numerical value x goes to a constant if it is taken over an increasingly large number of customers. We denote this constant by $W_q(x)$. So

$W_q(x) =$ the long-run fraction of the customers with a
waiting time of at most x units of time.

The numbers π_j and $W_q(x)$ can also be interpreted as limiting probabilities when the service times or the interarrival times are continuously distributed. Suppose that an outsider inspects the system's state at an arbitrary time when the system has been running for a long time. Then the probability that the outsider finds j customers in the system is equal to π_j, assuming that the outsider has no knowledge of the system's evolution up to the inspection time. Likewise, the probability that an arriving customer has a waiting time of at most x units of time is equal to $W_q(x)$ for a customer who arrives when the system has been running for a long time, assuming that the customer has no knowledge of the system's evolution up to the arrival time. The interpretation of π_j and $W_q(x)$ as probabilities is rather subtle; it assumes not only that the system has been running for a long time, but also that the outsider or arriving customer has no knowledge of the system's evolution up to then. This is why a more straightforward interpretation of π_j and $W_q(x)$ as long-term averages is often preferable. In operations research applications, this last interpretation is usually the most relevant.

9.4.1   *The M/M/1 Queue*

This model is the simplest queueing model and assumes the following:

- Customers arrive following a Poisson process with an average of λ customers per unit of time.
- A customer's service time is exponentially distributed with an expected value of $1/\mu$ units of time.
- There is a single server who can help only one customer at a time.
- There is an infinite waiting room.

The quantity ρ is defined as

$$\rho = \frac{\lambda}{\mu}.$$

The number ρ is the quotient of the average service time $1/\mu$ and the average interarrival time $1/\lambda$. Intuitively, it should be clear that if the number of waiting customers is not to grow indefinitely, the average service time must be less than the average interarrival time. Hence, we assume

$$\rho < 1.$$

The quantity ρ is called the *server utilization*. For the $M/M/1$ queue, the equilibrium distribution of the number of customers present is a geometric distribution:

$$\pi_j = (1 - \rho)\rho^j \quad \text{for } j = 0, 1, \ldots . \tag{9.2}$$

A proof of this result is given in Section 8.7. Since $\pi_0 = 1 - \rho$ gives the fraction of the time that there are no customers in the system, we find that

$$\text{the fraction of the time that the server is busy } = \rho, \tag{9.3}$$

which is why ρ is called the server utilization. Another consequence of the formula given above for the π_j is that the average number of customers in the system is equal to

$$L = \frac{\rho}{1 - \rho}. \tag{9.4}$$

This follows from the relation $L = \sum_{j=0}^{\infty} j\pi_j$ and the formula $\sum_{j=0}^{\infty} ja^j = a/(1-a)^2$ for $|a| < 1$. Once L is known, we find the quantities W, W_q, and L_q by applying Little's law $L = \lambda W$, the relation $W = W_q + 1/\mu$, and Little's law $L_q = \lambda W_q$. This gives the queueing formulas

$$W = \frac{1}{\mu(1 - \rho)}, \quad W_q = \frac{\rho}{\mu(1 - \rho)}, \quad \text{and} \quad L_q = \frac{\rho^2}{1 - \rho}. \tag{9.5}$$

Under the assumption that the customers are served in order of arrival, we have shown in Section 8.7 that the waiting time probability $W_q(x)$ is given by

$$W_q(x) = 1 - \rho e^{-\mu(1-\rho)x} \quad \text{for } x \geq 0. \tag{9.6}$$

In particular, the fraction of the customers who wait must equal the server utilization ρ. This conclusion also follows directly from equation (9.3) using the PASTA property. After all, the fraction of the customers who must wait is the fraction of the customers who find the server busy upon arrival. By the PASTA property, the latter is equal to the fraction of the time the server is busy.

Before giving an application of the formulas above, let us take Figure 9.1 to show what useful insight the formulas provide. Figure 9.1 shows the influence of the server utilization on the average queue length. In this figure, one can see that above a server utilization of about 0.8, the average queue length increases rapidly with every increase in the utilization. The phenomenon that, at a *high* load, the average queue length and the average waiting time per customer increase *strongly* with a small increase in the customer arrival rate holds in general for stochastic service systems.

Fig. 9.1   The queue length versus the server utilization.

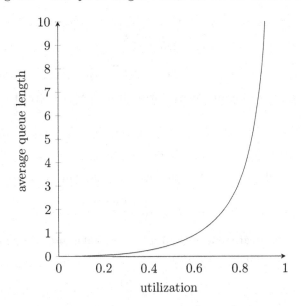

Example 9.1. At a port, the arrival of ships for unloading follows a Poisson process with an average of five ships per 24 hours. The unloading time per ship follows an exponential distribution. The port authority must decide whether to purchase a new unloading installation to replace the one in use. The question is what unloading rate the new installation must have. For an installation with unloading rate μ, the

average unloading time per ship is equal to $1/\mu$ hours, while the daily (depreciation) cost of such an installation is equal to $c_1\mu$ with $c_1 = 5000$. A ship's docking cost is proportional to the time the ship spends in the port (including unloading time); this cost equals $c_2 = 25$ per hour. The unloading installation, which is in operation day and night, can only unload one ship at a time.

What unloading rate must the new installation have for the average daily cost to be the lowest possible in the long run? For the chosen rate, calculate the fraction of the time that the installation is in use, the average number of ships waiting for unloading, and the probability that a ship must wait more than two hours.

Solution.   The different quantities must be expressed in the same unit of time. Take an hour as unit of time. The arrival rate is then equal to

$$\lambda = \frac{5}{24}$$

ships per hour. Suppose that an unloading installation with rate μ is purchased, so that the average unloading time per ship is equal to $1/\mu$ hours. The average docking cost per hour is equal to c_2 times the average number of ships present. Using equation (9.4), we then find that, in the long run, the average total cost per hour is equal to

$$g(\mu) = \frac{c_1}{24}\mu + c_2 L$$
$$= \frac{c_1}{24}\mu + c_2 \frac{\lambda/\mu}{1 - \lambda/\mu}.$$

The function $g(\mu)$ must be minimized for μ. If we set the derivative of $g(\mu)$ equal to zero, we find

$$\frac{c_1}{24} - \frac{c_2\lambda}{(\mu - \lambda)^2} = 0.$$

Because of the condition $\rho < 1$, we must have $\mu > \lambda$. This leads to the solution

$$\mu^* = \lambda + \sqrt{\frac{24c_2\lambda}{c_1}}.$$

If we fill in the numerical values $\lambda = 5/24$, $c_1 = 5000$, and $c_2 = 25$, then we find that an unloading installation with rate

$$\mu^* = 0.3664$$

must be purchased. For this installation, the average unloading time per ship is equal to 2.73 hours.

The installation's utilization ρ is equal to $\lambda/\mu^* = 0.569$. So, in the long run, the installation will be in use 56.9% of the time. It follows from the formula $L_q = \rho^2/(1-\rho)$ that, on average, 0.75 ships are waiting for unloading. Finally, it follows from equation (9.6) for $W_q(x)$ that the probability that an arriving ship must wait longer than two hours is equal to $0.569 \times e^{-0.316} = 0.41$.

9.4.2 The M/G/1 Queue

This model is a generalization of the $M/M/1$ queue and assumes the following:

- Customers arrive following a Poisson process with an average of λ customers per unit of time.
- A customer's service time has a general probability distribution with expected value $\mathbb{E}[B]$ and standard deviation $\sigma(B)$.
- There is a single server.
- There is an infinite waiting room.

The server utilization ρ is now defined by

$$\rho = \lambda\mathbb{E}[B],$$

and we assume

$$\rho < 1.$$

For the $M/G/1$ queue, we cannot give simple formulas for the probabilities π_j and $W_q(x)$. In fact, we need the complete probability distribution of the service time in order to give these probabilities. In our model description, we have assumed that only the expected value and standard deviation of the service time are known. These two key numbers suffice for several important performance measures. We have that

$$\text{the fraction of the time that the server is busy } = \rho \qquad (9.7)$$

and

$$\text{the fraction of the customers who must wait } = \rho.$$

The first formula is a particular case of Little's law from Section 9.3.1 for the average number of busy servers. The second formula follows from the first using the PASTA property. Of the service time, these formulas use only the expected value. The service time's standard deviation is needed for the famous *Pollaczek–Khintchine formula* for the average queue length. This formula is

$$L_q = \frac{1}{2}\left(1 + c_B^2\right)\frac{\rho^2}{1 - \rho}, \qquad (9.8)$$

where the service time's *coefficient of variation* c_B is defined by

$$c_B = \frac{\sigma(B)}{\mathbb{E}[B]}.$$

In the particular case of an exponentially distributed service time, we have $c_B = 1$, and the formula for L_q given here reduces to the formula for L_q given before for the $M/M/1$ queue.

Using Little's laws $L_q = \lambda W_q$ and $L = \lambda W$ and the relation $W = W_q + \mathbb{E}[B]$, we find $L = L_q + \rho$ and the *Pollaczek–Khintchine formula* for the average waiting time

$$W_q = \frac{1}{2}\left(1 + c_B^2\right)\frac{\rho\mathbb{E}[B]}{1 - \rho}. \qquad (9.9)$$

The Pollaczek–Khintchine formula is one of the most famous formulas from queueing theory. The insight this formula gives is, in fact, more important than the fact that the formula gives an explicit expression for the average queue length. The Pollaczek–Khintchine formula quantifies how the average queue length decreases as the service time's variability is reduced while the average service time remains the same. The decrease in the average queue length is determined by the decrease in the factor $1 + c_B^2$, where c_B is the service time's coefficient of variation. For example, $L_q = \rho^2/(1 - \rho)$ if the service time is exponentially distributed ($c_B^2 = 1$) and $L_q = \frac{1}{2}\rho^2/(1 - \rho)$ if the service time is constant ($c_B^2 = 0$). The difference is a factor of 2. If we denote by $L_q(\exp) = \rho^2/(1 - \rho)$ the average queue length in the particular case of exponentially distributed service times, then for the case of generally distributed service times, equation (9.8) for L_q can be rewritten as

$$L_q = \frac{1}{2}(1 + c_B^2)L_q(\exp).$$

Another interesting representation for L_q is (verify!)

$$L_q = (1 - c_B^2)L_q(\det) + c_B^2 L_q(\exp),$$

where $L_q(\det) = \frac{1}{2}\rho^2/(1 - \rho)$ is the average queue length in the particular case of deterministic service times.

A Heuristic Derivation of the Pollaczek–Khintchine Formula

An insightful heuristic derivation of the Pollaczek–Khintchine formula is as follows. This derivation can also be used for other situations. If the effect of the system's initial state has worn off and the system has reached steady state, then by the PASTA property, upon arrival a customer finds an average of L_q other customers in the queue. Moreover, the arriving customer has probability ρ of finding the server busy because the fraction of the time that the server is busy is ρ; see (9.7). This means that a customer's average waiting time W_q is equal to

$$W_q = L_q \mathbb{E}[B] + \rho \mathbb{E}[B_{res}],$$

where B_{res} is the residual service time of the customer being served when a new customer arrives. The Poisson arrival process has the property that the arrival times are, as it were, "randomly" placed on the time axis. The waiting time paradox from Section 9.1 tells us that

$$\mathbb{E}[B_{res}] = \frac{1}{2}(1 + c_B^2)\mathbb{E}[B].$$

By substituting this formula in the relation between L_q and W_q given above and applying Little's law $L_q = \lambda W_q$, we then find

$$L_q\left[\frac{1}{\lambda} - \mathbb{E}[B]\right] = \frac{1}{2}(1 + c_B^2)\rho\mathbb{E}[B].$$

This gives $L_q = \frac{1}{2}(1 + c_B^2)\rho^2/(1 - \rho)$, so that we have deduced the Pollaczek–Khintchine formula heuristically.

Why Is the Queue for the Ladies' Room so Much Longer?

At the beach, along a highway, or at a music festival, the queue for the ladies' room is always much longer than that for the men's room. According to a British study, women spend an average of 89 seconds in the restroom, compared to the average of 39 seconds for men. Although women spend a bit more than twice as long in the restroom as men, on average, the queue for the ladies' room is much more than twice as long as that for the men's room. The object of the study was how to explain this. An explanation can be given using the Pollaczek–Khintchine formula. We assume the following:

(1) On average, the same number of women as men arrive at the restrooms per unit of time.
(2) On average, a woman spends twice as long in the restroom as a man.
(3) The time a woman spends in the restroom is more variable than the time a man spends in the restroom.
(4) There is one separate toilet for women and one for men.
(5) The men's arrivals and the women's arrivals at the restrooms both follow a Poisson process.

We denote the women's arrival rate, the average time a woman stays in the restroom, and the coefficient of variation of the time a woman stays in the restroom by λ_w, m_w, and c_w, respectively. For the men, we use the notation λ_m, m_m, and c_m. The assumptions are that $\lambda_w = \lambda_m$, $m_w = 2m_m$, and $c_w \geq c_m$. Moreover, we assume $\lambda_w m_w < 1$. Applying the Pollaczek–Khintchine formula gives

the average length of the queue for the ladies' room
$$= \frac{1}{2}(1 + c_w^2)\frac{(\lambda_w m_w)^2}{1 - \lambda_w m_w} \geq \frac{1}{2}(1 + c_m^2)\frac{(\lambda_m 2m_m)^2}{1 - 2\lambda_m m_m}$$
$$\geq 4\frac{1}{2}(1 + c_m^2)\frac{(\lambda_m m_m)^2}{1 - \lambda_m m_m}.$$

That is,

the average length of the queue for the ladies' room
$$\geq 4 \times \text{(the average length of the queue for the men's room)}.$$

In fact, the difference increases by much more than a factor of 4 as the arrival rate $\lambda_w = \lambda_m$ increases.

Example 9.2. In the family business Multilever, the product Entabogen is produced on order. This high-quality product can only be made on one machine. This machine can also be used for other purposes but can be made available at any time for the production of Entabogen. The orders for Entabogen arrive following a Poisson process with an average of one order every two days. The production time of an order is fixed and equals one and a half days.

What fraction of the time is the machine in use for the production of Entabogen? What is the average number of orders either begin processed or waiting to be processed? What fraction of the orders must wait longer than five days before being put into production?

Solution. The $M/G/1$ queue with constant service times applies to this problem. Choose a day as unit of time. The parameter values are then

$$\lambda = \frac{1}{2}, \quad \mathbb{E}[B] = \frac{3}{2}, \quad \text{and} \quad \sigma(B) = 0.$$

It follows from equations (9.7)–(9.9) that

the fraction of the time that Entabogen is in production $= 0.75$

$$\text{the average number of orders being} = \frac{1}{2} \times \frac{(0.75)^2}{1 - 0.75} + 0.75$$

or waiting to be processed

$$= 1.875.$$

A fraction $1 - W_q(5)$ of the orders must wait longer than five days before production on the order begins. For the $M/G/1$ queue with constant service times, we have not given a formula for $W_q(x)$. Nevertheless, such a formula is known, although it is not simple. In the case of a constant service time, we have

$$1 - W_q(5) = 0.053.$$

For comparison, in the case of an exponentially distributed service time with the same expected value of one and a half days, equation (9.6) gives the waiting time probability $1 - W_q(5) = 0.326$. This shows how significant the effect of the service time's standard deviation is on the waiting time probability.

9.4.3   *The M/G/1 Queue with Priorities*

Let us consider Example 9.2 again for the case that, on average, 50% of the orders have a production time of one day and 50% of the orders have a production time of two days. Now, if in the order of completion, we distinguish between orders according to their production time, then we can reduce the average waiting time per order. This is easy to see from the situation where two orders A and B are waiting for production when the machine becomes available. Suppose that A arrives before B, where A has a production time of two days and B only needs one day. If the production is carried out in order of arrival, order A is carried out before order B and both orders are completed after $2 + 1 = 3$ days, where the total waiting time for both orders is equal to two days, for order B. If the order is changed and B is carried out before A, then once again, both orders are completed after $2 + 1 = 3$ days, but now the total waiting time of both orders is one day, for order A. Changing the order of the production of A and B therefore reduces the total waiting time by one day and has no effect on the waiting times of the orders that go into production after A and B. So, by prioritizing orders with a short production time over orders with a longer production time, one can reduce the average waiting time per order, taken over all orders.

A simple formula can be given for the effect a priority rule has on the average waiting time per customer. Consider the following situation:

- Customers arrive following a Poisson process with an average of λ customers per unit of time.
- Each arriving customer is of type 1 with probability r_1 and of type 2 with probability $r_2 = 1 - r_1$.
- The service time of a customer of type i has expected value $\mathbb{E}[B_i]$ and standard deviation $\sigma(B_i)$ for $i = 1, 2$.
- A single server is available.
- Customers of type 1 are given priority over customers of type 2, where the service time of a customer with a lower priority cannot be interrupted when a customer with a higher priority arrives.

This model is called the $M/G/1$ priority queue with non-preemptive priority. If we set

$$\lambda_i = \lambda r_i \quad \text{for } i = 1, 2,$$

then by the splitting property of the Poisson process, customers of type 1 arrive following a Poisson process with parameter λ_1 and customers of type 2 arrive following a Poisson process with parameter λ_2, where the two processes are independent of each other. Define the quantities

$$W_q^{(i)} = \text{average waiting time per customer taken}$$
$$\text{over the customers of type } i$$

for $i = 1, 2$ and

$$W_q = \text{average waiting time per customer taken over all customers.}$$

Then we have the formulas

$$W_q^{(1)} = \frac{\lambda_1 \mathbb{E}[B_1^2] + \lambda_2 \mathbb{E}[B_2^2]}{2(1 - \lambda_1 \mathbb{E}[B_1])} \tag{9.10}$$

$$W_q^{(2)} = \frac{\lambda_1 \mathbb{E}[B_1^2] + \lambda_2 \mathbb{E}[B_2^2]}{2(1 - \lambda_1 \mathbb{E}[B_1])\,(1 - \lambda_1 \mathbb{E}[B_1] - \lambda_2 \mathbb{E}[B_2])} \tag{9.11}$$

and

$$W_q = \frac{\lambda_1}{\lambda_1 + \lambda_2} W_q^{(1)} + \frac{\lambda_2}{\lambda_1 + \lambda_2} W_q^{(2)}. \tag{9.12}$$

The average queue length of type i customers satisfies $L_q^{(i)} = \lambda_i W_q^{(i)}$ for $i = 1, 2$. The heuristic derivation of the Pollaczek–Khintchine formula in the standard $M/G/1$ queue can easily be adapted to the $M/G/1$ queue with priorities. We leave out the details.

Application

Let us return to the modified Example 9.2 where, on average, 50% of the orders have a production time of one day and 50% have a production time of two days. First, we calculate the average waiting time per order if the orders are produced in order of arrival. If we use the random variable B to denote the production time of a randomly selected order, then

$$\mathbb{E}[B] = \frac{1}{2} \times 1 + \frac{1}{2} \times 2 = 1\frac{1}{2} \quad \text{and} \quad \mathbb{E}[B^2] = \frac{1}{2} \times 1^2 + \frac{1}{2} \times 2^2 = 2\frac{1}{2}.$$

It follows that $\sigma^2(B) = 1/4$ and therefore $c_B^2 = 1/9$. Applying the Pollaczek–Khintchine formula for the average waiting time in the $M/G/1$ queue, we find

$$W_q = \frac{1}{2} \times \frac{10}{9} \times \frac{0.75 \times 3/2}{1 - 0.75} = 2.5.$$

Hence, the average waiting time per order is $2\frac{1}{2}$ days if the orders are produced in order of arrival.

Now, assume that the orders with a production time of one day take precedence over the orders with a production time of two days. To find the average waiting time, we apply equations (9.10)–(9.12) for the $M/G/1$ priority queue with

$$\lambda = \tfrac{1}{2}, \quad r_1 = \tfrac{1}{2}, \quad r_2 = \tfrac{1}{2},$$

$$\mathbb{E}[B_1] = 1, \quad \mathbb{E}[B_1^2] = 1, \quad \mathbb{E}[B_2] = 2, \quad \mathbb{E}[B_2^2] = 4.$$

After some calculation, this gives

$$W_q^{(1)} = 0.833, \quad W_q^{(2)} = 3.333, \quad \text{and} \quad W_q = 2.083.$$

So, under the priority rule, the average waiting time per order is 2.083 days. This means a reduction of 16.7% in the average waiting time per order compared to the situation where orders are produced in order of arrival.

9.4.4 The $M/M/c$ Queue

In this model, we assume the following:

- Customers arrive following a Poisson process with an average of λ customers per unit of time.
- A customer's service time is exponentially distributed with an expected value of $1/\mu$ units of time.
- There are c servers, with one common queue.
- There is an infinite waiting room.

This model is also known as the *Erlang queue* or the *Erlang-C model*. For a model with several servers, the server utilization ρ is defined as

$$\rho = \frac{\lambda}{c\mu}.$$

We assume

$$\rho < 1.$$

By Little's law, in the long run, the average number of busy servers is equal to $\lambda \times \frac{1}{\mu} = c\rho$. This leads to the following interpretation of ρ:

the fraction of the time an individual server is busy $= \rho$. (9.13)

A measure for the service to customers is the fraction of the arriving customers who must wait. An explicit formula can be given for this fraction. For this, one needs the limiting distribution $\{\pi_j\}$ of the number of customers present at any given time. In the case of $c = 1$ server, we have derived this limiting distribution in Section 8.7 using the theory of continuous-time Markov chains. This derivation can easily be extended to the case of $c \geq 1$ servers. We leave out the details and give the final result. If we define the constant $C(c, \rho)$ by

$$C(c, \rho) = \frac{(c\rho)^c}{c!(1 - \rho)} \left[\frac{(c\rho)^c}{c!(1 - \rho)} + \sum_{j=0}^{c-1} \frac{(c\rho)^j}{j!} \right]^{-1}, \quad (9.14)$$

then the limiting distribution π_j is given by

$$\pi_j = \begin{cases} \dfrac{(c\rho)^j}{j!} \dfrac{c!(1 - \rho)C(c, \rho)}{(c\rho)^c}, & j = 0, \ldots, c - 1, \\ \dfrac{(c\rho)^c}{c!} \rho^{j-c} \pi_0, & j = c, c+1, \ldots. \end{cases} \quad (9.15)$$

So the quantity π_j gives the long-run fraction of the time that there are j customers in the system. This means that the average number of customers in the queue is equal to $L_q = \sum_{j=c}^{\infty}(j - c)\pi_j$, that is, after some calculation,

$$L_q = \frac{\rho}{1 - \rho} C(c, \rho). \quad (9.16)$$

The fraction of the time that all servers are busy simultaneously is $\sum_{j=c}^{\infty} \pi_j$. After some calculation, this implies that

the fraction of the time that all servers are busy at the same time $= C(c, \rho)$. (9.17)

The PASTA property gives that

the fraction of the customers who must wait $= C(c, \rho)$. (9.18)

This last formula for the delay probability is known as the *Erlang-C formula*.

Little's laws $L_q = \lambda W_q$ and $L = \lambda W$ and the relation $W = W_q + 1/\mu$ give

$$W_q = \frac{1}{c\mu(1 - \rho)} C(c, \rho), \quad (9.19)$$

$$L = L_q + c\rho. \quad (9.20)$$

Using the formulas given above for numerical calculations seems simpler than it is. For large values of c, due care must be taken when calculating the constant $C(c, \rho)$

to avoid numerical rounding errors. In the $M/M/c$ queue, it is possible to give an explicit formula for the waiting probability $W_q(x)$ if one assumes that the customers are served in order of arrival. As is done in Section 8.7 for the $M/M/1$ queue, one can deduce that

$$W_q(x) = 1 - C(c, \rho)e^{-c\mu(1-\rho)x} \qquad \text{for } x \geq 0. \tag{9.21}$$

When designing a queueing system, an important question is whether to have a *single* or *separate* queues. This is discussed in more detail in the following example.

Example 9.3. The post office in the town of Zeeburg has two counters, one for banking and one for postal services. Banking customers arrive following a Poisson process with an average of 15 customers per hour, and independently of that, customers for postal services arrive following a Poisson process with an average of 18 customers per hour. Each customer comes for only one of the two services. Every customer's service time is exponentially distributed with an expected value of three minutes. In the current setup, a customer for banking or postal services can only be served at the corresponding counter. The post office is considering opening both counters for both services, with a single queue. What is the effect of the new setup on the utilization of the counters and on the average number of customers at the post office?

Solution. Choose an hour as unit of time. First, we consider the current situation. In this situation, we are dealing with two separate $M/M/1$ queues. For the banking counter, an $M/M/1$ queue with $\lambda_g = 15$ and $1/\mu_g = 1/20$ applies, and for the postal counter, an $M/M/1$ queue with $\lambda_p = 18$ and $1/\mu_p = 1/20$. Equations (9.3) and (9.4) for the $M/M/1$ queue give that

     the fraction of the time the banking counter is busy $= \frac{15}{20}$,

     the fraction of the time the postal counter is busy $= \frac{18}{20}$,

     the average number of customers for banking $= \frac{15/20}{1-15/20} = 3$,

     the average number of customers for postal services $= \frac{18/20}{1-18/20} = 9$.

So in all, there are an average of $3 + 9 = 12$ customers at the post office.

Next, consider the proposed situation where both counters are open to both banking and postal services and there is a common queue. Then an $M/M/c$ queue applies with

$$\lambda = 15 + 18 = 33, \qquad \frac{1}{\mu} = \frac{1}{20}, \qquad \text{and} \quad c = 2.$$

Substituting the numerical values from the equations (9.13), (9.17), and (9.20) for the $M/M/c$ queue and some calculations give

     the fraction of the time each separate counter is busy $= 0.825$,

     the fraction of the time both counters are busy at the same time $= 0.746$,

     the average number of customers at the post office $= 5.17$.

So we see that combining the services more than halves the average number of customers present at the post office. As for the average waiting time per customer, in the system with separate counters, it is equal to $(3 - \frac{15}{20})/15 = 0.15$ hours for banking customers and $(9 - \frac{18}{20})/18 = 0.45$ hours for postal customers. In the system with combined counters with a single common queue, the average waiting time per customer is equal to $(5.17 - 1.65)/33 = 0.107$ hours for both types of customers. In this example, *pooling* leads to a decrease in the average waiting time for both types of customers. This is not always the case when pooling takes place in the presence of nonhomogeneous customer types with different service times. In general, the average waiting time per customer will decrease for one type of customer and increase for the other. As an example, consider the situation where customers of type 1 arrive following a Poisson process with an average of 50 customers per hour and a fixed service time of 1 minute, while customers of type 2 arrive following a Poisson process with an average of five customers per hour and a fixed service time of 10 minutes. It then follows from equation (9.9) for the $M/G/1$ queue that in the case of separate counters, the average waiting time per customer is equal to 2.5 minutes for customers of type 1 and to 25 minutes for customers of type 2. If the two servers are combined to a single common queue, a simulation shows that the average waiting time per customer is 6.14 minutes for both types of customers. In this example, pooling therefore results in a sharp increase in the average waiting time for most of the customers and a decrease for only a small percentage of the customers. Finally, we note that, in practice, the answer to the question whether or not pooling should take place depends not only on quantitative aspects but also on psychological considerations. Customers find it annoying when someone who arrives later is helped before them. On the other hand, separate queues may be chosen for psychological reasons, to influence customer behavior (for example, having specific lanes open only to carpoolers).

The cμ-Rule

Suppose that customers of types 1 and 2 arrive following independent Poisson processes with respective parameters λ_1 and λ_2. Let μ_i be the expected value of the service time of a customer of type i, for $i = 1, 2$. Suppose that for every customer of type i, a cost of c_i is incurred for every unit of time the customer is in the system. What (nonpreemptive) priority rule gives the lowest average cost per unit of time? The answer is given by the so-called $c\mu$-rule. If

$$\frac{c_1}{\mu_1} \geq \frac{c_2}{\mu_2},$$

give priority to customers of type 1; otherwise, give priority to customers of type 2. This rule's optimality can be seen intuitively by considering that c_i/μ_i can be interpreted as the cost per unit of work a customer of type i generates for the server. A service station would want to serve the relatively most expensive customers first. The result above can be easily deduced from the formulas in Section 9.4.3 using simple algebra (recall that $c_1 L^{(1)} + c_2 L^{(2)}$ gives the average cost per unit of time, where $L^{(i)}$ is the average number of customers of type i present). The $c\mu$-rule also applies when more than two types of customers arrive following independent Poisson processes.

One Fast Versus Two Slow Servers

An interesting question is how the $M/M/2$ queue with service rate μ for both servers compares to the $M/M/1$ queue in which the server has a service rate 2μ that is twice as high. If we denote the average queue length and the average number of customers in the system by $L_q^{(1)}$ and $L^{(1)}$ for the $M/M/1$ system with service rate 2μ and by $L_q^{(2)}$ and $L^{(2)}$ for the $M/M/2$ system with rate μ, then we have the following surprising result:

$$L^{(1)} < L^{(2)} \quad \text{but} \quad L_q^{(1)} > L_q^{(2)}.$$

This result can be easily deduced, using some algebra, from the formulas for L and L_q in the $M/M/c$ queue with $c = 1$ and $c = 2$. We leave out the details.

9.4.5   The M/G/c Queue

The $M/G/c$ queue arises by dropping the assumption of an exponentially distributed service time in the $M/M/c$ queue and assuming a general probability distribution instead. The utilization ρ is now defined as

$$\rho = \frac{\lambda \mathbb{E}[B]}{c},$$

and we have the assumption $\rho < 1$. The general $M/G/c$ queue is too complex to give a simple exact solution. However, a very useful approximation can be given for the average queue length. For this approximation, only the service time's expected value $\mathbb{E}[B]$ and standard deviation $\sigma(B)$ are needed. The approximation is a linear combination of formulas for the average queue lengths for the particular cases of exponentially distributed service times ($M/M/c$ queue) and deterministic service times ($M/D/c$ queue) with *the same* expected service time $\mathbb{E}[B]$. The weighting factor in the linear combination is the square of the service time's coefficient of variation. The approximation for the average queue length in the $M/G/c$ queue is

$$L_q^{\text{app}} = (1 - c_B^2)L_q(\det) + c_B^2 L_q(\exp), \tag{9.22}$$

where

$$
\begin{aligned}
c_B^2 &= \sigma^2(B)/[\mathbb{E}[B]]^2, \\
L_q(\exp) &= \text{average queue length in the } M/M/c \text{ queue}, \\
L_q(\det) &= \text{average queue length in the } M/D/c \text{ queue}.
\end{aligned}
$$

The approximation is exact for the case of $c = 1$ server; see Section 9.4.2. A formula for $L_q(\exp)$ is given by (9.16) in Section 9.4.4. For $L_q(\det)$, an exact formula can also be given, but that formula is rather laborious from a calculatory point of view.

We limit ourselves to giving the following simple but accurate approximation for $L_q(\det)$:

$$L_q^{\mathrm{app}}(\det) = \frac{1}{2}\left[1 + (1-\rho)(c-1)\frac{\sqrt{4+5c}-2}{16c\rho}\right]L_q(\exp). \qquad (9.23)$$

For the $M/G/c$ queue, let P_w be the long-run fraction of the customers who must wait in line. If we denote by $P_w(\exp)$ the probability P_w of having to wait for the particular case of exponentially distributed service times, then a good approximation for P_w is given by

$$P_w \approx P_w(\exp).$$

The probability $P_w(\exp)$ of having to wait is given by the Erlang-C formula (9.14) for $C(c,\rho)$ in Section 9.4.4.

9.4.6 The M/G/c/c Queue (the Erlang Loss Model)

In this model, which has many practical applications, we assume the following:

- Customers arrive following a Poisson process with an average of λ customers per unit of time.
- A customer's service time has a general probability distribution with expected value $\mathbb{E}[B]$ units of time.
- There are c servers.
- There is no waiting room; that is, a customer who finds all servers busy is lost.

A lost customer is assumed not to have any further influence on the system. This model is also called the *Erlang loss model*. The name Erlang is associated with this model for the following reason. The model was introduced at the beginning of the twentieth century by the Danish engineer A. K. Erlang in his pioneering studies of congestion in telephone systems. Erlang considered the situation of a telephone exchange with c lines for incoming calls, where an incoming call that finds all lines busy is not accepted. The question Erlang asked himself was how many lines are needed for the probability of an incoming call being rejected to be below a given value. Although the theory of stochastic processes was hardly developed at the beginning of the twentieth century, Erlang succeeded in deriving the probability distribution of the number of busy lines for the particular case of exponentially distributed call durations. Moreover, Erlang stated that he expected that the probability distribution he found also applies to the situation of generally distributed call durations in which only the expected duration of a call is important and not the form of the probability distribution. This famous insensitivity result was only proved many years after Erlang had stated his conjecture.

For the $M/G/c/c$ queue, the fraction of the time that j lines are busy satisfies

$$\pi_j = \frac{[\lambda\mathbb{E}[B]]^j/j!}{\sum_{k=0}^{c}[\lambda\mathbb{E}[B]]^k/k!} \quad \text{for } j = 0,1,\ldots,c. \qquad (9.24)$$

Of the service time, this formula uses only the expected value $\mathbb{E}[B]$. We will not prove the formula in its full generality. For the special case of exponentially distributed service times, the formula is derived in Section 8.7 using the theory of continuous-time Markov chains. By multiplying the numerator and denominator of the right-hand side of (9.24) by $e^{-\lambda\mathbb{E}[B]}$, one sees that the probability distribution $\{\pi_j\}$ for the Erlang loss model is a *truncated Poisson distribution*:

$$\pi_j = \frac{e^{-\lambda\mathbb{E}[B]}[\lambda\mathbb{E}[B]]^j/j!}{\sum_{k=0}^{c} e^{-\lambda\mathbb{E}[B]}[\lambda\mathbb{E}[B]]^k/k!} \quad \text{for } j = 0, 1, \ldots, c.$$

The PASTA property gives that

$$\text{the fraction of the customers who are lost} = \pi_c. \quad (9.25)$$

It then follows from Little's law that

$$\text{the average number of busy servers} = \lambda(1 - \pi_c)\mathbb{E}[B]. \quad (9.26)$$

In the literature, the probability π_c of losing a customer is commonly denoted by $B(c, u)$, where

$$B(c, u) = \frac{u^c/c!}{\sum_{k=0}^{c} u^k/k!} \quad \text{with } u = \lambda\mathbb{E}[B].$$

The formula for $B(c, u)$ is called the *Erlang loss formula* and gives the probability of losing a customer in the case of c servers and an offered load of u. The parameter u is dimensionless, and one often refers to an offered load of u Erlangs.

Example 9.4. The "Faxing Company" has decided to install a communication system for sending messages between two of its main offices. It is expected that during peak periods, an average of 100 messages per hour will be offered for transmission with an average transmission time of 3 minutes per message. The arrival of messages during peak periods can be described by a Poisson process. A message that finds all lines busy does not wait for a line to become available but is sent later during a quiet period. The wish is to dispose of sufficiently many transmission lines such that the probability of a message offered during a peak period finding all lines busy is no more than 0.5%. How many lines are needed? What is then the average number of busy lines?

Solution. The Erlang loss model applies. Take a minute as unit of time. The arrival rate λ and the expected value of the service time $\mathbb{E}[B]$ are given by $\lambda = 100/60$ and $\mathbb{E}[B] = 3$. Hence, the offered load is $u = (100/60) \times 3 = 5$ Erlangs. Therefore, we must determine the smallest value of c for which the loss probability satisfies $B(c, 5) \leq 0.005$. By trying different values of c, we find

$$B(10, 5) = 0.0184, \quad B(15, 5) = 0.0002,$$
$$B(12, 5) = 0.0034, \quad B(11, 5) = 0.0083.$$

Hence, the number of lines needed is 12. It follows from equation (9.26) that for 12 lines, an average of $5 \times (1 - 0.0034) = 4.98$ lines are busy during peak periods.

9.4.7 The $M/G/\infty$ Queue

Many real-world queueing situations can be closely approximated by an $M/G/\infty$ queue with an infinite number of servers. This queueing model assumes the following:

- Customers arrive following a Poisson process with an average of λ customers per unit of time.
- The number of available servers is unbounded, so that every customer is immediately assigned an idle server at arrival.
- The customer's service time B has a general probability distribution.

In the $M/G/\infty$ queue, the number of busy servers is therefore always equal to the number of customers present. The equilibrium distribution of the number of busy servers is given by

$$\pi_j = e^{-\lambda\mathbb{E}[B]}\frac{(\lambda\mathbb{E}[B])^j}{j!} \quad \text{for } j = 0, 1, \ldots; \tag{9.27}$$

that is, π_j is the long-run fraction of the time that j servers are busy. Heuristically, this formula can be seen by viewing the $M/G/\infty$ queue as a limiting case of the $M/G/c/c$ loss queue from Section 9.4.6. If we take $c \to \infty$ in equation (9.24), this gives equation (9.27). As in the Erlang loss model, in the $M/G/\infty$ queue, the *form* of the service time's probability distribution is not important for the equilibrium distribution; rather, only the *expected value* of the service time is relevant. This *insensitivity result* is extremely useful.

A Probabilistic Proof of the Insensitivity Property

The general validity of result (9.27) can be made plausible by giving an exact probabilistic proof in the case that the service time can take on only a finite number of values. First, consider the special case of a constant service time $D = \mathbb{E}[B]$. Fix a time t with $t > D$. Because the service time is a constant D, every customer in service at time $t - D$ has left the system at time t, and the customers in service at time t are exactly those who arrived between $t - D$ and t. In other words, the number of busy servers at time t is given by the number of customers who arrive during the time interval $(t-D, t)$. The number of customers who arrive during a time interval of length D has a Poisson distribution with expected value λD. This means that the number of busy servers at time t is Poisson distributed with expected value λD, thus proving the limiting result (9.27) for the specific case of a constant service time. This deduction can be directly extended to the case that the service time B can take on only a finite number of values $D_1, \ldots, D_m$ with respective probabilities $\pi_1, \ldots, \pi_m$ with $\sum_{k=1}^{m} \pi_k = 1$, in which case $\pi_1 D_1 + \pi_2 D_2 + \cdots + \pi_m D_m = \mathbb{E}[B]$. For convenience, we take $m = 2$. The trick is to refer to customers with the constant service time D_1 as type 1 customers and to customers with the constant service time D_2 as type 2 customers. Based on the Poisson process's splitting property (see

Appendix C), the type 1 and type 2 customers arrive following Poisson processes with respective parameters $\lambda_1 = \lambda\pi_1$ and $\lambda_2 = \lambda\pi_2$. Moreover, the remarkable property holds that the two Poisson processes are *independent* of each other. Now, fix t such that $t > D_1$ and $t > D_2$. The reasoning above tells us that at time t, the number of type 1 customers in service is Poisson distributed with expected value $\lambda_1 D_1$ and the number of type 2 customers in service is Poisson distributed with expected value $\lambda_2 D_2$. The sum of two independent Poisson-distributed random variables is again Poisson distributed. So the number of busy servers at time t is Poisson distributed with expected value $\lambda_1 D_1 + \lambda_2 D_2 = \lambda(\pi_1 D_1 + \pi_2 D_2) = \lambda\mathbb{E}[B]$. Thus, we have also proved the limiting result (9.27) for a discretely distributed service time that can take on only finitely many values. The validity of result (9.27) can then be made plausible by considering that every probability distribution can be approximated "arbitrarily closely" by a discrete distribution.

Example 9.5. In the picturesque town of Edam, on a beautiful summery day, tourist buses arrive from early morning to early evening following a Poisson process with an average of five buses per hour. There is sufficient parking space for all buses. The time a bus stays in Edam is normally distributed with an expected value of 45 minutes and standard deviation of five minutes with probability $\frac{2}{3}$, and normally distributed with an expected value of 90 minutes and standard deviation of 10 minutes with probability $\frac{1}{3}$. What is the probability that at four o'clock in the afternoon, there are more than seven tourist buses in Edam?

Solution.   Choose an hour as unit of time. The $M/G/\infty$ queue with

$$\lambda = 5 \quad \text{and} \quad \mathbb{E}[B] = \frac{2}{3} \times \frac{3}{4} + \frac{1}{3} \times \frac{3}{2} = 1$$

applies for the probability distribution of the number of buses present, where only the expected value of the time spent in Edam is relevant. The number of buses present at four o'clock in the afternoon on a summery day is therefore Poisson distributed with expected value $\lambda\mathbb{E}[B] = 5$. Consequently, the probability that there are more than seven tourist buses in Edam at four o'clock in the afternoon on a summery day is $1 - \sum_{j=0}^{6} e^{-5}5^j/j! = 0.133$.

9.4.8   *The $M/M/c/c + N$ Queue*

This highly useful queueing model assumes the following:

- Customers arrive following a Poisson process with an average of λ customers per unit of time.
- A customer's service time is exponentially distributed with an expected value of $1/\mu$ units of time.
- There are c servers, with a single common queue.
- There is a finite waiting room with the capacity for N customers.

A customer who finds all c servers busy and all N waiting places busy at arrival is not accepted and has no further influence on the system. The quantity ρ is again defined as $\rho = \lambda/c\mu$. There is now no need to assume $\rho < 1$ since the finite waiting room automatically prevents the number of customers in the system from increasing indefinitely.

In the $M/M/c/c + N$ queue, as before, the probability π_j gives the fraction of the time that there are j customers in the system, but the interpretation of $W_q(x)$ requires some further explanation. The reason is that some customers are not admitted into the system (and therefore have no waiting time), while other customers are admitted into the system and may need to wait until they are served. It is customary to consider only the waiting times of the admitted customers. So, the waiting time probability $W_q(x)$ can be interpreted as

> $W_q(x)$ = the fraction of the *admitted* customers with a waiting time of at most x units of time.

The equilibrium probabilities π_j are given by

$$\pi_j = \begin{cases} \frac{(c\rho)^j}{j!} \pi_0, & j = 0, \ldots, c-1, \\ \frac{(c\rho)^c}{c!} \rho^{j-c} \pi_0, & j = c, \ldots, c+N, \end{cases} \tag{9.28}$$

with

$$\pi_0 = \left[\sum_{j=0}^{c-1} \frac{(c\rho)^j}{j!} + \frac{(c\rho)^c}{c!} \sum_{j=c}^{c+N} \rho^{j-c} \right]^{-1}. \tag{9.29}$$

This result can be deduced the standard way using the equilibrium method of continuous-time Markov chains; cf. Chapter 8.

An important quantity is the probability π_{c+N}. Not only does this probability give the fraction of the time that the system is full, but based on the PASTA property, we also have that

$$\text{the fraction of the customers who are rejected } = \pi_{c+N}. \tag{9.30}$$

The average queue length L_q and the average number L of customers present are calculated as follows:

$$L_q = \sum_{j=c}^{c+N} (j-c)\pi_j \quad \text{and} \quad L = \sum_{j=1}^{c+N} j\pi_j. \tag{9.31}$$

The fraction of the customers who are rejected is equal to π_{c+N}; Little's law then gives that

$$\text{the average number of busy servers} = \lambda(1 - \pi_{c+N})\frac{1}{\mu}. \tag{9.32}$$

The number $\lambda(1 - \pi_{c+N})$ is called the effective arrival rate and gives the average number of *admitted* customers per unit of time. If the admitted customers are served in order of arrival, then the waiting time probability $W_q(x)$ can be calculated using the formula

$$W_q(x) = 1 - \frac{1}{1 - \pi_{c+N}} \sum_{j=c}^{c+N-1} \pi_j \sum_{k=0}^{j-c} e^{-c\mu x} \frac{(c\mu x)^k}{k!}, \quad x \geq 0. \tag{9.33}$$

The formulas above are not suitable for manual calculations. These rather laborious formulas are implemented in software packages.

Example 9.6. The software company "World Perfect" provides telephone support for its products. The company considers it of the utmost importance that it is easily reachable by telephone. It has decided to install an automated telephone system that automatically connects an incoming call to an information assistant if not all assistants are busy and otherwise keeps the call in a queue until an assistant becomes available. If all positions in the queue are filled, an incoming call is not accepted and the caller hears the busy tone. The calls' arrival process is assumed to be a Poisson process. This is a reasonable assumption since the company has a very large customer base. As long as the company is easily reachable by telephone, the Poisson assumption is not disturbed by the few callers who call again after having gotten the busy tone. When choosing the computerized system, the question is how many lines and how many queue positions must be created such that

(a) the fraction of the callers who get the busy tone is no more than 5%;
(b) at most 1% of the callers who are admitted wait more than 1 minute.

These goals must be achieved in a situation where, on average, 3 calls come in per minute and the average handling time per call is $2\frac{1}{2}$ minutes. The calls' handling time may be assumed to be exponentially distributed.

Solution.   Choose a minute as unit of time. In the $M/M/c/c+N$ queue, we then have $\lambda = 3$ and $1/\mu = 2.5$. To solve the problem, one typically requires tables or graphs in which the rejection probability π_{c+N} and the waiting time probability $1 - W_q(1)$ are given as functions of the number of lines c and the number of queue positions N. Table 9.1 gives these probabilities for several values of c and N, where c has been taken greater than λ/μ in view of the high service requirements. This table has been compiled using a software package for queueing models.

Table 9.1   The rejection and waiting time probabilities.

	$c = 8$		$c = 9$		$c = 10$	
	π_{c+N}	$1 - W_q(1)$	π_{c+N}	$1 - W_q(1)$	π_{c+N}	$1 - W_q(1)$
$N = 1$	0.163	0.008	0.109	0.004	0.069	0.002
$N = 2$	0.132	0.035	0.084	0.017	0.050	0.008
$N = 3$	0.110	0.081	0.065	0.041	0.036	0.019
$N = 4$	0.094	0.138	0.051	0.072	0.026	0.034
$N = 5$	0.081	0.199	0.041	0.105	0.019	0.050

An important conclusion can be drawn from the table:

For a given value of c, the rejection probability decreases as N increases, while the waiting time probability increases. The rejection probability decreases less quickly than the waiting time probability increases.

So caution is advised when choosing c and N. The table shows that the set targets $\pi_{c+N} \leq 0.05$ and $1 - W_q(1) \leq 0.01$ can be reached with 10 lines and 2 queue positions. It follows from equation (9.32) that for this configuration, the average number of busy lines is equal to 7.1. In other words, the fraction of the time that every individual information assistant is busy providing telephone support is 71%.

9.4.9 *The M/M/c Queue and Call Centers*

In this section, we discuss an approximation for Erlang's delay probability in the $M/M/c$ queue, based on the normal distribution. This normal approximation provides extremely useful insight for analyzing capacity and staffing problems such as those that occur in modern call centers. Queueing formulas for the $M/M/c$ queue have already been given in Section 9.4.4. In the $M/M/c$ queue, customers arrive following a Poisson process with arrival rate λ, the customers' service time has an exponential distribution with expected value $1/\mu$, and there are c servers. We denote the offered load by

$$R = \frac{\lambda}{\mu}.$$

The quantity R is dimensionless and represents the average amount of work coming in per unit of time for the c servers. In technical literature, it is common to speak of an offered load of R *Erlangs*. The queueing system reaches steady state only if the server capacity c is greater than the offered load R. Therefore, one assumes

$$\rho = \frac{R}{c} < 1.$$

The quantity ρ is called the server utilization. In Section 9.4.3, we have seen that, in the long run, the fraction of the time that a given server is busy is equal to ρ. In Section 9.4.1, we have also seen for the $M/M/1$ queue with one server that the offered load R should not be too close to the server capacity $c = 1$ to avoid excessively long queue lengths and waiting times. A rule of thumb for system design using the $M/M/1$ queue is that ρ should not be much larger than 0.8. A natural question is what happens in the $M/M/c$ queue with several servers when the offered load R approaches the server capacity c. Studying Table 9.2 closely helps us answer this. In this table, we give, for several values of R and c, the delay probability P_W, the average waiting time T_W per customer taken over the customers who must wait, and the 95% percentile $\eta_{0.95}$ of the waiting time distribution of customers who must wait. The service time in Table 9.2 is normalized such that $1/\mu = 1$. The formula for the delay probability P_W is

$$P_W = C(c, \rho),$$

where $C(c, \rho)$ is given by (9.14). It follows from

$$T_W = \frac{W_q}{P_W}$$

and equation (9.19) that

$$T_W = \frac{1}{c\mu(1 - \rho)}.$$

It follows from equation (9.21) that in steady state, a customer whom we know must wait has probability $e^{-c\mu(1-\rho)x}$ of having to wait longer than a time x. This means that the waiting time percentile η_p with $p = 0.95$ is determined by the solution to

$$e^{-c\mu(1-\rho)x} = 1 - p; \qquad (9.34)$$

that is,

$$\eta_p = -\frac{1}{c\mu(1 - \rho)} \ln(1 - p).$$

A queueing percentile of $\eta_{0.95} = 20$ seconds therefore means that, in the long run, 5% of the customers who must wait will wait longer than 20 seconds.

Table 9.2   Performance measures as functions of c and R.

	$\rho = R/c = 0.8$			$\rho = R/c = 0.95$			$\rho = R/c = 0.99$		
	P_W	T_W	$\eta_{0.95}$	P_W	T_W	$\eta_{0.95}$	P_W	T_W	$\eta_{0.95}$
$c = 1$	0.8	5	14.98	0.95	20	59.91	0.99	100	299.6
$c = 2$	0.711	2.5	7.49	0.926	10	29.96	0.985	50	149.8
$c = 5$	0.554	1	3.0	0.878	4	11.98	0.975	20	59.91
$c = 10$	0.409	0.5	1.5	0.826	2	5.99	0.964	10	29.96
$c = 25$	0.209	0.2	0.6	0.728	0.8	2.40	0.942	4	11.98
$c = 50$	0.087	0.1	0.3	0.629	0.4	1.20	0.917	2	5.99
$c = 100$	0.020	0.05	0.15	0.506	0.2	0.60	0.883	1	3.0
$c = 250$	3.9E-4	0.02	0.06	0.318	0.08	0.24	0.818	0.4	1.2
$c = 500$	8.4E-7	0.01	0.03	0.177	0.04	0.12	0.749	0.2	0.6

The following conclusion can be drawn from Table 9.2:

> *High values of the server utilization ρ go hand in hand with an acceptable service level for customers provided that the number of available servers is sufficiently large.*

The higher the number of servers, the higher the utilization ρ can be before the service level to the customers becomes poor. For large values of the numbers of servers, a relatively high delay probability P_W therefore does not necessarily mean that the service level is poor. For example, with $c = 100$ and $\rho = 0.95$, 50.6% of the customers must wait, but the average waiting time per customer who must wait is only $\frac{1}{5}$ of the average service time. Only 5% of the waiting customers must wait more than $\frac{3}{5}$ of the average service time.

The situation where there are many servers typically occurs at call centers. In practice, one often looks for the smallest number of servers such that a given service requirement is met. What is the least number c^* of servers necessary to ensure that the delay probability P_W is below a specified level α with, for example, $\alpha = 0.20$? Numerically, it is no problem at all to calculate the exact value of c^* by applying equation (9.14) for different values of c for a given value of the offered load R. For real-world situations, however, it is useful to have a formula that provides insight. Such a formula can be given using the normal probability distribution. The formula is an approximation and is known as the *square-root staffing rule*. This simple rule of thumb provides much useful information to managers of large call centers. The approximation formula is given by

$$c^* \approx R + k_\alpha \sqrt{R} \tag{9.35}$$

when the service requirement is that the delay probability P_W is not greater than a prescribed value α. The safety factor k_α is the solution to the equation

$$\frac{k\Phi(k)}{\varphi(k)} = \frac{1-\alpha}{\alpha}, \tag{9.36}$$

where $\Phi(x)$ is the probability distribution function of the standard normal distribution and $\varphi(x) = (1/\sqrt{2\pi})e^{-\frac{1}{2}x^2}$ is the corresponding probability density.

It is interesting to note that equation (9.35) is very similar to the famous formula for the reorder point s in the (s, Q) inventory model with service requirement; cf. Section 6.4.1. The safety factor k_α does not depend on R and, for every α, can easily be calculated from (9.36) using bisection. For $\alpha = 0.8, 0.5, 0.2, 0.1$, and 0.05, the respective values of k_α are 0.1728, 0.5061, 1.062, 1.420, and 1.740. The approximation formula clarifies the relationship between the queueing system's parameters and increases the manager's intuitive understanding of the system's dynamics. The approximation formula (9.35) also provides insight into the economies of scale that can be obtained by combining various call centers into a single common call center. As an illustration, consider two call centers, each with an offered load of R Erlangs and the same service requirement on the delay probability P_W. In the case of two separate call centers, a total of $2(R + k_\alpha \sqrt{R})$ servers are needed, whereas in the case of a single combined call center, $2R + k_\alpha \sqrt{2R}$ servers are required. This is a decrease of $(2 - \sqrt{2})k_\alpha \sqrt{R}$ servers. The quality of the approximation (9.35) is more than excellent. The value in (9.35) is rounded *up* to the nearest integer. Numerical experiments have shown that in almost all cases, the approximate value for the number of servers required is equal to the exact value. The difference between the two is never more than one. Table 9.3 shows the exact and approximate values of the least number of servers necessary to have such that $P_W \leq \alpha$ for different values of R and α. If R is large (and the number of servers needed is therefore also large), it is wise, from a cost perspective, to not take α too small in the service requirement $P_W \leq \alpha$. This was already noted in Table 9.2: for large c, a high value of the delay probability P_W does not mean that a customer's waiting time is unacceptably high. Suppose that an additional service requirement is that, in the long run, only $100(1 - p)\%$ of the waiting customers may have to wait more than f times the average service time for given values of p and f. Then it follows from equation (9.34) that the number c of servers must satisfy $-f(c - R) \leq \ln(1 - p)$; that is,

$$c \geq R - \frac{1}{f} \ln(1 - p).$$

Table 9.3   The exact values and approximations of c^*.

	$\alpha = 0.5$		$\alpha = 0.2$		$\alpha = 0.1$	
	exa	app	exa	app	exa	app
$R = 1$	2	2	3	3	3	3
$R = 5$	7	7	8	8	9	9
$R = 10$	12	12	14	14	16	15
$R = 50$	54	54	58	58	61	61
$R = 100$	106	106	111	111	115	115
$R = 250$	259	259	268	267	274	273
$R = 500$	512	512	525	524	533	532
$R = 1000$	1017	1017	1034	1034	1046	1045

A Derivation of the Approximation Formula

The derivation is not complicated and is based on an interesting relationship between the delay probability P_W in the $M/M/c$ queue and the loss probability P_{loss} in the Erlang loss model. In the Erlang loss model, every customer who finds all c servers busy upon arrival is lost. In Section 9.4.6, we have deduced that

$$P_{\text{loss}} = \frac{R^c/c!}{\sum_{k=0}^{c} R^k/k!}; \tag{9.37}$$

see equation (9.25). By doing some rewriting, we can use this formula and equation (9.18) for P_W to derive that

$$P_W = \frac{P_{\text{loss}}}{1 - \rho + \rho P_{\text{loss}}}. \tag{9.38}$$

For a fixed value of the offered load R, let X_R be a random variable that has a Poisson distribution with expected value R. If we multiply the numerator and denominator of the right-hand side of (9.37) by e^{-R}, we obtain

$$P_{\text{loss}} = \frac{\mathbb{P}(X_R = c)}{\mathbb{P}(X_R \leq c)}.$$

The Poisson distribution with expected value R has standard deviation $\sqrt{R}$. For R sufficiently large, the Poisson distribution of the random variable X_R is approximated by a normal distribution with expected value R and standard deviation $\sqrt{R}$. Take

$$c = R + k\sqrt{R}.$$

For R sufficiently large, it follows from the normal approximation that

$$\mathbb{P}(X_R \leq c) = \mathbb{P}\left(\frac{X_R - R}{\sqrt{R}} \leq k\right) \approx \Phi(k).$$

If we write $\mathbb{P}(X_R = c) = \mathbb{P}(c - 1 < X_R \leq c)$, then

$$\mathbb{P}(X_R = c) = \mathbb{P}\left(k - \frac{1}{\sqrt{R}} < \frac{X_R - R}{\sqrt{R}} \leq k\right)$$

$$\approx \Phi(k) - \Phi\left(k - \frac{1}{\sqrt{R}}\right) \approx \frac{1}{\sqrt{R}}\varphi(k).$$

This gives

$$P_{\text{loss}} \approx \frac{1}{\sqrt{R}} \frac{\varphi(k)}{\Phi(k)}.$$

If we substitute $\rho = R/c$ and $c = R + k\sqrt{R}$ in (9.38), we find

$$P_W = \frac{cP_{\text{loss}}}{c - R + RP_{\text{loss}}} = \frac{(R + k\sqrt{R})P_{\text{loss}}}{k\sqrt{R} + RP_{\text{loss}}}.$$

Next, we substitute $P_{\text{loss}} \approx \frac{1}{\sqrt{R}}\delta$ with $\delta = \varphi(k)/\Phi(k)$. This leads to

$$P_W \approx \frac{(R + k\sqrt{R})\delta}{kR + R\delta}.$$

For R sufficiently large, we have $k\sqrt{R} << R$, so that

$$P_W \approx \frac{R\delta}{kR + R\delta} = \frac{\delta}{k + \delta}$$

$$= \left(1 + \frac{k}{\delta}\right)^{-1} = \left[1 + \frac{k\Phi(k)}{\varphi(k)}\right]^{-1}.$$

This relation leads to equation (9.36) for $k = k_\alpha$ if we set the delay probability P_W equal to α. If k_α is determined by this equation, then the number of servers required in order to have $P_W \leq \alpha$ can be determined by using the approximation $c \approx R + k_\alpha\sqrt{R}$.

In the case of call centers, the $M/M/c$ queue is part of a broader problem, namely scheduling the numbers of servers depending on the expected load.

9.5   Networks of Queues

In this section, we give an introduction to the very important topic of networks of queues. Networks of queues have received much attention in recent decades because of their applications to computer and communication systems and to production systems. We first deal with the simplest case of a network of queues, namely a tandem queue. Then, we discuss a more general network of queues.

9.5.1   *Tandem Queues*

Consider a queueing system with two service stations 1 and 2 that are connected in series. Customers arrive at station 1 following a Poisson process with parameter λ and continue on to station 2 after being served at station 1. After the service at station 2, the customer leaves the system. This queueing system occurs, in particular, in production lines. We assume the following:

- Every service station has one server.
- A customer's service time at station i has an exponential distribution with expected value $1/\mu_i$.
- Every station has an infinite waiting room.

In addition to the assumption that the service times of the same customer at the different stations are independent of each other, we assume both $\rho_1 < 1$ and $\rho_2 < 1$, where

$$\rho_i = \frac{\lambda_i}{\mu_i} \quad \text{for } i = 1, 2.$$

Let $X_i(t)$ be the number of customers present at station i at time t; then the stochastic process $\{(X_1(t), X_2(t)), t \geq 0\}$ is a continuous-time Markov chain. We denote this Markov chain's equilibrium distribution by π_{nm}. In other words,

$$\pi_{nm} = \text{the long-run fraction of the time that } n \text{ customers are at}$$
$$\text{station 1 and } m \text{ customers are at station 2 simultaneously.}$$

For this probability distribution, we have the famous *product-form formula*:

$$\pi_{nm} = (1 - \rho_1)\rho_1^n(1 - \rho_2)\rho_2^m \quad \text{for } n, m = 0, 1, \dots.$$

An insightful heuristic explanation can be given for the product-form formula. This explanation is based on the following result.

Theorem 9.2 (Output theorem). *In an $M/M/c$ queue, in steady state, the customers' departure process is a Poisson process with the same rate as the Poisson arrival process.*

Let us explain this theorem for the case of $c = 1$ server. If, for the $M/M/1$ queue, we denote the arrival rate by λ and the service rate by μ, then in steady state, the probability distribution $\{\pi_j\}$ of the number of customers present is given by

$$\pi_j = (1 - \rho)\rho^j \quad \text{for } j = 0, 1, \dots,$$

where $\rho = \lambda/\mu$. In particular, $\pi_0 = 1 - \rho = 1 - \lambda/\mu$. In steady state, for Δt small, we have

$$\mathbb{P}(\text{a customer leaves during the interval } (t, t + \Delta t))$$
$$= \mathbb{P}(\text{at least one customer is present at time } t \text{ and this customer's}$$
$$\text{service ends during the interval } (t, t + \Delta t))$$
$$= (1 - \pi_0)\mu\Delta t = \tfrac{\lambda}{\mu}\mu\Delta t = \lambda\Delta t.$$

This argumentation, which can easily be extended to the $M/M/c$ case (verify) gives a heuristic explanation of the output theorem.

If we return to the tandem queue, then we see that not only station 1 behaves like an $M/M/1$ queue; in steady state, station 2 also behaves like an $M/M/1$ queue. The latter follows from the output theorem. In steady state, we therefore have

$$\mathbb{P}(n \text{ customers at station 1}) = (1 - \rho_1)\rho_1^n \quad \text{for } n = 0, 1, \dots,$$

$$\mathbb{P}(m \text{ customers at station 2}) = (1 - \rho_2)\rho_2^m \quad \text{for } m = 0, 1, \dots .$$

If, in steady state, the numbers of customers at stations 1 and 2 were independent of each other, we would have

$$\mathbb{P}(n \text{ customers at station 1, } m \text{ customers at station 2})$$
$$= \mathbb{P}(n \text{ customers at station 1})\mathbb{P}(m \text{ customers at station 2})$$
$$= (1 - \rho_1)\rho_1^n(1 - \rho_2)\rho_2^m \quad \text{for } n, m = 0, 1, \dots .$$

This is indeed the desired formula for the equilibrium probability π_{nm}. This can be verified mathematically through substitution in the balance equations

$$\lambda\pi_{00} = \mu_2\pi_{01}$$
$$(\lambda + \mu_1)\pi_{n0} = \mu_2\pi_{n1} + \lambda\pi_{n-1,0} \qquad\qquad n \geq 1$$
$$(\lambda + \mu_2)\pi_{0m} = \mu_2\pi_{0,m+1} + \mu_1\pi_{1,m-1} \qquad\qquad m \geq 1$$
$$(\lambda + \mu_1 + \mu_2)\pi_{nm} = \mu_2\pi_{n,m+1} + \mu_1\pi_{n+1,m-1} + \lambda\pi_{n-1,m} \quad n, m \geq 1.$$

These balance equations follow by applying the "rate out = rate in" principle from the theory of continuous-time Markov chains. The form of the solution π_{nm} and the heuristic arguments for this solution come in handy in the next section, where we consider a general open network of queues. There, too, we will see that the product form applies. In particular, this also holds for the tandem queue with c_1 servers at station 1 and c_2 servers at station 2, where we assume $\rho_1 = \lambda/(c_1\mu_1) < 1$ and $\rho_2 = \lambda/(c_2\mu_2) < 1$. In steady state, the stations 1 and 2 behave as independent $M/M/c$ queues with respective parameters (λ_1, μ_1, c_1) and (λ_1, μ_2, c_2); that is, for the equilibrium probabilities π_{nm}, we have

$$\pi_{nm} = \pi_n(\lambda, \mu_1, c_1)\pi_m(\lambda, \mu_2, c_2),$$

where $\pi_n(\lambda, \mu, c)$ is the equilibrium probability of having n customers in the $M/M/c$ queue; see equation (9.15). The following is a nice application.

Example 9.7. At an airport, passengers must first undergo a security check and then a baggage check. The times required per passenger for security screening and baggage screening are independent random variables that are exponentially distributed with an expected value of one and two minutes, respectively. The passengers arrive following a Poisson process with an average of 75 passengers per hour. In all, 10 officers are available to be assigned to either security or baggage checks. For what assignment is the average number of customers in the screening process the lowest?

Solution. Choose an hour as unit of time. If we assign c_1 officers to security checks and $c_2 = 10 - c_1$ officers to baggage checks, then to meet the stability condition for the $M/M/c$ queue, the values of c_1 and c_2 must satisfy

$$\frac{\lambda}{c_1\mu_1} < 1 \quad \text{and} \quad \frac{\lambda}{c_2\mu_2} < 1.$$

For the numerical values $\lambda = 75$, $\mu_1 = 60$, and $\mu_2 = 30$, this leads to $c_1 > 75/60$ and $c_1 < 225/30$. The admissible values for c_1 are therefore 2, 3, 4, 5, 6, and 7.

In Table 9.4, we give the average numbers L_1 and L_2 of passengers at the security check and baggage check, respectively, for various values of c_1. The averages L_1 and L_2 have been calculated by applying the equations (9.16) and (9.20). It follows from the table that the sum $L_1 + L_2$ is the least for $c_1 = 4$. It is therefore optimal to have four officers doing security checks and six officers doing baggage checks.

Table 9.4   The average numbers of passengers present.

c_1	L_1	L_2	$L_1 + L_2$
2	2.0513	2.5020	4.5533
3	1.3605	2.5086	3.8691
4	1.2692	2.5339	3.8031
5	1.2532	2.6304	3.8836
6	1.2505	3.0331	4.2836
7	1.2500	6.0111	7.2611

9.5.2   An Open Network of Queues

As a motivation for the open network model, we first consider the following practical situation. A hospital in a large city with many inhabitants has an emergency service open for emergencies between 20:00 and 8:00. Patients arrive at the emergency service following a Poisson process. The incoming patients are first assessed to gauge the severity of their complaints: on average, 30% of the incoming patients only need to pick up medication from the pharmacy, after which they leave the hospital, 20% are first sent to the X-ray department, 45% are first sent to the lab, and 5% require hospitalization. Of the patients who have been to the X-ray department, 30% must stay and are hospitalized, 10% go on to have a lab test done, and 60% need no further care and leave the hospital. Of the patients who have had lab work done, 10% must stay and are hospitalized, and 90% leave the hospital. Figure 9.2 summarizes the different patient flows.

The situation described in Figure 9.2 is a typical example of an open network of queues. We now consider the following general model for an open network of queues (*open Jackson network*):

- The network consists of K service stations numbered $j = 1, \ldots, K$.
- External arrivals of new customers at station j follow a Poisson process with parameter γ_j, for $j = 1, \ldots, K$.
- A customer's service time at station j is exponentially distributed with expected value $1/\mu_j$, for $j = 1, \ldots, K$.
- After being served at station i, a customer continues on to station j ($j = 1, \ldots, K$) with a given probability p_{ij} or leaves the system with probability $p_{i0} = 1 - \sum_{j=1}^{K} p_{ij}$.
- There is a single server at every station.
- There is an infinite waiting room at every station.

Fig. 9.2   An open network of queues.

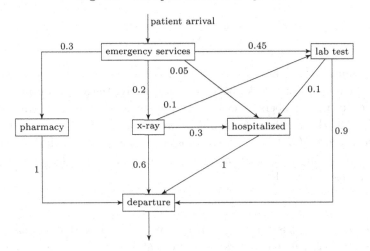

The assumption is made that the external arrival processes at the different stations are independent of one another, as are the service processes at the various stations. Moreover, we assume that the routing probabilities p_{ij} are such that every new incoming customer ultimately leaves the system with probability 1. Furthermore, to guarantee that the length of the queue at any of the stations does not explode, we must assume that at every station, the arrival rate of both external and internal customers is less than the service rate at the station.

Traffic Equations

Define the arrival rates $\lambda_1, \ldots, \lambda_K$ by

λ_j = the long-run average total number of arrivals at station j per unit of time,

where the arrivals consist of both external and internal ones. The λ_j are the unique solutions to the following system of linear equations:

$$\lambda_j = \gamma_j + \sum_{i=1}^{K} \lambda_i p_{ij} \quad \text{for } j = 1, \ldots, K. \tag{9.39}$$

These equations are called the *traffic equations*. This system is easy to understand by considering that, in the long run, the average number of customers who leave a particular station per unit of time is equal to average total number of customers who arrive at the station. Hence,

λ_i = the average number of customers who leave station i per unit of time.

A fraction p_{ij} of these customers go to station j. So the average number of internal arrivals at station j per unit of time is equal to $\sum_{i=1}^{K} \lambda_i p_{ij}$. If we add to this the term γ_j for the average number of external arrivals at station j per unit of time, we find the traffic equations. If we define the offered load ρ_j for the 1-server station j by

$$\rho_j = \frac{\lambda_j}{\mu_j} \quad \text{for } j = 1, \ldots, K,$$

then the assumption is that $\rho_j < 1$ for all j.

After the preparatory work above, we can now give the simultaneous probability distribution of the numbers of customers present at the different stations. If we define $X_j(t)$ as the number of customers present at station j at time t, then the stochastic process $\{(X_1(t), \ldots, X_K(t))\}$ is a continuous-time Markov chain with state space $I = \{(n_1, \ldots, n_K) : n_k \geq 0, \ k = 1, \ldots, K\}$. We denote this process's equilibrium distribution by

$$P(n_1, n_2, \ldots, n_K);$$

that is, $P(n_1, \ldots, n_K)$ is the long-run fraction of the time that n_1 customers are present at station 1, n_2 customers are present at station 2, ..., and n_K customers are present at station K at the same time. For this probability distribution, the *product-form formula* also applies:

$$P(n_1, n_2, \ldots, n_K) = P_1(n_1)P_2(n_2) \cdots P_K(n_K), \quad n_k \geq 0, \ k = 1, \ldots, K, \quad (9.40)$$

where for every fixed k,

$$P_k(n) = (1 - \rho_k)\rho_k^n \quad \text{for } n = 0, 1, \ldots, \quad (9.41)$$

that is $P_k(n)$ is the equilibrium probability π_n of an $M/M/1$ queue with load ρ_k. The so-called *product-form formula* (9.40) therefore gives the remarkable result that in steady state, the numbers of customers present at the various stations are independent of one another, and station k behaves as an isolated $M/M/1$ queue with arrival rate λ_k and service rate μ_k. This result is all the more remarkable since, in general, the arrival process at station j is not a Poisson process. A counterexample is given by the $M/M/1$ queue with *feedback*; that is, a customer who has been served returns directly to the queue with probability p and leaves the system with probability $1 - p$. Now, take the situation where the service rate is much higher than the arrival rate of new external customers. Then there is a clear dependence between consecutive interarrival times, so that there cannot be a Poisson arrival process.

From relations (9.40) and (9.41), we can deduce simple formulas for

$L = $ the average total number of customers in the system,

$W = $ the average time a newly arrived customer remains in the system.

For L, we have the formula

$$L = \sum_{k=1}^{K} L_k,$$

where L_k is the average number of customers present in an $M/M/1$ queue with arrival rate λ_k and service rate μ_k. Hence,

$$L_k = \frac{\rho_k}{1 - \rho_k}.$$

An expression for W then follows from Little's law:

$$L = \tilde{\lambda} W \quad \text{with } \tilde{\lambda} = \sum_{k=1}^{K} \gamma_k$$

(and *not* $\tilde{\lambda} = \sum_{k=1}^{K} \lambda_k$). This is easiest to see by directly applying the "money principle" from Section 9.3 (let every customer pay 1 euro per unit of time spent in the system and observe that the average number of new external customers per unit of time is equal to $\lambda^* = \gamma_1 + \cdots + \gamma_K$).

We may extend PASTA from single queues to open networks. This result is referred to as the arrival theorem for open networks.

Theorem 9.3 (Arrival theorem for open networks). *For the open network with exponential service times, in steady state, a customer who arrives at station k has probability $P(n_1, n_2, \ldots, n_K)$ of seeing n_1 other customers at station 1, n_2 other customers at station 2, ..., and n_K other customers at station K for all $(n_1, \ldots, n_K)$ with $n_k \geq 0$, $k = 1, \ldots, K$.*

Observe that PASTA implies that an external arrival observes the equilibrium distribution. The arrival theorem states that this is true for *all* arrivals (internal and external).

The BCMP Extension for the Product-Form Formula

The product-form formula (9.40) has been established under the assumption that the service times at the stations have an exponential probability distribution. It is also assumed that there is only one server at every station. In a famous paper (the BCMP paper), F. Baskett, K. M. Chandy, R. R. Muntz, and F. Palacios have shown that the product-form formula (9.40) remains valid if for each of the service stations, one of the following service disciplines holds:

(1) The service discipline is *first-come first-served*, where the service station may have several servers, and a customer's service time is *exponentially* distributed.

(2) The service discipline is *processor sharing*; that is, if n customers are present at the station, then every customer receives a fraction $1/n$ of the station's service capacity. A customer's service time may then have a *general* probability distribution.

(3) The service discipline is that every customer finds an idle server at the station in question (*infinitely many servers*). A customer's service time may then have a *general* probability distribution.

The product-form formula (9.40) remains valid, but the expression for the marginal probability distribution $\{P_k(n), n \geq 0\}$ then depends on the service discipline at a given station k:

Service discipline 1: The probability distribution $\{P_k(n), n \geq 0\}$ is the equilibrium distribution of the number of customers in an isolated $M/M/c$ queue with arrival rate $\lambda = \lambda_k$ and service rate $\mu = \mu_k$.

Service discipline 2: The probability distribution $\{P_k(n), n \geq 0\}$ is the geometric distribution $P_k(n) = (1 - \rho_k)\rho_k^n$ for $n = 0, 1, \ldots$ with $\rho_k = \lambda\mathbb{E}[S_k]$, where $\mathbb{E}[S_k]$ is the expected value of the service time a customer needs at station k (in the case of processor sharing, the same equilibrium distribution therefore holds as in the $M/M/1$ queue).

Service discipline 3: The probability distribution $\{P_k(n), n \geq 0\}$ is a Poisson distribution with expected value $\lambda_k\mathbb{E}[S_k]$, where $\mathbb{E}[S_k]$ is the expected value of the service time of a customer at station k.

9.5.3   *A Closed Network of Queues*

For the analysis of computer systems and production systems, it is often better to consider a network of queues with a *fixed* number of customers (jobs). In this type of system, a customer can leave the system but is then immediately replaced by another one. We consider the following simple model of a closed network of queues:

- The network consists of K service stations numbered $j = 1, \ldots, K$.
- The network has a constant number N of identical customers.
- Every service stations has one server, and a customer's service time at station j is exponentially distributed with expected value $1/\mu_j$.
- When a customer's service at station i is completed, the customer goes to station j with probability p_{ij}, where $\sum_{j=1}^{K} p_{ij} = 1$ for all i.

It is also assumed that the waiting room at every station is sufficiently large and that the service processes at the different stations are independent of one another. Moreover, the Markov matrix $\mathbf{P} = (p_{ij})$ is assumed to have the property that every station is accessible from every other station. By the assumptions above, the stochastic process that describes the numbers of customers present simultaneously at the stations is a continuous-time Markov chain. For all $n_1, \ldots, n_K \geq 0$ with $\sum_{j=1}^{K} n_j = N$, we once again define

$$P(n_1, \ldots, n_K)$$

as the long-run fraction of the time that there are, simultaneously, n_1 customers at station 1, n_2 customers at station 2, $\ldots$, and n_K customers at station K. For the closed network of queues under consideration, the equilibrium distribution $P(n_1, \ldots, n_K)$ also has the *product-form* solution. To give this, we need the so-called relative visit frequencies. First, define λ_j as

$\lambda_j =$ the long-run average number of arrivals at station j per unit of time.

Since λ_i also gives the average number of service completions at station i per unit of time, we find, as in the previous subsection, that the λ_j satisfy the system of linear equations

$$\lambda_j = \sum_{i=1}^{K} \lambda_i p_{ij} \quad \text{for } j = 1, \ldots, K.$$

However, this system does not have a unique solution; it follows from the theory of discrete-time Markov chains that the solution is determined uniquely up to a multiplicative constant. For a constant $\alpha > 0$, we have

$$\lambda_j = \alpha \hat{\pi}_j, \tag{9.42}$$

where $\{\hat{\pi}_j\}$ is the unique solution to the system of equations

$$\hat{\pi}_j = \sum_{i=1}^{K} \hat{\pi}_i p_{ij} \quad \text{for } j = 1, \ldots, K,$$
$$\sum_{j=1}^{K} \hat{\pi}_j = 1. \tag{9.43}$$

The $\hat{\pi}_j$ are nothing other than the equilibrium probabilities of a discrete-time Markov chain with the p_{ij} as one-step transition probabilities. The probability $\hat{\pi}_j$ can also be viewed as the relative frequency with which a customer visits station j.

The desired solution for the equilibrium probabilities $P(n_1, \ldots, n_K)$ is now given by the *product-form formula*

$$P(n_1, \ldots, n_K) = C_N P_1(n_1) \ldots P_K(n_K) \tag{9.44}$$

for a constant $C_N > 0$, where

$$P_k(n_k) = \left(\frac{\hat{\pi}_k}{\mu_k} \right)^{n_k} \quad \text{for } k = 1, \ldots, K. \tag{9.45}$$

In general, the product-form formula is less useful for practical calculations than it may seem. The reason is that this formula contains the normalization constant C_N. This constant can theoretically be calculated from the fact that the sum of all equilibrium probabilities is equal to 1:

$$C_N^{-1} = \sum_{\{n_1 + \cdots + n_K = N\}} \prod_{k=1}^{K} \left(\frac{\hat{\pi}_k}{\mu_k} \right)^{n_k}.$$

This requires the summation over all possible combinations $(n_1, \ldots, n_K)$. The number of possible combinations is $\binom{N+K-1}{N}$, which quickly becomes very large, leading to computational difficulties. Fortunately, there are several methods to circumvent this computational problem.

Buzen's algorithm

A well-known method to compute the normalizing constant C_N for a closed network of queues is Buzen's convolution method.

Algorithm 9.1 (Buzen's Algorithm). Define $G(m,k)$, $m = 0,\ldots,N$, $k = 1,\ldots,K$. Set

$$G(0,k) = 1, \quad k = 1,\ldots,K,$$
$$G(m,1) = \rho_1^m, \quad m = 0,\ldots,N.$$

For $k = 2,\ldots,K$, $m = 1,\ldots,N$, do

$$G(m,k) = G(m,k-1) + \rho_j G(m-1,k). \tag{9.46}$$

Then $C_N = G(M,K)^{-1}$.

Buzen's algorithm computes the normalizing constant C in order NK steps.

Mean-Value Analysis

If one only wants to know the average queue lengths and the average waiting times at the different service stations, the calculation of the normalization constant C can be avoided by applying the so-called *"mean-value"* algorithm. We briefly discuss this interesting approach. A basic component for this algorithm is the famous arrival theorem, which we give here without proof. To formulate this theorem, it is convenient to write the probability $P(n_1,\ldots,n_K)$ as $P_N(n_1,\ldots,n_K)$, where $N = \sum_{k=1}^{K} n_k$, to highlight the dependence on the fixed number N of customers.

Theorem 9.4 (Arrival theorem for closed networks). *For the closed network with exponential service times and a fixed number N of customers, in steady state, a customer who arrives at station k has probability $P_{N-1}(n_1, n_2, \ldots, n_K)$ of seeing n_1 other customers at station 1, n_2 other customers at station 2, $\ldots$, and n_K other customers at station K for all $(n_1,\ldots,n_K)$ with $n_1 + \cdots + n_K = N - 1$.*

In other words, in steady state, the probability distribution of the state seen by a customer arriving at a service station is the same as the probability distribution of the state at an arbitrary time in the closed network with one customer less. For $m = 1,\ldots,N$, define

$L_m(j) =$ the average number of customers at station j in the closed
network with a fixed number m of customers,

$W_m(j) =$ the average amount of time spent by the customer at station j
at each visit to the station in the closed network with a fixed
number m of customers.

Likewise, we define $\lambda_m(j)$ as the average number of arrivals at station j per unit of time when the network has a fixed number m of customers. The quantities we are interested in are, of course, $L_N(j)$ and $W_N(j)$. The following relations now apply:

$$W_m(j) = \frac{1}{\mu_j}\{1 + L_{m-1}(j)\}, \tag{9.47}$$

$$L_m(j) = \lambda_m(j)W_m(j), \qquad (9.48)$$

$$\lambda_m(j) = \alpha_m \hat{\pi}_j \qquad (9.49)$$

for some constant $\alpha_m > 0$. Here, the $\hat{\pi}_j$ are the unique solution to the linear system of equations (9.43). The relation (9.49) is nothing other than equation (9.42), while relation (9.48) follows directly from Little's law. The following heuristic reasoning gives relation (9.47). By the arrival theorem, in a closed network with m customers, a customer arriving at station j finds an average of $L_{m-1}(j)$ other customers at that station. This, together with the memoryless property of the exponential service time distribution, tells us that for a visit to station j, a customer's expected time spent at the station is equal to $[1 + L_{m-1}(j)] \times 1/\mu_j$. How do we calculate this? First, note that the unknown constant α_m in (9.49) can be written in terms of the $W_m(j)$. Since

$$\sum_{j=1}^{K} L_m(j) = m,$$

it follows from (9.48) and (9.49) that $m = \alpha_m \sum_{j=1}^{K} \hat{\pi}_j W_m(j)$, that is,

$$\alpha_m = m \left[\sum_{j=1}^{K} \hat{\pi}_j W_m(j) \right]^{-1}. \qquad (9.50)$$

Algorithm 9.2 (Mean value analysis).

Step 0. First, solve the $\hat{\pi}_j$ from the system of linear equations (9.43). Calculate $W_1(j) = 1/\mu_j$ for $j = 1, \ldots, K$. Let $m := 1$.

Step 1. Calculate α_m from (9.50) and then $\lambda_m(j)$ for $j = 1, \ldots, K$ from (9.49). Then, calculate $L_m(j)$ for $j = 1, \ldots, K$ from (9.48).

Step 2. Let $m := m+1$. Calculate $W_m(j)$ for $j = 1, \ldots, K$ from (9.47). If $m = N$, stop; otherwise, go back to Step 1.

The algorithm provides not only $L_N(j)$ and $W_N(j)$ for all j, but also the "throughput" $\lambda_N(j)$ as the average number of customers served at station j per unit of time.

Finally, it should be noted that the product-form formula (9.44) and the arrival theorem remain valid for a closed network of queues in which each service station is subject to one of the BCMP service disciplines for the open network mentioned in Section 9.5.2.

9.6 Exercises

9.1 At a port, a crane is currently available day and night to unload ships. The ships arrive in the port following a Poisson process with an average of four ships per 24 hours. The unloading time per ship is exponentially distributed with an average of $3\frac{1}{2}$ hours. The port manager is considering replacing the old crane by two smaller ones.

For each smaller crane, the unloading time per ship is also exponentially distributed, but with an average of seven hours. The two smaller ones are given one common queue.

   (a) What is the effect on the average waiting time per ship if the old crane is replaced by two smaller cranes?
   (b) Calculate, for both situations, the fraction of the ships that must wait longer than five hours to be unloaded

9.2 In a communication system with one transmission line, messages arrive following a Poisson process with an average of 250 messages per minute. A total of 1000 characters per second can be transmitted over the line. A message's length is exponentially distributed with an average of 200 characters. Suppose that there is a sufficiently large buffer to temporarily store arriving messages.

   (a) What is the average number of messages waiting for transmission in the buffer?
   (b) What is the average waiting time in the buffer per message?
   (c) What fraction of the time are there 10 or more messages awaiting transmission in the buffer?
   (d) What fraction of the messages must wait more than two seconds in the buffer?

9.3 Consider Exercise 9.2 again. Now, assume that one wishes to install a finite buffer such that the fraction of the incoming messages that find the buffer full and are therefore rejected is less than 1%.

   (a) What is the required buffer size?
   (b) Answer the questions from Exercise 9.2 again for this new buffer size.

9.4 Customers arrive at a taxi stop following a Poisson process with an average of 12 customers per hour. Every customer waits for a taxi. Taxis arrive following a Poisson process with an average of 15 per hour. A taxi only stops if there are passengers waiting; otherwise, the taxi drives on. Every taxi takes a single passenger. What is the average number of waiting customers?

9.5 An oil company employs seven teams for emergency repairs to pipelines. Calls for emergency repairs are made following a Poisson process with an average of one call per day. The repair time is exponentially distributed with an average of five days.

   (a) How long does it take, on average, before a requested repair can be started?
   (b) What is the probability that a requested repair will have to wait longer than a day for a repair team?

9.6 Tankers with crude oil arrive at a port following a Poisson process with an average of seven per 24 hours. There is only room for five tankers. A tanker that finds all berths busy continues on to a competing port. Three tankers can be unloaded simultaneously. A tanker's unloading time is exponentially distributed with an average of nine hours.

   (a) What is the average waiting time of a tanker mooring in the port?
   (b) What is the fraction of the tankers that continue on to another port?

9.7 At a maintenance base for planes, under the current overhaul policy, of each plane entering for maintenance, exactly one of the four engines is overhauled. The overhaul time of an engine has an expected value of $\frac{1}{2}$ days with a standard deviation of $\frac{1}{3}$ days. Only one engine can be overhauled at a time. Under the current revision policy, planes arrive for revision following a Poisson process with an average of three planes per two days. The base is considering changing the overhaul policy so that during a plane's maintenance, all four engines are successively overhauled. Under this maintenance policy, the frequency with which a plane comes in for maintenance is one fourth of what it was. Compare the two maintenance policies based on the average number of planes in the workshop.

9.8 At an airline reservation office, five agents are available to handle incoming telephone calls for flight reservations. In addition, three callers can be put in a queue until an agent is available. Potential customers calling when all agents and queue positions are busy will hear the busy tone. It is assumed that these callers turn to another company. The telephone calls come in following a Poisson process with an average of 40 calls per hour. The handling time of a reservation request is exponentially distributed with an expected value of four minutes.

    (a) What is the fraction of the callers who hear the busy tone?

    (b) What is the fraction of the callers who immediately get to speak to an idle agent?

    (c) What is the average number of busy agents?

9.9 Exactly every hour, a job arrives at a workstation. The processing time for a job is uniformly distributed between 15 and 45 minutes.

    (a) What is the fraction of the time that the workstation is busy?

    (b) What is the fraction of the jobs that must wait?

    (c) Also answer these questions if the jobs arrive following a Poisson process with an average of one job per hour.

9.10 Quotations are requested from a specialized contractor following a Poisson process with an average of two quotations per three working days. The agency has one employee who is responsible for writing quotations. This employee can work on only one quotation at a time. The time required to work out a quotation is normally distributed with an expected value of one day and a standard deviation of $\frac{1}{4}$ days.

    (a) What is the average time taken to answer a request for a quotation?

    (b) What is the fraction of the time that the employee is working on quotations?

9.11 A bicycle store has a workshop where bicycles are offered for repair following a Poisson process with an average of five bicycles per working day. The shop employs only one mechanic, who works eight hours a day. The repair time of each bicycle can be estimated very accurately upon arrival. Taken over all bicycles, the repair time turns out to be uniformly distributed between $\frac{1}{2}$ hours and $2\frac{1}{2}$ hours.

    (a) Determine the average time a bicycle waits for repair in the event that bicycles are repaired in order of arrival.

    (b) Determine the average time a bicycle waits for repair in the event that bicycles with a repair time of less than one hour are given priority over bicycles with a repair time of one hour or more.

9.12 In peak periods, customers arrive at a bank following a Poisson process with an average of two customers per minute. A customer's handling time is approximately exponentially distributed with an expected value of $2\frac{1}{2}$ minutes. The bank is considering redesigning the counter with a number of tellers working in parallel from a single common queue. How many tellers must work such that, on average, 95% of the customers do not have to wait longer than five minutes?

9.13 Consider Example 9.5 again. Now, suppose that there is only parking space for seven tourist buses. A bus that finds all parking spaces occupied upon arrival does not wait but drives on to Volendam.

(a) What is the average number of occupied parking spaces?

(b) What is the fraction of the buses that drive on to Volendam?

9.14 The "Video Palace" has the policy of having five units of a certain exclusive camcorder in stock. Every time a camcorder is sold, a replenishment order for one camcorder is placed with the supplier. The delivery time of the replenishment order is exactly one week. The demand for the camcorder follows a Poisson process with an average demand of two per week. A customer who wants the camcorder while the shop is out of stock buys it elsewhere.

  (a) What is the fraction of the customers who must look elsewhere?

  (b) What is the average number of camcorders in stock? (*Hint*: Use the Erlang loss model.)

  (c) Also answer the last question for the back-order case where a customer who finds the shop out of stock is willing to wait until his order is in, and also give the probability that a customer must wait for a back-order. (*Hint*: Use the $M/G/\infty$ queue.)

9.15 Customers bring cameras to an electronics shop for repair following a Poisson process with an average of five per week. Every camera goes directly into repair, and a camera's repair time is normally distributed with an expected value of two weeks and a standard deviation of one half week. Every customer who brings in a camera for repair is lent a spare camera until the repair is ready. There are sufficiently many spare cameras.

  (a) What is the equilibrium distribution of the number of spare cameras that have been lent out?

  (b) What is the fraction of the time that more than 15 cameras have been lent out?

  (c) Now, suppose that upon arrival, a broken camera's repair time is immediately determined and that spare cameras are lent out only if the repair time is more than two weeks. Answer the previous question for this situation.

9.16 Oil tankers bound for Rotterdam depart from ports in the Middle East following a Poisson process with an average of two tankers per day. The sailing time to Rotterdam has a gamma distribution with an expected value of eight days and a standard deviation of four days.

  (a) What is the probability that, at any given time, more than 25 tankers are on their way to Rotterdam?

  (b) Also answer this question if, independently of the tankers from the Middle East, oil tankers from Nigeria leave for Rotterdam following a Poisson process with an average of one tanker per day, where the sailing time is uniformly distributed between four and five days.

9.17 A call center provides product support in two languages. Requests for information are received by telephone. A sufficiently large number of bilingual operators are present to process every incoming request immediately. Requests from majority-language customers come in following a Poisson process with an average of λ_1 per hour, and, independently of that, requests from minority-language customers come in following a Poisson process with an average of λ_2 per hour. The handling time of a majority-language request is lognormally distributed with an expected value of α_1 minutes and a standard deviation of σ_1 minutes, while for a minority-language request, the handling time is lognormally distributed with an expected value of α_2 minutes and a standard deviation of σ_2 minutes. What is the limiting distribution of the total number of busy telephone operators?

9.18 Consider a production system with four machines, each of which operates as a one-server system. New external orders arrive at machine 1 following a Poisson process with an average of one order every two minutes. When a job at machine 1 is finished, the order is sent on to machine 2 with probability $\frac{1}{2}$ and to machine 3 with probability $\frac{1}{2}$. After the order is handled at machine 2 or 3, it continues on for the final handling at machine 4, after which the order is finished and leaves the system. The handling times at the various machines are independent of one another and exponentially distributed with expected values 48 seconds at machine 1, 144 seconds at machine 2, 180 seconds at machine 3, and 72 seconds at machine 4.

  (a) Determine the equilibrium distributions of the numbers of orders at the different machines.
  (b) Determine the utilization of each of the machines.
  (c) Determine the average time a new order takes to complete.

9.19 Consider a closed network with two customers moving between two servers. A customer's service time at station i is exponentially distributed with expected value $1/\mu_i$, for $i = 1, 2$. After a customer has been served at a station, the customer goes on to the other station with probability $\frac{1}{2}$ and joins the queue for the current station to be handled again with probability $\frac{1}{2}$.

  (a) Specify the simultaneous probability distribution of the numbers of customers at each service station.
  (b) What is the fraction of the time that service station i is busy?
  (c) What is the probability that a customer arriving at service station i finds it busy?

9.20 Consider a CPU model with one central processor and two disks. The processor is the central calculating unit, and the disks are used for backing storage. The closed system contains three jobs that alternately visit the central processor and one of the two disks. Assume that after handing by the processor, a job goes to disk 1 with probability $\frac{1}{3}$ and to disk 2 with probability $\frac{2}{3}$. The service times at the different stations are independent of one another and have an exponential probability distribution with respective expected values 5 ms for the central processor and 15 ms for each of the disks.

  (a) Specify the normalization constant in the simultaneous probability distribution of the numbers of jobs at the three stations.
  (b) What is the average number of jobs at each of the stations?

References

1. R. B. Cooper, *Introduction to Queueing Theory*, 2nd ed., Elsevier North Holland, Inc., New York, 1981.

2. R. W. Hall, *Queueing Methods for Services and Manufacturing*, Prentice Hall, Inc., Englewood Cliffs, New Jersey, 1991.

3. H. C. Tijms, *A First Course in Stochastic Models*, John Wiley & Sons Ltd., Chichester, England, 2003.

Chapter 10

Markov Decision Processes

10.1 Introduction

In Chapter 5, we discussed sequential decision problems with a finite number of decisions. In this chapter, we consider decision problems with an infinite horizon, in which the decision-making process continues indefinitely. Because the total expected reward (or cost) over an infinite horizon will generally not be finite, we must choose another optimization criterion for an infinite-horizon problem. The first option is to use the present value of the (expected) reward as criterion. We discuss this in more detail in Section 10.1.1. Another option is to choose the (expected) average reward per time period; roughly speaking, this is the total (expected) reward in n periods (after n decisions) divided by n, after which we take the limit as $n \to \infty$. To formulate a tractable infinite-horizon decision model, we will always assume that the reward (or costs) and any random variables that occur (for example, the demand during a period or the life span of machines) do not depend on the absolute time. This is called the *stationarity assumption*.

The Markov decision model is a versatile and powerful tool for analyzing probabilistic sequential decision problems with an infinite planning horizon. It has many applications in inventory control, maintenance, manufacturing and telecommunications, among others.

10.1.1 *Discounting and Present Value*

Discounting and present value are important elements of Markov decision models. If a capital K_0 is invested with interest at an *interest (rate)* of $i\%$ per year, then after 1 year, one receives an amount of $K_0 \frac{i}{100}$ in interest. Combining this with the initial capital, one then has $K_0(1 + \frac{i}{100})$. If this capital is reinvested with interest, one speaks of compound interest. If we assume that the interest rate is constant during t years, then after t years, the initial capital has grown to $K_t = K_0\left(1 + \frac{i}{100}\right)^t$. An amount U to be paid t years from now can be replaced by an amount K paid now that is such that with interest, it grows to U after t years. We must then have

$$K\left(1 + \frac{i}{100}\right)^t = U,$$

so

$$K = U \left(1 + \frac{i}{100}\right)^{-t} = \beta^t U,$$

where

$$\beta = \left(1 + \frac{i}{100}\right)^{-1}.$$

The factor β is called the *discount factor*. It is always assumed to be less than 1. The capital K is called the *present value* of U. By using present values, one can thus compare future payments and rewards.

Definition 10.1 (Present value). *Let $X = (x_0, x_1, \ldots)$ be a (finite or countable) sequence of amounts (cash flows), where x_t, for $(t = 0, 1, \ldots)$, is an amount paid (if $x_t < 0$) or received (if $x_t > 0$) in year t. The present value of X, denoted by $PV(X)$, is then*

$$PV(X) = \sum_{t=0}^{\infty} x_t \beta^t, \quad 0 < \beta < 1,$$

provided that the sum is finite.

If we choose the present value as decision rule, then we choose $X = (x_0, x_1, \ldots)$ over $Y = (y_0, y_1, \ldots)$ if $PV(X) > PV(Y)$ for a given discount factor β.

10.2   Markov Decision Processes

This chapter considers probabilistic sequential decision problems with an infinite planning horizon. Problems with a stochastic state have already been discussed in Section 5.8. Examples are the investment problem (see Section 5.8.4) and dice games (see Sections 5.8.2 and 5.8.5). However, these were problems with a finite horizon. In this chapter, we assume that the horizon is infinite. The associated theory is called *Markov decision theory*, and the decision problem is called a *Markov decision problem*.

A Markov decision problem involves a system whose state X_t at time $t \in T = \{0, 1, 2, \ldots\}$ is a random variable. The values X_t can take form the *state space* S. We will always assume that S is a countable set, so that we can speak of the state $i \in S = \{0, 1, 2, \ldots\}$. The process $\{X_t, t \geq 0\}$ is observed at every time $t \in T$, these moments are called *decision epochs*. If $X_t = i$, then decision $a \in D(i)$ is made. The *decision space* or *action space* $D(i)$ is assumed to be finite.

If $X_t = i$ and a decision $a_t \in D(i)$ is made, then we receive an immediate reward $r_t(i, a)$, and the process makes a transition to a new state $X_{t+1} = j$ according to the transition probabilities

$$\mathbb{P}(X_{t+1} = j \mid X_0 = x_0, B_0 = a_0; X_1 = x_1, B_1 = a_1; \ldots; X_t = i, B_t = a_t)$$
$$= \mathbb{P}(X_{t+1} = j \mid X_t = i, B_t = a_t),$$

where B_t is the decision taken at time t. This assumption makes that the decision process does not depend on the past of the decision process before time t. This process is called a Markov decision process. We make the following assumptions:

(i) we receive an *expected immediate reward r(i,a)*;
(ii) a transition to state X_{t+1} at time $t + 1$ occurs according to the transition probabilities

$$p(j|i,a) = \mathbb{P}(X_{t+1} = j \mid X_t = i, B_t = a).$$

It follows from (i) and (ii) that we assume *stationarity* as both $r(i,a)$ and $p(j|i,a)$ do not depend on t. This brings us to the following definition.

Definition 10.2 (Markov decision process). *A Markov decision process* $\{(X_t, B_t), t \in T\}$ *is fully defined by:*

- *decision epochs T,*
- *a state space S, with states $i \in S$,*
- *a decision space $D(i)$, with actions $a \in D(i)$,*
- *expected immediate rewards $r(i,a)$,*
- *transition probabilities $p(j|i,a)$.*

In a Markov decision process, decisions are made on the basis of a chosen *policy*.

Definition 10.3 (Decision rule). *If for each state $i \in S$, a decision is prescribed at time t, then the function $\delta_t : S \to S$ is called a decision rule.*

Definition 10.4 (Policy). *The infinite sequence $\pi = (\delta_0, \delta_1, \ldots, \delta_t, \ldots)$ is called a (deterministic) policy or strategy.*

Definition 10.5 (Stationary policy). *A policy $\pi = (\delta_0, \delta_1, \ldots)$ is stationary if at time t, the decision $\delta_t(i) \in D(i)$ in state $i \in S$ depends only on i, so that $\pi = (\delta, \delta, \ldots)$.*

A stationary policy has the property that in state $i \in S$ the same decision is made at any time. Mathematically, this means that the decision rule δ is a function $\delta : S \to D = \bigcup_{i \in S} D(i)$. Stationary policies form an important subclass of policies.

Suppose that the stationary policy $\pi = (\delta, \delta, \ldots)$ is followed. Since the decision $a = \delta(i) \in D(i)$ is then fixed for every state $i \in S$, the transition probabilities $p(j|i, \delta(i))$ are known for every state $i \in S$. The decision rule δ induces a matrix $\mathbf{P}(\delta) = (p(j|i, \delta(i)))$ of one-step transition probabilities for the state process $\{X_t, t \in T\}$. Hence, this process is a Markov chain with stationary transition probabilities.

Every policy generates an infinite sequence of rewards. To compare policies, we need a criterion that allows us to choose the best from two or more infinite sequences of rewards. The following two criteria are obvious choices:

(i) present value,
(ii) average reward per period.

These criteria translate to the following objective functions:

(i) the mathematical expectation of the present value of the rewards over the infinite horizon

$$V_\pi(i) = \mathbb{E}_\pi \left[\sum_{t=0}^{\infty} \beta^t r(X_t, B_t) \mid X_0 = i \right], \quad i \in S, \tag{10.1}$$

(ii) the mathematical expectation of the average rewards (per period)

$$\bar{V}_\pi(i) = \mathbb{E}_\pi \left[\lim_{n \to \infty} \frac{1}{n} \sum_{t=0}^{n-1} r(X_t, B_t) \mid X_0 = i \right], \quad i \in S. \tag{10.2}$$

The functions $V_\pi(\cdot)$ and $\bar{V}_\pi(\cdot)$ are called the *value functions* corresponding to the policy π; they are functions of the initial state $X_0 = i$ (see Example 10.1).

Definition 10.6. *Under the present value criterion (average value criterion), a policy π^* is better than a policy π if $V_{\pi^*}(i) \geq V_\pi(i)$ ($\bar{V}_{\pi^*}(i) \geq \bar{V}_\pi(i)$), for every initial state $i \in S$. A policy π^* is optimal if for every $i \in S$, we have $V_{\pi^*}(i) = V(i)$ ($\bar{V}_{\pi^*}(i) = \bar{V}(i)$).*

10.3   Markov Decision Processes: Discounted Rewards

In this section, we consider the general Markov decision problem with criterion the maximization of the rewards' expected present value. The discount factor is again denoted by β (with $0 < \beta < 1$). If C is the class of all possible policies, then the optimality criterion reads

$$\max_{\pi \in C} V_\pi(i) \quad \text{for all } i \in S.$$

The value of the objective function after maximization over all possible policies is called the *optimal value function* and is denoted by $V(i)$, $i \in S$, where i denotes the initial state of the decision process, so

$$V(i) = \max_{\pi \in C} V_\pi(i), \quad i \in S. \tag{10.3}$$

Example 10.1 (A network model). In this first example, we consider the special case with deterministic state transitions. In this case, a decision $\delta_t(i)$ leads to a known state, so we can also characterize the decision $\delta_t(i)$ by $\delta_t(i) = j \in S$. Then, the Markov decision model description with discounted rewards (present value) is equivalent to the following network description. The states in S correspond to the nodes of a directed graph $G = (S, E)$. In this graph, a decision $\delta_t(i) = j$ is represented by an arc $(i, j) \in E$. The set of arcs leaving i therefore represents all possible decisions in i. With every arc (that is, directed edge) $(i, j) \in E$ in G is associated a reward $r(i, j)$; this completes the network description. In this network, for every initial node (that is, initial state), a policy generates a path of infinite length (the length of a path is the number of arcs in the path) with corresponding sequence of rewards.

Fig. 10.1   A network with $|S| = 2$.

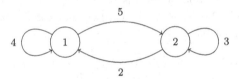

Consider the network in Figure 10.1, which represents a two-state system. We are looking for a policy that maximizes (for every initial state) the present value of the reward over an infinite horizon. We begin by calculating the value function of the policy $\tilde{\pi} = (\delta_0, \delta_1, \delta_2, \ldots)$, where

$$\delta_t(1) = 1, \quad t = 0, 2, 4, \ldots, \quad \delta_t(2) = 2, \quad t = 0, 2, 4, \ldots$$

and

$$\delta_t(1) = 2, \quad t = 1, 3, 5, \ldots, \quad \delta_t(2) = 1, \quad t = 1, 3, 5, \ldots$$

Starting in state 1, the policy $\tilde{\pi}$ generates the path $(1, 1, 2, 2, 1, 1, \ldots)$. The corresponding sequence of rewards is $(4, 5, 3, 2, 4, 5, \ldots)$, so that

$$
\begin{aligned}
V_{\tilde{\pi}}(1) &= 4 + 5\beta + 3\beta^2 + 2\beta^3 + 4\beta^4 + 5\beta^5 + 3\beta^6 + 2\beta^7 + \ldots \\
&= 4(1 + \beta^4 + \beta^8 + \ldots) + 5\beta(1 + \beta^4 + \beta^8 + \ldots) + 3\beta^2(1 + \beta^4 + \ldots) \\
&\quad + 2\beta^3(1 + \beta^4 + \ldots) \\
&= \frac{4 + 5\beta + 3\beta^2 + 2\beta^3}{1 - \beta^4},
\end{aligned}
$$

where we have used the summation formula for the geometric sequence, $\sum_{k=0}^{\infty} x^k = \frac{1}{1-x}$ for $|x| < 1$. Verify that

$$V_{\tilde{\pi}}(2) = \frac{3 + 2\beta + 4\beta^2 + 5\beta^3}{1 - \beta^4}.$$

The policy $\tilde{\pi}$ is an example of a *nonstationary* policy because the decision taken in a state is not the same at every point in time. There are four stationary policies:

$$
\begin{aligned}
\pi_1 &= (\delta^1, \delta^1, \ldots) \quad \text{with} \quad \delta^1(1) = 1 \quad \text{and} \quad \delta^1(2) = 1, \\
\pi_2 &= (\delta^2, \delta^2, \ldots) \quad \text{with} \quad \delta^2(1) = 1 \quad \text{and} \quad \delta^2(2) = 2, \\
\pi_3 &= (\delta^3, \delta^3, \ldots) \quad \text{with} \quad \delta^3(1) = 2 \quad \text{and} \quad \delta^3(2) = 2, \\
\pi_4 &= (\delta^4, \delta^4, \ldots) \quad \text{with} \quad \delta^4(1) = 2 \quad \text{and} \quad \delta^4(2) = 1.
\end{aligned}
$$

Let $V_k(i)$ be the present value of the sequence of rewards of policy π_k if $i \in S$ is the state (initial node) at time $t = 0$. It now follows from the definitions of the four stationary policies that (verify this)

$$
\begin{aligned}
V_1(1) &= \tfrac{4}{1-\beta}, & V_1(2) &= \tfrac{2+2\beta}{1-\beta}, \\
V_2(1) &= \tfrac{4}{1-\beta}, & V_2(2) &= \tfrac{3}{1-\beta}, \\
V_3(1) &= \tfrac{5-2\beta}{1-\beta}, & V_3(2) &= \tfrac{3}{1-\beta}, \\
V_4(1) &= \tfrac{5+2\beta}{1-\beta^2}, & V_4(2) &= \tfrac{2+5\beta}{1-\beta^2}.
\end{aligned}
$$

Let us now determine which of the four stationary policies is the best. In general, this will depend on the value of the discount factor β. Policy π_1 is better than policy π_2 if $V_1(1) \geq V_2(1)$ *and* $V_1(2) \geq V_2(2)$. Therefore, based on the calculated value functions, we must have

$$\frac{2 + 2\beta}{1 - \beta} \geq \frac{3}{1 - \beta}.$$

It follows that π_1 is better than π_2 if $\frac{1}{2} \leq \beta < 1$. One can easily check that for $\beta \geq \frac{1}{2}$, policy π_1 is also better than π_3 and π_4. In other words, π_1 is the best stationary policy for $\frac{1}{2} \leq \beta < 1$. Policy π_3 is better than π_1 if $V_3(1) \geq V_1(1)$ and $V_3(2) \geq V_1(2)$, that is, if

$$\frac{5 - 2\beta}{1 - \beta} \geq \frac{4}{1 - \beta} \quad \text{and} \quad \frac{3}{1 - \beta} \geq \frac{2 + 2\beta}{1 - \beta}.$$

Both inequalities give $\beta \leq \frac{1}{2}$, so that π_3 is better than π_1 for $\beta \leq \frac{1}{2}$. It even turns out that for $\beta \leq \frac{1}{2}$, policy π_3 is also better than π_2 and π_4, so π_3 is the best stationary policy for $\beta \leq \frac{1}{2}$. Thus, π_1 is the best stationary policy for $\frac{1}{2} \leq \beta < 1$ and π_3 is the best for $0 \leq \beta \leq \frac{1}{2}$. (If $\beta = \frac{1}{2}$, the two policies are equivalent.)

Remark 10.1. The result we found is not surprising because for small β (where future rewards hardly count), the first decision makes the largest contribution. In other words, follow the decision rule $\delta(1) = 2$ and $\delta(2) = 2$, with immediate rewards 5 and 3. (Compare this to policy π_3.) If β is close to 1, then future rewards contribute substantially. It should be clear, intuitively, that in this case, it is best to take advantage of the reward 4 $(4 + 4 + 4 + 4 > 5 + 2 + 5 + 2$, and so on) corresponding to the decision $\delta(1) = 1$ in state 1, which leads to $\delta(1) = 1$ and $\delta(2) = 1$; this corresponds to π_1.

We have now found the best stationary policy for the network model. For general Markov decision processes, we are obviously also interested in an optimal policy, that is, the best of all possible policies. Suppose that at time $t = 0$, in state $i \in S$, the decision $a \in D(i)$ is made, which leads to state $j \in S$ at $t = 1$. If an optimal policy is followed, then the maximum expected present value of the rewards from time $t = 1$ on is, by definition, equal to $V(j)$. By the dynamic programming principle of optimality, for an optimal policy, we must therefore have that for every initial state $i \in S$ and every decision $a \in D(i)$,

$$V(i) \geq r(i, a) + \beta \sum_{j \in S} p(j|i, a) V(j), \quad a \in D(i), \ i \in S.$$

This leads to the following *optimality equations* (note that $D(i)$ is a finite set):

$$V(i) = \max_{a \in D(i)} \{r(i, a) + \beta \sum_{j \in S} p(j|i, a) V(j)\}, \quad i \in S. \tag{10.4}$$

The difficulty in solving these equations is that the unknown value function $V : S \to \mathbb{R}$ occurs on both sides. The equations in (10.4) are examples of so-called *functional equations*. We present the following important result without proof.

Theorem 10.1. *If the expected immediate rewards are bounded, then the optimality equations (10.4) have a unique solution.*

We will now show that the best stationary policy is optimal for the network model.

Example 10.1 (Continued). Let $V^*(i)$, $i \in S$, be an optimal value function, which corresponds to an optimal policy. Now, consider state (node) 1. If in this state, the decision $\delta(1) = 1$ is chosen and after that, an optimal policy is followed, then the present value of the reward over an infinite horizon will be equal to $4 + \beta V^*(1)$. If we choose $\delta(1) = 2$, this gives $5 + \beta V^*(2)$. By the optimality principle of dynamic programming, we must therefore have

$$V^*(1) = \max \begin{cases} 4 + \beta V^*(1), & [\delta(1) = 1], \\ 5 + \beta V^*(2), & [\delta(1) = 2]. \end{cases} \tag{10.5}$$

Analogously, we find

$$V^*(2) = \max \begin{cases} 2 + \beta V^*(1), & [\delta(2) = 1], \\ 3 + \beta V^*(2), & [\delta(2) = 2]. \end{cases} \tag{10.6}$$

Together, equations (10.5) and (10.6) form the optimality equations for this infinite-horizon problem. To verify whether π_1 is indeed an optimal policy for $\frac{1}{2} \leq \beta < 1$, we substitute $V_1(1)$ and $V_1(2)$, respectively, for $V^*(1)$ and $V^*(2)$ on the right-hand side of (10.5). This gives

$$V^*(1) = \max \begin{cases} 4 + \beta \frac{4}{1-\beta} = \frac{4}{1-\beta}, & [\delta(1) = 1], \\ 5 + \beta \frac{2+2\beta}{1-\beta} = \frac{5-3\beta+2\beta^2}{1-\beta}, & [\delta(1) = 2]. \end{cases}$$

Since

$$4 \geq 5 - 3\beta + 2\beta^2 \iff \beta \in [\tfrac{1}{2}, 1),$$

we indeed have $V^*(1) = 4/(1 - \beta)$, so that $V_1(1)$ and $V_1(2)$ satisfy (10.5) with $\delta(1) = 1$ as optimal decision. Likewise, they satisfy equation (10.6) with $\delta(2) = 1$ as optimal decision; in other words, $V_1(\cdot)$ satisfies the optimality equations, so that π_1 is indeed an optimal policy for $\beta \in [\frac{1}{2}, 1)$. Likewise, one can show that π_3 is an optimal policy for $\beta \in [0, \frac{1}{2}]$. Note that for $\beta = \frac{1}{2}$, there are at least two optimal policies. (Are there more?)

In the example above, we found an optimal policy by first determining the value functions of all stationary policies and then showing that the best stationary policy's value function satisfies the optimality equations. For a large network, this method is inefficient. In the next three subsections, we discuss three methods to find the optimal value function and policy. To illustrate these methods, we will use the machine maintenance problem, which is a basic problem in Markov decision theory.

Example 10.2 (Machine maintenance). A machine can be in one of three conditions: good, reasonable, and bad. The machine is inspected at the beginning of every week. If it is in reasonable or bad condition, the machine can be sent out for maintenance. In that case, at the beginning of the next week, the machine is operational again and in good condition. Maintenance costs €90. The weekly revenue is €100, €70, and €20 if the machine is in good, reasonable, or bad condition, respectively. If no maintenance is carried out, then the machine's condition is described by a Markov chain with transition probabilities as in Table 10.1.

Table 10.1 Transition probabilities for the machine maintenance example.

Week: t	Good	Week: $t+1$ Reasonable	Bad
Good	0.7	0.2	0.1
Reasonable	0	0.7	0.3
Bad	0	0.1	0.9

If we translate the above into a general Markov decision model, we find

1. Decision epochs: week numbers, $T = \{0, 1, 2, \ldots\}$.
2. State space: $S = \{good, reasonable, bad\} = \{0, 1, 2\}$.
3. Actions: $0 =$ no maintenance, $1 =$ maintenance,
   $D(0) = \{0\}, D(1) = D(2) = \{0, 1\}$.
4. Immediate rewards $\{r(i, a)\}$:
   $r(0, 0) = 100$, $r(1, 0) = 70$, $r(2, 0) = 20$; $r(i, 1) = -90$ for $i = 1, 2$.
5. Transition probabilities $\{p(j|i, a)\}$:
$$p(0|0,0) = 0.7, \quad p(1|0,0) = 0.2, \quad p(2|0,0) = 0.1,$$
$$p(0|1,0) = 0, \quad p(1|1,0) = 0.7, \quad p(2|1,0) = 0.3,$$
$$p(0|2,0) = 0, \quad p(1|2,0) = 0.1, \quad p(2|2,0) = 0.9,$$
$$p(0|1,1) = 1.0, \quad p(0|2,1) = 1.0, \quad p(j|i,1) = 0, \quad (j \neq 0).$$

The optimality equations for this problem are as follows, where $0 < \beta < 1$:
$$V(0) = 100 + \beta\{0.7V(0) + 0.2V(1) + 0.1V(2)\}, \quad [\delta(0) = 0],$$
$$V(1) = \max \begin{cases} 70 + \beta\{0.7V(1) + 0.3V(2)\}, & [\delta(1) = 0], \\ -90 + \beta V(0), & [\delta(1) = 1], \end{cases}$$
$$V(2) = \max \begin{cases} 20 + \beta\{0.1V(1) + 0.9V(2)\}, & [\delta(2) = 0], \\ -90 + \beta V(0), & [\delta(2) = 1]. \end{cases}$$

Suppose that the solution $V(i)$, $i \in S$, of the optimality equations is known. The question now arises as to how to find an optimal policy. The answer can be found in the following theorem.

Theorem 10.2. *There exists an optimal stationary policy $\pi^* = (\delta, \delta, \ldots)$ with value function $V(i)$, $i \in S$, that satisfies (10.4), so that*
$$r(i, \delta(i)) + \beta \sum_{j \in S} p(j|i, \delta(i))V(j) = \max_{a \in D(i)} \{r(i, a) + \beta \sum_{j \in S} p(j|i, a)V(j)\}, \quad i \in S.$$

This theorem says that in any state $i \in S$, the optimal decision is the decision for which the maximum is attained in (10.4). Let us illustrate this using our example.

Example 10.2 (Continued). The solution of the optimality equations for $\beta = 0.8$ is $V(0) = 360.5701$, $V(1) = 267.3397$ and $V(2) = 198.4561$. To find the optimal decision in state 1, we consider

$$V(1) = \max \begin{cases} 70 + 0.8 \cdot \{0.7V(1) + 0.3V(2)\}, & [\delta(1) = 0], \\ -90 + 0.8 \cdot V(0), & [\delta(1) = 1]. \end{cases}$$

Since $70 + 0.8 \cdot \{0.7 \cdot 267.3397 + 0.3 \cdot 198.4561\} > -90 + 0.8 \cdot 360.5701 = 198.4561$, the optimal decision in state 1 is $\delta(1) = 0$. For state 2, we find $\delta(2) = 1$, so that the optimal stationary policy is to provide no maintenance when the machine is in good or reasonable condition, and provide maintenance when the condition is bad.

In the example above, the solution to the optimality equations is known. We now continue with methods that also determine the solution of the optimality equations.

10.3.1 *Value Iteration*

In this section, we discuss a method to approach the solution of (10.4) arbitrarily close. This method is known as *value iteration* or *successive approximation*.

Algorithm 10.1 (Value iteration algorithm).

Step 1. (Initialization) Specify the desired accuracy $\epsilon > 0$, set $n = 1$ and compute

$$V_1(i) = \max_{a \in D(i)} \{r(i,a)\} \text{ for all } i \in S.$$

Step 2. (Iteration) Set $n := n + 1$ and determine $V_n(i)$ using

$$V_n(i) = \max_{a \in D(i)} \{r(i,a) + \beta \sum_{j \in S} p(j|i,a)V_{n-1}(j)\} \text{ for all } i \in S. \tag{10.7}$$

Step 3. (Stop criterion) If $||V_n - V_{n-1}|| < \epsilon(1 - \beta)/(2\beta)$ then stop, else, return to step 2. Here, we use the maximum norm, that is, $||V_n - V_{n-1}|| = \max_{i \in S}\{|V_n(i) - V_{n-1}(i)|\}$.

If we view $V_n(i)$ as

$$V_n(i) = \text{maximum expected present value, beginning in state } i \in S,$$
$$\text{when } n \text{ additional decisions are to be made,}$$

then (10.7) is the dynamic programming recursion of a known *finite*-horizon problem. Intuitively, it is now obvious that as $n \to \infty$, we have $V_n(i) \to V(i)$ for all $i \in S$, so that V_n converges to the solution of the optimality equations. We will not prove this, but we will illustrate it for our machine maintenance example.

Example 10.2 (Continued). For our example, with $\beta = 0.8$, we get the following.

Step 1. The initialization is

$$V_1(0) = 100, \qquad [\delta_1(0) = 0],$$
$$V_1(1) = \max\{70, -90\} = 70, \quad [\delta_1(1) = 0],$$
$$V_1(2) = \max\{20, -90\} = 20, \quad [\delta_1(2) = 0].$$

Step 2. For $n \geq 2$, the recurrence relations (10.7) become

$$V_n(0) = 100 + 0.8\big(0.7V_{n-1}(0) + 0.2V_{n-1}(1) + 0.1V_{n-1}(2)\big),$$

$$V_n(1) = \max \begin{cases} 70 + 0.8\big(0.7V_{n-1}(1) + 0.3V_{n-1}(2)\big), \\ -90 + 0.8V_{n-1}(0), \end{cases}$$

$$V_n(2) = \max \begin{cases} 20 + 0.8\big(0.1V_{n-1}(1) + 0.9V_{n-1}(2)\big), \\ -90 + 0.8V_{n-1}(0). \end{cases}$$

Calculating the V_n gives the values in the table below (verify this for $n = 2$):

Table 10.2   Value iteration.

n	$V_n(0)$	$V_n(1)$	$V_n(2)$	$\|V_n - V_{n-1}\|$
1	100.0000	70.0000	20.0000	-
2	168.8000	114.0000	40.0000	68.800000
3	215.9680	143.4400	57.9200	47.168000
4	248.5261	164.2272	82.7744	32.558080
5	272.0729	181.8331	108.8209	26.046464
10	331.8429	238.6258	169.7237	9.027101
15	351.1580	257.9276	189.0439	2.353102
20	357.4859	264.2555	195.3719	0.771040
30	360.2389	267.0085	198.1249	0.082790
40	360.5345	267.3041	198.4205	0.008889
50	360.5663	267.3358	198.4522	0.000955
60	360.5697	267.3393	198.4556	0.000102
∞	360.5701	267.3397	198.4561	0.000000

Step 3. If we use, for example, $\epsilon = 0.001$, then we get $\epsilon(1 - \beta)/(2\beta) = 0.000125$ and we stop the algorithm after we have evaluated $n = 60$ in step 2. The greater the discount factor β is, the slower the convergence.

Theorem 10.3. *The sequence of functions $\{V_n : S \to \mathbb{R}, n = 1, 2, \ldots\}$ defined by (10.7) converges to the solution of the optimality equations (10.4).*

Proof. We prove Theorem 10.3 for the special case with deterministic state transitions. In this case, the optimality equations (10.4) simplify to

$$V(i) = \max_{(i,j) \in E} \{r(i, j) + \beta V(j)\}, \quad i \in S \tag{10.8}$$

and a similar simplification applies to (10.7):

$$V_n(i) = \max_{(i,j) \in E} \{r(i, j) + \beta V_{n-1}(j)\}, \text{ for all } i \in S. \tag{10.9}$$

Let $\delta(i) = j_i^*$ be an optimal decision in (10.8) for state $i \in S$. According to (10.8) and (10.9), for every $i \in S$, we have

$$V(i) = r(i, j_i^*) + \beta V(j_i^*),$$
$$V_n(i) \geq r(i, j_i^*) + \beta V_{n-1}(j_i^*),$$

so that
$$V(i) - V_n(i) \leq \beta\{V(j_i^*) - V_{n-1}(j_i^*)\} \leq \beta d_{n-1}$$
with
$$d_{n-1} = \max_{k \in S} |V(k) - V_{n-1}(k)|.$$

Let $\delta(i) = j_n(i)$ be an optimal decision in (10.9) for state $i \in S$. According to (10.8) and (10.9), for every $i \in S$, we have
$$V(i) \geq r(i, j_n(i)) + \beta V(j_n(i)),$$
$$V_n(i) = r(i, j_n(i)) + \beta V_{n-1}(j_n(i)),$$
so that
$$V(i) - V_n(i) \geq \beta\{V(j_n(i)) - V_{n-1}(j_n(i))\} \geq -\beta d_{n-1}.$$
Hence, for every $i \in S$, we have
$$|V(i) - V_n(i)| \leq \beta d_{n-1}.$$
Consequently,
$$\max_{i \in S} |V(i) - V_n(i)| = d_n \leq \beta d_{n-1},$$
so that $d_n \leq \beta^{n-1} d_1$, from which it then follows that
$$\lim_{n \to \infty} V_n(i) = V(i) \quad \text{for all } i \in S$$
because $0 < \beta < 1$ and $d_1 < \infty$. $\qquad\square$

We illustrate Theorem 10.3 by returning to our network model.

Example 10.1 (Continued). From Figure 10.1, we see that in this case, we have

Step 1.
$$V_1(1) = \max\{4, 5\} = 5,$$
$$V_1(2) = \max\{2, 3\} = 3.$$

Step 2. For $n = 2, 3, \ldots,$
$$V_n(1) = \max\{4 + \beta V_{n-1}(1), 5 + \beta V_{n-1}(2)\},$$
$$V_n(2) = \max\{3 + \beta V_{n-1}(2), 2 + \beta V_{n-1}(1)\}. \tag{10.10}$$

Take $\beta = 0.8$. Using (10.10), we consecutively find

$V_2(1) = 8$ $\qquad\qquad\qquad$ $V_2(2) = 6$

$V_3(1) = 10.4$ $\qquad\qquad\quad$ $V_3(2) = 8.4$

$V_4(1) = 12.32$ $\qquad\qquad\;$ $V_4(2) = 10.32$

$\qquad\vdots\qquad\qquad\qquad\qquad\qquad\vdots$

$$V_n(1) = 20 - 15 \cdot (0.8)^{n-1} \qquad V_n(2) = 18 - 15(0.8)^{n-1}, \qquad n = 1, 2, \ldots,$$

so that $\lim_{n \to \infty} V_n(1) = V(1) = 20$ and $\lim_{n \to \infty} V_n(2) = V(2) = 18$, giving the optimal value function for $\beta = 0.8$.

In the case of value iteration, we have a sequence of value functions, each corresponding to a finite-horizon problem. As the horizon grows, we obtain better and better approximations of the optimal value function corresponding to the infinite-horizon problem; that function is a solution of (10.4). The more accurate the solution must be, the more iterations are needed. So, theoretically, convergence requires infinitely many steps as may be observed from the example above.

10.3.2 *Policy Iteration*

Another method to find an optimal policy and the corresponding optimal value function uses an iteration method in which better and better *stationary* policies are found, known as *policy* or *strategy iteration*. If the state space S is finite, the method converges in finitely many steps.

Let $\pi = (\delta, \delta, \ldots, \delta, \ldots)$ be a stationary policy; that is, for any $i \in S$, the same decision $\delta(i) \in D(i)$ is made at every time $t \in T$. Let $V_\pi : S \to \mathbb{R}$ be the corresponding value function. It satisfies the following equations:

$$V_\pi(i) = r(i, \delta(i)) + \beta \sum_{j \in S} p(j|i, \delta(i)) V_\pi(j), \quad i \in S. \tag{10.11}$$

This is a system of linear equations with a unique solution; it is finite if $|S| < \infty$. The policy iteration method is based on the fact that the class of stationary policies contains an optimal policy provided that $|D(i)| < \infty$ for all $i \in S$.

Theorem 10.4. *Let $\pi^* = (\delta^*, \delta^*, \ldots)$ be a stationary policy that, for every $i \in S$, prescribes a decision $\delta^*(i) \in D(i)$ such that*

$$r(i, \delta^*(i)) + \beta \sum_{j \in S} p(j|i, \delta^*(i)) V(j) = \max_{a \in D(i)} \left\{ r(i, a) + \beta \sum_{j \in S} p(j|i, a) V(j) \right\}. \tag{10.12}$$

Then

$$V_{\pi^*}(i) = V(i) \quad \text{for all } i \in S$$

and π^* is optimal.

This theorem says that any stationary policy that prescribes decisions for which the maximum is reached in (10.4) is optimal.

Example 10.2 (Continued). There are four stationary policies for our problem:

$$\begin{aligned}
\pi_1 &= (\delta^1, \delta^1, \ldots) \quad \text{with} \quad \delta^1(0) = 0, \ \delta^1(1) = 0, \ \delta^1(2) = 0, \\
\pi_2 &= (\delta^2, \delta^2, \ldots) \quad \text{with} \quad \delta^2(0) = 0, \ \delta^2(1) = 1, \ \delta^2(2) = 0, \\
\pi_3 &= (\delta^3, \delta^3, \ldots) \quad \text{with} \quad \delta^3(0) = 0, \ \delta^3(1) = 0, \ \delta^3(2) = 1, \\
\pi_4 &= (\delta^4, \delta^4, \ldots) \quad \text{with} \quad \delta^4(0) = 0, \ \delta^4(1) = 1, \ \delta^4(2) = 1.
\end{aligned}$$

For each of these policies, the state of the machine is described by a Markov chain. Here are the matrices of one-step transition probabilities corresponding to two of the policies:

$$\mathbf{P}(\delta^1) = \begin{pmatrix} 0.7 & 0.2 & 0.1 \\ 0 & 0.7 & 0.3 \\ 0 & 0.1 & 0.9 \end{pmatrix},$$

$$\mathbf{P}(\delta^3) = \begin{pmatrix} 0.7 & 0.2 & 0.1 \\ 0 & 0.7 & 0.3 \\ 1.0 & 0 & 0 \end{pmatrix}.$$

With every policy is associated a value function, which can be calculated using (10.11). For policy π_3, with $\beta = 0.8$, the equations are as follows (see $\mathbf{P}(\delta^3)$):

$$V_3(0) = 100 + 0.8\big(0.7V_3(0) + 0.2V_3(1) + 0.1V_3(2)\big),$$
$$V_3(1) = 70 + 0.8\big(0.7V_3(1) + 0.3V_3(2)\big),$$
$$V_3(2) = -90 + 0.8\big(1.0V_3(0)\big),$$

with solution $V_3(0) = 360.5701$, $V_3(1) = 267.3397$, $V_3 = 198.4561$. The value functions for all four policies are given in the following table:

Table 10.3   The value functions for the four policies.

	π_1	π_2	π_3	π_4
$V_\pi(0)$	337.7622	303.8911	360.5701	316.1290
$V_\pi(1)$	234.6154	153.1128	267.3397	162.9032
$V_\pi(2)$	138.4615	115.1751	198.4561	162.9032

It follows from the table that π_3 is the best stationary policy. (*Question*: Why do we have $V_\pi(1) = V_\pi(2)$ for $\pi = \pi_4$?)

That π_3 is also optimal, that is, the best of all possible policies, follows from the fact that $V_{\pi_3} = V$; see Table 10.2. In other words, $V_{\pi_3}(i)$, $i \in S$, satisfies the optimality equations. As we have shown in Example 10.2 (continued) just before Section 10.3.1, we could also have found the optimal policy π_3 using Theorem 10.4 under the assumption that the optimal value function V, see Table 10.2 for $n = \infty$, is known.

In the example above, we have illustrated two methods to determine an optimal (stationary) policy. If the optimal value function V is known, Theorem 10.4 can be used to find an optimal policy. If V is unknown, then one determines the value functions of *all* stationary policies; the best stationary policy is then an optimal one. In practice, the latter method is very inefficient because it may be necessary to compare thousands of policies. It is important that there is an efficient method that determines both the optimal (stationary) policy and the optimal value function, namely the *policy iteration method*.

Let $\pi = (\delta, \delta, \ldots)$ be a stationary policy with value function $V_\pi(i)$, $i \in S$. This function then satisfies the equations (10.11). Solving those equations is called *policy evaluation* and amounts to solving a finite system of linear equations. (This system always has a unique solution.)

To obtain a better stationary policy, we define, for every initial state $i \in S$, the quantity $W(i)$ by

$$W(i) = \max_{a \in D(i)} \left\{ r(i,a) + \beta \sum_{j \in S} p(j|i,a) V_\pi(j) \right\}, \quad i \in S. \tag{10.13}$$

The value

$$W(i) = \text{the maximum present value of the rewards starting in state } i \in S$$
$$\text{if policy } \pi \text{ is followed from } t = 2 \text{ (that is, the second decision) on.}$$

The first decision in $i \in S$ may differ from the decision $\delta(i)$ that will be taken the next times the system is in state $i \in S$. Let $f(i) \in S$ be the decision that yields $W(i)$, so

$$W(i) = r(i, f(i)) + \beta \sum_{j \in S} p(j|i, f(i))V_\pi(j)$$

$$= \max_{a \in D(i)} \left\{ r(i, a) + \beta \sum_{j \in S} p(j|i, a)V_\pi(j) \right\}. \tag{10.14}$$

It immediately follows from (10.11) and (10.14) that $W(i) \geq V_\pi(i)$ for all $i \in S$. The function $W(i)$, $i \in S$, is the value function of the *nonstationary* policy $(f, \delta, \delta, \dots)$, which is a better policy than π unless $W(i) = V_\pi(i)$ *for all $i \in S$*. It seems obvious that if $f(i) \neq \delta(i)$ for some $i \in S$, we should always take the better decision $f(i)$ in i. If we do so for all states, then this results in a new stationary policy $\bar{\pi} = (f, f, \dots)$, known as the *stationary improvement* of π.

If $f(i) = \delta(i)$ for every $i \in S$, so that policy π is not changed, then we have $V_\pi(i) = W(i)$ for all $i \in S$. This means that, see (10.13),

$$V_\pi(i) = \max_{a \in D(i)} \left\{ r(i, a) + \beta \sum_{j \in S} p(j|i, a)V_\pi(j) \right\},$$

in which case V_π apparently satisfies the optimality equations (10.4). If (10.14) does not lead to a different decision in any state, then π is an optimal policy. This last property forms a natural termination criterion for the policy iteration algorithm.

Algorithm 10.2 (Policy iteration algorithm).

Step 1. (Initialization) Choose a stationary initial policy $\pi_0 = (\delta^0, \delta^0, \dots)$, and take $k := 0$.

Step 2. (Policy evaluation) Determine $V_k(i)$, $i \in S$, using

$$V_k(i) = r(i, \delta(i)) + \beta \sum_{j \in S} p(j|i, \delta(i))V_k(j), \quad i \in S.$$

Step 3. (Policy improvement) Determine a new policy $\pi_{k+1} = (\delta^{k+1}, \delta^{k+1}, \dots)$ by finding, for every state $i \in S$, a decision $\delta^{k+1}(i)$ using

$$W(i) = \max_{a \in D(i)} \left\{ r(i, a) + \beta \sum_{j \in S} p(j|i, a)V_k(j) \right\}.$$

Step 4. (Termination rule) Stop if $\delta^{k+1}(i) = \delta^k(i)$ for all $i \in S$ (in that case, π_k is optimal and V_k is the optimal value function). If $\delta^{k+1}(i) \neq \delta^k(i)$ for some $i \in S$, then set $k := k + 1$ and return to step 2.

Example 10.2 (Continued). Let us apply the policy iteration algorithm to our machine maintenance problem for $\beta = 0.8$.

Step 1. Take the initial policy $\pi_0 = (\delta^0, \delta^0, \dots)$ with $\delta^0(0) = \delta^0(1) = \delta^0(2) = 0$.

Step 2. The value function $V_0(i)$, $i \in S$, corresponding to π_0 follows from (10.11):

$$V_0(0) = 100 + 0.8\big(0.7V_0(0) + 0.2V_0(1) + 0.1V_0(2)\big),$$
$$V_0(1) = 70 + 0.8\big(0.7V_0(1) + 0.3V_0(2)\big),$$
$$V_0(2) = 20 + 0.8\big(0.1V_0(1) + 0.9V_0(2)\big).$$

The solution is $V_0(0) = 337.7622$, $V_0(1) = 234.6154$, $V_0(2) = 138.4615$.

Step 3. The decision in state 0 is always $\delta(0) = 0$. For state 1, we take

$$W(1) = \max \begin{cases} 70 + 0.8(0.7 \cdot 234.6154 + 0.3 \cdot 138.4615) = 234.6154, & \delta^1(1) = 0, \\ -90 + 0.8 \cdot 337.7622 = 180.2098, & \delta^1(1) = 1. \end{cases}$$

We find that $\delta^1(1) = \delta^0(1) = 0$, so there are no changes.
For state 2, we have

$$W(2) = \max \begin{cases} 20 + 0.8(0.1 \cdot 234.6154 + 0.9 \cdot 138.4615) = 138.4615, & \delta^2(2) = 0, \\ -90 + 0.8 \cdot 337.7622 = 180.2098, & \delta^2(2) = 1. \end{cases}$$

We find that $\delta^1(2) = 1 \neq \delta^0(2)$, so there is a change.

Step 4. Since $\delta^1(2) \neq \delta^0(2)$, the policy π_0 is changed to policy $\pi_1 = (\delta^1, \delta^1, \ldots)$ with $\delta^1(0) = \delta^1(1) = 0, \delta^1(2) = 1$.

Step 2. Using (10.11), we calculate that the value function corresponding to π_1 satisfies $V_1(0) = 360.5701, V_1(1) = 267.3397, V_1(2) = 198.4561$.

Step 3. For state 1, we now take

$$W(1) = \max \begin{cases} 70 + 0.8(0.7 \cdot 267.3397 + 0.3 \cdot 198.4561) = 267.3397, & \delta^2(1) = 0, \\ -90 + 0.8 \cdot 360.5701 = 198.4561, & \delta^2(1) = 1; \end{cases}$$

consequently, $\delta^2(1) = \delta^1(1) = 0$, so there are no changes.
For state 2, we take

$$W(2) = \max \begin{cases} 20 + 0.8(0.1 \cdot 267.3397 + 0.9 \cdot 198.4561) = 184.2756, & \delta^2(2) = 0, \\ -90 + 0.8 \cdot 360.5701 = 198.4561, & \delta^2(2) = 1; \end{cases}$$

consequently, $\delta^2(2) = \delta^1(2) = 1$, so there are no changes.

Step 4. Since $\delta^2(i) = \delta^1(i)$ for $i = 0, 1, 2$, policy π_1 is optimal.

In general, the policy iteration algorithm converges in finitely many steps if $|S| < \infty$ and $|D(i)| < \infty$ for every $i \in S$ because there are then only finitely many stationary policies and every iteration leads to a strictly better policy.

10.3.3 *Linear Programming*

Lastly, we show how to find an optimal policy and the optimal value function using linear programming. We begin by considering the optimality equations once more:

$$V(i) = \max_{a \in D(i)} \{r(i, a) + \beta \sum_{j \in S} p(j|i, a) V(j)\}, \quad i \in S.$$

This is equivalent to the system of inequalities

$$V(i) \geq r(i, a) + \beta \sum_{j \in S} p(j|i, a) V(j), \quad \text{for all } a \in D(i), \text{ for all } i \in S, \qquad (10.15)$$

where for every $i \in S$, there is equality for at least one decision $a \in D(i)$. The decisions where we have equality form the optimal stationary policy. A solution of (10.15) with this property can be found using the LP problem

$$\text{Minimize} \quad \sum_{i \in S} x_i$$

$$\text{subject to} \quad x_i - \beta \sum_{j \in S} p(j|i,a)x_j \geq r(i,a), \quad a \in D(i),\ i \in S,$$

in which the inequalities correspond to (10.15) and the chosen objective function ensures that for every $i \in S$, there is, indeed, at least one $a \in D(i)$ such that the associated inequality is an equality. The latter is easily verified. Suppose that for some i, we have

$$x_i > r(i,a) + \beta \sum_{j \in S} p(j|i,a)x_j \quad \text{for all } a \in D(i),$$

then x_i can be decreased regardless of the values of the x_k for $k \neq i$. This follows from the equation

$$x_k \geq r(k,a) + \beta \sum_{j \in S} p(j|k,a)x_j.$$

So the objective function ensures that for every $i \in S$, there is an $a \in D(i)$ for which the associated constraint in the LP holds with equality.

Example 10.2 (Continued). The LP model for our problem (with $\beta = 0.8$) is

$$\begin{aligned}
\text{Minimize} \quad & x_0 + x_1 + x_2 \\
\text{subject to} \quad & x_0 = 100 + 0.8\big(0.7x_0 + 0.2x_1 + 0.1x_2\big), && [\delta(0) = 0], && \text{(a)} \\
& x_1 \geq 70 + 0.8\big(0.7x_1 + 0.3x_2\big), && [\delta(1) = 0], && \text{(b)} \\
& x_1 \geq -90 + 0.8x_0, && [\delta(1) = 1], && \text{(c)} \\
& x_2 \geq 20 + 0.8\big(0.1x_1 + 0.9x_2\big), && [\delta(2) = 0], && \text{(d)} \\
& x_2 \geq -90 + 0.8x_0, && [\delta(2) = 1]. && \text{(e)}
\end{aligned}$$

The optimal basic solution is $x_0 = 360.5701$, $x_1 = 267.3397$, and $x_2 = 198.4561$, and the binding constraints (in which there is equality) are (a), (b), and (e). This corresponds to the optimal decisions $\delta(0) = \delta(1) = 0$ and $\delta(2) = 1$ found earlier.

Remark 10.2. We would find the same optimal solution if we changed the criterion to minimizing $b_1 x_1 + b_2 x_2 + b_3 x_3$ for $b_1 > 0$, $b_2 > 0$, and $b_3 > 0$. This property holds in general. In the LP, too, we can replace the objective function

$$\min \sum_{i \in S} x_i \quad \text{with} \quad \min \sum_{i \in S} b_i x_i, \text{ where } b_i > 0.$$

Closing Remark

In this section, all formulations have been focused on *maximizing* the expected present value of *rewards*. If we would rather *minimize* the expected present value of *costs*, then *min* must be replaced by *max* everywhere, and vice versa, and $r(i,a)$

must be viewed as the expected immediate costs, which are usually denoted by $c(i, a)$. The optimality equations (10.4) then become

$$V(i) = \min_{a \in D(i)} \left\{ c(i, a) + \beta \sum_{j \in S} p(j|i, a) V(j) \right\}, \quad i \in S,$$

and the corresponding LP problem becomes

Maximize $\sum_{j \in S} x_j$

subject to $x_i - \beta \sum_{j \in S} p(j|i, a) x_j \leq c(i, a), \quad a \in D(i), \ i \in S.$

10.4 Markov Decision Processes: Average Rewards

In this section, we consider the general Markov decision problem with criterion the maximization of the expected average rewards (per period). If C is the class of all possible policies, then the optimality criterion reads

$$\max_{\pi \in C} \bar{V}_\pi(i) \quad \text{for all } i \in S.$$

The value of the objective function after maximization over all possible policies is called the *optimal value function* and is denoted by $\bar{V}(i)$, $i \in S$, where i denotes the initial state of the decision process, so

$$\bar{V}(i) = \max_{\pi \in C} \bar{V}_\pi(i), \quad i \in S. \tag{10.16}$$

For this criterion, there also exist a value iteration method, a policy iteration method, and an LP formulation. We will only discuss the latter briefly.[1]

The state process $\{X_t, t \geq 0\}$ in a Markov decision process is a Markov chain if we apply a stationary policy. Let $\pi = (\delta, \delta, \ldots)$ be such a stationary policy and $\mathbf{P}(\delta)$ the matrix of one-step transition probabilities. We assume that for every stationary policy $\pi = (\delta, \delta, \ldots)$, the Markov chain underlying policy π has no two or more disjoint closed sets, which implies that the stationary probability distribution corresponding to $\mathbf{P}(\delta)$ exists. If $\{q_i, i \in S\}$ is this equilibrium distribution, then we have (see Theorem 7.4)

$$q_j = \sum_{i \in S} q_i p(j|i, \delta), \quad j \in S,$$

$$\sum_{j \in S} q_j = 1. \tag{10.17}$$

The equilibrium probability q_i can be interpreted as the long-run fraction of time that the process is in state $i \in S$.

The (expected) immediate reward of decision $\delta(i) \in D(i)$ in state $i \in S$ is $r(i, \delta(i))$. The following result is clear from the second interpretation of q_i.

---

[1]The interested reader is referred to Chapter 1 in reference 1 to this chapter for the alternative methods of value iteration and policy iteration.

Theorem 10.5. *Let $\{q_i, i \in S\}$ be the equilibrium distribution of the Markov chain corresponding to the policy $\pi = (\delta, \delta, \ldots)$. The expected value of the average reward per period is given by*

$$\mathbb{E}_\pi \left[\lim_{t \to \infty} \frac{1}{t} \sum_{n=0}^{t-1} r(X_n, \delta(X_n)) \right] = \sum_{i \in S} r(i, \delta(i)) q_i. \tag{10.18}$$

Example 10.2 (Continued). We have already seen that there exist four stationary policies for our maintenance problem (denoted by π_i with $i = 1, 2, 3, 4$). We will now determine the expected average reward per period of policy $\pi_1 = (\delta, \delta, \ldots)$ with $\delta(0) = \delta(1) = \delta(2) = 0$.

The corresponding transition probability matrix is

$$\mathbf{P}(\delta) = \begin{pmatrix} 0.7 & 0.2 & 0.1 \\ 0 & 0.7 & 0.3 \\ 0 & 0.1 & 0.9 \end{pmatrix}.$$

(Verify that state 0 is *transient* and that states 1 and 2 form an irreducible set; see Section 7.4.1). The equilibrium distribution is obtained from (10.17):

$$q_0 = 0.7q_0,$$
$$q_1 = 0.2q_0 + 0.7q_1 + 0.1q_2,$$
$$q_2 = 0.1q_0 + 0.3q_1 + 0.9q_2,$$
$$q_0 + q_1 + q_2 = 1.$$

The solution is $q_0 = 0$, $q_1 = 0.25$, $q_2 = 0.75$.

Since $\delta(0) = \delta(1) = \delta(2) = 0$, the immediate rewards are given by $r(0, \delta(0)) = 100$, $r(1, \delta(1)) = 70$, $r(2, \delta(2)) = 20$ (if $\delta(2) = 1$, then $r(2, \delta(2)) = -90$). According to (10.18), the expected average reward is

$$\sum_{i=0}^{2} r(i, \delta(i)) q_i = 0.25 \cdot 70 + 0.75 \cdot 20 = 32.5.$$

Verify that for π_2, π_3, and π_4, we find the values 13.5, 60.8, and 56.2, respectively. Apparently, for the criterion we are now considering, π_3 is also the best stationary policy (we will see later that it is even optimal). This illustrates a phenomenon that occurs more often, namely that the optimal policy for the average reward criterion is also optimal for the present value criterion provided that the discount factor β is close enough to 1.

10.4.1   *Linear Programming*

For $i \in S$ and $a \in D(i)$, define

$x(i, a) = $ probability that at an arbitrary time, the state is $i \in S$
         and the decision $a \in D(i)$ *is made*.

It follows from this definition that

$\sum_{a \in D(i)} x(i, a) = $ probability that the state is $i \in S$ at an arbitrary time, so that

$$\sum_{i \in S} \sum_{a \in D(i)} x(i, a) = 1.$$

The probability that at time $t - 1$, for $t \to \infty$, the system is in state $i \in S$ and decision $a \in D(i)$ is made, and that, moreover, at time t, the system is in state $j \in S$, is equal to $x(i, a)p(j|i, a)$. If we take the sum of these expressions over all $a \in D(i)$ and all $i \in S$, then we find the probability that the system is in state $j \in S$ at time t, so that

$$\sum_{i \in S} \sum_{a \in D(i)} x(i, a)p(j|i, a) = \sum_{a \in D(j)} x(j, a).$$

Since $r(i, a)$ is the reward in state $i \in S$ if decision $a \in D(i)$ is made, the sum

$$\sum_{i \in S} \sum_{a \in D(i)} r(i, a)x(i, a)$$

gives the expected reward per period. The LP model (that can be used to find an optimal policy and the optimal value function) is therefore

$$
\begin{aligned}
\text{Maximize} \quad & \sum_{i \in S} \sum_{a \in D(i)} r(i, a)x(i, a) \\
\text{subject to} \quad & \sum_{i \in S} \sum_{a \in D(i)} x(i, a)p(j|i, a) = \sum_{a \in D(j)} x(j, a), \quad j \in S \\
& \sum_{i \in S} \sum_{a \in D(i)} x(i, a) = 1
\end{aligned}
$$

(10.19)

and $\qquad x(i, a) \geq 0, \quad a \in D(i), \ i \in S.$

One can show that under certain conditions regarding $\mathbf{P}(\delta)$, this LP problem has an optimal basic solution such that for every $i \in S$, there is a unique $a = \delta(i)$ such that $x(i, \delta(i)) > 0$, in other words, such that there exists an optimal deterministic stationary policy.

Example 10.2 (Continued). If, for the sake of brevity, we write x_{ia} for $x(i, a)$, then the LP model for our problem is

$$
\begin{aligned}
\text{Maximize} \quad & 100x_{00} - 90x_{01} + 70x_{10} - 90x_{11} + 20x_{20} - 90x_{21} \\
\text{subject to} \quad & 0.7x_{00} + x_{01} + x_{11} + x_{21} = x_{00} + x_{01} \qquad (j = 0) \\
& 0.2x_{00} + 0.7x_{10} + 0.1x_{20} = x_{10} + x_{11} \qquad (j = 1) \\
& 0.1x_{00} + 0.3x_{10} + 0.9x_{20} = x_{20} + x_{21} \qquad (j = 2) \\
& x_{00} + x_{01} + x_{10} + x_{11} + x_{20} + x_{21} = 1
\end{aligned}
$$

and $\qquad x_{ia} \geq 0 \qquad\qquad\qquad\qquad\qquad\qquad (i = 0, 1, 2; a = 0, 1).$

The optimal solution is given by $x_{00} = 0.5085$, $x_{10} = 0.3390$, $x_{21} = 0.1525$, and all other x_{ij} equal to zero. The optimal value is 60.85. It follows from this solution that $\delta(0) = \delta(1) = 0$ and $\delta(2) = 1$, which corresponds to the optimal stationary policy found before.

Note that we can immediately deduce from the LP problem that $x_{01} = 0$ (why?); in other words, the probability that the system is in state 0 ($=$ "good") and the decision $a = 1$ ("send out for maintenance") is taken is zero, which we have always assumed by taking $D(0) = \{0\}$.

10.5 Exercises

10.1 At the beginning of each day a piece of equipment is inspected to reveal its actual working condition. The equipment will be found in one of the working conditions $i = 1, ..., N$, where the working condition i is better than the working condition $i + 1$. The equipment deteriorates in time. If the present working condition is i and no repair is done, then at the beginning of the next day the equipment has working condition j with probability q_{ij}. It is assumed that $q_{ij} = 0$ for $j < i$ and $\sum_{j \geq i} q_{ij} = 1$. The working condition $i = N$ represents a malfunction that requires an enforced repair taking two days. For the intermediate states i with $1 < i < N$ there is a choice between preventively repairing the equipment and letting the equipment operate for the present day. A preventive repair takes only one day. A repaired system has the working condition $i = 1$. The cost of an enforced repair upon failure is C_f and the cost of a preemptive repair in working condition i is C_i. We wish to determine a maintenance rule which minimizes the long-run average repair cost per day. Formulate this in terms of a Markov decision model.

10.2 An electricity plant has two generators $j = 1$ and 2 for generating electricity. The required amount of electricity fluctuates during the day. The 24 hours of a day are divided into six consecutive periods of 4 hours each. The amount of electricity required in period k is d_k kWh for $k = 1, ..., 6$. Also the generator j has a capacity of generating c_j kWh of electricity per period of 4 hours for $j = 1, 2$. An excess of electricity produced during one period cannot be used for a next period. At the beginning of each period k it has to be decided which generators to use for that period. The following costs are involved. An operating cost of r_j is incurred for each period in which generator j is used. Also, a setup cost of S_j is incurred each time generator j is turned on after having been idle for some time. Formulate this as a Markov decision problem.

10.3 A student is concerned about her car and does not like dents. When she drives to school, she can choose to park it on the street in one space, park it on the street and take up two spaces, or put it in a parking lot. If she parks on the street in one space, her car gets dented with probability $1/10$. If she takes up two spaces, the probability of a dent is $1/50$, and the probability of a \$15 fine is $3/10$. Using a parking lot costs \$5, but the car will not get dented. If her car gets dented, she can have it repaired, in which case it is out of commission for one day and costs her \$50 in fees and cab fares. She can also drive her car dented, but she feels that the resulting loss of value and pride is equivalent to a cost of \$9 per school day. She wishes to determine the optimal policy for where to park and whether to repair the car when dented in order to minimize her (long-run) expected average cost per school day.

    (a) Formulate this problem as a Markov decision problem by identifying the states, actions, immediate rewards, and transition probabilities.

    (b) Carry out one iteration of the policy iteration algorithm, starting with the decision rule in which the student always parks on the street in one space and is not bothered about getting dents repaired. Interpret the new decision rule.

    (c) Show that it is optimal to park in one space if the car is not dented and to have the car repaired if it is dented. Also determine the minimal expected average cost.

    (d) Formulate the linear program that corresponds to this Markov decision problem. (You need *not* solve it.)

    (e) How can you determine an optimal strategy from the optimal solution of the dual linear program?

10.4 There are three drill installations at an oil platform, each of which can have a failure. When the failure is beneath the water surface, a specialized diving team has to come by helicopter. The costs for such an operation consist of 50 000 euros travel costs plus 25 000 euros per installation. The capacity of the diving team is such that all broken installations are repaired after exactly one week.

For safety reasons, all drill installations should be turned off during repair activities. Therefore, it can be advantageous to repair multiple installations at the same time. On the other hand, there are costs for installations that are not working. These costs are 50 000 euros per installation per week. At the beginning of each week, the decision of calling the diving team or not should be made based on the number of broken installations.

The time between two consecutive failures of a drill installation is memoryless with a mean of 5 weeks. In other words, the probability that a random installation breaks down in a random week is 20%, provided that this installation is in use.

  (a) This problem can be modeled as a Markov decision problem. What do you choose as states i, decisions d, and optimal value function $V(i)$?
  (b) Determine the direct costs $c(i, d)$ and transition probabilities $p(j|i, a)$ for each state i and decision a.
  (c) Formulate the optimality equations with a discount factor of $\beta = 0.95$ per week.

At the moment, the oil platform has the following strategy: if at the beginning of a week, one or more installations are out of order, all installations are shut down and the diving team is called to repair the broken installation(s).

  (d) Calculate the current expected discounted costs with an infinite horizon if at $t = 0$, all drill installations are in use.
  (e) Is the oil platform's current strategy optimal? If not, give at least one better strategy.
  (f) Give an LP formulation that you can use to determine the optimal strategy.
  (g) How can you determine the optimal strategy from the solution of the LP problem?

10.5 Each week, a computer manufacturer's product either sells well or sells poorly. If the product sells well in a given week and the computer manufacturer takes no further action, he earns €6000 during that week and the product will still sell well the next week with probability 0.5 (that is, the product will sell poorly the next week with probability 0.5). Alternatively, the computer manufacturer can choose to advertise the product when it sells well. Advertising costs €2000 but increases the probability that the product will sell well the next week to 0.8. If the product sells poorly in a given week and the computer manufacturer takes no further action, he "earns" −€3000, and the product will sell well the next week with probability 0.4. Alternatively, the computer manufacturer can choose to do research to improve his product when it sells poorly. Doing research costs €2000 but increases the probability that the product will sell well the next week to 0.7. The computer manufacturer aims to maximize his long-term discounted earnings, using a weekly discount factor of 0.9.

  (a) Model this problem as a Markov decision problem. What do you choose as states, decisions, and optimal value function?
  (b) Determine the direct rewards and transition probabilities for each state and decision.
  (c) Formulate the corresponding optimality equations explicitly for this specific problem (that is, not in generic form).

The computer manufacturer currently advertises when the product sells well and does research when the product sells poorly.

(d) Use the policy iteration algorithm to determine whether the computer manufacturer's current policy is optimal.

(e) Formulate the linear program that the computer manufacturer could have used to determine the optimal policy.

(f) In the optimal solution to this linear program, which constraints are binding (that is, have no slack), and what is the objective function value?

10.6 In a store that sells furniture, there are at most 2 black couches in the showroom. There are no other black couches in stock, so the maximum number of black couches in the store is 2. At the end of the week, the number of black couches is counted, and the owner decides whether to order new black couches and if so, how many. The order is delivered immediately. The costs of an order consist of a fixed price of 80 euros plus 80 euros per ordered black couch. The demand per week is stochastic: 0 with probability $\frac{1}{4}$, 1 with probability $\frac{1}{2}$, and 2 with probability $\frac{1}{4}$. If the demand in a week is greater than the number of black couches in stock, the shortage is delivered directly from the production facility at a cost of 240 euros per black couch. Inventory costs are 0 euros regardless of the number of black couches in stock. The store starts out with 0 black couches in stock.

The owner of the store wonders what the optimal inventory and order strategy is. She wants to minimize the expected discounted costs over an infinite horizon. The discount factor is 0.8 per week.

(a) Determine the states, decisions in each state, direct costs as a function of the states and decisions, and transition probabilities of this Markov decision problem.

(b) Formulate the optimality equations for the optimal value function $V(i)$.

(c) Carry out the initialization and two additional iterations of the value iteration algorithm. In each iteration, determine the corresponding best strategy.

(d) What are the expected discounted costs when the starting inventory is 0 and strategy $(2, 0, 0)$ is used? This means when there are 0 black couches in stock, you order 2 black couches. When you have 1 or 2 black couches in stock, you do not order any new black couches.

(e) What do you have to do to determine whether the strategy from (d) is optimal using policy iteration? (You do not have to calculate these things.)

10.7 An intensively used lease car is inspected every weekend (after being used for a week). The inspection report is summarized in a grade ranging from 1 to 5. The higher the grade, the worse the condition of the car. A perfect car receives grade 1. The grade 5 indicates that the car cannot be used anymore. For the intermediate grades (2 up to 4), there is the possibility to have the car repaired preventively. If this is done, the car cannot be used for a week. After the preventive repair, the car is in perfect condition. A car with grade 5 should be replaced by a new one, but a new car (that is always in perfect condition) takes two weeks to arrive. The table below shows the probabilities $p(i, j)$ that a car with grade $i = 1, 2, 3, 4$ that is not being repaired receives grade $j = 1, 2, 3, 4, 5$ after the next week: A car with grade 1, 2, 3, or 4 that is not being repaired generates a profit of 1. A car that is repaired preventively generates no profit that week. When the grade is 5, there are no profits for two weeks (lead time). The goal is to maximize the long-term discounted profit, using a weekly discount factor of 0.85.

Table 10.4 Probability that a car with grade i, without repair, receives grade j after the next week.

	$j = 1$	$j = 2$	$j = 3$	$j = 4$	$j = 5$
$i = 1$	0.75	0.20	0.05	0	0
$i = 2$	0	0.50	0.20	0.20	0.10
$i = 3$	0	0	0.50	0.25	0.25
$i = 4$	0	0	0	0.30	0.70

(a) Formulate this problem as a Markov decision problem by identifying the states, actions, immediate rewards, and transition probabilities. (Hint: Introduce an extra state 0 to indicate that a new car has already been on order for a week.)

(b) Suppose that the leasing company decides that the car should always be repaired preventively if its grade equals 2, 3 or 4. Calculate the discounted profit of this policy when starting with a new car.

(c) Carry out one iteration of the policy iteration algorithm, starting with the decision rule in (b).

(d) Carry out the initialization and two additional iterations of the value iteration algorithm.

10.8 The owner of a racehorse wants to maximize his horse's (discounted) returns. The (daily) discount factor is $\frac{2}{3}$. It is possible to participate in a race every day, but after participating, the horse may not be fit the next day. If the horse is fit, the expected returns from participating in a race are 2 million euros. If the horse is tired, the expected returns from participating in a race are only 1 million euros. Participating in a race is free. If the horse is fit and participates in a race, it is fit again the next day with probability $\frac{2}{3}$, and it is still tired the next day with probability $\frac{1}{3}$. If the horse is fit and does not participate in a race, it will still be fit the next day. If the horse is tired and participates in a race, it will still be tired the next day. If a tired horse rests for a day, it will be fit the next day with probability $\frac{1}{2}$ and will still be tired the next day with probability $\frac{1}{2}$.

(a) Model this problem as a Markov decision problem. What do you choose as states, decisions, and optimal value function?

(b) Determine the direct rewards and transition probabilities for each state and decision.

(c) Formulate the corresponding optimality equations explicitly for this specific problem (that is, not in generic form).

(d) Use the policy iteration algorithm to determine an optimal policy as well as the optimal discounted rewards.

Suppose that, instead of maximizing the discounted returns of his horse, the owner wishes to maximize the long-term average rewards.

(e) Formulate the linear program that the owner could use to determine an average optimal policy.

(f) What is the interpretation of the decision variables in your linear program?

(g) In the optimal solution, how many decision variables will have a (strictly) positive value? Motivate your answer.

10.9 Consider a queueing system with room for two customers. We assume that a customer's service can only start or end at, respectively, the beginning or end of an hour and that the system can serve only one customer at a time. The number of customer arrivals during an hour is zero with probability $p_0 = 0.3$, and it is one with probability $p_1 = 0.7$. Customers finding the system full depart without attempting to enter later.

The system offers two kinds of service, fast and slow, with respective cost $c_f = 6$ and $c_s = 3$ per hour. Service can be switched between fast and slow at the beginning of each hour. If there are no customers in the system at the beginning of an hour, the server can take a break and service costs are 0 during that hour. With fast service, a customer in service at the beginning of an hour will complete service at the end of the hour with probability $q_f = 0.9$ independently of the number of hours the customer has been in service. With slow service, that probability is $q_s = 0.5$. There is a cost $r(i) = 2i$ for each hour during which there are i customers in the system. The system operator, who controls the service speed, aims to minimize the long-term discounted costs with an hourly discount factor of 0.99.

(a) Model this problem as a Markov decision problem. What do you choose as states, decisions, and optimal value function?
(b) Determine the direct rewards and transition probabilities for each state and decision.
(c) Formulate the corresponding optimality equations explicitly for this specific problem (that is, not in generic form).
(d) Perform the initialization and two additional iterations of the value iteration algorithm.
(e) Use the policy iteration algorithm to determine whether the policy you found in (d) is optimal.
(f) The optimal solution to the *average* cost linear program is $\pi_0^* = 0.0604, \pi_{1,s}^* = 0.2819, \pi_{1,f}^* = 0.0000, \pi_{2,s}^* = 0.6577, \pi_{2,f}^* = 0.0000$. What is the optimal policy for minimizing the long-term average costs, and what are the corresponding minimum expected average costs?

10.10 A local football club has several sponsors that donate money every year. The club divides the sponsors into two classes, low (L) and high (H), based on their donation behavior in the current year. Every year, the club wishes to determine how to discuss next year's donation with each sponsor. Either the club management can take a sponsor to a fancy dinner, which costs the club €1500, or the club's treasurer can simply make a phone call to a sponsor at no costs. From a sponsor in class L, the club expects to receive €2000 next year if it takes the sponsor to dinner and €1000 if the treasurer calls the sponsor. From a sponsor in class H, the club expects to receive €3500 next year if it takes the sponsor to dinner and €2500 if the treasurer calls the sponsor.

Whether a sponsor is taken to dinner also affects the sponsor's classification in the subsequent year. If a sponsor is in class L this year, then the probability that it is in class L next year is 0.3 if it is taken to dinner and 0.5 if the treasurer calls it. If a sponsor is in class H this year, then the probability that it remains in class H next year is 0.8 if it is invited for dinner and 0.4 if the treasurer calls it. Club management wishes to maximize the long-term (discounted) profit from sponsoring, using a yearly discount factor of 0.9.

(a) Model this problem as a Markov decision problem. What do you choose as states, decisions, and optimal value function?
(b) Determine the direct rewards and transition probabilities for each state and decision.
(c) Formulate the corresponding optimality equations explicitly for this specific problem (that is, not in generic form).
(d) Currently, the club management takes every sponsor in class H to a fancy dinner, while the treasurer calls every sponsor in class L. What is the most efficient

procedure to determine whether the club's current strategy is optimal? Give the name of the procedure and explain how it can be used for this purpose. (You do not have to carry it out!)

(e) Perform the initialization and two additional iterations of the value iteration algorithm.

(f) When this Markov decision problem is solved using linear programming, the optimal solution is $V^*(L) = 16192.66055; V^*(H) = 17568.80734$. Use this information to determine the club's optimal policy.

10.11 The electricity company of a large developing country exploits an extremely old-fashioned electricity plant. In good shape, the plant delivers energy with a daily value of €2 000 000. However, the old plant overheats quite often due to the sudden rise in energy demand around 8:30 a.m. When the plant is in good shape, the probability that the plant overheats in the morning is $\frac{1}{3}$. The plant remains in good shape with the complementary probability $\frac{2}{3}$. In order to cool an overheated plant, it must be shut down during the remainder of the day, rendering no energy that day. If the plant is not shut down, it remains overheated and only delivers daily energy worth €1 000 000. The plant is inspected each morning at 9:00 a.m., and if overheated, the company may decide to shut it down for that day at a cost of €1 000 000 because of possible customer damage. (The value of the energy delivered before 9:00 a.m. may be neglected in all cases.) The plant never overheats the morning after having been shut down for a day. The objective of the electricity company is to maximize the total discounted (net) rewards. The daily discount factor is $\frac{3}{5}$.

(a) Formulate this problem as a discounted Markov decision problem with infinite horizon by identifying the states, actions, immediate rewards, and transition probabilities. The state is observed at the 9:00 a.m. inspections.

(b) Formulate the corresponding optimality equations explicitly for this specific problem (that is, not in generic form).

(c) Carry out the initialization and two additional iterations of the value iteration algorithm.

We now consider the stationary policy where the plant is always shut down at 9:00 a.m. if it is overheated and not shut down otherwise.

(d) Compute the discounted rewards of this strategy for each initial state s.

(e) Check whether this strategy is optimal.

10.12 A factory has a tank for temporarily storing chemical waste. The tank has a capacity of 4 m^3. Each week the factory produces k m^3 chemical waste with probability p_k for $k = 0, ..., 3$ with $p_0 = 1/8$, $p_1 = 1/2$, $p_2 = 1/4$ and $p_3 = 1/8$. If the amount of waste produced exceeds the remaining capacity of the tank, the excess is specially handled at a cost of \$30 per m^3. Upon the end of the week a decision has to be made as to whether or not to empty the tank. There is a fixed cost of \$25 to empty the tank and a variable cost of \$5 for each m^3 of chemical waste that is removed. Compute an average cost optimal policy by linear programming.

10.13 Use linear programming for the average reward criterion to calculate an optimal policy for Problem 10.1 with the numerical data $N = 5$, $C_f = 10$, $C_2 = C_3 = 7$ and $C_4 = 5$. The deteriorating probabilities are $q_{11} = 0.9$, $q_{12} = 0.1$, $q_{22} = 0.8$, $q_{23} = 0.1$, $q_{24} = q_{25} = 0.05$, $q_{33} = 0.7$, $q_{34} = 0.1$, $q_{35} = 0.2$, and $q_{44} = q_{45} = 0.5$.

10.14 Use linear programming for the average reward criterion to calculate an optimal policy for Problem 10.2 with the numerical data $d_1 = 20$, $d_2 = 40$, $d_3 = 60$, $d_4 = 90$, $d_5 = 70$, $d_6 = 30$, $c_1 = 40$, $c_2 = 60$, $r_1 = 1000$, $r_2 = 1100$, $S_1 = 500$ and $S_2 = 300$.

References

1. R. J. Boucherie, N. M. van Dijk, editors, *Markov Decision Processes in Practice*, International Series in Operations Research and Management Science. Vol. 248, Springer, 2017.
2. M. L. Puterman, *Markov Decision Processes: Discrete Stochastic Dynamic Programming*, John Wiley & Sons, Inc., Hoboken, New Jersey, 2005.

Chapter 11

Simulation

11.1 Introduction

In this chapter, we discuss how simulation can be used for stochastic models from operations research.[1] A model is a description of an (often complicated) real-world system aimed at gaining insight into the system's functioning. Almost every model is a simplification of reality. A model's formulation is strongly determined by the quantities to be studied and the questions to be answered. The model must, of course, reflect the essence of the system, but, in general, one should be careful not to immediately build a highly detailed model. Experience shows that insight into the system's functioning is usually best obtained by beginning with relatively simple models. Modeling a real-world situation is often a matter of trial and error. Most models are formed in an ad hoc manner, tested, adjusted, retested, and so on until a satisfactory result is obtained.

To solve the model, one can use analytical methods and/or computer simulation. An analytical method provides a mathematical formula into which the system's characteristics can be substituted, while simulation imitates the system's behavior on the computer. Wherever possible, an analytical solution method is generally preferable to a simulation approach. The technique of computer simulation is less straightforward than it may seem at first sight. Setting up a properly functioning computer program is full of pitfalls and can be very time-consuming. One also should not think lightly about drawing conclusions from the computer program's output. In most cases, the model is stochastic, and drawing reliable conclusions from the simulation results requires thorough statistical knowledge.

For many complex systems, there is, of course, no other option than a simulation model, but sometimes it is possible to use a simulation model in combination with a

---

[1]In statistics, the power of computer simulation has led to the so-called *bootstrap method*, a revolutionary new method for making statistical statements about *small* sample sets. Traditional statistical methods—developed before powerful computers were available—often require a larger sample to apply an approximation based on the normal probability distribution. In the bootstrap method, however, this is not necessary, and the information contained in the sample is fully used. The idea is to carry out a simulation that is directly based on the available data ("let the data speak for itself").

less realistic but simpler analytical model. The analytical model can provide insight into the form of the solution, and simulation can further refine this. For example, the famous square root formula from inventory theory comes from a remarkably simple analytical model. Simulation can be used to find out how this square root formula works for more complex inventory systems and how the formula could be adapted if necessary.

Simulation has both advantages and disadvantages.

Advantages of Simulation:

(a) Simulation models allow us to study systems that are too complex for an analytical solution.

(b) Simulation models are often more convincing and understandable for managers than analytical models.

(c) Simulation allows the user to gain experience in a short time instead of the years it would in real-life.

(d) Simulation can quickly answer "What if ... ?" questions.

Disadvantages of Simulation:

(a) Simulation models can be expensive and can require a long development time.

(b) Simulation often generates an enormous amount of numbers from which it is sometimes difficult to draw meaningful conclusions.

(c) Simulation poses the danger of creating a model for the real-world system that is too detailed, causing the user to become swamped.

In the design of a simulation program for a real-world system, modeling usually goes hand in hand with *verification* and *validation*. Verification means checking whether the computer program is error free and consistent with the model, while validation means checking whether the simulation model, if implemented correctly, gives results that are useful for the actual system. A recipe for validation is difficult to give. Some guidelines can be given for verification:

(1) Modularize and test every subroutine of the program individually.

(2) Where possible, use analytical solutions for special cases of the model (for example, in queueing systems, an analytical solution is often possible for the particular case of exponential probability distributions).

(3) Vary one parameter or select specific extreme values for the parameters and check whether the model's behavior is understandable.

What else is this chapter about? First, we discuss the commonly used method of discrete-event simulation, which starts out with a well-defined (stochastic) model. Our intention is for the reader to put the discrete-event simulation into practice by writing a computer program (in C++) for a specific application. A good understanding of simulation and especially of the pitfalls of simulation is best obtained by writing one's own computer program for a specific simulation application and experimenting with it. To write a simulation program, in addition to programming experience, only an elementary knowledge of probability is required. This knowledge is necessary for generating samples from probability distributions. We briefly discuss the basic principle of generating random numbers between 0 and 1 using the random number generator on the computer; the uniform (0,1) random numbers enable us to draw random samples from almost all probability distributions. Next, we pay the necessary attention to the so-called output analysis of the simulation results, that is, the construction of point estimates and confidence intervals. The importance of an adequate statistical analysis of the output cannot be sufficiently emphasized. Due to statistical fluctuations in the simulation output, it is extremely dangerous to draw conclusions based on a single run or on a too short run of an otherwise correct simulation program. In stochastic simulations, a confidence interval is essential. In the output analysis, we distinguish between so-called short-term simulation and long-term simulation. Short-term simulation involves simulating a performance measure that involves a clearly defined time period; in this case, one is, for example, interested in the expected time until a production process breaks down for the first time. Long-term simulation involves a time period that is, in principle, unlimited; for example, one may be interested in the long-run fraction of the time that the production process is in working order. Finally, we discuss procedures for generating random samples from specific probability distributions.

11.2 Discrete-Event Simulation

The principle of discrete-event simulation can best be explained using a concrete example.

Example 11.1. A bank opens at 10 o'clock in the morning. There are c tellers who are equally fast and can serve any of the incoming customers. Suppose that data analysis shows that during the bank's business hours, customers arrive following a Poisson process with a rate of λ customers per hour; in other words, the customer interarrival times are independent of one another and exponentially distributed with expected value $1/\lambda$. The incoming customers line up in a single queue in front of the counter, and as soon as a teller is idle, the first customer from the queue is served by that teller. The customers' service times are independent random variables that are also independent of the arrival process. Suppose that data analysis shows that each customer's service time is exponentially distributed with an expected value of $1/\mu$ hours. The bank closes at 5 o'clock in the afternoon but still serves the customers who are already inside.

How does this system behave? For example, we could be interested in the effect of varying the number of available tellers or increasing the service rate. The problem formulation is as follows.

Problem Formulation

For given values of c, λ, and μ, determine

    a. the average queue length during the day,
    b. a customer's average waiting time,
    c. the fraction of the time that a teller is busy.

    Although the problem above has been highly simplified by, among others, the special assumptions concerning the stochastic arrival process and the service process, it is representative of more complex stochastic systems. The problem is known in the literature as the $M/M/c$ queue (see also Section 9.4.4).

States

The key to building a good model is the choice of the system's state. The system's state must contain sufficient information about the system's past so that, given the current state, the past is statistically irrelevant for predicting the system's future behavior. Obviously, the choice of the state variables is determined by the system quantities to be studied.
    In the bank model, the state variables are as follows:

- the statuses of the tellers (busy or idle and, if busy, the time at which a teller becomes idle),
- the number of customers in the queue,
- the arrival times of the customers in the queue (needed to determine the average waiting time per customer).

    Note that the same state variables would be used if we had assumed different probability distributions for the customers' interarrival times and for the service times.

Events

Events are occurrences that may change the model's state. For discrete-event simulation, it is essential that events can only take place at discrete times, so that the model's state does not continuously change. In the bank model, the events are customer arrivals or departures.
    The assumption that the events only take place at discrete points in time allows us to *compress the actual time scale in a simulation and only advance the simulation clock at moments when an event takes place and then adjust the state*. The successive steps of the simulation clock are usually of different sizes. In discrete-event simulation, time intervals during which irrelevant changes take place are skipped. This avoids unnecessary simulation efforts.

Statistical Counters

Statistical counters are variables that are used to store statistical information that can be used to calculate the requested performance measures at the end of the simulation. In the bank model, we need the following statistical counters:

- the number of customers who have entered so far,
- the total waiting time of all customers who have entered so far,
- the total time a teller has been busy so far,
- the total area below the graph of the number of customers in the queue so far.[2]

The statistical counters are, of course, not updated continuously; like the state variables, they are only updated at the discrete times when an event takes place.

The last statistical counter in the list, needed for the average length of the queue during the day, requires some explanation. If we set

$$Q(t) = \text{number of customers in the queue at time } t,$$

then $Q(t)$ is a step function that can only change value when an event (arrival or departure) takes place. If, for convenience, we take the number of tellers equal to $c = 1$, then the graph in Figure 11.1 could represent $Q(t)$.

Fig. 11.1   A possible realization of $Q(t)$.

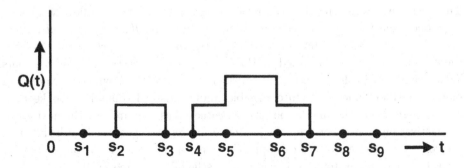

Here, s_i is the time of the ith event (in the realization above, s_1, s_2, s_4, s_5, and s_9 are arrival times, and s_3, s_6, s_7, and s_8 are departure times). The average number of customers in the queue after a simulation of T time units is estimated by

$$\widehat{Q}(T) = \frac{1}{T} \int_0^T Q(t)dt$$

(in the simulation program, you might take T equal to the length of the time interval between 10:00 and the first time after 17:00 when the bank is empty). Although

---

[2]A customer's waiting time does not include service time, and the queue does not include customers in service.

the integrand is a continuous time variable, for calculating the integral, it suffices to update the statistical counters at the discrete times when an event takes place. The integral can simply be calculated as the sum of the areas of a number of rectangles. The areas of the same rectangles also allow us to calculate the total waiting time per customer. The total area under the graph up to time T divided by T gives the average length of the queue, while the average waiting time per customer follows by dividing the total area under the graph up to time T by the number of customers who have arrived at the bank up to time T.[3]

Important observation. The numerical value that we find for the average queue length (average waiting time, fraction of the time a teller is busy) after a single simulation run is a realization of a random variable. *In general, that numerical value will fluctuate considerably per run.* One should not underestimate the size of the fluctuations that can occur, in particular when the queueing system is busy. Because of these fluctuations, a simulation that consists of only one observation does not suffice. A high number of runs are needed to even out the fluctuations. A careful statistical analysis of the output is an absolute necessity for arriving at useful conclusions. We return to this aspect later.

Computer Program

The question now is what the computer program looks like. Earlier, we already mentioned that the basis for the simulation design is that *the simulation clock skips the time interval between two successive events, considering it irrelevant.* As far as the simulation logic is concerned, nothing happens between two successive events. This means that in the program, we consider only those times when an event takes place. At those times, the state variables and statistical counters (bookkeeping) are adjusted. So in the simulation, one repeatedly has to search for the next event. The activation of an event typically causes future related events to be scheduled. In the computer program of, for example, the bank model, when a customer arrival is activated, the next customer's arrival time is scheduled. Moreover, if a desk teller is idle at this just-scheduled time, then the departure time of the customer who just arrived is scheduled.[4] The simulation uses an event list, that is, a list of events that have already been scheduled along with the times at which they are scheduled to take place. It is important that this list be managed efficiently. This means that a good data structure must be chosen for this list. If the possible number of events on

---

[3]Little's law, a sort of law of nature in queueing theory, can be substantiated heuristically based on this. Little's law states that if a queueing system is observed over an *unlimited* length of time, the average queue length is equal to the average number of customer arrivals per unit of time multiplied by the average waiting time per customer. Little's law $L = \lambda W$ is also very useful for the verification of simulation programs for queueing systems.

[4]There are several possibilities for the moment when a customer service time is generated. The service time can already be generated when the customer arrives; another possibility is to wait until the customer goes into service.

the list is not too high (say, ≤ 100), then a linked list is recommended, where events are sorted by order of occurrence. In this case, it is wise to go through the list from back to front when a new event must be added. If the possible number of events is high, a more advanced data structure is needed, for example, a binary search tree or heaps. We do not go into these matters here; instead, we refer the reader to one of the many existing good books about data structures. When writing a simulation program for the $M/M/c$ bank model, it is useful to first make a flowchart of the model; see Figure 11.2. Of course, when writing the program, one must follow the usual rules for good programming. This means, in particular, *modularization*, that is, using a separate module for every event routine, for every procedure for generating random samples, and so on. A further explanation of the procedure for generating random samples from the exponential probability distribution for the customer interarrival times and service times is given in Section 11.3.

For the reader who has mastered a programming language, we now give a rather nice assignment to set up a discrete-event simulation model.

Assignment 11.1. Write a simulation program for a single run of the following reliability model. An electronic system has K components that are all in operation. The running times of the components are independent of one another and have a common Weibull distribution with distribution function $F(x) = 1 - \exp(-(\lambda x)^\alpha)$ for $x \geq 0$; see Section 11.3 for how to generate a sample from this probability distribution. A broken component is immediately repaired. The repair takes a units of time with probability p and b units of time with probability $1 - p$. After a repair, a component is as good as new. The repair times are independent of one another and also independent of the running times. The electronic system shuts down completely when none of the components work. The problem is to determine the expected time until the system shuts down for the first time. Take $K = 4$, $\alpha = 1.01$, $\lambda = 0.249$, $a = \frac{1}{2}$, $b = \frac{3}{2}$, and $p = \frac{1}{2}$.

11.3   Random Number Generator

In the computer program for the $M/M/c$ queue in Section 11.2, samples must be generated from a probability distribution, namely the exponential distribution. The question is how to do this. The answer is that it suffices to have access to a so-called random number generator that randomly chooses numbers between 0 and 1. The latter means that every subinterval of the unit interval must have the same probability of containing the chosen number as any other interval of the same length. So the probability that a number is chosen from a given subinterval of the unit interval is equal to the subinterval's length. The uniform distribution on (0,1) is the probability distribution that assigns to every subinterval of (0,1) a probability mass equal to the interval's length. We call a random variable with this distribution a U(0,1) random variable.

Every U(0,1) random number U directly determines a random number X from

Fig. 11.2   Flowchart of the $M/M/c$ bank model.

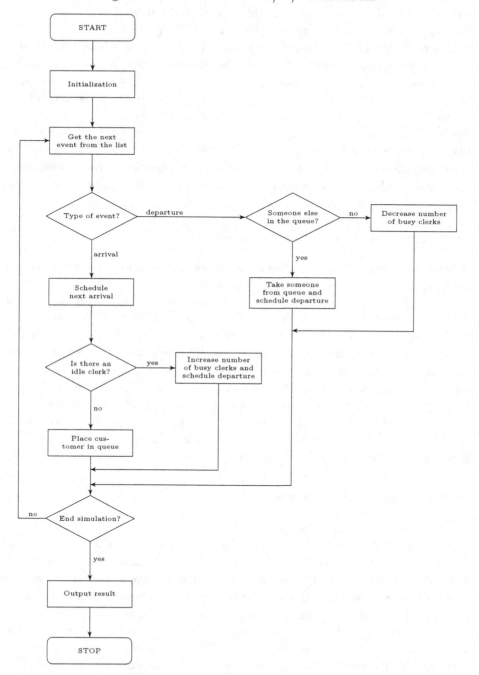

any interval (a, b) via $X := a + (b - a)U$. In fact, a random number generator that produces U(0,1) random numbers suffices to generate samples from almost every conceivable probability distribution. We briefly explain this remarkable result using two examples. In Section 11.6, we go into more detail about generating samples from probability distributions.

First, consider the particular case that the random variable X has the discrete two-point distribution

$$\mathbb{P}(X = a) = p \quad \text{and} \quad \mathbb{P}(X = b) = 1 - p.$$

If we now generate a U(0,1) random number U, we obtain a random sample for the random variable X by giving the variable X the value a if $U \leq p$ and the value b otherwise. This immediately follows from the fact that $\mathbb{P}(U \leq u) = u$ for $0 \leq u \leq 1$.

Next, consider the specific case that X is a continuously distributed random variable with a strictly increasing distribution function. A basic method for generating a sample from X is the so-called inversion sampling method.

Algorithm 11.1 (Inversion Sampling Method).

Step 1. Generate a random number u between 0 and 1.
Step 2. Calculate x as the solution to $F(x) = u$; that is, $x = F^{-1}(u)$.
Step 3. Output $X := x$.

The inversion sampling method is only practical when the inverse function $F^{-1}(u)$ can be quickly calculated numerically. This is, for example, the case when $F(x)$ is a Weibull probability distribution function, that is,

$$F(x) = 1 - e^{-(\lambda x)^{\alpha}}, \quad x \geq 0,$$

where α and λ are given positive numbers. For the special case $\alpha = 1$, the Weibull distribution reduces to the exponential distribution. The inverse function $F^{-1}(u)$ can then be trivially calculated. The equation $F(x) = u$ has the explicit solution

$$x = \frac{1}{\lambda} [-\ln(1 - u)]^{1/\alpha}.$$

We can do one fewer calculation by replacing $1 - u$ with u. This is allowed because $1 - U$ is also uniformly distributed between 0 and 1 if U is uniformly distributed between 0 and 1.

Pseudorandom Numbers

In the remainder of this section, we give an idea of how a random number generator on the computer works. Procedures for generating U(0,1) distributed numbers are present on all platforms.

To generate numbers that are uniformly distributed between 0 and 1, one can, of course, take a hat with 10 lots numbered $0, 1, \ldots, 9$ and then successively pull a number of lots out of the hat at random, with replacement. The resulting sequence of numbers with a decimal point in front of it can be seen as a U(0,1) random number. However, such a procedure is not the way computers generate U(0,1) random numbers. Such an approach is not only impractical but also has the disadvantage that one cannot reproduce calculations that have been carried out unless all generated random numbers are stored in the computer memory. Storing a vast number of random numbers in the computer memory is practically impossible. An ingenious solution has been found for this obstacle. Instead of generating truly random numbers, the computer generates so-called *pseudorandom* numbers by iterating a particular function f. This is done as follows. Choose a function f, the random number generator, and an initial number z_0, called the seed. Then, let

$$z_1 = f(z_0), \ z_2 = f(z_1), \ldots, \ z_n = f(z_{n-1}),$$

thus generating a deterministic sequence $\{z_i\}$ of numbers. The function f is chosen such that, statistically, the sequence $\{z_i\}$ cannot be distinguished from a sequence of truly random numbers. This means that the output of the function f must be able to withstand a battery of statistical "randomness" tests.

The first pseudorandom number generator that can be used for real-world applications is the so-called *linear congruential generator* developed in the 1950s by the American number theorist D. Lehmer. This generator determines the ith integer z_i in the sequence of pseudorandom numbers generated by the recursion

$$z_i = az_{i-1} + c \ (\text{modulo } m),$$

where a, c, and m are fixed positive integers and z_0 is a positive start value that can be chosen.[5] This recursion always produces one of the numbers $1, \ldots, m$. As U(0,1) random numbers u_i, we take

$$u_i = \frac{z_i}{m}.$$

The number $m > 0$ is called the modulus, the number a with $0 < a < m$ is called the multiplier, the number c with $0 \leq c < m$ is called the increment, and the integer z_0 with $0 \leq z_0 < m$ is called the seed. Of course, the sequence of numbers generated this way always ends up in a loop; after at most m different numbers, the sequence repeats. The length of a cycle is called the generator's *period*. Obviously, one requires a long period, so that m must be chosen very large. Because of the computer's word size, it is efficient to choose m equal to 2^β or $2^\beta - 1$ for some β.

The choice of the parameters a, c, and m determines the statistical quality of the random number generator. If $c = 0$, then we speak of a *multiplicative* generator. Such a generator ($c = 0$) has the advantage that the calculations can be done very efficiently. Moreover, generators with $c = 0$ generally have better statistical properties. The multiplicative generator with $m = 2^{31} - 1$ and $a = 630\,360\,016$ is widely used. This generator has period $m - 1$ and generates each of the numbers

---

[5]The number a (modulo m) gives the least positive remainder from the division of a by m. For example, 17 (modulo 5) = 2.

$1, \ldots, m-1$ exactly once in a cycle (so, if one chooses one of the numbers $1, \ldots, m-1$ as the seed z_0, this will result in a cycle of length $m-1$ every time).

A good pseudorandom number generator of course has a long period. This is also necessary because simulation projects can require hundreds of thousands or millions of random numbers. For most operations research applications, pseudorandom number generators with $m = 2^{31} - 1$ or $m = 2^{32}$ suffice. However, for simulations in, among others, physics and telecommunication, generators with even longer periods are needed nowadays. In the 1990s, with a completely different approach than that of the multiplicative congruential generator, the American mathematician G. Marsaglia developed pseudorandom number generators with incredibly long periods, including the so-called Christopher Columbus generator with a period of about 2^{1492} (to give an idea of how long it takes for this generator to repeat itself: at one million random numbers per second, this takes more than 10^{435} years). The Christopher Columbus generator is based on technical manipulations with modified Fibonacci-like number sequences, is fast, and contains no multiplications or divisions. A highly technical pseudorandom number generator that also does not use any multiplications or divisions is the Mersenne twister generator developed by the Japanese researchers M. Matsumoto and T. Nishimura in the late 1990s. This fast generator has a period of length 2^{19937}, gives very high-quality pseudorandom numbers, and is, nowadays, the first choice of generator for most Monte Carlo simulations.

One should not think too lightly of designing a good pseudorandom number generator. It is incredibly unwise to just choose some values for a, c, and m for the linear congruential generator or to invent a "crazy" algorithm by, for example, combining a number of existing algorithms. That something is strange or complicated does not have to mean that it is random. A random number generator must always be supported by a strong theoretical basis. A natural question is how to determine whether a pseudorandom number generator is good. A first requirement is that the generator produce sufficiently many different numbers in the interval $(0,1)$ before the sequence repeats itself (long period). Another requirement is that nonoverlapping number combinations $(u_1, u_2, \ldots, u_d)$, $(u_{d+1}, \ldots, u_{2d}), \ldots$ are sufficiently uniformly distributed over the d-dimensional unit cube, at least for small values of d. To check this condition, one needs to design a great number of statistical tests, of which the spectral test is one of the most important. This test has caused many pseudorandom number generators used in the past to be rejected. We do not further discuss the statistical tests for pseudorandom number generators.

Nowadays, one may assume that a computer's compiler has a good pseudorandom number generator (this was not always the case in the past). Finally, we would like to emphasize once more that a great advantage of a pseudorandom number generator is that a generated sequence of pseudorandom numbers can be *reproduced* by taking the same seed for the generator. This is important for two reasons:

(a) The detection of programming errors in simulation programs is facilitated when the run can be repeated under the same experimental conditions.

(b) A comparison between different decisions rules in, for example, an inventory system, is more straightforward when the decision rules are exposed to the same experimental conditions.

The modern random number generators such as the Christopher Columbus and Mersenne twister generators have the option of using simultaneously several independent streams of random numbers for different purposes such as, for example, for generating interarrival times and service times.

11.4   Short-Term Simulation

Short-term simulation is simulation for a system with a finite time horizon. For example, in the bank model in which the bank closes at 17:00, the simulation is a short-term one. However, in a short-term simulation, the length of the finite time horizon may also be stochastic, for example if the simulation ends when a certain state has been reached. An example of this is given in Assignment 11.1 in Section 11.2. In a short-term simulation, the simulation runs are typically *independent*. This allows one to use classical statistics directly. The statistical analysis of the computer output is discussed in Section 11.4.1.

11.4.1   *Statistical Analysis*

The output of a single simulation run for a stochastic model is a single number for each performance measure under consideration. It should be clear that little meaning can be given to one number. Statistical analysis is required, and therefore multiple observations.

Now, assume that, through simulation, we have obtained n independent observations $X_1, \ldots, X_n$ for a specific stochastic performance measure X in the model under consideration. Independent runs for a short-term simulation are automatically obtained by simply letting the random number generator go on once it has been initialized (the generator's period must, of course, be sufficiently long). How can we estimate the unknown value $\theta = \mathbb{E}[X]$, and how can we indicate the quality of this estimate? The answer is given by two fundamental results from probability theory and statistics. The first pillar of simulation is the *strong law of large numbers*. This law tells us that the *sample mean*

$$\overline{X}(n) = \frac{1}{n} \sum_{k=1}^{n} X_k$$

becomes arbitrarily close to the unknown value θ when n is very large. More precisely, if the experiment is carried out an infinite number of times, then regardless of the individual results, the limit of the sample mean is equal to the desired value θ *with certainty*. In practice, the number of times an experiment is carried out is, of course, finite. The quality of the approximation $\overline{X}(n)$ for a large but fixed value of

n can be determined by another statistical pillar of simulation, namely the *central limit theorem*. If we set $\sigma^2 = \text{var}(X)$, then

$$\lim_{n\to\infty} \mathbb{P}\left(\frac{X_1 + \cdots + X_n - n\theta}{\sigma\sqrt{n}} \leq x\right) = \Phi(x) \text{ for all } x,$$

where $\Phi(x)$ is the standard $N(0,1)$ distribution function. In other words,

$$\lim_{n\to\infty} \mathbb{P}\left(\frac{\overline{X}(n) - \theta}{\sigma/\sqrt{n}} \leq x\right) = \Phi(x) \text{ for all } x.$$

The problem is that, in general, we do not know σ (otherwise, we would probably also know θ). This problem can be solved by replacing σ^2 with its estimator

$$S^2(n) = \frac{1}{n-1} \sum_{k=1}^{n} \left[X_k - \overline{X}(n)\right]^2.$$

This quantity is called the *sample variance*. The following modified version of the central limit theorem applies:

$$\lim_{n\to\infty} \mathbb{P}\left(\frac{\overline{X}(n) - \theta}{\sqrt{S^2(n)}/\sqrt{n}} \leq x\right) = \Phi(x) \text{ for all } x.$$

This means that for n fixed, the quantity $(\overline{X}(n) - \theta)/(\sqrt{S^2(n)/n})$ is approximately $N(0,1)$ distributed provided that n is sufficiently large. For a given value of α (for example, $\alpha{=}0.05$), let $z_{1-\alpha/2}$ be the $(1 - \alpha/2)$-percentile of the standard normal distribution. In other words,

$$\mathbb{P}\left(-z_{1-\alpha/2} \leq Z \leq z_{1-\alpha/2}\right) = 1 - \alpha$$

with Z a standard $N(0,1)$ distributed random variable. So for n fixed and sufficiently large, we find

$$\mathbb{P}\left(-z_{1-\alpha/2} \leq \frac{\overline{X}(n) - \theta}{\sqrt{S^2(n)}/\sqrt{n}} \leq z_{1-a/2}\right) \approx 1 - \alpha;$$

that is,

$$\mathbb{P}\left(\overline{X}(n) - \frac{z_{1-\alpha/2}\sqrt{S^2(n)}}{\sqrt{n}} \leq \theta \leq \overline{X}(n) + \frac{z_{1-\alpha/2}\sqrt{S^2(n)}}{\sqrt{n}}\right) \approx 1 - \alpha.$$

So, in addition to the point estimate $\overline{X}(n)$ for the desired value of θ, we have the following result.

Approximate $100(1 - \alpha)\%$ confidence interval:

$$\left[\overline{X}(n) - \frac{z_{1-\alpha/2}\sqrt{S^2(n)}}{\sqrt{n}}, \ \overline{X}(n) + \frac{z_{1-\alpha/2}\sqrt{S^2(n)}}{\sqrt{n}}\right]. \tag{11.1}$$

One says that the desired value of θ lies in this interval with approximately a $100(1-\alpha)\%$ degree of confidence. This should not be interpreted as the probability that θ lies in the interval. The quantity θ is not a random variable, but a number (unknown to us) that either lies in the interval or does not. Every time we carry out a simulation, we find a confidence interval with other endpoints. The confidence interval (11.1) can be interpreted as follows. If we carry out n observations many times in a row, independently of one another and, every time, determine the corresponding confidence interval (11.1), then in approximately $100(1 - \alpha)\%$ of the cases, the confidence interval (11.1) will include the actual value of θ.

The approximate covering of θ may differ significantly from the actual covering if applying the central limit theorem is not justified. How large one must choose n for $\overline{X}(n)$ to be approximately normally distributed depends largely on the form of the underlying probability distribution of the performance measure X. In general, one should expect n to be in the tens of thousands rather than in the hundreds. When a sample is generated from a symmetric probability distribution, the sample mean is much more quickly approximately normally distributed than in a draw from an asymmetric probability distribution. It is difficult to make a quantitative statement on the minimum size of n. We restrict ourselves to the pragmatic remark that in a short-term simulation, the number of observations must quickly run into thousands or tens of thousands if the resulting confidence interval is not to be too wide. For such an order of magnitude of the number of observations, applying the central limit theorem is practically always justified. The confidence interval is narrower the greater the number n of simulation runs. We would prefer to choose the number of simulation runs in such a way that a prescribed desired margin of error is obtained in the estimate for θ. However, this is easier said than done. What is often done in practice is to first carry out a pilot simulation with a value of n that is not too large, say $n = k$. This gives a confidence interval with half-width $z_{1-\alpha/2}\sqrt{S^2(k)}/\sqrt{k}$. For increasing n, the term $S^2(n)$ will not change because $S^2(n)$ is an estimator for the (unknown) constant σ^2. Based on the term $1/\sqrt{n}$, a value of n can then be estimated for which the resulting confidence interval has approximately the desired width. The fact that the width of the confidence interval is proportional to $1/\sqrt{n}$ implies the following.

Rule of thumb: *Four times as many observations are needed for a twice as narrow confidence interval.*

We illustrate the above with a concrete example.

Example 11.2. An electronic system has one spare for a critical system component that is subject to stochastic failure. So the system consists of one working component and another component on standby. The standby component is turned on when the working component breaks down. The defective component is immediately repaired, after which it is put on standby. The system shuts down completely when the working component breaks down while the other component is still in repair. How long does it take before the system is down for the first time when both components are in perfect condition at time 0? To answer this question, we must know the distributions of the running time and repair time. Suppose that the running time has the Weibull distribution with distribution function $F(x) = 1 - \exp(-(\lambda x)^{\alpha})$ and that the repair time has constant value R. As an illustration, we give a simulation program (in C++) for this problem.

Simulation Program for Example 11.2

```
// confidence model
// author: Stefan Boer

#include <iostream.h>
#include <stdlib.h>
#include <math.h>


    const double t = 1.96; // normal distribution for
                           // sufficiently many runs

    const SEED    = 12345;// every time the program runs, SEED
                          // must be changed to obtain a differ-
                          // -ent sequence of random numbers

// global variables

    int runs;            // number of simulation runs
    double R;            // replacement time

    double alpha;        // form parameter Weibull distribution

    double lambda;       // scale parameter Weibull distribution
    double sum, sumsq;

// functions

double draw_weibull(double alpha, double lambda) {
    long int dummy;

    dummy = rand(RAND_MAX)+1;       // random from 1..RAND_MAX+1

    return
    (1/lambda)*pow(-log(((float)dummy)/(RAND_MAX+2)),1/alpha);
    // Inverse Trafo Method
};  // draw_weibull
```

```
void input() {

    cout << "R     = "; cin >> R;
    cout << "alpha = "; cin >> alpha;
    cout << "lambda = "; cin >> lambda;
    cout << '\n';
    cout << "Number of runs: "; cin >> runs;
}; // input


void simulate() {
    double simtime;
    double lifetime;

    sum = 0;
    sumsq = 0;
    for (int i=0; i<runs; i++) {
        simtime = draw_weibull(alpha,lambda);
        do {
            lifetime = draw_weibull(alpha, lambda);
                simtime += lifetime;
            } while (lifetime >= R);
        sum    += simtime;
        sumsq += (simtime*simtime);
    }; // for i
}; // simulate


void report() {
    double mean, stderror;

    mean    = sum/runs;
    stderror = sqrt((sumsq-double(runs)*mean*mean)/
                (double(runs-1)*double(runs)));
    cout << endl;

    cout << "How long does it take, on average, before" <<
            << "the system no longer works: ";
    cout << mean << endl;
    cout << "95% bti: [";
    cout << (mean-t*stderror);
    cout << ",";
```

```
        cout << (mean+t*stderror);
        cout << "]" << endl;
}; // report


bool one_more_time() {
    char test;

    cout << "\nAgain? (y/n)";
    do
        cin >> test;

    while ((test != 'Y') && (test != 'y') && (test != 'N') &&
        (test != 'n'));
    return bool((test == 'Y') || (toets == 'y'));
}; // one_more_time


// main program

int main() {

    srand(SEED); // the generator is only initialized
                 // at the beginning of the first run
    do {
        input();
        simulate();
        report();
    } while (one_more_time());
}; // main
```

Table 11.1   Some simulation results from Example 11.2.

n	$\mathbb{E}[X]$	$\mathbb{P}\{X > 2\}$	$\mathbb{P}\{X > 10\}$
625	32.65($\pm$2.49)	0.941($\pm$0.019)	0.738($\pm$0.035)
2500	32.84($\pm$1.27)	0.939($\pm$0.009)	0.745($\pm$0.017)
10000	32.64($\pm$0.65)	0.947($\pm$0.004)	0.735($\pm$0.009)

This simulation program therefore considers the random variable

$$X = \text{time until the system shuts down for the first time.}$$

The program determines a point estimate for the expected value $\theta = \mathbb{E}[X]$ and constructs an approximate 95% confidence interval for this point estimate. The notation 32.65 ($\pm$2.49) means, of course, a confidence interval $[32.65 - 2.49, 32.65 +$

2.49]. In Table 11.1, we also give point estimates and 95% confidence intervals for the probabilities $\mathbb{P}(X > t_0)$ for $t_0 = 2$ and $t_0 = 10$. The table nicely illustrates the rule of thumb: four times as many simulations are needed for a twice as narrow confidence interval. The calculation of the probability $\mathbb{P}(X > t_0)$ is not included in the program above but can easily be added (we leave this to the interested reader). After all, suppose that we have n independent observations $X_1, \ldots, X_n$ for the time X until the system shuts down for the first time. Now, for a given value of t_0, defined the derived observation X_i' by

$$X_i' = \begin{cases} 1 \text{ if } X_i > t_0, \\ 0 \text{ otherwise.} \end{cases}$$

For the random variable X' defined by $X' = 1$ if $X > t_0$ and $X' = 0$ otherwise, we have

$$\mathbb{E}[X'] = 1 \times \mathbb{P}(X > t_0) + 0 \times \mathbb{P}(X \leq t_0)$$
$$= \mathbb{P}(X > t_0).$$

In the same way as $\theta = \mathbb{E}[X]$ is estimated, the probability $\mathbb{P}(X > t_0)$ is estimated from the derived observations $X_1', \ldots, X_n'$. The confidence interval for the estimate of a probability has a simple form.

Confidence Interval for a Probability

Suppose that the observations $X_1, \ldots, X_n$ can only take on the values 0 and 1. In that case, we are actually simulating an unknown probability. The sample variance $S^2(n)$ then has the simple form $S^2(n) = \frac{n}{n-1}\overline{X}(n)[1 - \overline{X}(n)]$ (verify). The approximate $100(1 - \alpha)\%$ confidence interval for the estimate $\overline{X}(n)$ of the probability then has the simple form

$$\left[\overline{X}(n) - \frac{z_{1-\alpha/2}}{\sqrt{n-1}}\sqrt{\overline{X}(n)[1 - \overline{X}(n)]}, \overline{X}(n) + \frac{z_{1-\alpha/2}}{\sqrt{n-1}}\sqrt{\overline{X}(n)[1 - \overline{X}(n)]}\right]$$

for n large. This formula also gives insight into the simulation efforts for *very small* probabilities. Suppose, for example, that the probability that is to be simulated is in the order of 10^{-6} and that one wants a 95% confidence interval whose half-width is less than $f \times 10^{-6}$ for some f with $0 < f < 1$. Roughly speaking, the simulation length n must then satisfy $(1.96/\sqrt{n})10^{-3} \leq f \times 10^{-6}$; that is, n is in the order of $(1/f)^2 \times 4 \times 10^6$. For $f = 0.1$, this is 400 million. So the simulation time becomes very large when one wants to accurately determine a very small probability.

Verification

Although the simulation program for the previous example is relatively simple, it is still wise to do some extra checks to be sure that the program is working correctly. How could one do that in this particular case? One could take the specific case of an exponentially distributed running time and an exponentially distributed repair time. In that particular case, an analytic solution is fairly easy to deduce. Instead of developing this verification approach further, we give another approach.

A mathematically profound but practically very useful result from probability theory is the following. For stochastic processes, under general conditions, the time until a "rarely occurring" event occurs the first time is approximately exponentially distributed. In other words, if

$\theta =$ the expected time until the rare event occurs for the first time,

then the stochastic time T at which this rare event occurs for the first time satisfies

$$\mathbb{P}(T > t) \approx e^{-t/\theta} \text{ for all } t \geq 0.$$

In Example 11.2, we have chosen the numerical values of the parameters such that $\mathbb{E}[\text{repair time}] << \mathbb{E}[\text{running time}]$. Because of this, it is reasonable to consider the occurrence of the event "the system shuts down" a rare event. Indeed, the numerical results in the table can be used to check that the following is a good approximation:

$$\mathbb{P}(X > t_0) \approx e^{-t_0/\mathbb{E}[T]}.$$

Such a verification gives confidence in the correct functioning of the computer code.

Assignment 11.2. Consider Assignment 11.1 from Section 11.2. Expand the simulation program to find a point estimate with associated confidence interval for the expected value of the time X until the electronic system shuts down for the first time. Also ensure that the program can simulate the probability $\mathbb{P}(X > t_0)$ for a given value of t_0 (for example, $t_0 = 100$ and $t_0 = 400$). Verify the correctness of your computer program.

11.5 Long-Term Simulation

Long-term simulation is also called steady-state simulation. This term is a good indication of the purpose of long-term simulation, namely to analyze the system's behavior when the effect of the initial state has worn off and, statistically, the system has reached an equilibrium situation. In other words, we are interested in the system's average behavior when it has been running for a long time. In the bank model in Section 11.2 where the bank closes at the end of the afternoon and reopens the next morning, there is, of course, no question of the system reaching steady state. Now, suppose that the bank model concerns a production hall that is in operation day and night. The incoming customers are production orders, and the tellers are the machines on which the orders are processed. We can model such a situation as a queueing model with an infinite time horizon. Long-term performance measures for such a model are:

(a) the average number of production orders in the queue per unit of time over a very long period of time,

(b) the average waiting time per production order over a considerable number of production orders,

(c) the long-run average occupancy rate of a machine.

Long-term simulation has its own problems. First, the simulation cannot continue forever. The choice of the simulation's duration is already a problem. How should the duration be chosen such that the warm-up period before the system reaches an equilibrium does not influence the results? Just as big a problem is the statistical analysis of the simulation's output. From a mathematical point of view, statistical analysis for long-term simulation is much more difficult than it is for short-term simulation, where one "automatically" has independent observations. However, for practical purposes, a useful solution to this problem exists that we discuss below (the "batch means" method).

    Partly in view of the difficulties outlined above, one may wonder whether, in practice, there are systems that qualify for long-term simulation. In reality, every system has a finite lifespan. Of course, a long-term simulation is of no use if, for example, one needs to simulate an inventory with a planned lifespan of 5 years for more than 50 years to obtain accurate steady-state results. However, in practice, there are sufficiently many situations in which a long-term simulation is useful. These are often situations where extremely many state changes occur in a relatively short period of time. For example, in the analysis of a control rule for routing messages in a communication network, long-term simulation is typically applicable because this type of system receives high numbers of messages in the time period during which we are interested in the system's average behavior. An hour is a very long time period when a state change occurs every microsecond.

    Let us explain how a long-term simulation can be carried out using the $M/M/c$ queueing model for the production hall described above that runs day and night. Suppose that we want to use simulation to estimate the value of the average waiting time per production order over a substantial number of orders and also to construct a confidence interval for these estimates. How should we proceed? We first give a fundamental result from probability theory. Define the random variable

$$D_k = \text{the } k\text{th order's waiting time (excluding processing time)},$$

where we assume that the orders are processed in order of arrival (FIFO rule). The random variables D_n are, of course, correlated and therefore not independent of one another. Nevertheless, the strong law of large numbers applies in this situation. Regardless of the values the individual variables D_k take on, in the limit, we have

$$\lim_{k \to \infty} \frac{1}{k} \sum_{m=1}^{k} D_m = W_q \quad \text{with probability 1} \qquad (11.2)$$

for a constant W_q. This constant is the desired long-term average waiting time per order. This result holds not only for the $M/M/c$ queue but also in general for queueing processes that can be described by a *regenerative* stochastic process. A stochastic process is called regenerative if a certain state repeats itself from time to time, after which, from a probabilistic point of view, the process begins again, as it were, independently of the system's evolution up to that point. For a queueing process, a typical regeneration state is the state corresponding to the arrival of a customer who finds the system empty. The remarkable thing about the result (11.2) is that the average waiting time per order becomes less stochastic the greater the number of orders over which the average is taken, with limit a constant. This is a significant result that is the basis for the simulation approach.

The Batch Means Method

Suppose that we have carried out one long simulation run for the production hall and have obtained observations $D_1, \ldots, D_n$ for the waiting times for the first n production orders with n *very large*. Based on (11.2), we can, of course, take the average $\frac{1}{n} \sum_{k=1}^{n} D_k$ as a point estimate for the desired value W_q. How can we construct a confidence interval for this point estimate? We cannot do this directly because the observations $D_1, \ldots, D_n$ are (strongly) correlated. The trick to circumvent this problem is to aggregate the observations in such a way that the aggregated observations are approximately independent of one another. The run consisting of n observations is divided into r subruns of b observations each with b *sufficiently large*. The value for r is usually chosen between 25 and 50. The length b of the subruns is $b = n/r$. Define the aggregate observations $\overline{D}_1, \ldots, \overline{D}_r$ by

$$\overline{D}_1 = \frac{1}{b} \sum_{k=1}^{b} D_k, \ \overline{D}_2 = \frac{1}{b} \sum_{k=b+1}^{2b} D_k, \ldots, \overline{D}_r = \frac{1}{b} \sum_{k=(r-1)b+1}^{rb} D_k.$$

Then

$$\frac{1}{r} \sum_{\ell=1}^{r} \overline{D}_\ell = \frac{1}{n} \sum_{k=1}^{n} D_k.$$

After all,

$$\frac{1}{r} \sum_{\ell=1}^{r} \overline{D}_\ell = \frac{1}{rb} \left\{ \sum_{k=1}^{b} D_k + \sum_{k=b+1}^{2b} D_k + \cdots + \sum_{k=(r-1)b+1}^{rb} D_k \right\} = \frac{1}{n} \sum_{k=1}^{n} D_k.$$

To calculate a point estimate for W_q, it does not matter whether we take the original observations $D_1, \ldots, D_n$ or the aggregate observations $\overline{D}_1, \ldots, \overline{D}_r$. However, if we choose b sufficiently large, then we can use the aggregate observations to construct a confidence interval. If b is sufficiently large, then the subruns are sufficiently long to ensure that the observations $\overline{D}_1, \ldots, \overline{D}_r$ are approximately *independent of one another*. Moreover, for b sufficiently large, for practical purposes, one may assume that the aggregate observations $\overline{D}_1, \ldots, \overline{D}_r$ are approximately *normally distributed*. A standard result from statistics now tells us that the point estimate

$$\bar{D}(r) = \frac{1}{r} \sum_{l=1}^{r} \bar{D}_l$$

for the desired value W_q has the *approximate* $100(1 - \alpha)\%$ confidence interval given below.

Approximate $100(1 - \alpha)\%$ *confidence interval:*

$$\left[\overline{D}(r) - \frac{t_{r-1,1-\alpha/2}\sqrt{\overline{S}^2(r)}}{\sqrt{r}}, \ \overline{D}(r) + \frac{t_{r-1,1-\alpha/2}\sqrt{\overline{S}^2(r)}}{\sqrt{r}} \right],$$

where

$$\overline{S}^2(r) = \frac{1}{r-1} \sum_{k=1}^{r} \left[\overline{D}_k - \overline{D}(r) \right]^2.$$

The number $t_{r-1,1-\alpha/2}$ is the well-known $(1 - \alpha/2)$-percentile of Student's t-distribution with $r - 1$ degrees of freedom. For every r, the number $t_{r-1,1-\alpha/2}$ is greater than the percentile $z_{1-\alpha/2}$ of the standard normal distribution. For example, for $\alpha = 0.05$, we have $z_{1-\alpha/2} = 1.96$ and $t_{r-1,1-\alpha/2} = 2.064$ for $r = 25$. Because of the (almost) normality of the $\overline{D}_k$, an important aspect of this confidence interval for normally distributed observations is that it holds for every value of r, including small values such as $r = 25$.

The Length of Subruns and the Warm-up Period

The question of how to choose the size of n and b remains. Empirical studies show that correlation is the main obstacle to establishing a confidence interval. This means that particular attention should be paid to the choice of b (= the length of the subrun). For n (= the number of subruns), a value between 25 and 50 is usually chosen; in practice, $n = 25$ is a commonly used value. The question of how to choose b is very difficult to answer and depends largely on the problem. If we take, for example, the queueing model for the production hall, then b must be chosen greater as the production hall's load increases. When the hall is busy, the correlation between the waiting times $D_1, D_2, \ldots$ of course lasts much longer than when it is less busy. In general, b will have to be chosen in the order of thousands, tens of thousands, or even more. In practice, choosing b is often a matter of common sense and trial and error. It is better to be on the safe side than to choose b too conservatively, especially now that faster and faster computers are available. In the batch means simulation, we also have to deal with warm-up phenomena. The initial state is usually not representative of the state of the system in steady state. This means that we need to simulate for some time before the warm-up phenomena disappear. It makes sense to discard the initial observations associated with the warm-up period. A very long run that is divided into a number of subruns has the advantage that the observations in the warm-up period are only discarded once. After that, we automatically generate representative initial states for the subruns because a subrun's final state is taken as the next subrun's initial state. The question of how long the warm-up period should be is difficult to answer. In practice, very heuristic choices are made for the point up to which the observations must be discarded. But here too, it is nowadays so that one can choose on the safe side because of the enormous increase in the processing speed of computers.

Earlier, we described how to estimate the long-run average for a stochastic process with a discrete time parameter. The approach for a stochastic process with a continuous time parameter is entirely analogous. Suppose that in the queueing model for the production hall, we consider the continuous-time process $\{L_q(t), t \geq 0\}$, where the random variable $L_q(t)$ indicates the numbers of orders in the queue at time t. Suppose that we are interested in the long-run average

$$\lim_{t \to \infty} \frac{1}{t} \int_0^t L_q(u) du.$$

By the strong law of large numbers for regenerative stochastic processes, this limit exists and the average $(1/t) \int_0^t L_q(u) du$ always converges to the same constant regardless of the realization of the stochastic process $\{L_q(t), t \geq 0\}$. This constant is denoted by L_q and is the desired average queue length. How can we find L_q using simulation? To do this, we carry out a single long simulation run in which the production hall is simulated during T units of time, and we divide this run into r subruns of $t_0 = T/r$ units of time each. Here, T and t_0 are large (r between 25 and 50). For the resulting realization of the stochastic process $\{L_q(t)\}$, define the aggregate observations

$$\overline{L}_1 = \frac{1}{t_0} \int_0^{t_0} L_q(u) du, \; \overline{L}_2 = \frac{1}{t_0} \int_{t_0}^{2t_0} L_q(u) du, \ldots, \overline{L}_r = \frac{1}{t_0} \int_{(r-1)t_0}^{rt_0} L_q(u) du.$$

From

$$\frac{1}{r} \sum_{k=1}^r \overline{L}_k = \frac{1}{rt_0} \left[\int_0^{t_0} L_q(u) du + \int_{t_0}^{2t_0} L_q(u) du + \cdots + \int_{(r-1)t_0}^{rt_0} L_q(u) du \right]$$

and $rt_0 = T$, it follows that

$$\frac{1}{r} \sum_{k=1}^r \overline{L}_k = \frac{1}{T} \int_0^T L_q(u) du.$$

The aggregate observations $\overline{L}_1, \ldots, \overline{L}_r$ allow us to construct a confidence interval for the point estimate $(1/T) \int_0^T L_q(u) du$. If we choose t_0 sufficiently large, then we may assume that, by approximation, the aggregate observations $\overline{L}_1, \ldots, \overline{L}_r$ are independent and can be viewed as samples from a normal distribution. This leads to the following approximate $100(1-\alpha)\%$ confidence interval for the point estimate $\overline{L}(r) = \frac{1}{r} \sum_{k=1}^r \overline{L}_k$ of the average queue length L_q:

$$\left[\overline{L}(r) - \frac{t_{r-1,1-\alpha/2} \sqrt{\overline{S}^2(r)}}{\sqrt{r}}, \; \overline{L}(r) + \frac{t_{r-1,1-\alpha/2} \sqrt{\overline{S}^2(r)}}{\sqrt{r}} \right],$$

where

$$\overline{S}^2(r) = \frac{1}{r-1} \sum_{k=1}^r \left[\overline{L}_k - \overline{L}(r) \right]^2.$$

Regenerative Simulation Method

The big problem with the batch means method is the correlation between the observations. An alternative method for long-term simulation exists that does not involve this problem at all. This is the so-called *regenerative method*, which automatically generates independent observations. Starting from the assumption that the stochastic process to be simulated is regenerative, the method generates independent observations by simulating successive cycles of the process. A cycle is the time interval between two successive regeneration times of the process. At a regeneration time, the process starts over, as it were, independently of what has happened before. So the observations obtained from the successive cycles are independent of one another. However, this comes at a price. The biggest practical problem is that the (usually unknown) length of the cycle can be enormous so that obtaining sufficiently many observations becomes very expensive. The statistical analysis for the regenerative method requires a *large* number of observations. In the statistical analysis, so-called ratio estimators are needed. For example, in a queueing model, the average waiting time per customer is estimated by dividing the total waiting time of all customers in one cycle by the total number of customers arriving in that cycle. The confidence intervals for ratio estimators are surrounded by more uncertainty than those of traditional estimators. The disadvantages mentioned above are the main reason that, in practice, the less subtle batch means method is preferred to the mathematically elegant regenerative method.

Verification

Above, we illustrated long-term simulation with the queueing model for the production hall. We have shown how to use simulation to estimate the average waiting time per customer ($= W_q$) and the average queue length ($= L_q$). As a verification of the computer program's correctness, we can use Little's law:

$$L_q = \lambda W_q,$$

where λ is the long-run average number of customers (production orders) admitted into the system per unit of time. This famous law applies to virtually every conceivable queueing system; see Section 9.3.

Assignment 11.3. Adjust the computer program for the reliability problem in Assignment 11.1 from Section 11.2 so that it becomes suitable for using long-term simulation to estimate the fraction of the time that the system is down as well as a 95% confidence interval. Observe that the system is only down when all K units are in repair; as soon as one of these K units comes out of repair, the system is up again until all K units fail again. Use simulation experimentally to verify that in the long run, the fraction of the time that the system is down does not depend on the form of the repair time's probability distribution as long as the expected value of the repair time remains the same (cf. Exercise 8.18 in Chapter 8). To do this, consider two probability distributions for the repair time, namely an exponential distribution with parameter λ and a uniform distribution on (a, b), where λ and a, b are chosen such that $1/\lambda = (a + b)/2$.

11.6 Random Samples from Probability Distributions

From almost every conceivable probability distribution, random samples can be generated using U(0,1) random numbers. We already discussed the inversion sampling method for sampling from a continuous probability distribution with a strictly increasing distribution function in Section 11.3. This method can only be used if the inverse of the distribution function can be calculated quickly. The inversion sampling method has the pleasant property that it needs only one U(0,1) random number and gives a *monotonous* output (that is, the generated random sample is bigger the higher the U(0,1) random number is that has been used). This can be useful for the application of variance-reduction techniques.

In this section, we discuss the fundamental principles of some generally applicable methods that underlie customized procedures for specific probability distributions. We do not discuss those procedures. In software tools such as Matlab and R, and also on the internet, one can find balanced and fast codes for specific distributions such as the binomial distribution, the hypergeometric distribution, the Poisson distribution, the (multivariate) normal distribution, the gamma distribution, and the beta distribution, to name a few.

11.6.1 *The Acceptance-Rejection Method*

The acceptance-rejection method is in fact a generalization of the "hit-or-miss method" used to generate a random point in a "jagged-edged" region G in the plane. This is done by circumscribing the region with a rectangle and repeatedly generating a random point in the rectangle until the point that is generated falls into the region G.[6] The hit-or-miss method can, of course, also be applied to determine the area of a region in a higher-dimensional space (hence, in particular, also to determine higher-dimensional integrals). We describe the acceptance-rejection method for the case of a continuous probability density $f(x)$, but exactly the same approach can be used for a discrete distribution. The idea of the method is simple. To generate a sample from a "difficult" density $f(x)$, we choose a probability density $g(x)$ from which it is easy to generate a sample and for which there exists a constant c such that

$$f(x) \le cg(x) \quad \text{for all } x.$$

---

[6]For example, generating a random point in the closed unit disk $\{(x, y) : x^2 + y^2 \le 1\}$ is done as follows. Take the rectangle $R = \{(x, y) : -1 \le x, y \le 1\}$ to circumscribe the disk. Repeatedly generate a random point $(v_1 = -1 + 2u_1, v_2 = -1 + 2u_2)$ in the rectangle R using two random numbers u_1 and u_2 from $(0, 1)$ until the point (v_1, v_2) satisfies $v_1^2 + v_2^2 \le 1$, in which case a random point in the disk has been found.

The graph of $f(x)$ is therefore enclosed by the graph of $cg(x)$. Note that the constant c satisfies $c \geq 1$ because of the identity $1 = \int f(x)dx \leq c \int g(x)dx = c$. The idea is now to generate a random point under the graph of $cg(x)$ and only accept those samples that fall under the graph of $f(x)$.

Algorithm 11.2 (Acceptance-rejection method).

Step 1. Generate a sample Y from the probability distribution with density $g(y)$ and generate a random number U from $(0,1)$.

Step 2. If $U \leq f(Y)/cg(Y)$, accept the sample Y as a sample from the probability density $f(x)$. Otherwise, repeat Step 1.

Intuitively, the method generates a random point $(Y, U \times cg(Y))$ under the graph of $cg(y)$ and accepts this point only if it lies under the graph of $f(y)$, that is, if $U \times cg(Y) \leq f(Y)$.

The proof that the algorithm indeed generates a sample from the probability distribution of a random variable X with continuous probability density $f(x)$ goes as follows. Let $G(y)$ be the distribution function associated with the probability density $g(y)$. If the random variable Y has probability density $g(y)$, then the conditional probability density of Y given $Y \leq x$ is given by $g(y)/G(x)$. The probability that a given iteration of the algorithm results in a value of Y that is both less than or equal to x and accepted is equal to

$$\mathbb{P}(Y \leq x)\mathbb{P}(\text{sample } Y \text{ is accepted} \mid Y \leq x)$$

$$= G(x) \int_{-\infty}^{x} \mathbb{P}(\text{sample } Y \text{ is accepted} \mid Y = y) \frac{g(y)}{G(x)} dy$$

$$= \int_{-\infty}^{x} \frac{f(y)}{cg(y)} g(y)dy = \frac{1}{c} \int_{-\infty}^{x} f(y)dy.$$

If we let $x \to \infty$, then we find that $\mathbb{P}$ (sample Y is accepted) $= 1/c$. The number of required iterations is therefore geometrically distributed with expected value c. The ultimately generated sample $X = Y$ satisfies

$$\mathbb{P}(X \leq x) = \sum_{n=1}^{\infty} \mathbb{P}(X \leq x \text{ and } X \text{ is generated during iteration } n)$$

$$= \sum_{n=1}^{\infty} \left(1 - \frac{1}{c}\right)^{n-1} \frac{1}{c} \int_{-\infty}^{x} f(y)dy = \int_{-\infty}^{x} f(y)dy,$$

thus proving the statement.

As an illustration, suppose that the random variable X is concentrated on the finite interval (a, b) with continuous probability density $f(x)$. Take

$$g(x) = \frac{1}{b-a} \quad \text{for } a < x < b;$$

that is, $g(x)$ is the probability density of the uniform distribution on (a, b). Choose m as small as possible such that $f(x) \leq m$ for $a < x < b$. For $c = (b-a)m$, we then have $cg(x) \geq f(x)$ for $a < x < b$. Generating a sample from the uniform distribution on (a, b) is extremely simple: if the random variable Z is uniformly distributed on (a, b), draw a random number U between 0 and 1 and take $Z := a + (b - a)U$ for the random sample from the probability distribution of Z.

Nonhomogeneous Poisson Process

Many simulation applications involve a Poisson arrival process with a time-dependent arrival rate. Such a process is called a nonhomogeneous Poisson arrival process and is characterized by the arrival rate $\lambda(t)$ for $t \geq 0$; cf. Appendix C. If $\lambda(t) = \lambda$ for all $t \geq 0$, then the Poisson process is homogeneous. The arrival rates in a homogeneous Poisson process with parameter λ are easy to generate because the times between successive arrivals are independent of one another and exponentially distributed with expected value $1/\lambda$. Generating a sample from the exponential probability distribution is easy to do using the inversion sampling method; see Section 11.3.

For the simulation of a nonhomogeneous Poisson process, one can use the so-called *thinning method*, in fact a type of acceptance-rejection method. Fix a number λ_0 with $\lambda_0 \geq \lambda(t)$ for all $t \geq 0$. The idea is to first generate arrival times in a homogeneous Poisson process with parameter λ_0 and then thin out these times by accepting an arrival at time τ only with probability $\lambda(\tau)/\lambda_0$.

Suppose that the time until the next arrival after a given time s must be generated.

Algorithm 11.3 (Thinning Algorithm).

Step 1. Let $\overline{\lambda} = \max\{\lambda(t) \mid t \geq s\}$.
Step 2. Generate a U(0,1) random number U_1, and let $T := -\ln(U_1)/\overline{\lambda}$.
Step 3. Generate a U(0,1) random number U_2. If $U_2 \leq \lambda(s + T)/\overline{\lambda}$, let the next arrival after time s take place at time $s + T$; otherwise, set $s := s + T$ and go back to Step 2.

11.6.2 *The Inversion Sampling Method for Discrete Densities*

Suppose that the random variable X can only take on a finite number of possible values, with

$$\mathbb{P}(X = x_i) = p_i \quad \text{for } i = 1, \ldots, N.$$

The inversion sampling method can be applied to generate a sample from the probability distribution $\{p_i\}$. To do this, generate a U(0,1) random number U and determine the index ℓ for which

$$\sum_{i=1}^{\ell-1} p_i < U \leq \sum_{i=1}^{\ell} p_i \tag{11.3}$$

with the convention that $\sum_{i=1}^{\ell-1} = 0$ for $\ell = 1$. Output $X := x_\ell$. After all, the probability that a U(0,1) distributed random variable U takes on a value in the interval $(\sum_{i=1}^{\ell-1} p_i, \sum_{i=1}^{\ell} p_i)$ is equal to the length p_ℓ of this interval.

The inversion sampling method stands or falls with a quick search procedure for the index ℓ that satisfies (11.3). This is, in particular, a problem when N is large. A binary search procedure may then be better than a linear search procedure but still not fast enough for practical purposes. An extremely effective method of finding the index ℓ that satisfies (11.3) is given below.

The Ahrens–Kohrt method

The idea is to split the total probability mass 1 of the points $x_1, \ldots, x_N$ into m equal portions of mass $1/m$ each, where m is a fixed large number, for example $m = 2N$. Take m buckets numbered $1, \ldots, m$. In these buckets, store a probability mass of $1/m$ and one or more points x_i with parts of their unused probability mass chosen such that those parts add up to $1/m$. Every bucket is allocated only a few x_i because m has been chosen large. Next, randomly choose one of the m buckets. In this chosen bucket, determine the element x_ℓ for which (11.3) holds. This element can be found very quickly using a linear search method because the bucket contains only a few elements.

How does one implement the idea above? We first need some preliminary work. If necessary, first renumber the possible values x_i of the variable X such that $x_1 < x_2 < \cdots < x_N$. Then,

- calculate the array elements $F[i] = \sum_{j=1}^{i} p_j$ for $i = 1, \ldots, N$;
- for every bucket $k = 1, \ldots, m$, calculate the indices

$$m_k = \min\left\{ i \mid F[i] > \frac{k-1}{m} \right\} \text{ and } M_k = \min\left\{ i \mid F[i] \geq \frac{k}{m} \right\};$$

- in bucket k, store the elements x_j with $j = m_k, \ldots, M_k$.

We illustrate this with the following numerical example:

$$N = 4, \; p_1 = 0.3, \; p_2 = 0.2, \; p_3 = 0.35, \; p_4 = 0.15.$$

Suppose that we choose $m = 5$. It then follows from $F[1] = 0.3$, $F[2] = 0.5$, $F[3] = 0.85$, and $F[4] = 1$ that

$$(m_1, M_1) = (1, 1), \; (m_2, M_2) = (1, 2), \; (m_3, M_3) = (2, 3),$$

$$(m_4, M_4) = (3, 3), \text{ and } (m_5, M_5) = (3, 4).$$

In bucket 1, x_1 is stored with probability mass 0.2 from $p_1 = 0.3$; in bucket 2, x_1 is stored with probability mass 0.1 and x_2 with probability mass 0.1; in bucket 3, x_2 is stored with probability mass 0.1 and x_3 with probability mass 0.1; in bucket 4, x_3 is stored with probability mass 0.2; and in bucket 5, x_3 is stored with probability mass 0.05 and x_4 with probability mass 0.15.

Fig. 11.3   Ahrens–Kohrt method with $N = 4$ and $m = 5$.

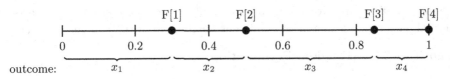

It is important to note that due to the order $x_1 < x_2 < \cdots < x_N$, every bucket contains only adjacent x_i, so that every bucket k can unambiguously be represented by the index m_k. After our preliminary work, we can give a fast procedure for generating a random sample from the probability distribution $\{p_i\}$. This procedure becomes faster the larger m is (the preliminary work, on the other hand, increases as m increases; however, those calculations only need to be done once). In practice, m is usually chosen between N and $3N$.

Algorithm 11.4 (Ahrens–Kohrt method).

Step 1. Generate a U(0,1) random number U.

Step 2. (Randomly choose one of the m buckets.) Let $k := 1 + \lfloor mU \rfloor$.

Step 3. (In bucket k, look for the element x_ℓ for which $F[\ell - 1] < U \leq F[\ell]$.) Let $\ell := m_k$. If $m_k = M_k$, go to Step 5; otherwise, go to Step 4.

Step 4. While $F[\ell] < U$, set $\ell := \ell + 1$. As soon as $F[\ell] \geq U$, go to Step 5.

Step 5. Output $X := x_\ell$.

In Step 3, one can find the element x_ℓ such that $F[\ell - 1] < U \leq F[\ell]$ in bucket k because $k := 1 + \lfloor mU \rfloor$ is equivalent to $\frac{k-1}{m} < U \leq \frac{k}{m}$. Applying this method to the numerical example above, suppose that the random number $u = 0.82137 \ldots$ is generated. Then one searches bucket 5 and the output is x_3 from that bucket.

The Array Method

For the special case that every p_i is, for example, of the form $0.01k_i$ with k_i an integer, a method exists that is related to the Ahrens–Kohrt method and is even faster. This is the so-called *array method*. We explain this method by means of a numerical example. Suppose $p_1 = 0.25$, $p_2 = 0.10$, $p_3 = 0.45$, and $p_4 = 0.20$. Form an array $A[i]$, with $i = 1, \ldots, 100$, with $A[i] = 1$ for $i = 1, \ldots, 25$, $A[i] = 2$ for $i = 26, \ldots, 35$, $A[i] = 3$ for $i = 36, \ldots, 80$, and $A[i] = 4$ for $i = 81, \ldots, 100$. Drawing a sample from the probability distribution $\{p_i\}$ now goes as follows. Generate a U(0,1) random number U. Let $k := 1 + \lfloor 100U \rfloor$ and $\ell := A[k]$. Output $X := x_\ell$.

Random Permutation

In many applications, either a random permutation of the numbers $1, \ldots, n$ is requested, or a random sample of r different numbers from the numbers $1, \ldots, n$.

Algorithm 11.5 (Random Permutation).

Step 1. Initialize $t := n$ and $a[j] := j$ for $j = 1, \ldots, n$.

Step 2. Generate a random number u between 0 and 1.

Step 3. Set $k := 1 + \lfloor tu \rfloor$ (random sample from $1, \ldots, t$). Interchange the current values of $a[k]$ and $a[t]$.

Step 4. Set $t := t - 1$. If $t > 1$, return to Step 2; otherwise, stop, and the random permutation is given by $(a[1], \ldots, a[n])$.

A random sample of r different numbers from $1, \ldots, n$ is obtained by only doing the first r steps of this algorithm and taking $a[n], \ldots, a[n - r + 1]$.

11.7   Variance-Reduction Methods

In the previous sections, we have seen that in a simulation, a high number of runs may be required to obtain a reliable estimate for the performance measure under consideration. However, in some applications, statistical tricks can be used to reduce the variance of the estimators without the need for more runs. Several variance-reduction methods have been developed. Which method can be used depends largely on the application in question. Note that the variance-reduction technique usually requires additional programming.

The most commonly used variance-reduction method is that of *common random numbers*. This method is often used when comparing two or more alternative system designs. In addition to this method, we discuss the variance-reduction method of *conditional Monte Carlo*. We do not discuss the variance-reduction methods of antithetic variables or of control variables; we refer the interested reader to the literature. In practice, these latter two methods are rarely used because of implementation difficulties and the fact that the success of these methods is not sure beforehand. Looking back, the variance reduction that is achieved often does not outweigh the additional programming efforts.

11.7.1   *Common Random Numbers*

This variance-reduction method is typically used in practice when comparing two or more alternative system designs. For example, an airline may want to compare the system with separate check-in desks with each its own queue to the system of pooled check-in desks with one common queue. The underlying idea is that, as much as possible, different system designs should be compared under the same experimental conditions so that any differences in the system performance are the result of the differences in the design and not of random fluctuations in the experimental conditions. In simulation, the experimental conditions are determined by the U(0,1) random numbers produced by the random number generator. However, for a pseudorandom number generator, these numbers can be reproduced by choosing the same start number (seed) for the generator. This is the basis for the method of common random numbers. When implementing this method, *synchronization* is essential; that is, a U(0,1) random number that is used for a particular purpose in one system must be used for the same purpose in any other system; for example, it should not be used for the arrival time in one system and for the service time in the other system. An essential step for realizing synchronization is that it is possible, when programming, to use separate generators for the different purposes (for example, arrival times and service times) with *disjoint* streams of U(0,1) random numbers. We return to the question of how to do this in practice further on. In some applications, it can be challenging to program in such a way that full synchronization is achieved and one settles for partial synchronization, for example only synchronization of the arrival times. Complete synchronization can easily be realized in, for example, the following problem. Suppose that at a port, a particular unloading installation must be replaced, either by one of the same type or by one of a new type that unloads twice as fast. Based on a given stochastic arrival time of the ships and a given probability distribution of the unloading time, the port manager wishes to compare the average time a ship spends in the port for the two situations. In this problem, full synchronization is easy to realize. In other cases, it may be tough to program. Take, for example, the situation of a production hall in which different types of jobs arrive at stochastic times and must be handled by different machines, where one wants to compare different rules for routing the jobs to the machines based on the average total processing time per job. In this situation, one will probably only synchronize the job arrival times.

In situations where *two* alternative systems are compared based on a given performance measure, it is recommended to do so based on an estimate and a confidence interval for the *difference* between the performance measure for one system and the performance measure for the other system. Considering the differences of the observations in general allows one to better distinguish between the systems, especially when the performance measures of the two systems do not differ by much. Again, it can be expected that the variance of the estimator for the difference is significantly reduced when using the method of common random numbers. This can also be seen using the formula

$$\text{var}(X - Y) = \text{var}(X) + \text{var}(Y) - 2\text{cov}(X, Y),$$

where X is the performance measure of one system and Y is the same performance measure of the other system. If both systems are exposed to, as much as possible, the same experimental conditions, then X and Y are (strongly) positively correlated, and it follows from $\text{cov}(X, Y) > 0$ that $\text{var}(X - Y) < \text{var}(X) + \text{var}(Y)$.

We illustrate the above with an inventory model.

Example 11.3. Suppose that customers for a specific product arrive at a store following a Poisson process with a rate of λ customers per day. Each customer's demand is approximately normally distributed with expected value μ and standard deviation σ, where the demand sizes of the different customers are independent of one another. If the on-hand stock is insufficient to fully meet a customer's demand, the customer's entire demand is lost ("all-or-nothing demand"). The store manager places a replenishment order with the factory when the inventory level gives reason to do so. The lead time of a replenishment order is equal to L_1 days with probability p and equal to L_2 days with probability $1 - p$. The replenishment strategy is based on the economic inventory, that is, the on-hand inventory plus the quantity already ordered at the factory. A replenishment has an (s, S) strategy when the economic inventory is replenished to S as soon as it drops to s, and the economic inventory is not replenished as long as it is more than s.

The aim is to compare the long-term behavior of two different (s, S) strategies based on, on the one hand, the service level and, on the other hand, the total cost. The service level is defined as the fraction of the customers whose demand is not lost. The total cost refers to the average daily holding and ordering costs, where a linear holding cost of $h > 0$ per unit of stock per day is incurred and a fixed ordering cost of $\dot{K} > 0$ is incurred for every replenishment order. For the numerical data $\lambda = 75$, $\mu = 4$, $\sigma = \frac{1}{2}$, $L_1 = \frac{1}{2}$, $L_2 = 1\frac{1}{2}$, $p = \frac{1}{2}$, $K = 100$, and $h = 1$, we want to compare the ordering strategies $(s, S) = (425, 675)$ and $(s, S) = (420, 670)$.

In this example, it seems obvious to synchronize the customer interarrival times and demand sizes in both (s, S) inventory systems. The synchronization of stochastic lead times is not easy because the frequencies with and times at which the replenishment orders are placed do not, in general, coincide for two different (s, S) policies. The most useful seems to be to generate independent samples for the lead times in the two (s, S) models. However, numerical experiments show that in the case of nonfixed lead times, the effect of variance reduction may be small if the lead times are not synchronized as much as possible using common random numbers. How to use the same random number generator to produce different streams of random numbers is discussed at the end of Section 11.3. Using the setup described above, we have simulated the inventory model until both $N = 150\,000$ customers have arrived and a time $T = 2000$ has elapsed (T has been chosen such that the number of arrivals during T units of time is equal to $N = 150\,000$). In the simulation with common random numbers, we found a 95% confidence interval $(0.0035, 0.0063)$ for the difference in the service levels of the two policies $(s, S) = (425, 675)$ and $(420, 670)$ and a confidence interval $(3.476, 5.088)$ for the difference in the average costs per unit of time. The individual point estimates and 95% confidence intervals for the service levels and the average costs per unit of time are $0.972\,(\pm 0.003)$ and $367.58\,(\pm 6.83)$ for the $(425, 675)$ policy and $0.967\,(\pm 0.004)$ and $372.30\,(\pm 6.79)$ for the $(420, 670)$ policy. It is interesting to see how great the variance reduction is.

To determine this, we have also simulated the differences in the service level and average cost by doing two independent runs for the two (s, S) policies. In the simulation with independent runs, we found the 95% confidence intervals $(-0.0033, 0.0057)$ and $(-4.245, 13.415)$, respectively, for these differences. If we compare the differences when using common random numbers with these differences, we see that a substantial variance reduction has been achieved. The confidence intervals of the differences for the independent runs do not give a clear answer to the difference between the two policies, while the confidence intervals of the differences when using common random numbers do.

11.7.2 *Conditional Monte Carlo*

This method depends largely on the model and can only be applied when the stochastic performance measure considered in the model can be naturally related to another random variable such that the performance measure's *conditional* expected value can be calculated *exactly* given the observed value of the other random variable. If one determines the conditional expected value by an exact formula instead of simulation, then a source of stochastics is decreased, hopefully lowering the variance of the final estimator. In general, one cannot be absolutely certain that this will happen. Only a pilot study can determine whether there is indeed a significant reduction in the sample variance.

Before we illustrate this method with a concrete example, we first give the mathematical motivation for this approach. Let Z be the random variable we are interested in. Suppose that the random variable Y is related to Z in a natural way and that we can calculate the conditional expected value of Z given the observed value y of Y exactly for every possible value y. For convenience, we abbreviate the conditional expected value $\mathbb{E}[Z \mid Y = y]$ as $w(y)$. If we define the random variable W to be $W = w(Y)$ as a function of the random variable Y, then we know from probability theory that

$$\mathbb{E}[W] = \mathbb{E}[Z] \quad \text{and} \quad \text{var}(W) \le \text{var}(Z). \tag{11.4}$$

In a simulation model, the situation is more complicated than that described by (11.4). For example, in a specific simulation model, Z could be an individual customer's waiting time, while we ultimately want to estimate the average waiting time per customer. The fact that the reduction in variance applies to a customer's individual waiting time does not mean that it applies to the average waiting time per customer (the reason is that covariances also occur when one takes the average). Nevertheless, it is reasonable to hope that reduction will occur in the variance of the estimator for the average waiting time per customer. The alternative estimator is based on the values of $\mathbb{E}[Z \mid Y]$ associated with the observations for Y.

We now illustrate the above with a concrete example. For this, we take the inventory model from Example 11.3, where we focus on estimating the fraction of the customers whose demand is lost for a given (s, S) policy. Define Z_k by

$$Z_k = \begin{cases} 1 \text{ if the } k\text{th customer's demand is lost,} \\ 0 \text{ otherwise.} \end{cases}$$

Then $\frac{1}{N} \sum_{k=1}^{N} Z_k$ is the classic estimator for the fraction of the customers whose demand is lost, where N is the number of customers in the simulation run. We can construct a conditional Monte Carlo estimator using the random variable

$$Y_k = \text{the on-hand inventory before the } k\text{th customer arrives.}$$

This random variable is related to Z_k in a natural way, and the conditional expectation $\mathbb{E}[Z_k \mid Y_k = y]$ is easy to calculate. After all, it follows from

$$\mathbb{E}[Z_k \mid Y_k = y] = 1 \cdot \mathbb{P}(Z_k = 1 \mid Y_k = y) + 0 \cdot \mathbb{P}(Z_k = 0 \mid Y_k = y)$$
$$= \mathbb{P}(\text{the } k\text{th customer's demand is greater than } y)$$

that $\mathbb{E}[Z_k \mid Y_k = y] = 1 - \Phi(\frac{y-\mu}{\sigma})$, where $\Phi(z)$ is the standard $N(0, 1)$ distribution function. Now, suppose that from the observations $Y_1, \ldots, Y_N$ in the simulation model, we calculate the derived observations

$$W_k = 1 - \Phi\left(\frac{Y_k - \mu}{\sigma}\right), \quad k = 1, \ldots, N. \tag{11.5}$$

Then $(1/N) \sum_{k=1}^{N} W_k$ is an alternative estimator for the fraction of the customers whose demand is lost. The sample variance of this conditional estimator will be smaller than that of the classic estimator $(1/N) \sum_{k=1}^{N} Z_k$.

11.7.3 *Importance Sampling*

In telecommunication problems and risk analysis, it is common for the probability of a rare event (buffer overflow, subsystem failure, and so on) to have to be estimated; these are very small probabilities, in the order of 10^{-6} or smaller. To estimate very small probabilities with a standard simulation approach generally requires a great deal of computing time; the amount of computing time needed to find a confidence interval with a given relative precision is inversely proportional to the value of the probability that is to be estimated.

In the article P. Heidelberger, *Fast simulation of rare events in queueing and reliability models*, ACM Transactions on Modeling and Computer Simulation 5, 43–85, 1995, an overview is given of special variance-reduction techniques for estimating very small probabilities in queueing and reliability problems. These "fast-simulation" methods use the theory of large deviations from probability theory. *Importance sampling* is an important concept for this. Suppose that X is a random variable with probability density $f(x)$ and that one wants to estimate $\mathbb{E}[a(X)]$ for a given function $a(x)$. Suppose that direct simulation from the density $f(x)$ is inefficient because $\text{var}(a(X))$ is very large with respect to $\mathbb{E}[a(X)]$. The idea is then to sample from another probability density $g(y)$ to achieve variance reduction. This idea is based on the simple observation that $\mathbb{E}[a(X)]$ can be represented as

$$\int a(x)f(x)dx = \int \left[a(y)\frac{f(y)}{g(y)}\right]g(y)dy = \mathbb{E}_Y\left[a(Y)\frac{f(Y)}{g(Y)}\right],$$

where the random variable Y is distributed following the probability density $g(y)$. If one produces a significant number of independent samples $y_1, \ldots, y_M$ from the density $g(y)$, then by the law of large numbers, one can estimate $\mathbb{E}[a(X)]$ by

$$\frac{1}{M}\sum_{k=1}^{M} w_k a(y_k) \quad \text{with } w_k = \frac{f(y_k)}{g(y_k)}.$$

The trick is to choose $g(y)$ in such a way that the variance of the random variable $a(Y)f(Y)/g(Y)$ is small so that variance reduction occurs.

In some situations, the density $f(x)$ is given up to a multiplicative constant, as is typically the case in applications of Bayesian statistics. In this case, one uses the alternative estimate

$$\frac{\sum_{k=1}^{M} w_k a(y_k)}{\sum_{k=1}^{M} w_k}.$$

The alternative estimate follows from the previous estimate by observing that $(1/M)\sum_{k=1}^{M} w_k \approx \int [f(y)/g(y)]g(y)\,dy = \int f(y)\,dy$ by the law of large numbers; that is, $(1/M)\sum_{k=1}^{M} w_k \approx 1$ for M large. In the alternative estimate, it suffices to know the w_k up to a multiplicative constant, so that this estimate can be used when $f(x)$ is given up to a multiplicative constant.

To estimate a (small) probability $\mathbb{P}(X \in A)$, apply the above with $a(x) = 1$ for $x \in A$ and $a(x) = 0$ for $x \notin A$, in which case $\mathbb{E}[a(X)] = \mathbb{P}(X \in A)$. For this choice of $a(x)$, the estimator's variance is determined by

$$\mathbb{E}_Y\left[\left(\frac{a(Y)f(Y)}{g(Y)}\right)^2\right] = \int_A \left(\frac{f(y)}{g(y)}\right)^2 g(y)dy.$$

This means that for variance reduction, $f(y)/g(y)$ must be small for $y \in A$; that is, $g(y)$ must be large for $y \in A$ because $f(y)$ is typically small on A for small values of the probability $\mathbb{P}(X \in A)$. In other words, the new probability density $g(x)$ must be such that it has a greater mass on the set A than the old probability density $f(x)$, so that the event A occurs more often under the new probability density. The choice of $g(x)$ is a subtle matter and depends largely on the problem; the probability density g should not put too much mass on the set A because, otherwise, the estimator's variance would go up instead of down. A common choice is

$$g(x) = e^{tx}f(x)/M(t)$$

for an appropriate $t > 0$, where $M(t) = \mathbb{E}\left[e^{tX}\right]$ is the moment-generating function of X.

Example 11.4. In certain statistical applications, an important problem is the determination of the probability that a random walk with a negative drift is more likely to exceed a value of b than to drop below a value of $-a$, where $a, b > 0$ are given. Suppose that $X_1, X_2, \ldots$ are independent random variables with the same density $f(x)$, namely the $N(\mu, 1)$ density with $\mu < 0$. Let θ be defined as the probability that the random walk $\{X_1 + \cdots + X_n, n = 1, 2, \ldots\}$ exceeds b before it drops below $-a$. How can importance sampling be used to estimate the small probability θ better than with standard simulation? For this, we take the $N(-\mu, 1)$ density for $g(x)$. This density puts more mass on (a, ∞) than the original density $f(x)$. Let $Y_1, Y_2, \ldots$ be a sequence of independently distributed random variables with density $g(y)$. Since $-\mu$ is positive, for the random walk $\{Y_1 + \cdots + Y_n, n = 1, 2, \ldots\}$, the probability of exceeding the level b before dropping below the level $-a$ will be greater than for the original random walk. Now, set the function $h(v_1, \ldots, v_n)$ equal to 1 for $v_1 + \cdots + v_n > b$ and equal to 0 otherwise. Define the random variable N as the smallest n for which $Y_1 + \ldots + Y_n$ is either less than $-a$ of greater than b. The importance sampling for θ is then given by

$$h(Y_1, \ldots, Y_N) \prod_{i=1}^{N} \frac{f(Y_i)}{g(Y_i)} = \delta(Y_1 + \cdots + Y_N - b)e^{2\mu(Y_1 + \cdots + Y_N)/\sigma^2},$$

where $\delta(x) = 1$ for $x > 0$ and $\delta(x) = 0$ otherwise. The expected value of this estimator is θ. The variance of this estimator is less than the variance of the direct estimator without importance sampling. For example, for the numerical example with $\mu = -3$, $a = 4$, and $b = 7$, doing $n = 10^4$ simulation runs leads to the 95% confidence intervals 0.011 (± 0.002) and 0.00996 ($\pm 7.4 \times 10^{-5}$) for the desired probability θ using the direct approach and the importance-sampling approach, respectively. The latter leads to a great reduction in the variance of the estimator for θ.

11.8 Exercises

11.1 A beer company brings a new beer with the brand name Babarras to the market and prints one of the letters of this brand name underneath each bottle cap. Each of the letters A, B, R, and S must be collected a certain number of times in order to get a nice beer glass. The quota for the letters A, B, R, and S are 3, 2, 2, and 1. These letters appear with probabilities 0.15, 0.10, 0.40, and 0.35, where the letters underneath the bottle caps are independent of each other. Simulate the expected value and the probability mass function of the number of bottles that must be purchased in order to form the word Babarras.

11.2 An opaque bowl contains 11 envelopes in the colors red and blue. You are told that there are four envelopes of one color each containing $100 and seven empty envelopes of the other color, but you cannot see the envelopes in the bowl. The envelopes are taken out of the bowl, one by one and in random order. Each time an envelope has been taken out, you must decide whether or not to open this envelope. Once you have opened an envelope, you get the money in that envelope (if any) and the process stops. Your stopping rule is to open the envelope drawn as soon as four or more envelopes of each color have been taken out of the bowl. Simulate your probability of winning $100.

11.3 A retired gentleman considers participation in an investment fund. An adviser shows him that with a fixed yearly return of 14%, which was realized in the last few years, he could withdraw $15,098 from the fund at the end of each of the coming 20 years. This is music to the ears of the retired gentleman, and he decides to invest $100,000. However, the yearly return fluctuates. If the return was $r\%$ for the previous year, then for the coming year the return will remain at $r\%$ with a probability of 0.5, will change to $0.8r\%$ with a probability of 0.25, and will change to $1.2r\%$ with a probability of 0.25. Simulate a probability histogram of the invested capital after 15 years when at the end of each of those 15 years $15,098 is withdrawn with the stipulation that the entire amount remaining in the fund is withdrawn if the remaining capital is less than $15,098.

11.4 You play repeatedly a game with 11 cards: ace, two, three, ..., nine, ten, and joker. Each card counts for its face value, where the ace counts for 1. The randomly ordered cards are turned over one by one. Your score is the sum of the face values of the cards turned over as long as the joker has not shown up. You can stop any moment you wish. If the joker appears, the game is over and your end score is zero. You stop as soon as your score is s or more. Use simulation with common random numbers to find the average score per game and the probability of getting a score zero as function of the stopping level s.

11.5 Consider the investment problem from Section 5.8.4 in which, during 52 weeks, an investor invests 5/24 of her current capital in a dot-com company. Use simulation to determine a probability histogram of the investor's final capital at the end of the 52 weeks. For $c = 2, 3, 4$, and 5, simulate the probability that at some time during the 52 weeks, the investor's capital is c or more times as large as her starting capital of ten thousand euros.

11.6 The Spiderweb Plan of Argus Investments is a 60-month contract in which the customer agrees to pay a fixed amount, say 100 euros, at the beginning of every month. Upon entering into the contract, Argus immediately deposits 150 times the amount of 100 euros in an investment fund. The total amount of 15 thousand euros stays in the fund for five years, and at the end of the five years, Argus pays out the value of this investment minus 15 thousand euros to the customer. The Spiderweb Plan is launched when the fund's return over the previous year was 10%, as was the average annual return over a number of previous years (the customer is paid $(1 + 0.1)^5 \times 15000 - 15000 = 9158$ euros if the return is exactly 10% during the next five years). Suppose that the investment fund's return fluctuates from year to year according to the following probability model: if the return was $r\%$ in the past year, then the coming year's return is $r\%$ with probability p_0, decreases to $(1 - f_d)r\%$ with probability p_1, and increases to $(1 + f_u)r\%$ with probability p_2, where $p_0 + p_1 + p_2 = 1$ and $0 < f_d, f_u < 1$. For the numerical values $p_0 = 0.5$, $p_1 = p_2 = 0.25$, and $f_u = f_d = 0.2$, simulate the probability histogram of the amount the customer receives after five years.

11.7 In many TV shows, the wheel of fortune is used to determine the winner among the remaining candidates. Numbers are spaced evenly along the outer edge of the wheel, for example the numbers 1, 2, ..., 100. Every candidate spins the wheel of fortune once or twice; in the case of a second turn, it must immediately follow the first one. At every spin, the wheel stops randomly at one of the hundred numbers. The candidate's final score is the sum of the results provided that this is not more than one hundred. If it is more, the candidate gets zero points. The final winner is the candidate with the most points, and if several candidates have the same score, then the winner is the one who spun the wheel first. For the case of two candidates A and B, with candidate A spinning the wheel first, player A's optimal strategy is to stop after the first spin only if it gives more than 53 points (the probability of player A being the final winner is then the highest, namely 0.4596). Now, consider the case of three candidates A, B, and C, who play in this order. Use simulation with common random numbers to determine an optimal stopping point for the first player A under the assumption that players B and C play optimally. Also determine the winning probabilities of all players with the corresponding 95% confidence intervals.

11.8 A supermarket has a refrigeration system with three cooling elements. The current policy is to replace only the broken element when a cooling element breaks down. The management is considering changing to another policy, namely replacing all three cooling elements as soon as one of them breaks down. The reason is that the fixed cost for shutting down the cooling system is quite high, namely $K = 1000$ euros. The replacement cost is $v = 650$ euros for each cooling element. The lifespans of the three cooling elements are independent of one another, and each has a Weibull distribution with an expected value of $\mu = 150$ days and a standard deviation of $\sigma = 50$ days. Use the method of common random numbers to compare the average total cost per day for both replacement policies.

11.9 At a currency exchange office that is open from 8:00 to 20:00, customers arrive following a nonhomogeneous Poisson process with arrival rate

$$\lambda(t) = \begin{cases} 5 + 10t & \text{for } 0 \leq t \leq 2, \\ 23 + t & \text{for } 2 \leq t \leq 7, \\ 33.5 - 0.5t & \text{for } 7 \leq t \leq 10, \\ 53.5 - 2.5t & \text{for } 10 \leq t \leq 12. \end{cases}$$

Take an hour as unit of time and let 8:00 correspond to $t = 0$ and 20:00 to $t = 12$. From 20:00 on, no more customers are admitted, but all customers who have arrived before then are processed. The currency exchange office has two separate counters: one for foreign currency and one for traveler's cheques. A customer comes for foreign currency with probability 3/4 and for traveler's cheques with probability 1/4. The processing time is Erlang-2 distributed with an expected value of 3 minutes for a customer for foreign currency and lognormally distributed with an expected value of 5 minutes and a standard deviation of 8 minutes for a customer for traveler's cheques. Use common random numbers to compare the average waiting time per customer for both types of customers and the average occupancy rate of each counter for the system with separate counters where customers can only go to the counter corresponding to their specific demand and for the system with a common queue where every counter helps both types of customers.

11.10 The Hubble Space Telescope is a laboratory in space for making astronomical observations. The telescope contains a large number of instruments, including six gyroscopes to ensure the telescope's stability. The six gyroscopes have been arranged in such a way that three working gyroscopes suffice for stability. The gyroscopes' lifespans are independent random variables that each have an exponential distribution with an expected value of $1/\lambda = 10$ years. When a fourth gyroscope fails, the telescope puts itself into "safe mode." In safe mode, no more observations are carried out. It takes a fixed duration of 0.24 days to reach safe mode. As soon as the telescope has reached this state, a signal is sent to the base station on earth. After such a signal is received, the base prepares a shuttle mission to space to carry out the necessary repairs. The time needed to reach the telescope and carry out the repairs has a normal probability distribution with an expected value of 2.4 months and a standard deviation of 0.5 months. After repairs are carried out, all six gyroscopes are back in perfect condition. However, it may happen that during the time needed to carry out the repair mission, the remaining two working gyroscopes also break down. If this happens, the telescope crashes and is lost forever. Use simulation to answer the following questions:

(a) What is the probability distribution of the number of shuttle missions to be carried out in the next 10 years?
(b) Suppose that in two years' time, a major astronomical event must be photographed for six months. What is the probability that during those six months, the telescope functions, that is, has at least three functioning gyroscopes, at least 95% of the time?
(c) What is the probability that the telescope crashes in the next ten years?

11.11 A specialized national courier service has purchased $C = 250$ delivery cars. These cars must be distributed among five regional centers that operate independently of one another and each have their own customer base. Every customer order requires one delivery car, and one delivery car cannot carry out more than one customer order at a time. In each region, transport takes place from a central point. At the regional center i, customer orders arrive following a Poisson process with a rate of λ_i orders per hour, where $\lambda_1 = 5$, $\lambda_2 = 10$, $\lambda_3 = 10$, $\lambda_4 = 50$, and $\lambda_5 = 37.5$. The processing time of a customer order from region i is a random variable S_i (the processing time is the time between the beginning of the delivery and the moment the car that is used is available for a new job). For each region i, the processing time S_i has a lognormal probability distribution with expected value μ_i and standard deviation σ_i, where $\mu_1 = 2$, $\mu_2 = 2.5$, $\mu_3 = 3.5$, $\mu_4 = 1$, $\mu_5 = 2$ and $\sigma_1 = 1.5$, $\sigma_2 = 2$, $\sigma_3 = 3$, $\sigma_4 = 1$, and $\sigma_5 = 2.7$. The national management wants to distribute the $C = 250$ cars over the five regional centers in such a way that approximately the same service is delivered by the five centers, where the service is measured using the probability that an incoming customer order must wait for an available car.

(a) Use simulation to compare the service provided at the different centers for each of the distributions ($c_1 = 16$, $c_2 = 34$, $c_3 = 46$, $c_4 = 63$, $c_5 = 91$) and ($c_1 = 17$, $c_2 = 35$, $c_3 = 46$, $c_4 = 62$, $c_5 = 90$).
(b) For each of these two distributions, determine, for every center, the probability that a customer order that must wait stays in the queue longer than half the order's average processing time.

11.12 An airline receives calls for information about flights following a Poisson process with a rate of 25 calls per hour. The airline has both domestic and international flights. The probability that an incoming call concerns a domestic flight is $2/3$. The processing time of a call concerning a domestic flight has a lognormal probability distribution with an expected value of 3 minutes and a standard deviation of 2 minutes, while the processing time of a call concerning an international flight has a lognormal probability distribution with an expected value of 5 minutes and a standard deviation of 5 minutes. Incoming calls that find all agents busy wait. The service requirement is, on the one hand, that the probability that an incoming call must wait should not be more than 50% and, on the other hand, that a call that must wait has a probability of at most 5% of waiting longer than 2 minutes. There are two options: (1) separate agents for domestic and international calls; (2) pooled agents for both types of phone calls.

(a) Use simulation to determine for both options how many agents are needed to satisfy the service requirement.

(b) In a graph, indicate the simulated complementary probability distribution function of the waiting time of a domestic call that must wait. Do the same for international calls.

11.13 A beach club has separate restrooms for men and women. On a summery day, women arrive at the restroom following a Poisson process with arrival rate λ_1 per minute and men following a Poisson process with arrival rate λ_2 per minute. The time a woman spends in the restroom is lognormally distributed with expected value μ_1 minutes and a standard deviation of σ_1 minutes, while the time a man spends in the restroom is lognormally distributed with expected value μ_2 minutes and a standard deviation of σ_2 minutes. Suppose $\mu_2 = 2\mu_1$ and $\sigma_1/\mu_1 \geq \sigma_2/\mu_2$. The club's owner is considering allowing members of one gender to use the other gender's restroom if it is empty (that is, they are not allowed to queue for the other gender's restroom). Choose numerical values for μ_1, σ_1, and σ_2 yourself.

(a) Determine the average waiting times for both the women and the men in both the old and the new situations.

(b) For both situations, simulate the graph of the complementary probability distribution function of the waiting time of women (respectively, men) who must wait.

11.14 At a call center, phone calls come in between 8:00 and 20:00. The calls arrive following a nonhomogeneous Poisson process with arrival rate function

$$
\begin{aligned}
&\lambda(t) = 5 + 5t \quad &&(0 < t \leq 1), \quad &&\lambda(t) = 20t - 10 \quad &&(1 < t \leq 2),\\
&\lambda(t) = 20t - 10 \quad &&(2 < t \leq 3), \quad &&\lambda(t) = 15t + 5 \quad &&(3 < t \leq 4),\\
&\lambda(t) = 125 - 15t \quad &&(4 < t \leq 5), \quad &&\lambda(t) = 10t \quad &&(5 < t \leq 6),\\
&\lambda(t) = 60 \quad &&(6 < t \leq 7), \quad &&\lambda(t) = 60 \quad &&(7 < t \leq 8),\\
&\lambda(t) = 20t - 100 \quad &&(8 < t \leq 9), \quad &&\lambda(t) = 350 - 30t \quad &&(9 < t \leq 10),\\
&\lambda(t) = 250 - 20t \quad &&(10 < t \leq 11), \quad &&\lambda(t) = 140 - 10t \quad &&(11 < t \leq 12),
\end{aligned}
$$

where t is measured in hours and $t = 0$ corresponds to 8:00, $t = 12$ to 20:00, and so on. The number of available operators varies from hour to hour and is equal to c_i during the time interval $(i - 1, i]$, for $i = 1, 2, \ldots, 12$. An operator who works until time $t = i$ finishes any call he started before $t = i$. The values of $c_1, \ldots, c_{12}$ are

$$
c_1 = 10,\ c_2 = 25,\ c_3 = 40,\ c_4 = 60,\ c_5 = 60,\ c_6 = 60,
$$
$$
c_7 = 65,\ c_8 = 65,\ c_9 = 70,\ c_{10} = 75,\ c_{11} = 45,\ c_{12} = 25.
$$

A call's handling time has a lognormal probability distribution with an expected value of $\mu = 1$ minute and a standard deviation of $\sigma = 1$ minute. A customer who must wait too long before speaking to an operator becomes impatient and hangs up. The time a customer is willing to wait patiently is Erlang-2 distributed with an expected value of $\theta = 2$ minutes; this time does not depend on the customer's position in the queue (customers cannot see one another).

(a) For each of the twelve time periods of one hour, determine what percentage of the incoming customers must wait and the average number of busy operators.

(b) Over the full period of twelve hours, determine, out of the customers who must wait and decide to stay in the queue, the percentage of customers who must wait longer than x. Plot this percentage in a graph as a function of x.

11.15 A train maintenance site has a number of parallel dead-end tracks. Each of these tracks has room for two trains. The position on the track where the track stops is called the closed side, and the other one is called the open side. Trains can only leave the track via the open side. This means that at the end of its maintenance, a train may be temporarily blocked by a train that came for maintenance on the same track but has not been fully processed yet. There are two maintenance teams for each track. A train arriving for maintenance is assigned to the closed side if that is free on one of the tracks. If all closed and open sides are occupied, then the train waits in the switching yard until a spot becomes available. Suppose that trains arrive for maintenance following a Poisson process with a rate of λ trains per day. A train's service time is lognormally distributed with an expected value of μ days and a standard deviation of σ days. In all, there are s tracks and therefore $2s$ maintenance teams.

(a) Determine the average time a train spends at the site, the probability that a train must wait in the switching yard, and the probability that a train becomes blocked.

(b) In a graph, show the simulated complementary probability distribution function of the time a train spends at the site.

11.16 An information center has three counters A, B, and C for three different services. Customers for the services A, B, and C arrive following independent Poisson processes with a rate of $\lambda_A = 0.8$ customers per minute for service A, $\lambda_B = 0.16$ customers per minute for service B, and $\lambda_C = 0.08$ customers per minute for service C. In principle, every counter can handle all services, but first and foremost, a customer for a specific service goes to the corresponding counter. If a counter becomes vacant while there is no queue at that counter, one customer from another queue may go to the vacated counter. Customers for service A have priority over customers for service B, and customers for service B have priority over customers for service C. A customer's service time has a lognormal distribution whose expected value and standard deviation are given by $\mu_A = 1$ minute and $\sigma_A = 1.2$ minutes for service A, $\mu_B = 5$ minutes and $\sigma_B = 4$ minutes for service B, and $\mu_C = 10$ minutes and $\sigma_C = 8$ minutes for service C. Use computer simulation to determine the average waiting time for each of the three types of services and the probability that a customer must wait longer than half the expected service time. Compare the results for three strictly separated queues to those for a common queue for the three counters. Use common random numbers.

11.17 A national emergency service has four regional centers 1, 2, 3, and 4. The four centers receive telephone calls following Poisson processes with a rate of λ_i calls for help per hour at center i, where $\lambda_1 = 70$, $\lambda_2 = 100$, $\lambda_3 = 50$, and $\lambda_4 = 80$. A call's service time has a lognormal distribution with an expected value of μ_i minutes and a standard deviation of σ_i for processing by the regional center i, where $\mu_1 = \mu_2 = 3$, $\sigma_1 = \sigma_2 = 2.5$, $\mu_3 = \mu_4 = 3.5$, and $\sigma_3 = \sigma_4 = 3$. The number of available operators at center i is equal to c_i, where $c_1 = 5$, $c_2 = 8$, $c_3 = 4$, and $c_4 = 7$. The organization aims to have at least 95% of the incoming calls picked up by an idle operator within 30 seconds. To see how this goal can be achieved, use computer simulation to compare two systems:

(a) The centers work independently of one another.

(b) If possible, overflow takes place after 28 seconds; that is, if at a particular center, a call is in the queue for 28 seconds or more and a line frees up at another center, then those waiting in the queue become eligible to be diverted to the idle line at the other center. The priority of requests for help in the queue is determined by the current waiting time. If several centers have a free operator, then the order of preference for call forwarding is first center 1, then center 2, then center 3, and finally center 4.

11.18 Consider a two-server facility with heterogeneous servers. The fast server is always available and the slow server is activated for assistance when too many customers are waiting. Batches of customers arrive according to a Poisson process with rate $\lambda = 1.4$. The batch size is geometrically distributed with mean 8. The service facility has ample waiting room. The service times of the customers are independent of each other and have an exponential distribution. The mean service time is $1/\mu_1$ when service is provided by the fast server and is $1/\mu_2$ for the slow server, where $\mu_1 = 9$ and $\mu_2 = 3$. The slow server can only be turned off when it has completed a service. Each server can handle only one customer at a time. Service is non-preemptive, that is, the fast server cannot take over a customer from the slow server. A fixed cost of $K = 10$ is incurred each time the slow server is turned on and there is an operating cost of $r = 20$ per unit of time the slow server is on. Also, a holding cost of $h = 1$ per time unit is incurred for each customer in the system. The control rule is a so-called hysteretic (m, M) rule under which the slow server is turned on when the number of customers present is $M = 11$ or more and the slow server is switched off when this server completes a service and the number of customers left behind in the system is below $m = 4$. Simulate the long-run average cost per unit time.

References

1. A. M. Law, *Simulation Modeling and Analysis*, 4th ed., McGraw Hill, New York, 2007.

2. S. M. Ross, *Simulation*, 5th ed., Academic Press, New York, 2013.

Appendix A

Complexity Theory

An optimization problem can either be solved via an *exact algorithm* or via a *heuristic*. An exact algorithm determines the optimal solution, but often uses much computation time. A heuristic determines a feasible solution that is as good as possible, in as little time as possible. A natural question for a discrete (combinatorial) optimization problem is: can one endeavor to create an exact algorithm for it, or will the computation time be so large that one should pursue developing a heuristic? Complexity theory provides a mathematical framework to answer this question. In other words, complexity theory is a mathematical framework to study the difficulty of combinatorial optimization problems.

Let us first introduce some definitions. With *problem* we refer to a generic description of the problem, while an *instance* I of the problem is the problem for a specific set of numerical parameters. For example, the traveling salesman *problem* (TSP) is: given n cities and the distances between each pair of cities, what is the shortest possible route that visits each city exactly once and returns to the origin city? An example instance of TSP would be 8 cities and the accompanying distance matrix. The *size* $|I|$ of an instance is the length of the string necessary to specify the input data. For TSP, we have $|I| = n$, with n the number of cities.

Algorithm Running Time

The *algorithm running time* is an upper bound on the number of calculation steps an algorithm needs, as a function of the size of the instance. Algorithm running time is expressed in *big-O notation*.

Definition A.1 (Algorithm running time). *The running time of an algorithm is $O(f(n))$ if there exist a constant $c > 0$ and an integer n_0 such that, for an instance with size $n = |I|, n \geq n_0$, the number of steps needed is bounded by $cf(n)$.*

For example, if an algorithm needs $7n^3 + 230n + 10\log(n)$ steps on an instance of size n, then the algorithm running time is $O(n^3)$ because $7n^3$ is the expression's dominant component for large n.

Example A.1. Suppose we have an unordered list of n integers, $i_1, i_2, \ldots, i_n$, and we want to find k, the value of the largest integer. We use the following simple algorithm.

Algorithm A.1.

Step 0. Set $k := i_1$.
Step 1. For j from 2 to n do: if $i_j > k$ then $k := i_j$.

In terms of calculation steps, this algorithm performs $n - 1$ comparisons between i_j and k, and it adjusts the value of k at most n times. Hence, the algorithm performs at most $2n - 1$ calculation steps, so its algorithm running time is $O(n)$.

The following rules apply to algorithms consisting of sub-algorithms or containing subroutines.

Rule A.1 (Addition rule). *If an algorithm consists of two sub-algorithms with running times $O(f(n))$ and $O(g(n))$, then the running time of the algorithm is $O(\max\{f(n), g(n)\})$.*

Rule A.2 (Product rule). *If an algorithm consists of executing $O(f(n))$ times a subroutine with running time $O(g(n))$, then the running time of the algorithm is $O(f(n) \cdot g(n))$.*

If, for a problem with instances of size $|I| = n$, we have an algorithm with a running time that is polynomial in n, we say that we have a *polynomial algorithm* for the problem.

Classes $\mathcal{P}$ and $\mathcal{NP}$

Each combinatorial optimization problem has a corresponding *decision problem* by introducing a threshold for the objective value. For example, for TSP the optimization problem is to find a tour with minimum length, while the decision problem is: does a tour exist with length smaller than or equal to k? Thus, a decision problem has a 'yes' or 'no' answer. We classify problem instances according to this answer, that is, every instance that results in the answer 'yes' to the decision problem is called a *'yes' instance*. For our TSP example, a 'yes' instance is an instance for which there indeed exists a tour with length smaller than or equal to k.

A decision problem is called *polynomial solvable* if a polynomial algorithm exists which solves the problem. Such problems are said to belong to complexity class $\mathcal{P}$.

Definition A.2 (Class $\mathcal{P}$). *The complexity class $\mathcal{P}$ contains all decision problems which are polynomial solvable.*

There are also decision problems for which no polynomial algorithm has been developed until now, and the conjecture among mathematicians is that it is not even possible to do so. To formally classify such problems, an additional definition is required: a *concise certificate* is a representation of a candidate solution. For our TSP decision problem, a candidate solution would be the representation of a tour, for example as a list with the same city at the beginning and the end (the origin) and every other city appearing only once.

Definition A.3 (Class $\mathcal{NP}$). *The complexity class $\mathcal{NP}$ contains all decision problems for which a concise certificate exists that can be verified by a polynomial algorithm.*

Clearly, the class $\mathcal{P}$ is a subset of the class $\mathcal{NP}$. A famous open problem is the question whether $\mathcal{P}$ equals $\mathcal{NP}$. The general belief among mathematicians is that this is not the case.

Finally, we introduce $\mathcal{NP}$-*completeness*, which is defined based on the concept of a *polynomial reduction*.

Definition A.4 (Polynomial reduction). *A decision problem R polynomially reduces to a problem Q (notation: $R \propto Q$), if a polynomial function g exists that transforms instances of R to instances of Q such that I is a 'yes' instance of R if and only if $g(I)$ is a 'yes' instance of Q.*

Example A.2. In this example, we polynomially reduce the partition problem to the two-machine scheduling (decision) problem with the makespan criterion. We first introduce these two problems.

In the partition problem, we are given n positive integers $s_1, \ldots, s_n$. The question is whether we can divide this set into two subsets with the same sum. Formally, if we define

$$b = \frac{1}{2} \sum_{j=1}^{n} s_j \tag{A.1}$$

does there exist a subset $J \subset I = \{1, \ldots, n\}$ such that

$$\sum_{j \in J} s_j = b = \sum_{j \in I \setminus J} s_j \ ?$$

In the two-machine scheduling problem with the makespan criterion, abbreviated as $P2||C_{max}$, we are given a set of jobs j with processing times p_j. A job should be produced on either machine 1 (M1) or machine 2 (M2) and each machine can produce one job at a time. The *makespan*, C_{max}, of a schedule is the completion time of the last job. In the optimization problem, the objective is to construct a schedule with minimum makespan, while the decision problem is: does there exist a schedule with a makespan smaller than or equal to a certain threshold?

To polynomially reduce the partition problem to $P2||C_{max}$, we first show how to transform instances of the partition problem to instances of $P2||C_{max}$. We take the positive integers $s_1, \ldots, s_n$ from the partition problem and construct an instance of $P2||C_{max}$ with n jobs having processing times $p_j = s_j$. Moreover, we set the threshold in $P2||C_{max}$ equal to the value b from (A.1).

Now, suppose we have a 'yes' instance of the partition problem. We need to show that, via the transformation above, we then also have a 'yes' instance of $P2||C_{max}$. Since we have a 'yes' instance of the partition problem, we know that there exists a subset $J \subset I$ with

$$\sum_{j \in J} s_j = b.$$

We schedule all jobs from J on M1 and all jobs from $I \backslash J$ on M2. This gives

$$C_{max} = \max\left\{ \sum_{j \in J} p_j, \sum_{j \in I \backslash J} p_j \right\} = \max\{b, b\} = b,$$

so we indeed have a 'yes' instance of $P2||C_{max}$.

Finally, suppose we have a 'yes' instance of $P2||C_{max}$. We need to show that, via the transformation above, we also have a 'yes' instance of the partition problem. We define J as all jobs on M1. Since we know that $C_{max} \leq b$, we have

$$\sum_{j \in J} s_j \leq b.$$

Similarly, we get

$$\sum_{j \in I \backslash J} s_j \leq b$$

since these jobs are scheduled on M2. By definition (see (A.1)), we have

$$\sum_{j \in I} s_j = 2b$$

which, in combination with the above, leads to the conclusion that

$$\sum_{j \in J} s_j = b,$$

so we indeed have a 'yes' instance of the partition problem. Concluding, we have shown that partition $\propto P2||C_{max}$.

Definition A.5 ($\mathcal{NP}$-completeness). *A decision problem $R \in \mathcal{NP}$ is called $\mathcal{NP}$-complete if all problems from the class $\mathcal{NP}$ polynomially reduce to R.*

Definition A.6 ($\mathcal{NP}$-hardness). *An optimization problem is called $\mathcal{NP}$-hard if the corresponding decision problem is $\mathcal{NP}$-complete.*

To show that an optimization problem is $\mathcal{NP}$-hard, one thus has to show that the corresponding decision problem is $\mathcal{NP}$-complete. To do so, one first needs to show that the decision problem belongs to the class $\mathcal{NP}$, by showing that there exists a concise certificate that can be verified in polynomial time (see Definition A.3). Subsequently, one needs to show that the decision problem polynomially reduces to a decision problem that is already known to be $\mathcal{NP}$-complete.

Note that solving one $\mathcal{NP}$-complete problem polynomially would imply that $\mathcal{P} = \mathcal{NP}$.

Appendix B

Useful Formulas for the Normal Distribution

This appendix lists some useful facts for the standard normal distribution. The formulas for the standard normal density function and the standard normal distribution function are

$$\phi(x) = \frac{1}{\sqrt{2\pi}} e^{-\frac{1}{2}x^2} \quad \text{and} \quad \Phi(x) = \frac{1}{\sqrt{2\pi}} \int_{-\infty}^{x} e^{-\frac{1}{2}y^2} \, dy.$$

The calculation of $\Phi(x)$ seems complicated but does not give any problems in practice. The function can be calculated to every desired degree of accuracy using a rational approximation (a rational function is the quotient of two polynomials). An approximation that is sufficiently accurate for practical purposes is the following:

$$\Phi(x) \approx 1 - \frac{1}{2}(1 + d_1 x + d_2 x^2 + d_3 x^3 + d_4 x^4 + d_5 x^5 + d_6 x^6)^{-16}, \quad x \geq 0,$$

where

$$d_1 = 0.0498673470 \quad d_4 = 0.0000380036$$
$$d_2 = 0.0211410061 \quad d_5 = 0.0000488906$$
$$d_3 = 0.0032776263 \quad d_6 = 0.0000053830.$$

The symbol $a \approx b$ means that a is approximately equal to b. The absolute error of the approximation is less than 1.5×10^{-7}. The formula above can be applied directly only for $x \geq 0$. It can also be applied to calculate $\Phi(x)$ for $x < 0$ by using the relation $\Phi(x) = 1 - \Phi(-x)$ for $x < 0$.

A common problem is that of solving the equation

$$\Phi(k) = \alpha$$

with α a given number between 0 and 1. The percentile k can be found through a straightforward calculation by using a rational approximation for the inverse function $\Phi^{-1}(\alpha)$. An approximation that is useful in practice is

$$k \approx w - \frac{c_0 + c_1 w + c_2 w^2}{1 + d_1 w + d_2 w^2 + d_3 w^3}, \quad 0.5 \leq \alpha < 1,$$

where

$$w = \sqrt{-\ln(1-\alpha)^2}$$

$$c_0 = 2.515517 \qquad\qquad d_1 = 1.432788$$
$$c_1 = 0.802853 \qquad\qquad d_2 = 0.189269$$
$$c_2 = 0.010328 \qquad\qquad d_3 = 0.001308.$$

The absolute error of this approximation is less than 4.5×10^{-4}. If $0 < \alpha < 0.5$, then the solution for $\Phi(k) = \alpha$ can be found by applying the formula above for the inverse to the equation $\Phi(-k) = 1 - \alpha$.

The *normal loss function* $I(z)$ is defined by

$$I(z) = \frac{1}{\sqrt{2\pi}} \int_z^\infty (x - z)e^{-\frac{1}{2}x^2} \, dx.$$

This integral plays an important role in, for example, inventory management theory. The function $I(z)$ can be calculated using the relation

$$I(z) = \phi(z) - z\{1 - \Phi(z)\} \quad \text{for every } z.$$

A basic problem in inventory management is that of solving the equation

$$I(k) = \beta,$$

where β is a given number between 0 and 1. In real-world situations, the solution k satisfies $-4 \le k \le 4$. The solution k can be obtained through a straightforward calculation by using the rational approximation

$$k \approx \frac{a_0 + a_1v + a_2v^2 + a_3v^3}{b_0 + b_1v + b_2v^2 + b_3v^3 + b_4v^4},$$

where

$$v = \sqrt{\ln(25/\beta^2)} \quad b_0 = 1.0000000$$
$$a_0 = -5.3925569 \quad b_1 = -7.2496485 \times 10^{-1}$$
$$a_1 = 5.6211054 \quad b_2 = 5.07326622 \times 10^{-1}$$
$$a_2 = -3.8836830 \quad b_3 = 6.69136868 \times 10^{-2}$$
$$a_3 = 1.0897299 \quad b_4 = -3.29129114 \times 10^{-3}.$$

The absolute error of this approximation is less than 2×10^{-4} provided that $-4 \le k \le 4$.

Appendix C

The Poisson Process

A stochastic process that is inextricably connected with the Poisson distribution is the Poisson process. This is a counting process that counts the number of occurrences of a particular event over time. The event can be of all kinds: the arrival of customers at a bank, the occurrence of severe earthquakes, the receipt of calls at a telephone exchange, the occurrence of outages at a power plant, the receipt of phone calls at a general emergency number, and so on. In the remainder of this section, we use the terminology of customer arrivals for the occurrence of events over time. When is a counting process a Poisson process? For this, we must assume that the population of potential customers is *infinitely large* and that customers behave independently of one another.

Definition C.1 (Poisson process). *Let the random variable $N(t)$ be the number of customers arriving at a service station up to time t. The counting process $\{N(t),\ t \geq 0\}$ is called a Poisson process with rate λ, $\lambda > 0$, if it has the following properties:*

(a) $N(0) = 0$.
(b) The process has independent increments, that is, the numbers of arrivals in disjoint time intervals are independent of each other.
(c) The number of arrivals in any interval of length t is Poisson distributed with mean λt. That is, for all $s, t \geq 0$

$$\mathbb{P}(N(t+s) - N(s) = n) = e^{-\lambda t}\frac{(\lambda t)^n}{n!}, \quad n = 0, 1, \ldots.$$

Property (c) implies that

$$\lambda = \text{expected value of the number of arrivals}$$
$$\text{during a time interval of unit length.}$$

The number λ is called the *arrival rate* of the Poisson process.

The Poisson process provides a good model description in many real-world situations. The explanation for this lies in the following, roughly formulated, result. Suppose that at the micro level, there are *numerous* stochastic processes that are *independent* of one another and that each generate arrivals *sparsely*. Then one can show that at the macro level, the resultant of all these processes is approximately a Poisson process. This explains, for example, why the arrival process of customers at a post office can often be described by a Poisson process. The population of potential customers is very large: every individual behaves according to a specific pattern, but the combination of all these typically independent patterns leads to an unpredictable whole that can be described by a Poisson process.

Property (b), which assumes the absence of aftereffects in the arrival process, is the most characteristic property of the Poisson process and can only be satisfied if the customer population is infinitely large. Intuitively, it should be clear that this property implies that for every given time t, the probability distribution of the time until the next customer arrives does not depend on how long ago the last customer came in. This property is known as the *memoryless property* of the Poisson process. The Poisson process with the memoryless property is the complete opposite of a deterministic arrival process in which customers arrive at fixed intervals.

In a Poisson arrival process, the number of arrivals during a given time interval has a *discrete* distribution, namely the Poisson distribution, while the time between two consecutive arrivals has a *continuous* distribution, namely the exponential distribution.[1] This can be seen as follows:

$\mathbb{P}$(the time between two consecutive arrivals is greater than y)

$= \mathbb{P}$(there are no arrivals during an interval of length y)

$= e^{-\lambda y}$ for every $y > 0$.

In a Poisson arrival process with rate λ, the stochastic time T between two consecutive arrivals is continuously distributed with distribution function

$$\mathbb{P}(T \leq y) = 1 - e^{-\lambda y} \quad \text{for } y \geq 0.$$

This continuous distribution function has the exponential density function: $\lambda e^{-\lambda y}$. The expected value of the exponential distribution is $1/\lambda$. An expected value of $1/\lambda$ for the interarrival time is consistent with the significance of the arrival rate λ. Because of property (b) of the Poisson process, it should not come as a surprise that the interarrival times of consecutive customers are independent of one another.

Example C.1. Shared taxis are waiting at a train station. A shared taxi leaves as soon as it has four passengers or 10 minutes have passed since the first passenger got in. Assume that the passenger arrivals follow a Poisson process with an average of one passenger every 3 minutes.

1. What is the probability that the first passenger to get in must wait 10 minutes until the taxi leaves?

---

[1]The Poisson process (with its arrival rate) and the Poisson distribution (with its expected value) should not be mixed up. The Poisson process is a stochastic process, that is, a collection of random variables indexed by a time variable. Every individual variable represents the cumulative number of arrivals up to a given time t and is Poisson distributed with expected value λt if λ is the Poisson process's arrival rate.

2. The first passenger to get in has already been in the taxi for 5 minutes, during which time two more passenger have gotten in. What is the probability that the taxi does not leave for 5 more minutes?

The answer to Question 1 is based on the observation that the first passenger must wait 10 minutes only if the number of passengers who arrive in the next 10 minutes is less than three. If we choose a minute as unit of time, then the time the first passenger must wait is Poisson distributed with an expected value of 10λ with $\lambda = \frac{1}{3}$. Hence,

$\mathbb{P}$(the first passenger waits 10 minutes)

$= \mathbb{P}$(zero, one, or two passengers get in during the next 10 minutes)

$$= e^{-10\lambda} + e^{-10\lambda}\frac{10\lambda}{1!} + e^{-10\lambda}\frac{(10\lambda)^2}{2!} = 0.3528.$$

The answer to Question 2 is based on the memoryless property of the Poisson process. The waiting time until the arrival of the next passenger is exponentially distributed with expected value three minutes, regardless of how long ago the first passenger got into the taxi. The probability that the first passenger must wait another five minutes is equal to the probability $e^{-5\lambda} = 0.1889$ that no new passenger arrives during that time.

Example C.2. A passenger arrives at a bus stop at a random time between 17:00 and 17:15 to take the first bus home. He can take either bus 1 or bus 3. Bus 1 leaves on the hour and every 15 minutes, while the departure times of bus 3 are Poisson distributed with an average of one bus every 15 minutes. What is the probability that the passenger takes bus 1?

Under the condition that, according to the schedule, the passenger arrives x minutes before bus 1 leaves, the probability of the passenger taking bus 1 is equal to the probability of no bus 3 leaving in the next x minutes. This probability is equal to $e^{-\lambda x}$, regardless of how long ago the last bus 3 left. Here, $\lambda = 1/15$. Taking the average over the uniformly distributed time before bus 1 leaves gives

$$\mathbb{P}(\text{the passenger takes bus 1}) = \frac{1}{15}\int_0^{15} e^{-\lambda x}\,dx = 1 - \frac{1}{e},$$

which is at least $1/2$.

Alternative Definitions of the Poisson Process

We have seen that properties of the Poisson process imply that the interarrival times have an exponential distribution. This property of the interarrival times turns out to also *define* a Poisson process. The following is an alternative definition of the Poisson process.

Definition C.2.   *Let $T_1, T_2, \ldots$ be the times between the successive arrivals of customers. For any $t > 0$, define $N(t)$ as the largest nonnegative integer n such that $\sum_{k=0}^{n} T_k \leq t$ with $T_0 = 0$. If $T_1, T_2, \ldots$ are independent random variables each having an exponential distribution with mean $\frac{1}{\lambda}$, then the counting process $\{N(t), t \geq 0\}$ is a Poisson process with arrival rate λ.*

Definition C.2 emphasizes that the Poisson process has both a discrete component (the Poisson distribution of the number of arrivals) and a continuous component (the exponential distribution of the interarrival times). This is also illustrated by the following situation. It is well known from physics experiments that the emission of alpha particles from radioactive material can be described by a Poisson process: the number of particles that are emitted during a given time interval is Poisson distributed, and the times between the emission of consecutive particles are independent and exponentially distributed.

   From Definition C.2, we derive another alternative definition that is often the most useful one for applications. To do this, we need the concept of the "failure rate" of a positive random variable. The term is easiest to explain if we assume that the positive random variable X represents the lifetime of a machine, where we, moreover, assume that X is exponentially distributed with expected value $1/\mu$. Then, regardless of the age x of the machine, we have

$\mathbb{P}(\text{a machine of age } x \text{ fails in the upcoming short interval of length } \Delta x)$

$\approx \mu \Delta x + o(\Delta x) \qquad \text{as } \Delta x \to 0,$

where $o(\Delta x)$ is the general notation for a rest term that is negligibly small compared to Δx when Δx itself is very small, that is, $o(\Delta x)/\Delta x \to 0$ as $\Delta x \to 0$C. A machine with an exponential lifespan therefore has a *constant* failure rate μ ("used is as good as new"). The derivation of this result is as follows. Since $\mathbb{P}(X \leq x) = 1 - e^{-\mu x}$ for $x > 0$, we have

$$\mathbb{P}(X \leq x + \Delta x \mid X > x) = \frac{\mathbb{P}(x < X \leq x + \Delta x)}{\mathbb{P}(X > x)}$$

$$= \frac{e^{-\mu x} - e^{-\mu(x+\Delta x)}}{1 - e^{-\mu x}} = 1 - e^{-\mu \Delta x}.$$

It follows from the Taylor series expansion $e^{-u} = 1 - u + u^2/2! - u^3/3! + \cdots$ that

$$1 - e^{-\mu \Delta x} = \mu \Delta x - \frac{(\mu \Delta x)^2}{2!} + \frac{(\mu \Delta x)^3}{3!} - \cdots = \mu \Delta x + o(\Delta x).$$

Based on the above, we can now give the second alternative definition of the Poisson process.

Definition C.3. *The process $\{N(t),\ t \geq 0\}$ counting the number of arriving customers is called a* Poisson process *with rate λ, $\lambda > 0$, if it has the following properties:*

(a) $N(0) = 0$.
(b) The numbers of arrivals in disjoint intervals are independent of one another.
(c) For every $t \geq 0$, we have $\mathbb{P}(N(t + \Delta t) - N(t) = 1) = \lambda \Delta t + o(\Delta t)$ as $\Delta t \to 0$;
(d) for every $t \geq 0$, we have $\mathbb{P}(N(t + \Delta t) - N(t) \geq 2) = o(\Delta t)$ as $\Delta t \to 0$.

Conditions (c) and (d) state that in a very short time interval of length Δt, the number of customer arrivals is exactly one with probability $\lambda \Delta t$, zero with probability $1 - \lambda \Delta t$, and two or more with a probability that is negligible compared to Δt. This definition has the additional advantage that it can be directly extended to the situation where the customers' arrival rate is time dependent.

Merging and Splitting Poisson Processes

In applications, one often needs to merge two Poisson processes or split (or thin) a Poisson process. We explain the merging of two independent Poisson processes with the following example. Suppose that a call center handles information requests for two completely different companies. Calls for company A come in following a Poisson process with rate λ_A, and, independently of that, calls for company B come in following a Poisson process with rate λ_B. Merging these two processes gives the combined arrival process of the requests. One can prove that the merged process is a Poisson process with rate $\lambda_A + \lambda_B$. For example, this means that the probability that no call comes in during the upcoming t units of time is equal to $e^{-t(\lambda_A + \lambda_B)}$. The merged Poisson arrival process moreover has the property that a random call has probability $\lambda_A/(\lambda_A + \lambda_B)$ of being for company A and $\lambda_B/(\lambda_A + \lambda_B)$ of being for company B. For example, suppose that the expected value of the call handling time is m_A for company A and m_B for company B; then the average handling time over all calls is equal to $\frac{\lambda_A}{\lambda_A + \lambda_B} m_A + \frac{\lambda_B}{\lambda_A + \lambda_B} m_B$.[2]

As an example of splitting a Poisson process, consider the situation where the occurrence of earthquakes in a given area is described by a Poisson process with rate λ. Suppose that the magnitudes of the earthquakes are independent of one another, where the earthquake is classified as major with probability p and minor with probability $1 - p$. Then one can prove that, over time, the occurrence of major earthquakes is described by a Poisson process with rate λp and that of minor earthquakes as a Poisson process with rate $\lambda(1 - p)$. This result is not surprising. What is surprising is that the two Poisson processes that describe major and minor earthquakes are *independent* of each other. Another illustration of this result is given in the following example.

---

[2]By the law of total expectation, we have $\mathbb{E}$[call handling time]=$\mathbb{E}$[call handling time for A | call is for A]$\mathbb{P}$(call is for A)+$\mathbb{E}$[call handling time for B | call is for B]$\mathbb{P}$(call is for B).

Example C.3. A piece of radioactive material emits α-particles following a Poisson process with an average rate of 0.84 particles per second. A Geiger counter records every emitted particle with probability 0.95, each independently of the other particles. In a 10-second period, 12 particles have been registered. What is the probability that more than 15 particles were emitted during that period?

The process that describes the emission of unregistered particles is a Poisson process with rate $0.84 \times 0.05 = 0.0402$ particles per second, and this process is independent of the process that describes the emission of registered particles. So, the number of particles missed by the Geiger counter during the 10-second period is Poisson distributed with expected value $10 \times 0.0402 = 0.402$, regardless of the number of emitted particles that were registered. The desired probability is the probability that, during a given period of 10 seconds, more than three unregistered particles were emitted and is therefore equal to $1 - \sum_{j=0}^{3} e^{-0.402}(0.402)^j/j! = 0.00079$.

Clustering of Arrival Times

A Poisson process has the characteristic property that the arrival times of customers tend to cluster. This is clearly shown in Figure C.1. This figure gives the simulated arrival times in the time interval $(0, 45)$ for a Poisson process with arrival rate $\lambda = 1$. The mathematical explanation of the phenomenon of clustered arrivals is as follows. The time between two consecutive arrivals has probability density $f(y) = \lambda e^{-\lambda y}$. The probability that the interarrival time takes on a value between y and $y + \Delta y$ is approximately equal to $f(y)\Delta y$ for Δy small. The density $f(y) = \lambda e^{-\lambda y}$ is the greatest for $y = 0$ and is decreasing from $y = 0$ on. This means that *short* interarrival times occur relatively frequently; that is, the arrival times cluster. Another explanation lies in the relationship between the Poisson process and the uniform distribution: given that a Poisson arrival process has exactly n arrivals in a given time interval $(0, T)$, the joint distribution of the n arrival times is the same as the joint distribution of the order statistics of n independent random variables that are uniformly distributed on $(0, T)$. This is a very useful result in queueing theory. Numbers drawn randomly from an interval also tend to cluster. If the numbers were evenly distributed over the interval, they would not be randomly drawn.

The phenomenon of clustered arrival times also sheds a different light on a series of murders in Florida that caused a great deal of commotion at the time. In the period between October 1992 and October 1993, nine foreign tourists were murdered in Florida. The murders were attributed to the fact that foreign tourists were recognizable as such by their rental cars. This may well explain the explosion of killings, but it is also quite possible that this is an unexceptional probability event when one considers a longer period of time. Suppose that on any given day, the probability that a foreign tourist is killed somewhere in Florida is 1%. The stochastic process that describes the occurrence of murders of foreign tourists in Florida over time can reasonably well be modeled as a Poisson process with rate 3.65 murders per year. What is the probability that within, say, 10 years, there is a

Fig. C.1   Clustering property of the Poisson process.

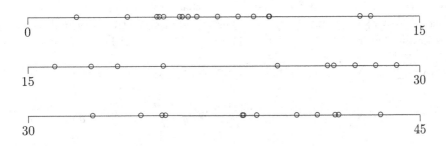

12-month period with 9 or more such murders? There is no simple formula for this probability, but it can be easily found using a computer simulation. The probability is about 36%. Over a period of 20 years, the probability of such a series of murders even increases to around 60%.

A similar event that caused much commotion in Amsterdam is the following. In the first five months of the year 2000, seven fatal accidents occurred in Amsterdam involving people who were hit by a tram while distracted. In previous years, an average of 3.7 people a year died as a result of this type of accident. Is the high number of fatal tram accidents bad luck or something else? A reasonable modeling assumption is that, over time, fatal tram accidents occur following a Poisson process. Here too, the Poisson process's clustering property explains the many accidents in a short period of time. If we use a computer simulation to estimate the probability that within ten years, there is a continuous period of five months with at least seven fatal tram accidents, we find a value of about 10.5%.

The Inspection Paradox

Assume that buses arrive at a bus stop with exponentially distributed interarrival times with mean $1/\lambda$. Let $N(t)$ be the number of buses that have arrived by time t, then $N(t)$ is a Poisson process with rate λ. We are interested in the probability distribution of the length of the interval between two consecutive bus arrivals. To this end, we fix some time t and observe the total interarrival time between the last bus that arrived before time t and the first bus that arrives after time t. It seems reasonable that the length of the interarrival interval covering the inspection time t is exponentially distributed with mean $1/\lambda$. However, it turns out that the interarrival interval covering the inspection time t tends to have a larger length than the ordinary interarrival interval. This is the *inspection paradox*.

A mathematical proof of the inspection paradox is as follows. Let $T_1, T_2, \ldots$ be the interarrival times in the Poisson arrival process of buses. Define $\delta(t)$ as the time elapsed since the last arrival before time t and $\gamma(t)$ as the waiting time from time t

until the next arrival.[3] By the memoryless property of the Poisson process, $\gamma(t)$ and $\delta(t)$ are independent, and $\gamma(t)$ is exponentially distributed with mean $1/\lambda$. To obtain the distribution of $\delta(t)$, first observe that $\mathbb{P}(\delta(t) = t) = \mathbb{P}(T_1 > t) = e^{-\lambda t}$ since $\{\delta(t) = t\}$ may only occur if no bus has arrived by time t. Conditioning on the Erlang density $\lambda^n y^{n-1} e^{-\lambda y}/(n-1)!$ of $X_1 + \cdots + X_n$,[4] For $0 \leq x < t$ we have

$$\mathbb{P}(\delta(t) \leq x) = \sum_{n=1}^{\infty} \mathbb{P}(t - x \leq X_1 + \cdots + X_n \leq t < X_1 + \cdots + X_{n+1})$$

$$= \sum_{n=1}^{\infty} \int_{t-x}^{t} e^{-\lambda(t-y)} \frac{\lambda^n y^{n-1}}{(n-1)!} e^{-\lambda y}\, dy = \int_{t-x}^{t} \lambda e^{-\lambda(t-y)}\, dy = 1 - e^{-\lambda x}.$$

Thus $\mathbb{E}[\delta(t)] = te^{-\lambda t} + \int_0^t \lambda x e^{-\lambda x}\, dx = \frac{1}{\lambda}(1 - e^{-\lambda t})$. This leads to

$$\mathbb{E}[\delta(t) + \gamma(t)] = \frac{2}{\lambda} - \frac{e^{-\lambda t}}{\lambda} \qquad \text{for any } t > 0,$$

proving the inspection paradox saying that for t large the particular interarrival time covering the fixed epoch t is in expectation about twice as large as the interarrival time between two consecutive arrivals (more generally, $\lim_{t \to \infty} \mathbb{E}[\delta(t) + \gamma(t)] = (1 + c_X^2)\mathbb{E}[T]$ when the interarrival time T has a continuous distribution with coefficient of variation $c_T = \sigma(T)/\mathbb{E}[T]$).

The inspection paradox can be explained by the fact that you have a higher probability of arriving at the bus stop during a long interarrival time than during a short one. The inspection paradox is a common source of confusion in many situations of public service in transportation.

Nonhomogeneous Poisson Process

In many real-world situations, the customer arrival rate is time dependent. The alternative definition of the Poisson process in which arrivals are described by considering very short time intervals can directly be extended to the case of time-dependent arrivals. A counting process $\{N(t), t \geq 0\}$, where $N(t)$ indicates the total number of arrivals up to time t, is called a *nonhomogeneous Poisson process* with arrival rate function $\lambda(t)$ if

(a) the numbers of arrivals in disjoint intervals are independent of one another;
(b) for every $t \geq 0$, we have $\mathbb{P}(N(t + \Delta t) - N(t) = 1) = \lambda(t)\Delta t + o(\Delta t)$ as $\Delta t \to 0$;
(c) for every $t \geq 0$, we have $\mathbb{P}(N(t + \Delta t) - N(t) \geq 2) = o(\Delta t)$ as $\Delta t \to 0$.

Here, $o(h)$ is the general notation for a function $f(h)$ with the property that $f(h)$ goes to zero faster than h itself as h goes to zero, that is, $\lim_{h \to 0} f(h)/h = 0$ as $h \to 0$ (for example, $f(h) = h^2$ is an $o(h)$-function).

---

[3]In literature, $\delta(t)$ is referred to as age, and $\gamma(t)$ as excess life.
[4]The sum $T_1 + \cdots + T_n$ is a sum of independent exponentially distributed random variables T_i, $i = 1, \ldots, n$. This sum has an Erlang distribution (also known as gamma distribution) with n phases and rates λ per phase.

For a nonhomogeneous Poisson arrival process $\{N((t), t \geq 0\}$, one can show that for all $t_1, t_2 \geq 0$ with $t_2 > t_1$, the total number of arrivals during the time interval (t_1, t_2) has a *Poisson distribution* with expected value $M(t_2) - M(t_1)$, where

$$M(t) = \int_0^t \lambda(x)dx \quad \text{for } t \geq 0.$$

The function $M(t)$ therefore gives the expected number of arrivals during the interval $(0, t)$.

Example C.4. At an ATM (cash dispenser), customer arrivals follow a nonhomogeneous Poisson process. If we take 6:00 as $t = 0$ and an hour as unit of time, then between 6:00 and 12:00, customers arrive with rate $\lambda(t) = 7.5t$ for $0 \leq t \leq 6$, and between 12:00 and 6:00 the next day, they arrive with rate $\lambda(t) = 45 - 2.5(t-6)$ for $6 \leq t \leq 24$. What is the probability distribution of the total number of persons, per 24 hours, that withdraw cash? The answer is that this number follows a Poisson distribution with expected value

$$\int_0^{24} \lambda(t)dt = \int_0^6 7.5t\,dt + \int_6^{24} \left(45 - 2.5(t-6)\right)dt = 540.$$

Suppose that every customer withdraws 50 euros with probability $\frac{3}{4}$ and 100 euros with probability $\frac{1}{4}$. What is the expected value of the total amount withdrawn per 24 hours? We call a customer who withdraws 50 euros a type 1 customer, and one who withdraws 100 euros a type 2 customer. The type 1 customers arrive following a nonhomogeneous Poisson process with rate $\frac{3}{4}\lambda(t)$, and the type 2 customers arrive following a nonhomogeneous Poisson process with rate $\frac{1}{4}\lambda(t)$ (why?). The total number of times per 24 hours that 50 euros is withdrawn is then Poisson distributed with expected value $\frac{3}{4} \times 540 = 405$, and the total number of times that 100 euros is withdrawn is Poisson distributed with expected value $\frac{1}{4} \times 540 = 135$. The total expected amount that is withdrawn per 24 hours therefore has expected value $\mu = 50 \times 405 + 100 \times 135 = 33\,750$ euros.

If the arrival rate function $\lambda(t)$ is bounded for $t \geq 0$, an interesting alternative description of the nonhomogeneous Poisson arrival process can be given. Choose a fixed number λ with $\lambda \geq \lambda(t)$ for all $t \geq 0$. Now, construct a counting process as follows:

1. Arrivals follow a Poisson process with rate λ.
2. An arrival at time s is accepted with probability $\lambda(s)/\lambda$ and rejected otherwise.

The counting process $\{N(t), t \geq 0\}$ with $N(t)$ the number of accepted arrivals up to time t is then a nonhomogeneous Poisson process with rate function $\lambda(t)$. This can be seen as follows. The probability that the Poisson process generates an arrival during the period $(t, t+\Delta t)$ is $\lambda\Delta t + o(\Delta t)$ for Δt small, and the probability that this arrival is accepted is then $\lambda(t)/\lambda$; in other words, the probability that an accepted arrival comes in during the period $(t, t+\Delta t)$ is equal to $(\lambda\Delta t) \times (\lambda(t)/\lambda) + o(\Delta t) = \lambda(t)\Delta t + o(\Delta t)$. This construction of a nonhomogeneous Poisson process is useful for simulating arrivals in such an arrival process.

Poisson Process in Higher Dimensions

The Poisson process on the real line can be generalized to a Poisson process in the plane or in higher dimensions. This generalization gives a probabilistic description of randomly occurring points in the plane or in higher dimensions. For clarity, we limit ourselves to a definition of the Poisson process in the plane. This process is useful, among other things, when describing the positions of mobile phones in a particular region. A stochastic process consisting of randomly occurring points (called entities) in the plane is called a *two-dimensional Poisson process* with parameter λ if

(a) the number of entities in a given bounded region A has a Poisson distribution with expected value $\lambda(A) = \lambda \times area(A)$;
(b) the numbers of entities in disjoint bounded regions are independent of one another.

A direct consequence of this definition is that if Q is a given point in the plane, the distance R to the nearest entity has probability density $2\lambda\pi r e^{-\lambda\pi r^2}$. After all, the probability that there is no entity within a circle around Q with radius r is $\mathbb{P}(R > r) = e^{-\lambda\pi r^2}$.

Another interesting result is the following. Under the condition that a given bounded region A contains r Poisson entities, the positions of these r entities are uniformly distributed over the region A and therefore have the same probability distribution as r points randomly chosen independently of one another in the region A. This result also provides a method to simulate Poisson entities in a given bounded region A:

1. Do a random draw X from a Poisson distribution with expected value $\lambda(A) = \lambda \times area(A)$.
2. If $X = r$, generate r independent random points in the region A.

The *hit-or-miss* method is typically used to generate random points in a region A: circumscribe a rectangle around A, and in that rectangle, randomly generate points until r points fall into the region A. This approach can be directly extended to a nonhomogeneous Poisson process in the plane.[5]

Compound Poisson Process

The compound Poisson process is an extension of the Poisson process and has applications in, among other things, risk theory and inventory management. What is the probability distribution of the total amount claimed from an insurance company during a given time period if claims arrive following a Poisson process and the size of each claim is a random variable? What is the probability distribution

---

[5]In a nonhomogeneous Poisson process with rate function $\lambda(x, y)$, change $\lambda(A)$ in Step 1 to $\lambda(A) = \iint_A \lambda(x, y)\, dx\, dy$.

of the total demand for a product during a given time period if customer arrivals follow a Poisson process and the demand sizes are random variables? Suppose that $\{N(t), t \geq 0\}$ is a Poisson process with parameter λ and $\{Y_1, Y_2, \ldots\}$ is a sequence of random variables that do not depend on the Poisson process. For every $t \geq 0$, define the random variable $X(t)$ by

$$X(t) = \sum_{i=1}^{N(t)} Y_i.$$

The stochastic process $\{X(t)\}$ is called a *compound Poisson process*. The following basic results hold for this process. The expected value and standard deviation of $X(t)$ are given by

$$\mathbb{E}[X(t)] = \lambda t \mathbb{E}[Y_1] \quad \text{and} \quad \text{var}[X(t)] = \lambda t \mathbb{E}[Y_1^2].$$

If the Y_i have a discrete probability distribution $\{a_j, j = 0, 1, \ldots\}$, then for fixed $t = t_0$, the probabilities $p_k = \mathbb{P}(X(t_0) = k)$ can be calculated recursively from

$$p_0 = e^{-\lambda t_0 (1 - a_0)} \quad \text{and} \quad p_k = \frac{\lambda t_0}{k} \sum_{j=0}^{k-1} (k - j) a_{k-j} p_j \quad \text{for } k = 1, 2, \ldots.$$

This scheme is called Panjer's recursion in actuarial sciences. Alternatively, the p_k can be efficiently calculated by applying the Fast Fourier Transform method to the generating function of the p_k. This generating function is explicitly given by $\sum_{k=0}^{\infty} p_k z^k = e^{-\lambda t_0 [1 - A(z)]}$, where $A(z) = \sum_{j=0}^{\infty} a_j z^j$.

Reference

1. H. C. Tijms, *A First Course in Stochastic Models*, John Wiley & Sons Ltd., Chichester, England, 2003.

Appendix D

Answers to Selected Exercises

Chapter 1

1.1 The maximum profit per week is 198 euros and is obtained by using 25 liters each of ingredients A and B for Spicy Gonzales and 15 liters of ingredient A and 5 liters of ingredient B for Cool Gringo.

1.4 Give out 35, 77, and 98 3500-euro loans in weeks 1, 2, and 3, respectively, and 144 and 102 15000-euro loans in weeks 1 and 2, respectively. The solution is found by rounding down the continuous LP solution.

1.5 Take 10 standard rolls and cut each up into three rolls of width 40 cm, and take 20 standard rolls and cut each up into one roll of width 40 cm and three rolls of width 25 cm. The length of the trim loss is 100 cm.

1.7 The net production of mozzarella for the retailers is 1050 kg and that for the pizzeria is 200 kg, while the net production of ricotta is 450 kg for the retailers and 300 kg for the feed company. The maximum revenue is 5649.33 euros.

1.8 The optimal solution (after rounding off) is to sell 323 packages of 250 g of Jake mix, 507 packages of 250 g of Party mix, 619 packages of 250 g of Lulu mix, and 200 packages of 250 g of Elite mix. The net profit is 9115 euros.

1.9 (a) $y = 0.0744x^2 - 0.0357x + 0.413$. (b) $y = 0.51 + 5.7\log(x)$.

1.11 (a) Bet on red with probability $\frac{4}{12}$, on white with probability $\frac{5}{12}$, and on blue with probability $\frac{3}{12}$. (b) The optimal strategy for Fatima is to choose the number 1 with probability 0.2901, the number 2 with probability 0.2290, and the number 3 with probability 0.4809. (c) If player A chooses one of the numbers 11, 17, 28, 44, and 67 with probability 0.2 each, then player A has probability at least 0.6 of winning regardless of player B's strategy.

1.12 In units of one thousand euros, set aside 15 054.2600 at the end of year 0 and invest in bonds of type 1 for a nominal value of 5490.8997 (this costs 0.985×5490.8997 at the end of year 0), in bonds of type 2 for a nominal value of 5839.8997, and in bonds of type 3 for a nominal value of 2325.5814; at the beginning of year 1, put 768.0574 into the savings account.

1.15 (a) 406.25 euros. (b) The selling price must increase by at least 46.875 euros.

1.17 (a) 11.5%. (b) 1%. (c) 175 euros. (d) The set of feasible solutions changes, and in fact increases. The problem must be solved again.

1.19 (a) The net profit of the company increases by 509.09 euros for each ton that is sold over the first 100 (up to 50 tons more). (b) The optimal values of the decisions variables remain the same, while the net profit decreases by 5454.54 euros. (c) If the right-hand-side coefficient has the new value 2.95, then the net profit decreases by exactly 181.82 euros, and if the right-hand-side coefficient has the new value 2.2, the net profit decreases by at least 2909.09 euros. (d) The net profit decrease by exactly 136.36 euros. (e) It follows from the 100% rule that the optimal values of the decision variables do not change. The net profit does decrease, by 2000 euros. (f) The "economic value" of alloy 5 is 309.09 euros. If the cost of alloy 5 is more than $1000 - 309.09 = 690.91$ euros, it is not worthwhile to use it.

1.20 (a) 120 euros. (b) $125 + 500 = 625$ euros. (c) 750 euros. (d) It follows from the 100% rule that the optimal values of the decision variables do not change. (e) The economic cost per hectare of hemp is $225(< 260)$ euros, so it *may* be worthwhile to cultivate hemp.

1.22 (a) Let x_{ij} be the amount of coal (in thousands of tons) supplied by coal company i to location j. The optimal supply schedule is $x_{12} = 150$, $x_{13} = 25$, $x_{14} = 100$, $x_{21} = 75$, $x_{23} = 225$, $x_{34} = 200$. The minimum cost is 12 400 euros. (b) The costs increase by $20 \times 15\,000 = 300\,000$ euros. (c) The optimal values of the decision variables do not change. The costs, however, do increase by 775 000 euros.

1.23 (a) The maximum net profit is 248 250 euros. (b) Factory 3, with an increase in the net profit of 7000 euros. (c) 4500 euros. (d) The 100% rule gives $\frac{2}{10} + \frac{2}{5} + \frac{2}{5} \leq 1$, and therefore the optimal values of the decision variables do not change; the net profit does decrease, by 110 euros.

1.24 (a) Make 15 units of product 1 at the high price and 10 units at the low price. For product 2, these numbers are 12 and 9. The maximum profit is 633 euros. (b) No, the profit would have increased by at most 180 euros. (c) The 100% rule says that the optimal production plan remains the same as long as the reduction does not exceed 4.7%.

1.26 (a) The participation in projects 1 and 2 is 71.43% and 63.72%, respectively; there is no participation in project 3. The maximum profit is 9.5815, with unit 100 thousand euros. (b) For the periods 1 and 4, it is 3.5%; for period 2, it is 27.19%; for period 3, it is 25.9%; and for periods 5 and 6, it is 3%. (c) It is now optimal to participate in project 1 for 71.21% and in project 2 for 66.07%, and not participate in project 3; one-year loans are taken out in periods 1 and 2.

Chapter 2

2.1 At most 8 matches can take place, namely matches 1 through 6 and matches 9 and 10.

2.2 (a) The maximum profit is 4320 euros and is obtained by carrying 2 units of cargo 1, 2 units of cargo 5, and 3 units of cargo 6. (b) The maximum profit is 4260 euros and is obtained by carrying 11 units of cargo 1 and 1 unit of cargo 7.

2.4 (a) The minimum total travel time is 57 minutes and is obtained by sending 7 officers from the Marnixstraat to the Prinsengracht and 3 from the Marnixstraat to the Keizersgracht, sending 5 officers from the Nieuwe Zijds to the Spui, and sending 1 officer from the Warmoesstraat to the Keizersgracht and 7 from the Warmoesstraat to the Spui. (b) In three minutes, all officers can be on the scene by sending 6 officers from the Marnixstraat to the Prinsengracht and 4 from the Marnixstraat to the Keizersgracht, sending 1 officer from the Nieuwe Zijds to the Prinsengracht and 4 from the Nieuwe Zijds to the Spui, and sending 8 officers from the Warmoesstraat to the Spui (the solution given in (a) is also optimal for the second criterion).

2.7 (a) The optimal solution is 26 quarters, 26 guilders, and 27 rijksdaalders. (b) In this case, the optimal solution is 24 quarters, 29 guilders, and 26 rijksdaalders.

2.8 Person 1 is given gold coins 1 and 2, 4 through 8, 14, 15, 17, and 20.

2.9 The optimal solution is to produce 175, 200, and 75 tons at the beginning of periods 1, 3, and 6, respectively.

2.11 On Monday, Tuesday, Wednesday, Friday, and Saturday, 8, 4, 1, 2, and 5 bus drivers begin without overtime; the number of bus drivers who begin with overwork is 5 on Wednesday and 1 on Friday.

2.12 The minimum is 1, namely 1 guard is free on Tuesday and Sunday, 12 guards are free on Sunday and Monday, 5 are free on Tuesday and Wednesday, 10 on Thursday and Friday, and 2 guards are free on Saturday and Sunday.

2.15 Cut 475 meters of standard roll according to combination 1, 75 meters according to combination 2, 125 meters according to combination 3, 75 meters according to combination 4, and 75 meters according to combination 5. The total trim loss is 290 square meters.

2.17 The maximum total profit is 27 915 euros and is obtained by making 444 units of product 1 and 417 units of product 2.

2.20 The greatest angle insertion heuristic gives the tour $A - B - C - G - D - F - E - A$ with length 73.74. Applying the 2-opt heuristic to this tour gives the improved tour $A - B - C - D - G - F - E - A$ with length 73.07. This last tour is also optimal.

2.22 Clarke and Wright's heuristic gives the two routes O-D-A-E-B-G-O and O-C-F-O with a total travel time of 328.

Chapter 3

3.1 The shortest paths are $1 \to 4 \to 2$, $1 \to 4 \to 2 \to 3$, $1 \to 4$, $1 \to 4 \to 6 \to 7 \to 5$, $1 \to 4 \to 6$, $1 \to 4 \to 6 \to 7$, and $1 \to 4 \to 6 \to 7 \to 8$, with shortest distances 4, 13, 2, 14, 5, 7, and 13.

3.2 The shortest path is $1 \to 2 \to 5 \to 6 \to 7$ with cost 331 thousand euros.

3.5 (a) The minimum number of steps is seven. (b) The fastest possible time is 17 minutes.

3.7 The safest path is $1 \to 2 \to 5$ with probability 0.8742 of arriving safely.

3.8 The optimal transfer is $A \to B \to D$ with 96.03% of the liquid reaching tank D.

3.9 The optimal order is $1 \to 4 \to 2 \to 5 \to 3$ with probability 0.8493 of avoiding interception.

3.11 On the optimal route, Rotterdam – Gouda – Utrecht – Arnhem – Zutphen – Hengelo – Enschede, he sits at most 37 minutes in one train.

3.12 The network can accommodate at most 2200 cars.

3.13 Checkpoints should be set up on the roads (A, C), (A, D), and (B, E), with total cost 84.

3.15 Swimmer 2 should do the butterfly stroke, swimmer 3 the freestyle, swimmer 4 the breaststroke, and swimmer 5 the backstroke.

3.17 Airline A should get 35 flights to Bangkok and 10 to Nairobi, airline B should get 30 flights to Rio, 15 to Singapore, and 15 to Nairobi, and airline C should get 20 flights to Sydney and 25 flights to Singapore. The total cost is 225 500.

3.18 The tour is $A - C - B - C - B - A - D - B - C - A$ with 9 crossings.

3.19 The cycle is $A - B - D - E - B - E - A - F - D - C - E - C - F - A$ with total cost 87.

Chapter 4

4.1 The first time, do not switch doors. The second time, switch.

4.3 The bet is unfavorable for the player. The player's probability of winning is $\frac{1}{3}$.

4.5 The odds are $\frac{0.15}{0.85} \times \frac{0.8}{0.2} = \frac{12}{17}$, so the probability is $\frac{12}{29} = 0.4138$.

4.7 The odds are $\frac{1/3}{2/3} \times \frac{1/2}{1/4} = \frac{6}{7}$, so the probability is $\frac{6}{13} = 0.4615$.

4.8 The odds are $\frac{1}{10}$, so the probability is $\frac{1}{11} = 0.091$.

4.10 The contestant should draw three marbles from the vase and guess the color that is dominant among these three marbles. He then wins 8500 euros with probability 0.7407.

4.12 Hollandia should not accept the 5 million euros compensation offer. If Imperia makes a new compensation offer of 7.5 million euros, Hollandia should accept; otherwise, Hollandia should let it go to trial. The expected value of the settlement is 7.25 million euros.

Chapter 5

5.1 The optimal path is $(0,0) \rightarrow (0,1) \rightarrow (1,1) \rightarrow (2,1) \rightarrow (2,2)$, with minimum cost 22.

5.2 The optimal path is $(0,0) \rightarrow (1,0) \rightarrow (2,0) \rightarrow (2,1) \rightarrow (2,2)$, with $f(0,0) = 5$.

5.3 Define the value function $f(i)$ as the minimum cost for building a pipeline from node i to the end node 7. This leads to $f(1) = 331$ thousand euros and $1 \rightarrow 2 \rightarrow 5 \rightarrow 6 \rightarrow 7$ as the optimal route for the pipeline.

5.5 The minimum time is 17 minutes: first persons 1 and 2 to go the northern bank with 1 turning back, then persons 3 and 4 go the northern bank with 2 turning back, and finally persons 1 and 2 go to the northern bank.

5.7 The optimal schedule is to use generator 2 in periods 1, 2, and 3, use both generators in periods 4 and 5, and use generator 1 in period 6, with minimum total daily cost 9300 euros.

5.10 Node i corresponds to a remaining debt of i cents, and an arc from i to $i - w_k$ means that at node i, one pays a single coin of type k. Every arc is attributed cost 1.

5.12 The minimum cost is 19 000 euros and is obtained by buying a new machine at the beginning of years 1 and 4 or at the beginning of years 1 and 5.

5.15 This problem is a minimum-cost problem in an acyclic network. The maximum profit is 130 and is obtained by transporting 2 units of cargo 5.

5.17 Define the value function $f_k(b)$ as the maximum profit obtained from a budget b that is still available for the products k through n. The recursion is $f_k(b) = \max\{f_{k+1}, a_k + f_{k+1}(b - b_j)\}$ with $f_k(i) = -\infty$ if $i < 0$.

5.19 Define the value function $f_k(j)$ as the minimum cost to transport j passengers using only airplane types k through 3. The minimum cost is $f_1(575) = 33$. Use two planes of type 1, none of type 2, and two of type 3.

5.20 Define the value function $g_k(i)$ as the least possible value of the greatest difference over the districts k through 5 if there are still i seats to be allocated over those districts. For $R = 26$, the numbers of seats allocated to the five districts are 9, 7, 5, 4, and 1, with $g_1(26) = 0.6810$; for $R = 27$, the allocations are 9, 8, 6, 3, and 1, with $g_1(27) = 0.5449$; and for $R = 28$, the allocations are 10, 8, 6, 3, and 1, with $g_1(28) = 0.5743$. The allocation is not house monotonous.

5.22 Define f_k as the maximum expected number of euros to be won if the candidate has just given a correct answer to question k $(k = 1, \ldots, N)$. The optimal strategy is to stop after question 4 if the candidate is still playing, with maximum expected gain $p_1 f_1 = 19.07$.

5.24 The maximum probability of reaching the goal is 0.303.

5.27 **(a)** Define the value function $f_k(i)$ as the maximum expected profit when the player still has k tosses to go and has tossed heads i times so far. For $K = 10, 25, 50, 100$, and 1000, $f_K(0)$ has the values 0.7437, 0.7679, 0.7780, 0.7839, and 0.7912. **(b)** The maximum expected yield is 2.6245.

5.28 If the gang has N members, the optimal strategy is to let the first $s_N - 1$ pass and then arrest the first member who is taller than all the previous ones, where s_N has values 4, 10, 19, and 38 for $N = 10, 25, 50$, and 100, respectively, with corresponding arrest probabilities 0.3987, 0.3809, 0.3743, and 0.3710.

5.31 The maximum probability has values 0.2198, 0.4654, 0.8322, and 0.9728 for $N = 7$, 10, 15, and 20, respectively.

5.32 **(a)** The minimum expected number of turns is 16.923. **(b)** A one-step look-ahead policy is: continue if $\frac{10}{36} \times k + \frac{1}{36} \times (i + k) < \frac{25}{36} \times 8$; otherwise, stop. Using this heuristic, the expected number of turns needed to obtain 100 points is equal to 17.164.

5.33 **(a)** The maximum probability that the opponent wins is 0.5548. **(b)** The maximum probability has values 0.1914, 0.4240, 0.8004, and 0.9631 for $N = 7, 10, 15$, and 20, respectively.

5.34 The maximum probability of winning is very small, equal to 0.00362. For $m = 1, 5$, and 10, the probabilities of winning are 0.0100, 0.0625, and 0.1724, respectively.

5.35 The minimum expected value is 6.254.

Chapter 6

6.1 The optimal order quantity is $Q^* = 216$ with an average cost of 777.69 euros. The deviation is 0.03% for $Q = 200$ and 1.07% for $Q = 250$. If the lead time is half a month, the order should be placed when the stock has dropped to 100.

6.3 Without discount, order 700 units of 100 grams when the stock has dropped to 50 units. With discount, the optimal order quantity is 782.6 units.

6.5 Buy 300 units of 100 grams at the beginning of week 1 and 760 units at the beginning of week 3.

6.7 The optimum is 7 solar panels.

6.9 Buy 886 ski jackets.

6.11 Agree to 400 repairs. The probability is approximately 50%.

6.12 Produce 170 spare parts.

6.14 (b) The expected profit is at its maximum for $Q_1 = Q_2 = 109$, with value 729.05 euros.

6.15 (a) The average profit is maximized for $Q = 163$.

6.19 The reorder point is $s = 231$.

6.21 The order quantity is $Q = 6450$, and the reorder point is $s = 8144$ with implied penalty cost $b = 12.40$.

Chapter 7

7.2 A Markov chain with six states is needed, where a state describes both where the ride is going and where the driver's license is.

7.4 The machines are identical, so that a Markov chain with seven states suffices. State $(0,0)$ means that both machines are available, state $(0,b)$ with $1 \leq b \leq 3$ means that only one of the machines is available and that the other machine is still in maintenance for b days, while state (a,b) with $1 \leq a < b \leq 3$ means that neither machine is available, where one machine still has a days of maintenance to go and the other b days.

7.5 If the particle begins at point 12, then the probability of being back at point 12 after eight jumps is 0.2627, while the probability is $\frac{1}{12}$ when the initial point is chosen at random.

7.6 (b) The probability that it will be sunny five days from now is 0.7440, while the probability that it will be sunny five days from now and the day after that is 0.6426. The expected number of sunny days in the next 14 days is 10.18. The long-run fraction of the time that it is sunny is equal to 0.7912.

7.7 A Markov chain with three states suffices. The expected number of individuals in stage i five time units from now is 125.15 for $i = 1$, 205.96 for $i = 2$, and 668.89 for $i = 3$. In the long run, the numbers of individuals in stages 1, 2, and 3 have multinomial probability distributions with parameters $N = 1000$ and $p_1 = 0.125$, $p_2 = 0.175$, and $p_3 = 0.7$.

7.8 The desired probability is 0.5760 and can be calculated using an absorbing Markov chain with six states.

7.9 The probabilities are 0.2937, 0.5237, 0.7833, 0.9552, and 0.9981 for the same number three times in a row, and 0.0539, 0.1113, 0.2159, 0.3895, and 0.6270 for a six three times in a row.

7.10 The probabilities are $\frac{7}{8}$, $\frac{3}{4}$, $\frac{2}{3}$, and $\frac{2}{3}$.

7.11 An approximation of the probability is $1 - (1 - 0.011797)^{49} = 0.4409$. A simulation with 150 thousand runs gives the value 0.4424.

7.12 The probability of Joe reaching his goal is 0.5819 and the expected number of bets is 2.0323.

7.13 The probabilities are 0.6 and 0.5. The expected lengths are 6.80 and 8.41.

7.14 The professor has his driver's license with him 42.86% of the time. Production is stopped 2.44% of the time.

7.15 The game is fair.

7.17 The long-run frequencies of HTH and HTT are $\frac{1}{8}$ and $\frac{1}{12}$, respectively.

7.18 The average on-hand inventory is 4.387, the order frequency is 0.5005, and the average demand lost per week is 0.0938.

7.20 The fraction of the time that the panel works is 0.9814, and the average cost per week is 52.46.

7.23 The proportion of time the student is eating at home is 0.2110.

7.24 The long-run fraction of weekends that Linda and Bob visit the same venue is $\frac{1}{19}$.

7.26 The loss probability is 0.0467.

Chapter 8

8.1 Take state $(i, 0)$ for i waiting passengers and no available sherut, and take state $(i, 1)$ for i waiting passengers and one available sherut. The continuous-time Markov chain has $8 + 7 = 15$ states.

8.3 A continuous-time Markov chain with five states is needed, including the state that describes the situation where station 1 is blocked and station 2 is occupied.

8.5 (a) The average number of cars at the gas station is $L = 1.839$. (b) The fraction of the time that the gas station is full is 0.0323, and by the PASTA property, the fraction of the cars that decide not to stop is also equal to 0.0323.

8.8 Let states $(i, 0)$ and $(i, 1)$ correspond to the situation where i trains are at the unloading area and the crane is out of service and in working order, respectively. The average number of trains at the unloading area is $L = \sum_{i=1}^{N} i(\pi(i, 0) + \pi(i, 1))$. The average number of trains that are unloaded per unit of time is $\mu \sum_{i=1}^{N} \pi(i, 1)$. This is also the average number of trains that arrive at the unloading area per unit of time. The fraction of the trains that are unloaded immediately upon arrival is $\pi(0, 1)$ by the PASTA property. The fraction of the time that the crane is out of service is $\sum_{i=1}^{N} \pi(i, 0)$. The average amount of time W a train spends at the unloading area follows from Little's formula $L = \lambda^* W$ with $\lambda^* = \mu \sum_{i=1}^{N} \pi(i, 1)$.

8.10 (b) It is a Poisson distribution with expected value λ/μ.

8.11 Let state i correspond to the situation where the inventory level is i. The average inventory level is $\sum_{i=1}^{Q} i\pi_i$. The average number of replenishments per unit of time is the average number of transitions per unit of time from state 1 to state Q and is therefore equal to $(\lambda + \mu)\pi_1$.

8.13 (a) The state j gives the economic inventory. The equilibrium probabilities are $\pi_j = 1/Q$ for $j = s+1, \ldots, s+Q$. (b) Because of the fixed delivery time $L > 0$ of a replenishment order, the net inventory at time t is equal to the economic inventory at time $t - L$ minus the total demand during the interval $(t - L, t]$. Let $d_k(t)$ be the probability that the net inventory at time t is equal to k. The probability $d_k = \lim_{t\to\infty} d_k(t)$ is

$$d_k = \sum_{j=\max(k,s+1)}^{s+Q} \frac{1}{Q} e^{-\lambda L} \frac{(\lambda L)^{j-k}}{(j-L)!} \quad \text{for } k = s+Q, s+Q-1, \dots .$$

(c) The average on-hand inventory is $\sum_{k=1}^{s+Q} k d_k$, and the average delay in delivery is $\sum_{k=-\infty}^{0} -k d_k$. The fraction of the demand that is delivered late is $\sum_{k\le 0} d_k = 1 - \sum_{k=1}^{s+Q} d_k$.

8.15 The necessary capacity is $c = 18$ (loss probability 0.0071).

8.18 Let the state be the number of working units. The fraction of the time that the system is down is π_0. For this so-called Engset model, in which one never needs to wait for repair, the remarkable result is that the equilibrium probabilities π_j do not depend on either the form of the running time's probability distribution or that of the repair time's probability distribution; they depend only on the expected values of these distributions (this can be checked experimentally with a computer simulation).

Chapter 9

9.1 Old situation: $M/M/1$ model with $W_q = 4.9$ hours and $1 - W_q(5) = 0.3217$. New situation: $M/M/2$ model with $W_q = 3.61$ hours and $1 - W_q(5) = 0.2370$.

9.3 The $M/M/c/c + N$ model with $c = 1$ applies. The required size is $N = 15$, and the quantities are $L_q = 3.36$, $W_q = 0.817$ seconds, and $1 - W_q(2) = 0.1101$.

9.4 The $M/M/1$ model applies with $\lambda = 12$ and $\mu = 15$ (the unit of time is an hour). The average number of waiting customers is $L_q = 3.2$.

9.6 The average waiting time per tanker is 5.12 hours, and the fraction of the tankers that continue on is 0.076.

9.7 Under the current maintenance policy, there are, on average, 3.375 planes in the workshop, and under the new maintenance policy, there are 2.25 planes ($M/G/1$ model).

9.8 The fraction of the callers who hear the busy tone is 0.0116, and the fraction of the callers who are immediately helped is 0.7062. The average number of occupied agents is 2.636.

9.12 Take $c = 7$; then $1 - W_q(5) = 0.0059$ (for $c = 6$, we have $1 - W_q(5) = 0.0795$).

9.14 Without back-ordering, the Erlang loss model gives the probability 0.0367 of losing a customer and the value 3.07 for the average number of video recorders in stock. For the back-order model, the average inventory level is 3.02, and a customer's probability of having to wait for a back-order is 0.0527.

9.15 The number of spare cameras that are lent out is Poisson distributed with expected value 10 in situation 1 and expected value 5.997 in situation 2. The probability of having lent out more than 15 cameras is 0.0487 in situation 1 and 0.0005 in situation 2.

9.17 The limiting distribution is a Poisson distribution with expected value $\lambda_1 \sigma_1 + \lambda_2 \sigma_2$.

Chapter 10

10.1 This problem can be put in the framework of a Markov decision model. Since an enforced repair takes two days and the state of the system has to be defined at the beginning of each day, we need an auxiliary state for the situation in which an enforced repair is in progress already for one day. Thus the set of possible states of the system is taken as $S = \{1, 2, \ldots, N, N+1\}$. State i with $1 \leq i \leq N$ corresponds to the situation in which an inspection reveals working condition i, while state $N+1$ corresponds to the situation in which an enforced repair is in progress already for one day. The actions are defined by $a = 0$ if no repair is done, $a = 1$ if a preventive repair is done, and $a = 2$ if an enforced repair is done. The set of possible actions in state i is taken as $\{0\}$ for $i = 1$, $\{0,1\}$ for $1 < i < N$, and $\{2\}$ for $i = N$ and $i = N + 1$. The one-step transition probabilities $p(j \mid i, a)$ are $p(j \mid i, 0) = q_{ij}$ for $1 \leq i < N$, $p(1 \mid i, 1) = 1$ for $1 < i < N$, $p(N + 1 \mid N, 2) = p(1 \mid N + 1, 2) = 1$, and the other $p(j \mid i, a) = 0$. The one-step rewards $r(i, a)$ are given by $r(i, 0) = 0$, $r(i, 1) = -C_i$, $r(N, 2) = -C_f$ and $r(N + 1, 2) = 0$.

10.2 This problem is an example of a Markov decision model with deterministic state transitions. A decision has to be made at the beginning of each period of 4 hours. In view of the fluctuating demand for electricity and the fixed cost for turning an idle generator on, we choose as state space $S = \{(k, m) : k = 1, \ldots, 6 \text{ and } m = 1, 2, 3\}$. Here the first state variable k indicates the current period k and the second state variable m indicates which generators are currently on, where $m = 1$ (2) means that only generator 1 (2) is on and $m = 3$ means that both generators are on. The possible actions are $a = 1$ if only generator 1 is used, $a = 2$ if only generator 2 is used, and $a = 3$ if both generators are used. Clearly, the action a is only feasible in state (k, m) if the amount of electricity produced under this action is at least as large as d_k. Note that the demand process for electricity generates itself every 24 hours so that period 1 succeeds period 6. Hence, for $1 \leq k \leq 5$, $p(s \mid (k, m), a)$ is 1 if $s = (k+1, a)$ and is 0 otherwise. Also, $p(s \mid (6, m), a)$ is 1 if $s = (1, a)$ and is 0 otherwise. Further, the one-step rewards (negative costs) are given by $r(s, a) = -c(s, a)$ with

$$c((k, 1), 1) = r_1, \quad c((k, 1), 2) = S_2 + r_2, \quad c((k, 1), 3) = S_2 + r_1 + r_2,$$
$$c((k, 2), 1) = S_1 + r_1, \quad c((k, 2), 2) = r_2, \quad c((k, 2), 3) = S_1 + r_1 + r_2,$$
$$c((k, 3), 1) = r_1, \quad c((k, 3), 2) = r_2, \quad c((k, 3), 3) = r_1 + r_2.$$

10.4 (a) Decision epochs: start of each week. States: number of broken installations: $S = \{0, 1, 2, 3\}$. Decisions: $a \in \{0, 1\}, 0 = $ do not repair, $1 = $ repair. (b) Expected immediate costs: $c(0, 0) = 15, c(1, 1) = 225, c(1, 0) = 60, c(2, 1) = 250, c(2, 0) = 105, c(3, 1) = 275$. Transition probabilities: $p(0|0, 0) = 0.512, p(1|0, 0) = 0.384, p(2|0, 0) = 0.096, p(3|0, 0) = 0.008, p(1|1, 0) = 0.64, p(2|1, 0) = 0.32, p(3|1, 0) = 0.04, p(2|2, 0) = 0.8, p(3|2, 0) = 0.2, p(0|1, 1) = p(0|2, 1) = p(0|3, 1) = 1$. (d) 1666.71. (e) The oil platform's current strategy is not optimal, it can be improved by not repairing when only one installation is broken.

10.5 Decision epochs: weekly. States: $i \in \{W \text{ (sells well)}, P \text{ (sells poorly)}\}$. Decisions: $D(W) \in \{A, N\}, D(P) \in \{R, N\}$ where $A = $ advertise, $R = $ do research, $N = $ take no action. Immediate rewards: $r(W, A) = 4000, r(W, N) = 6000, r(P, R) = -5000, r(P, N) = -3000$. Transition probabilities: $p(W|W, A) = 0.8, p(W|W, N) = 0.5, p(W|P, R) = 0.7, p(W|P, N) = 0.4$. Optimal policy: $\delta(W) = A, \delta(P) = R$.

10.6 (a) Decision epochs: end of each week. States: number of couches in stock: $S =$

$\{0, 1, 2\}$. Decisions: number of couches to order: $D(0) = \{0, 1, 2\}$, $D(1) = \{0, 1\}$, $D(2) = \{0\}$. Expected immediate costs: $c(0, 0) = 240, c(0, 1) = 220, c(0, 2) = 240, c(1, 0) = 60, c(1, 1) = 160, c(2, 0) = 0$. Transition probabilities: $p(0|0, 0) = 1, p(0|0, 1) = p(0|1, 0) = \frac{3}{4}, p(1|0, 1) = p(1|1, 0) = \frac{1}{4}, p(0|0, 2) = p(0|1, 1) = p(0|2, 0) = \frac{1}{4}, p(1|0, 2) = p(1|1, 1) = p(1|2, 0) = \frac{1}{2}, p(2|0, 2) = p(2|1, 1) = p(2|2, 0) = \frac{1}{4}$. (c) $V_3(0) = 308, V_3(1) = 204, V_3(2) = 68$, with strategy $\delta = (2, 0, 0)$. (d) 740.

10.8 Decision epochs: daily. States: $i \in \{F \text{ (fit)}, T \text{ (tired)}\}$. Decisions: $a \in \{0, 1\}, 0 = $ do not race, $1 = $ race. Immediate rewards: $r(F, 0) = 0, r(F, 1) = 2M, r(T, 0) = 0, r(T, 1) = 1M$. Transition probabilities: $p(F|F, 0) = 1, p(F|F, 1) = \frac{2}{3}, p(F|T, 0) = \frac{1}{2}, p(F|T, 1) = 0$. Optimal policy: participate in a race every day. The optimal discounted rewards are 4.8 (3) million euros for starting with a fit (tired) horse.

10.10 (a) Decision epochs: yearly. States: $i \in \{L, H\}$. Decisions: $a \in \{0, 1\}, 0 = $ call, $1 = $ take to dinner. (b) Immediate rewards: $r(L, 0) = 1000, r(L, 1) = 500, r(H, 0) = 2500, r(H, 1) = 2000$. Transition probabilities: $p(L|L, 0) = 0.5, p(L|L, 1) = 0.3, p(L|H, 0) = 0.6, p(L|H, 1) = 0.2$. (e) $V_3(L) = 3949.75, V_3(H) = 5329.1$.

10.12 The minimal average cost per week is 17.7687.

10.13 The minimal average cost per day is 0.4338.

10.14 The minimal average cost per period of 24 hours is 1516.67.

Index

Printed in the United States
by Baker & Taylor Publisher Services